HYDROLOGY FOR ENGINEERS

McGRAW-HILL SERIES IN WATER RESOURCES AND ENVIRONMENTAL ENGINEERING

Ven Te Chow, Rolf Eliassen, and Ray K. Linsley, Consulting Editors

Chanlett: Environmental Protection
Graf: Hydraulics of Sediment Transport
Hall and Dracup: Water Resources Systems Engineering
James and Lee: Economics of Water Resources Planning
Linsley and Franzini: Water Resources Engineering
Linsley, Kohler, and Paulhus: Hydrology for Engineers
Metcalf and Eddy, Inc.: Wastewater Engineering: Collection Treatment, Disposal
Rich: Environmental Systems Engineering
Walton: Groundwater Resource Evaluation
Wiener: The Role of Water in Development: An Analysis of Principles of Comprehensive Planning

McGRAW-HILL
BOOK COMPANY
New York
St. Louis
San Francisco
Düsseldorf
Johannesburg
Kuala Lumpur
London
Mexico
Montreal
New Delhi
Panama
Paris
São Paulo
Singapore
Sydney
Tokyo
Toronto

RAY K. LINSLEY, JR.
Professor of Hydraulic Engineering
Stanford University
Chairman, Hydrocomp, Inc.

MAX A. KOHLER
Consulting Hydrologist
Formerly Associate Director, Hydrology
U.S. National Weather Service

JOSEPH L. H. PAULHUS
Consulting Hydrometeorologist
Formerly Chief, Water Management Information Division
U.S. National Weather Service

Hydrology for Engineers

SECOND EDITION

This book was set in Times New Roman.
The editors were B. J. Clark and M. E. Margolies;
the cover was designed by Pencils Portfolio, Inc.;
the production supervisor was Dennis J. Conroy.
New drawings were done by J & R Services, Inc.
Kingsport Press, Inc., was printer and binder.

Library of Congress Cataloging in Publication Data

Linsley, Ray K.
Hydrology for engineers.

(McGraw-Hill series in water resources and environmental engineering)
Includes bibliographies.
1. Hydrology. I. Kohler, Max Adam, date joint author. II. Paulhus, Joseph L. H., joint author. III. Title.
GB661.L53 1975 551.4′8 74-9922
ISBN 0-07-037967-X

HYDROLOGY FOR
ENGINEERS

5 6 7 8 9 0 KP KP 7 9 8 7

To the memory of
MERRILL BERNARD
friend and colleague
whose enthusiasm was a source
of inspiration to the authors

CONTENTS

PREFACE

The first edition of "Hydrology for Engineers" was published in 1958, and has been widely used as a textbook in both senior and graduate courses. In the intervening years there have been many important developments in the science of hydrology, and techniques now available are generally superior to those available in 1958. This second edition represents an extensive revision of the earlier text. Chapters on hydrologic simulation, stochastic hydrology, and morphology of river basins have been added, and extensive changes have been made throughout the balance of the text. The importance of the digital computer as a tool for hydrologic analysis is stressed, but recognizing that not everyone has access to computers, we still discuss the older methods, though in somewhat less detail. The basic processes of hydrology continue to be stressed, in a belief that an understanding of these processes is essential to the application of any of the tools of hydrology.

Since most nations of the world now use metric units of measurement while the United States has barely begun a conversion from the English system, both systems of units are included in text, tables, and figures. In the chapters dealing with meteorological topics, where metric units are in general use with relatively few exceptions, metric units are usually given first with

English equivalents in parentheses. In other chapters the reverse system is followed. It is hoped that this arrangement will serve the countries where the metric system is in use and ease the transition in those countries where the change is yet to occur. Problems are included in both systems of units.

The student should find hydrology an interesting subject but one very different from most of his engineering courses. The natural phenomena with which hydrology is concerned do not lend themselves to such rigorous analyses as are possible in engineering mechanics. There is therefore a greater variety of methods, more latitude for judgment, and a seeming lack of accuracy in problem solutions. Actually, the accuracy of sound hydrologic solutions compares favorably with other types of engineering computations. Uncertainties in engineering are frequently hidden by use of factors of safety, rigidly standardized working procedures, and conservative assumptions regarding properties of materials.

The authors gratefully acknowledge the helpful suggestions, data, and other assistance from their colleagues in the National Weather Service, Stanford University, Hydrocomp, Inc., and elsewhere. Special mention should be made of Professor Stephen Burges for his careful review of the chapter on stochastic methods.

RAY K. LINSLEY, JR.
MAX A. KOHLER
JOSEPH L. H. PAULHUS

LIST OF SYMBOLS AND ABBREVIATIONS

SYMBOLS

A = area
a = coefficient
B = width
b = coefficient
C = Chézy coefficient; a coefficient
C_p = synthetic-unit-hydrograph coefficient of peak
C_t = synthetic-unit-hydrograph coefficient of lag
c = coefficient; concentration
D = depth; overland-flow detention volume; degree-days
d = diameter; coefficient
E = evaporation; sediment washed from impervious areas
E_a = a reference evaporation rate
E_T = evapotranspiration
e = vapor pressure; base of napierian logarithms
e_s = saturation vapor pressure
F = fall; first cost; force; infiltration volume

f = relative humidity
$f(\ )$ = function of
f_c = final infiltration capacity
f_i = infiltration rate
f_0 = initial infiltration capcity
f_p = infiltration capacity
G = safe yield of a groundwater basin; gully-erosion rate
G_i = bed-load transport
g = gage height; acceleration of gravity
H_v = latent heat of vaporization
h = height; head; Hurst coefficient
I = inflow; antecedent-precipitation index; interload
i = rainfall intensity
i_s = supply rate (rainfall less retention)
J = probability
j = probability (exponent)
K = Muskingum storage constant; frequency factor; compaction coefficient; permeability coefficient; hydraulic conductivity
K_r = recession constant
k = coefficient; number
L = length; lower-zone moisture-storage index
L_c = distance from outlet to center of basin
L_0 = length of overland flow
M = snowmelt rate
m = coefficient or exponent
N = normal precipitation; number
n = Manning roughness coefficient; coefficient or exponent; number
O = outflow; operating cost
O_g = subsurface seepage
P = precipitation
P_e = precipitation excess
P_r = radar returned power
p = pressure; porosity; probability
$\mathrm{p}F$ = logarithm of capillary potential in centimeters of water
Q = volume of discharge or runoff
Q_a = net longwave radiation
Q_{ar} = reflected longwave radiation
Q_e = energy used for evaporation
Q_g = groundwater-flow volume
Q_h = sensible-heat transfer
Q_{ir} = incident minus reflected radiation
Q_n = net radiant energy

Q_0 = emitted longwave radiation
Q_r = reflected shortwave radiation
Q_S = volume of surface streamflow
Q_s = shortwave radiation; suspended sediment load
Q_v = advected energy
Q_θ = change in energy storage
q = discharge rate
q_b = base-flow discharge
q_d = direct-runoff discharge
q_e = equilibrium flow rate
q_h = specific humidity
q_o = overland-flow rate
q_p = peak discharge
R = hydraulic radius; Bowen's ratio; soil splash
R_g = gas constant
R_I = runoff index
R_n = range of a series
R_s = sediment residue on ground surface
r = radius, range
S = storage; volume of surface retention; transport of sediment
S_c = storage constant of an aquifer
S_d = depression-storage capacity
S_g = groundwater storage
S_i = interception storage
S_I = season index
S_L = lower-zone moisture storage
S_s = surface storage
S_U = upper-zone moisture storage
s = slope
s_b = slope of channel bottom
T = temperature; transmissibility; time base of unit hydrograph
T_L = lag time
T_d = dewpoint temperature
T_r = return period or recurrence interval
T_w = wet bulb temperature
t = time
t_e = time to equilibrium
t_p = basin lag
t_R = duration of rain
t_r = unit duration for synthetic unit hydrograph
U = unit hydrograph ordinate; upper-zone moisture storage index
u = wave celerity; a factor in well hydraulics

V_e = volume of surface detention at equilibrium
V_i = interception storage
V_s = depression storage
V_0 = volume of surface detention when $i = 0$
v = velocity
v_s = settling velocity
v_* = friction velocity
W = infiltration index
W_p = precipitable water
$W(u)$ = well function of u
w = specific weight
w_r = mixing ratio
X = a variable
$\bar{X}$ = the mean of X
x = distance; constant or exponent
Y = a variable
$\bar{Y}$ = the mean of Y
y = a vertical distance; a reduced variate in frequency analysis
y_n = a factor in frequency analysis
Z = drawdown in a well; drop-size function; a variable
z = a vertical distance
α = a ratio; evaporation portion of advected energy
β = constant
Δ = slope of vapor pressure-temperature curve; an increment
ε = a mixing coefficient; emissivity
θ = an angle
Λ = total potential
μ = absolute viscosity; mean
ν = kinematic viscosity
γ = Bowen's ratio coefficient
π = 3.1416 . . .
ρ = density; correlation coefficient
$\sum$ = summation
σ = standard deviation; Stefan-Boltzmann constant
τ = shear
Υ = du Boy's coefficient
Φ = infiltration index; bed-load function
ψ = capillary potential; function of ρ; bed-load function

ABBREVIATIONS

Å = angstrom (10^{-10} m)
acre-ft = acre-foot

atm = atmosphere
Btu = British thermal unit
°C = degrees Celsius
cal = calorie
cm = centimeter (10^{-2} m)
cfs = cubic feet per second
csm = cubic feet per second per square mile
d = day
D = darcy
deg = degree
°F = degrees Fahrenheit
ft = foot
g = gram
gal = gallon
h = hour
hm = hectometer (10^2 m)
Hg = mercury
in. = inch
K = kelvin
km = kilometer (10^3 m)
kn = knot
l = liter
lat = latitude
lb = pound
ln = logarithm to the base *e*
log = logarithm to the base 10
Ly = langley
m = meter
mi = mile
mb = millibar
min = minute
mm = millimeter (10^{-3} m)
mgd = million gallons per day
nmi = nautical mile
ppm = parts per million
s = second
sfd = second-foot-day
y = year
μm = micrometer (10^{-6} m)

1

INTRODUCTION

"*Hydrology* treats of the waters of the Earth, their occurrence, circulation, and distribution, their chemical and physical properties, and their reaction with their environment, including their relation to living things. The domain of hydrology embraces the full life history of water on the Earth" [1].† Engineering hydrology includes those segments of the field pertinent to design and operation of engineering projects for the control and use of water. The boundaries between hydrology and other earth sciences such as meteorology, oceanography, and geology are indistinct, and no good purpose is served by attempting to define them rigidly. Likewise, the distinctions between engineering hydrology and other branches of applied hydrology are vague. Indeed, the engineer owes much of his present knowledge of hydrology to agriculturists, foresters, meteorologists, geologists, and others in a variety of fields.

† Numbered references will be found at the end of each chapter.

1-1 The Hydrologic Cycle

The concept of the *hydrologic cycle* is a useful, if academic, point from which to begin the study of hydrology. This cycle (Fig. 1-1) is visualized as beginning with the evaporation of water from the oceans. The resulting vapor is transported by moving air masses. Under the proper conditions, the vapor is condensed to form clouds, which in turn may result in precipitation. The precipitation which falls upon land is dispersed in several ways. The greater part is temporarily retained in the soil near where it falls and is ultimately returned to the atmosphere by evaporation and transpiration by plants. A portion of the water finds its way over and through the surface soil to stream channels, while other water penetrates farther into the ground to become part of the earth's groundwater supply. Under the influence of gravity, both surface streamflow and groundwater move toward lower elevations and may eventually discharge into the ocean. However, substantial quantities of surface and underground water are returned to the atmosphere by evaporation and transpiration before reaching the oceans.

This description of the hydrologic cycle and the schematic diagram of Fig. 1-1 are enormously oversimplified. For example, some water which enters surface streams may percolate to the groundwater, while in other cases groundwater is a source of surface streamflow. Some precipitation may remain on the ground as snow for many months before melting releases the water to streams or groundwater. The hydrologic cycle is a convenient means for a rough delineation of the scope of hydrology as that portion between precipitation on the land and the return of this water to the atmosphere or ocean. The hydrologic cycle serves also to emphasize the four basic phases of interest to the hydrologist: precipitation, evaporation and transpiration, surface streamflow, and groundwater. These topics are the subject of much more detailed discussion later in this text.

If the discussion of the hydrologic cycle gives any impression of a continuous mechanism through which water moves steadily at a constant rate, this impression should be dispelled. The movement of water through the various phases of the cycle is erratic, both in time and area. On occasion, nature seems to work overtime to provide torrential rains which tax surface-channel capacities to the utmost. At other times it seems that the machinery of the cycle has stopped completely and, with it, precipitation and streamflow. In adjacent areas the variations in the cycle may be quite different. It is precisely these extremes of flood and drought that are often of most interest to the engineering hydrologist, for many hydraulic engineering projects are designed to protect against the ill effect of extremes. The reasons for these climatic extremes are found in the science of meteorology and should be

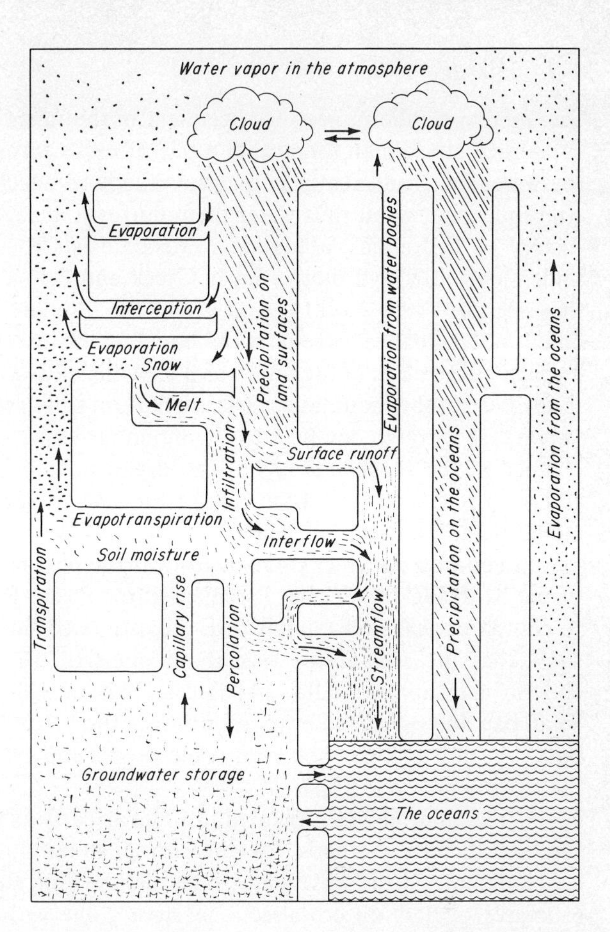

FIGURE 1-1
The hydrologic cycle.

understood, in broad detail at least, by the hydrologist. This aspect of hydrology is discussed in the following chapters.

The hydrologist is interested in more than obtaining a qualitative understanding of the hydrologic cycle and measuring the quantities of water in transit in this cycle. He must be able to deal quantitatively with the interrelations between the various factors so that he can accurately predict the influence of man-made works on these relationships. He must also concern himself with the frequency with which various extremes of the cycle may occur, for this is the basis of economic analysis, which should be an important determinant for all hydraulic projects. The final chapters of this text deal with these quantitative problems.

1-2 History

The first hydraulic project has been lost in the mists of prehistory. Perhaps some prehistoric man found that a pile of rocks across a stream would raise the water level sufficiently to overflow the land which was the source of his wild food plants and thus water them during a drought. Whatever the early history of hydraulics, abundant evidence exists to show that the builders understood little hydrology. Early Greek and Roman writings indicate that these people could accept the oceans as the ultimate source of all water but could not visualize precipitation equaling or exceeding streamflow [2]. Typical of the ideas of the time was a view that seawater moved underground to the base of the mountains. There a natural still desalted the water, and the vapor rose through conduits to the mountain tops, where it condensed and escaped at the source springs of the streams. Marcus Vitruvius Pollio (ca. 100 B.C.) seems to have been one of the first to recognize the role of precipitation as we accept it today.

Leonardo da Vinci (1452–1519) was the next to suggest a modern view of the hydrologic cycle, but it remained for Pierre Perrault [3] (1608–1680) to compare measured rainfall with the estimated flow of the Seine River to show that the streamflow was about one-sixth of the precipitation. The English astronomer Halley [4] (1656–1742) measured evaporation from a small pan and estimated evaporation from the Mediterranean Sea from these data. As late as 1921, however, some people still questioned the concept of the hydrologic cycle [5].

Precipitation was measured in India as early as the fourth century B.C., but satisfactory methods for measuring streamflow were a much later development. Frontinus, water commissioner of Rome in A.D. 97, based estimates of flow on cross-sectional area alone without regard to velocity. In the United States, organized measurement of precipitation started under the Surgeon General of the Army in 1819, was transferred to the Signal Corps in 1870, and finally, in 1891, to a newly organized U.S. Weather Bureau, renamed the National Weather Service in 1970. Scattered streamflow measurements were made on the Mississippi River as early as 1848, but a systematic program was not started until 1888, when the U.S. Geological Survey undertook this work. It is not surprising, therefore, that little quantitative work in hydrology was done before the early years of the twentieth century, when men such as Horton, Mead, and Sherman began to explore the field. The great expansion of activity in flood control, irrigation, soil conservation, and related fields which began about 1930 gave the first real impetus to organized research in hydrology, as need for more precise design data became evident. Most of today's concepts of hydrology date from 1930.

1-3 Hydrology in Engineering

Hydrology is used in engineering mainly in connection with the design and operation of hydraulic structures. What flood flows can be expected at a spillway or highway culvert or in a city drainage system? What reservoir capacity is required to assure adequate water for irrigation or municipal water supply during droughts? What effect will reservoirs, levees, and other control works exert on flood flows in a stream? These are typical of questions the hydrologist is expected to answer.

Large organizations such as federal and state water agencies can maintain staffs of hydrologic specialists to analyze their problems, but smaller offices often have insufficient hydrologic work for full-time specialists. Hence, many civil engineers are called upon for occasional hydrologic studies. It is probable that these civil engineers deal with a larger number of projects and greater annual dollar volume than the specialists do. In any event, it seems that knowledge of the fundamentals of hydrology is an essential part of the civil engineer's training.

1-4 Subject Matter of Hydrology

Hydrology deals with many topics. The subject matter as presented in this book can be broadly classified into two phases: data collection and methods of analysis. Chapters 2 to 6 deal with the basic data of hydrology. Adequate basic data are essential to any science, and hydrology is no exception. In fact, the complex features of the natural processes involved in hydrologic phenomena make it difficult to treat many hydrologic processes by rigorous deductive reasoning. One cannot always start with a basic physical law and from this determine the hydrologic result to be expected. Rather, it is necessary to start with a mass of observed facts, analyze these facts, and from this analysis establish the systematic pattern that governs these events. Thus, without adequate historical data for the particular problem area, the hydrologist is in a difficult position. Most countries have one or more government agencies with responsibility for data collection [6]. It is important that the student learn how these data are collected and published, the limitations on their accuracy, and the proper methods of interpretation and adjustment.

Typical hydrologic problems involve estimates of extremes not observed in a small data sample, hydrologic characteristics at locations where no data have been collected (such locations are much more numerous than sites with data), or estimates of the effects of man's actions on the hydrologic characteristics of an area. Generally, each hydrologic problem is unique in that it deals with a distinct set of physical conditions within a specific river baisn. Hence, quantitative conclusions of one analysis are often not directly transferable to another problem. However, the general solution for most problems can be developed from application of a few relatively basic concepts. Chapters 6 to 14 describe these concepts and explain how they are utilized to solve specific phases of a hydrologic problem.

REFERENCES

1. Federal Council for Science and Technology, "Scientific Hydrology," Washington, June 1962.
2. For a survey of early literature on hydrology see A. K. Biswas, "History of Hydrology," 2d ed., North-Holland, Amsterdam, 1972.
3. P. Perrault, "De l'Origine des fontaines," Paris, 1674, trans. by A. LaRocque, Hafner, New York, 1967.
4. E. Halley, An Account of the Evaporation of Water, *Phil. Trans. R. Soc. Lond.*, vol. 18, pp. 183–190, 1694.
5. P. Ototsky, Underground Water and Meteorological Factors, *Q. J. R. Meteorol. Soc.*, vol. 47, pp. 47–54, 1921.
6. World Meteorological Organization, Organization of Hydrometeorological and Hydrological Services, *Rep. WMO/IHD Proj.* 10, Geneva, 1969.

BIBLIOGRAPHY

AMERICAN SOCIETY OF CIVIL ENGINEERS: Hydrology Handbook, *Man.* 28, 1949.

BRUCE, J. P., and CLARK, R. H.: "Introduction to Hydrometeorology," Pergamon, New York, 1966.

CHOW, V. T. (ed.): "Handbook of Hydrology," McGraw-Hill, New York, 1964.

EAGLESON, P.: "Dynamic Hydrology," McGraw-Hill, New York, 1970.

JOHNSTONE, DON, and CROSS, W. P.: "Elements of Applied Hydrology," Ronald, New York, 1949.

KALININ, G. P.: "Problemi Globolnoy Gidrologii" ("Problems of Global Hydrology"), in Russian, Gidrometeoizdat, Leningrad, 1963.

KAZMANN, R. G.: "Modern Hydrology," 2d ed., Harper & Row, New York, 1972.

LINSLEY, R. K., and FRANZINI, J. B.: "Water-Resources Engineering," 2d ed., McGraw-Hill, New York, 1972.

———, KOHLER, M. A., and PAULHUS, J. L. H.: "Applied Hydrology," McGraw-Hill, New York, 1949.

MEAD, D. W.: "Hydrology," 1st ed., McGraw-Hill, New York, 1919; 2d ed., rev. by H. W. Mead, 1950.

MEINZER, O. E. (ed.): "Hydrology," vol. 9, Physics of the Earth Series, McGraw-Hill, New York, 1942; reprinted by Dover, New York, 1949.

MEYER, A. F.: "Elements of Hydrology," 2d ed., Wiley, New York, 1928.
WISLER, C. O., and BRATER, E. F.: "Hydrology," 2d ed., Wiley, New York, 1959.

PROBLEMS

1-1 List the agencies in your state which have responsibilities of a hydrologic nature. What is the special responsibility of each agency?

1-2 Repeat Prob. 1-1 for federal agencies.

1-3 List the major hydraulic projects in your area. What specific hydrologic problems did each project involve?

2

WEATHER AND HYDROLOGY

The hydrologic characteristics of a region are determined largely by its geology and geography, climate playing a dominant part. Among climatic factors that establish the hydrologic features of a region are the amount and distribution of precipitation; the occurrence of snow and ice; and the effects of wind, temperature, and humidity on evapotranspiration and snowmelt. Hydrologic problems in which meteorology plays an important role include determination of probable maximum precipitation and optimum snowmelt conditions for spillway design, forecasts of precipitation and snowmelt for reservoir operation, and determination of probable maximum winds over water surfaces for evaluating resulting waves in connection with the design of dams and levees. Obviously, the hydrologist should have some understanding of the meteorological processes determining a regional climate. The general features of climatology are discussed in this chapter.

SOLAR AND EARTH RADIATION

2-1 Solar and Earth Radiation

Solar radiation, the earth's chief source of energy, determines the world's weather and climates. Both earth and sun radiate essentially as *blackbodies*; i.e., they emit for every wavelength almost the theoretical maximum amount of radiation for their temperatures.

Radiation wavelengths are usually given in micrometers (μm)† (10^{-6} m) or in angstroms (Å) (10^{-10} m). Maximum energy of solar radiation is in the visible range of 0.4 to 0.8 μm, and that of earth radiation is about 10 μm. Radiation from the sun is shortwave radiation, and that from the earth is longwave.

The rate at which solar radiation reaches the upper limits of earth's atmosphere on a surface normal to the incident radiation and at earth's mean distance from the sun is called the *solar constant*. Measurements have ranged from 1.89 to 2.05 Ly/min, with most of the uncertainty resulting from corrections for atmospheric effects rather than from fluctuations in solar activity, which are considered relatively small. (The abbreviation Ly is for langley; 1 Ly = 1 cal/cm^2.) High-altitude observations with airborne instruments, which minimize atmospheric effects, indicate a range of 1.91 to 1.95 Ly/min, and a value of 1.94 Ly/min is often used as the solar constant.

2-2 Solar Radiation at Earth's Surface

A large part of the solar radiation reaching the outer limits of the atmosphere is scattered and absorbed in the atmosphere or reflected from clouds and the earth's surface. Scattering of radiation by air molecules is most effective for the shortest wavelengths. With sun overhead and clear sky, over half the radiation in the blue range (short wavelengths about 0.45 μm) is scattered, thus accounting for the blue sky. Very little radiation in the red range (about 0.65 μm) is scattered. Estimates of radiation scattered to space average about 8 percent of the incident solar radiation (*insolation*).

Clouds reflect much incident solar radiation to space. The amount reflected depends on the amount and type of clouds and their albedo.‡ The

† Formerly called the micron (μ).

‡ *Reflectivity* is the ratio of the amount of electromagnetic radiation (within any specified range of wavelength) reflected by a body to the amount incident upon it, commonly expressed as a percentage. *Albedo* is the ratio of the amount of solar radiation (or sometimes visible radiation) reflected by a surface to the amount incident upon it, also expressed as a percentage. For example, the reflectivity of fresh snow for infrared wavelengths (terrestrial radiation) is nearly zero, but the albedo is about 85 percent. The albedo is the reflectivity for the solar or visible range of radiation.

albedo (and absorption) of clouds varies greatly with thickness and liquid-water content and inversely with solar elevation. A high, thin overcast may reflect less than 20 percent of incident radiation, while a 2000-ft (600-m) layer of stratus or stratocumulus clouds may reflect over 80 percent.

About half the incident radiation at the outer limits of the atmosphere eventually reaches earth's surface. Much of it is absorbed, but some is reflected to the atmosphere and to space. The albedo of earth's surface varies widely, depending on solar altitude and type of surface. It is less for moist soil surfaces than for dry and tends to be less for high solar altitudes than for low. The albedo (in percent) ranges from 10 to 20 for green forests; 15 to 30 for grass-covered plains; 15 to 20 for marshy lands; 15 to 25 for crop-covered, cultivated fields; 10 to 25 for dark, bare soils when dry, and 5 to 20 when moist; 20 to 45 for dry, light, sandy soils; 40 to 50 for old, dirty snow; and 60 to 95 for pure, white snow, the highest albedo being for fresh, clean, dry snow and low solar altitude.

The albedo of ocean surfaces depends on surface roughness as well as on solar altitude. The albedo (in percent) of a calm sea is 2 to 3 with solar altitude between 90 and 50°, increases to 12 with sun at 20°, and is about 40 with sun at 5°. The albedo of a rough sea is greater than that of a calm sea for solar altitudes greater than about 45° and is less for lower altitudes. Estimates of average albedo for the total of all ocean surfaces range from 6 to 8 percent.

The mean weighted albedo of the earth's surface has been estimated at about 14 percent. Estimates of the albedo for the planet as a whole, i.e., including atmosphere, clouds, etc., have ranged from 35 to 43 percent.

The above discussion generally treats radiation scatter, reflection, and absorption in the mean. Very little of earth's surface is normal to incident solar radiation; and the greater the angle of the surface from normal, the less the radiation intensity. Thus, less solar radiation reaches high latitudes than reaches low latitudes. Differences in insolation are one of the primary factors in determining the general circulation of the earth's atmosphere.

2-3 Heat Balance of Earth's Surface and Atmosphere

Since the surface area of a sphere is 4 times that of a great circle, solar radiation intercepted by the planet earth averages, for the globe as a whole, one-fourth the solar constant, or about 0.5 Ly/min. This amount is arbitrarily set at 100 units in Table 2-1, which presents estimates of the various components of earth's heat balance. On the scale of Table 2-1, solar radiation absorbed by the earth's surface is evaluated at 44 units.

The earth's surface radiates almost as a blackbody at a mean temperature of about 15°C (59°F).† This emission is rated in Table 2-1 as 116 units, which is roughly $2\frac{1}{2}$ times the 44 units of solar radiation absorbed. Net loss of heat is prevented and a heat balance maintained because the atmosphere reflects back to the surface about 85 percent of the emitted radiation. Were it not for this *greenhouse effect*, the mean temperature of the earth would be about −40°C (−40°F).

Figure 2-1 shows the average annual values of the radiation balance of

† There are two temperature scales in common use. These are the *Celsius scale* (formerly called centigrade) with freezing point of water at 0° and boiling point at 100°, and the *Fahrenheit scale*, with freezing point at 32° and boiling point at 212°. Temperatures on one scale are converted to the other by the relationships $F = \frac{9}{5}C + 32$ and $C = \frac{5}{9}(F - 32)$. Hence, −40° on both scales represents the same degree of heat. The Celsius scale is the recommended standard scale for international use and is generally used for meteorological and hydrologic purposes.

Table 2-1 COMPONENTS OF EARTH'S AVERAGE ANNUAL HEAT BALANCE IN PERCENTAGE UNITS

A Shortwave radiation		
1 Solar radiation at upper boundary of atmosphere		100
2 Reflected to space by clouds and atmospheric scattering		34
3 Absorbed by atmosphere (includes absorption by ozone, clouds, dust, and water vapor)		15
4 Direct solar and diffuse radiation reaching earth's surface:		
a Reflected	7	
b Absorbed	44	
		51
B Longwave radiation		
1 Emitted by atmosphere:		
a To space	51	
b To earth's surface	99	
		150
2 Emitted by earth's surface:		
a To atmosphere	108	
b To space	8	
		116
3 Net longwave radiation of earth's surface (116 − 99)		17
C Nonradiative heat transfers		
1 Turbulent heat transfer from earth's surface to atmosphere		5
2 Latent heat transfer (evaporation and condensation) from earth to atmosphere		22
D Heat balances		
1 Of atmosphere:		
Heating components = 15 + 108 + 5 + 22 = 150		
Cooling component = 150		
2 Of earth's surface:		
Heating component = 44 + 99 = 143		
Cooling component = 116 + 5 + 22 = 143		

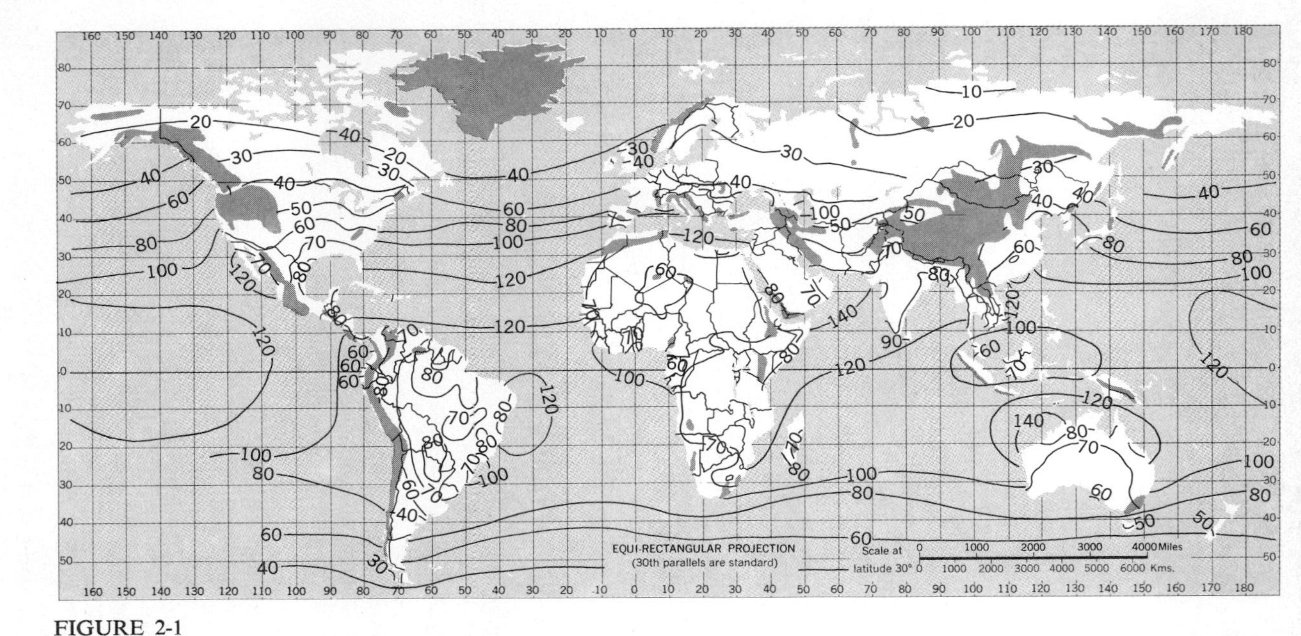

FIGURE 2-1
Mean annual heat balance at earth's surface, in kilocalories per square centimeter. (*From M. I. Budyko, N. A. Yefimova, L. I. Aubenok, and L. A. Strokina, The Heat Balance of the Surface of the Earth*, *Soviet Geog.: Rev. and Translation, vol. 3, pp.* 3–16, *May* 1962.)

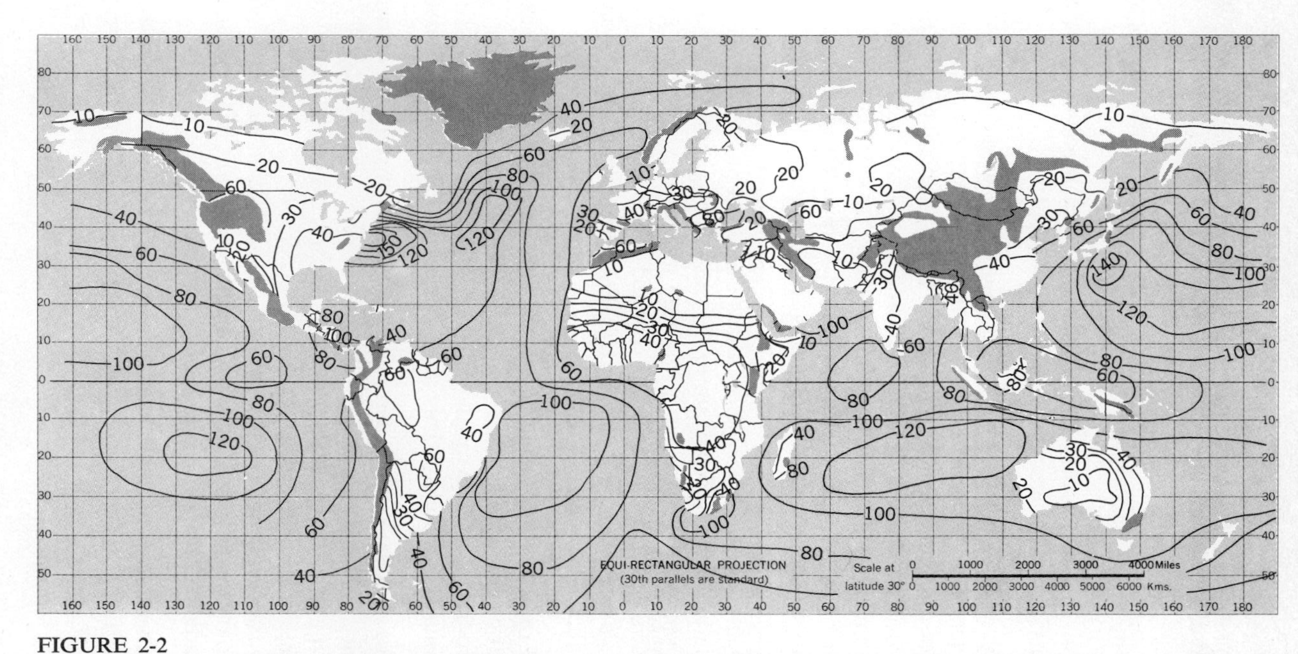

FIGURE 2-2
Mean annual amount of heat utilized by evaporation, in kilocalories per square centimeter. (*From M. I. Budyko, N. A. Yefimova, L. I. Aubenok, and L. A. Strokina, The Heat Balance of the Surface of the Earth, Soviet Geog.: Rev. and Translation, vol. 3, pp.* 3–16, *May* 1962.)

the earth's surface, i.e., the difference between the absorbed shortwave radiation and the effective (or net) longwave radiation. Figure 2-2 shows the average amount of heat utilized by evaporation.

2-4 Measurement of Radiation

Actinometer and *radiometer* are general names for instruments used to measure intensity of radiant energy. There are five types:

Pyrheliometer For measuring intensity of direct solar radiation
Pyranometer For measuring global† radiation, i.e., the combined intensity of direct solar radiation and diffuse sky radiation (radiation reaching earth's surface after being scattered from direct solar beam by molecules and suspensoids in the atmosphere)
Pyrgeometer For measuring hemispherical longwave radiation, used face up to measure atmospheric radiation or inverted to measure terrestrial and reflected atmospheric radiation
Pyrradiometer, or *total hemispherical radiometer* For measuring all-wave radiation flux, used face up to measure hemispherical longwave radiation plus global radiation or inverted to measure terrestrial and reflected atmospheric radiation plus reflected solar radiation
Net Pyrradiometer, or *net radiometer* For measuring net all-wave radiation flux

Details of radiation-measuring instruments can be found in texts listed in the Bibliography at the end of this chapter.

In order to obtain reliable measurements from radiometers, convective heat exchange must be excluded. This can be accomplished in several ways:

1 Blowing a stream of air across the receiver to overcome natural convection
2 Covering the receiver with a material transparent only to radiation components to be measured
3 Using two receivers equally exposed to convection but with one shaded from radiation and heated to compensate for the heat exchange of the irradiated receiver

Net radiometers (net pyrradiometers) are not generally used for network observations because their measurements are applicable only to the type of ground surface at the installation site and their area of representativity is thus

† Although the term "global radiation" is widely used, it is a misnomer. As used here the term refers to hemispherical shortwave radiation.

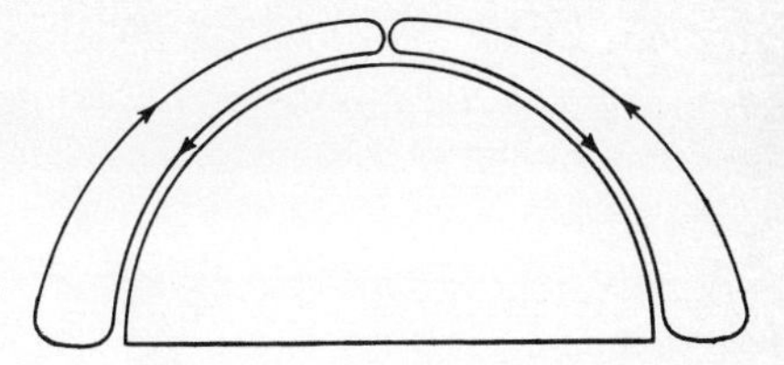

FIGURE 2-3
Simple thermal circulation on nonrotating earth (Northern Hemisphere).

limited. Greatest use of radiometers in hydrology is in studies of evaporation and snowmelt. For most studies of evaporation, incident all-wave radiation data are adequate since the reflectivity of water is nearly constant. The reflectivity of snow, however, differs widely for shortwave, or solar, radiation compared with longwave, or atmospheric, radiation. Hence, both incident shortwave and longwave radiation data are required. Because of the sparsity of all-wave radiation measurements, application of the energy-budget technique in snowmelt computations frequently requires that incident longwave radiation be computed. Several procedures, some empirical and some based on theoretical considerations, have been developed [1–4] for computing longwave radiation from regularly observed surface and upper-air data, such as temperature, vapor pressure, cloudiness, and incident solar radiation.

THE GENERAL CIRCULATION

2-5 Thermal Circulation

If the earth were a nonrotating sphere, a purely thermal circulation (Fig. 2-3) would result. The equator receives more solar radiation than the higher latitudes. Equatorial air, being warmer, is lighter and tends to rise. As it rises, it is replaced by cooler air from higher latitudes. The only way the air from the higher latitudes can be replaced is from above by the poleward flow of air rising from the equator. The true circulation differs from that of Fig. 2-3 because of the earth's rotation and the effects of land and sea distribution and landforms.

2-6 Effects of Earth's Rotation

The earth rotates from west to east, and a point at the equator moves at about 1670 km/h (1040 mi/h) while one at 60° lat moves at one-half this speed. From the principle of conservation of angular momentum, it follows

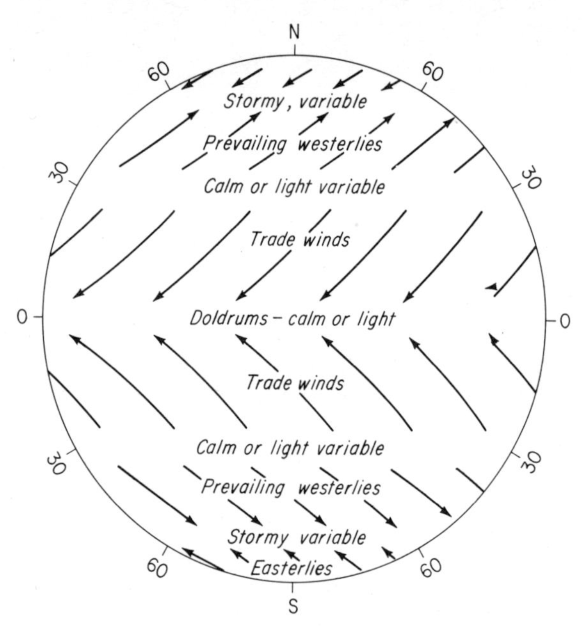

FIGURE 2-4
Idealized circulation at earth's surface assuming a smooth surface of uniform composition.

that a parcel of air at rest relative to earth's surface at the equator would attain a theoretical eastward velocity of 2505 km/h (1560 mi/h), relative to earth's surface, if moved northward to 60° lat. Conversely, if a parcel of air at the North Pole were moved southward to 60° lat, it would reach a theoretical westward velocity of 835 km/h (520 mi/h). However, wind speeds of this magnitude are never observed in nature because of friction. The force that would be required to produce these changes in velocity is known as the *Coriolis force.* This apparent force always acts to the right in the Northern Hemisphere and to the left in the Southern Hemisphere.

An oversimplified pattern of the general circulation at the surface is shown in Fig. 2-4. The wind and temperature distributions are shown for the Northern Hemisphere winter season in Figs. 2-5 and 2-6. The physical reasons for these patterns are only partly known. Early attempts to determine a natural mechanism for the general circulation were based on the idea of a convectionally driven meridional circulation. Eddy transport of angular momentum is now considered most important.

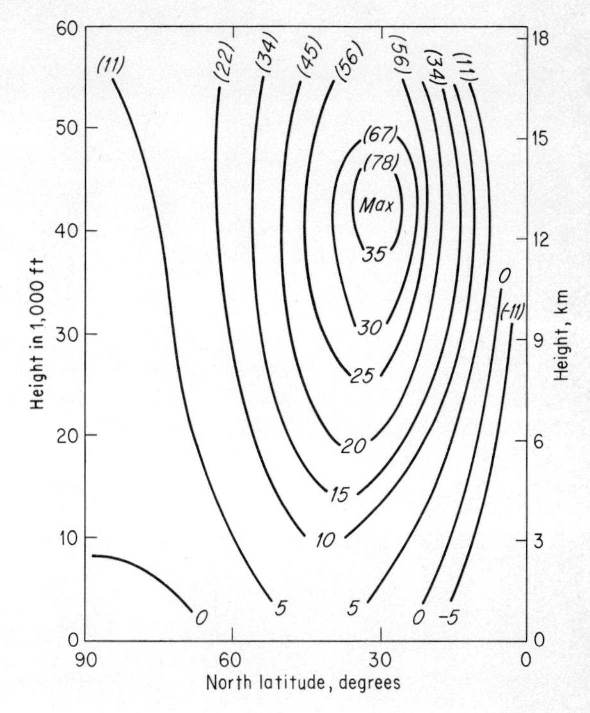

FIGURE 2-5
Vertical north-south cross section of average speed, in meters per second and (miles per hour), of eastward component of wind in Northern Hemisphere in winter. (Negative value indicates westward component.) Maximum wind center marks mean location of jet-stream core.

2-7 Jet Streams

Jet streams are a prominent feature of the general circulation. They are caused by air masses being brought into motion by strong pressure-gradient forces resulting from steep meridional temperature gradients and by angular momentum imparted by rotation of the earth's surface. *Jet streams* are quasi-horizontal, sinuous, undulating bands, or ribbons, of air traveling near the tropopause at speeds ranging from about 30 m/s (70 mi/h) to over 135 m/s (300 mi/h). The *tropopause*, the boundary between troposphere and stratosphere, ranges in height from about 8 km (5 mi) at the poles to about 16 km (10 mi) at the equator (Fig. 2-6). The *troposphere*, which extends from earth's surface to the tropopause, is characterized by generally decreasing temperature with height, considerable vertical wind motion, most of the water

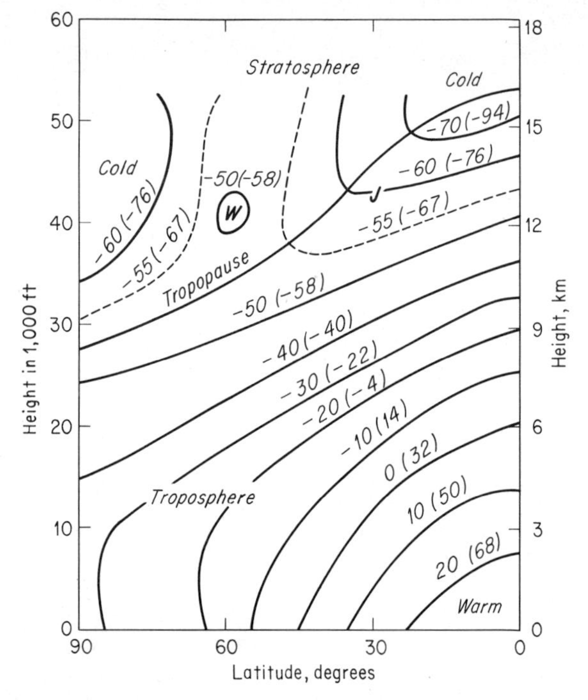

FIGURE 2-6
Vertical north-south cross section of mean temperature, in degrees Celsius (and Fahrenheit), over the Northern Hemisphere in winter. Tropopause marks boundary between troposphere, where temperature generally decreases with height, and stratosphere, which is relatively isothermal. *J* marks the mean position of the jet-stream core. (*Adapted from H. Riehl, "Introduction to the Atmosphere, 2d ed. Copyright* 1972, *McGraw-Hill Book Company. Used with permission of McGraw-Hill Book Company.*)

vapor in the atmosphere, and weather. The *stratosphere* is the relatively isothermal layer extending from the tropopause to about 20 to 25 km (12 to 16 mi), where temperature begins to increase with height.

Being located near the tropopause, which slopes downward toward the poles, jet streams are positioned where the horizontal temperature gradient reverses. In westerly jets of the Northern Hemisphere, warmer air is found to the south below the jet-core level and to the north at heights above the jet core. The reverse applies to easterly jets. Jet streams often appear to produce a break, or discontinuity, in the tropopause, and a transfer of cold, stratospheric air to the troposphere is believed to take place through the gap.

Jet streams apparently provide the mechanism for generating low-level

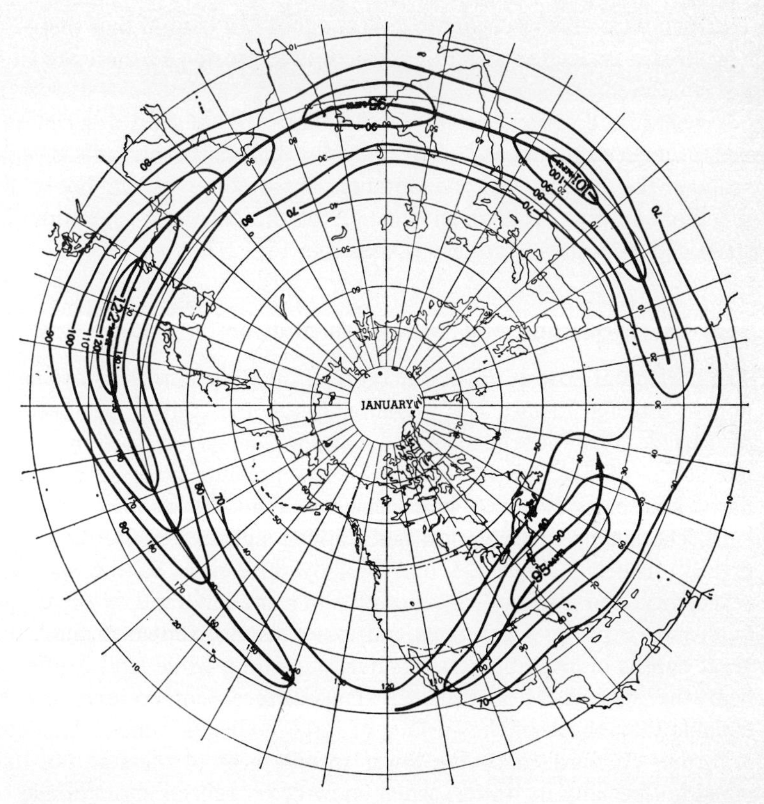

FIGURE 2-7
Mean location of Northern Hemisphere tropospheric jet streams and average speeds, in miles per hour, in January. (*From J. Namias and P. F. Clapp, Confluence Theory of the High Tropospheric Jet Stream, J. Meteorol., vol.* 6, *pp.* 330–336, *October* 1949. *Used with permission of American Meteorological Society.*)

pressure systems that determine the weather. Convergence and divergence of air entering or leaving the maximum-wind centers of the jet result in the addition or removal of air in the jet layer which must be compensated by an opposite effect. This is provided by the layers below jet level, where the thermal structure is more favorable for vertical motion. Thus, convergence of air at jet level results in a piling of air reflected as high pressure and accompanying divergence at the earth's surface. Similarly, divergence at jet level results in removal of air that is compensated by low pressure and convergence at the surface. Paradoxically, the low-pressure systems of the lower levels appear to provide the energy for jet streams. Research has not yet disclosed which is the originator; although it can be said that there is always a jet stream associated with any major general storm [5], there are jet streams not so related.

There are several types of jet streams. A detailed description of the meteorological factors associated with the different jet streams is outside the scope of this text but can be obtained from some of the books listed in the Bibliography. The mean location of the Northern Hemisphere tropospheric jet streams in January is shown in Fig. 2-7.

2-8 Effect of Land and Water Distribution

The horizontal flow of air in any layer of the atmosphere always has a component directed toward low pressure. Thus, the converging flows of air in the surface layer at the equator and about 60° lat, as indicated by the idealized circulation of Fig. 2-4, imply belts of low pressure at these latitudes. Similarly, high pressure would be expected at about 30° lat and at the poles.

The idealized circulation and implied simple pressure distribution are greatly distorted (Figs. 2-8 and 2-9) by differences in the specific heats, reflectivity, and mixing properties of water and land and by the existence of barriers to airflow. Heat gains and losses are distributed through relatively great depths in large bodies of water by mixing, while land is affected only near the surface. Consequently, land-surface temperatures are far less equable than those of the surface of large bodies of water. This condition is further emphasized by the lower specific heat of the soil and its higher albedo, especially in winter, when snow cover reflects most of the incident radiation back to space. In winter there is a tendency for the accumulation of cold dense air over land masses and warm air over oceans. In summer, the situation is reversed.

The pressure and wind patterns of Figs. 2-8 and 2-9 show the effects of the earth's rotation, land and water distribution, and seasonal changes. They also show that winds blow clockwise around high pressure and counterclockwise around low pressure in the Northern Hemisphere and vice versa in the Southern Hemisphere.

2-9 Migratory Systems

The semipermanent features of the general, or mean, circulation (Figs. 2-8 and 2-9) are statistical and at any time may be distorted or displaced by transitory, or migratory, systems. Both semipermanent and transitory features are classified as cyclones or anticyclones. A *cyclone* is a more or less circular area of low atmospheric pressure in which the winds blow counterclockwise in the Northern Hemisphere. *Tropical cyclones* form at low latitudes

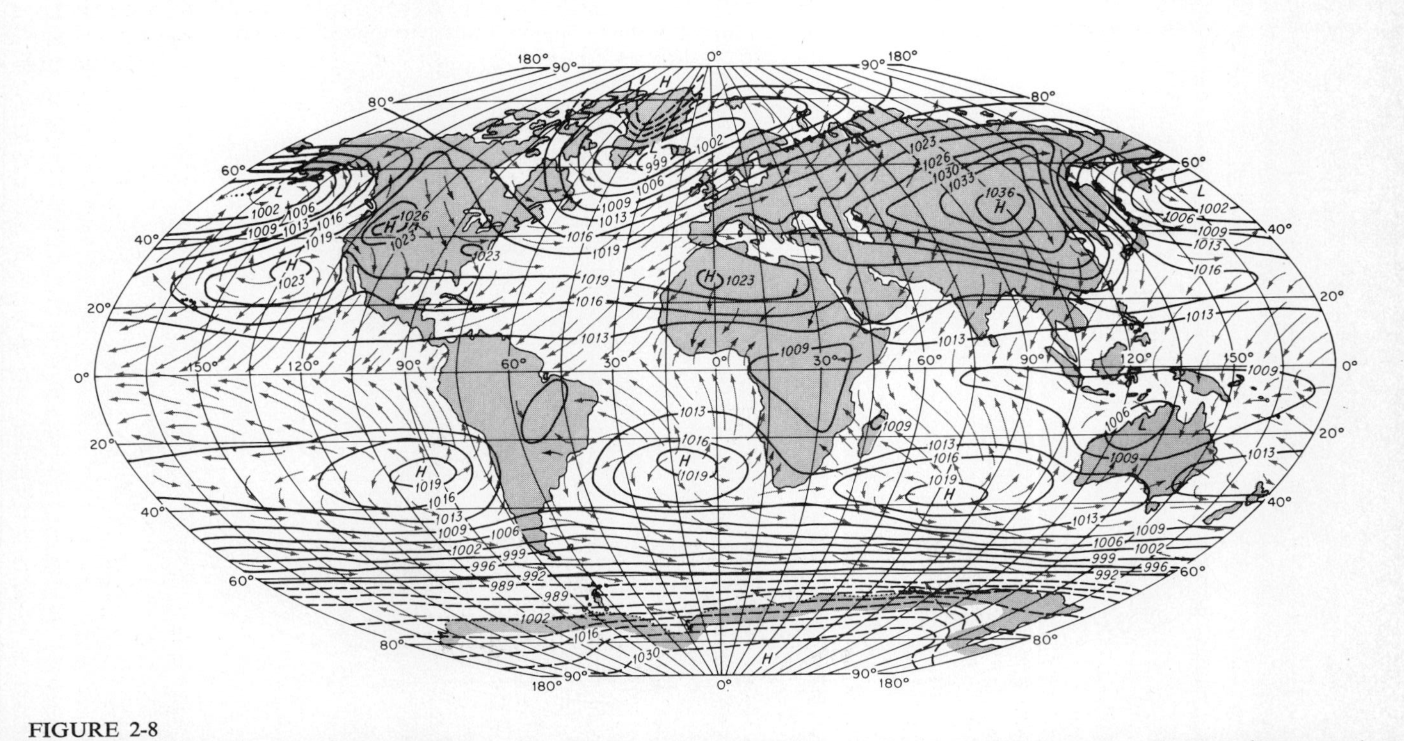

FIGURE 2-8
January mean sea-level pressure, in millibars, and prevailing winds. (*From V. C. Finch, G. T. Trewartha, A. H. Robinson, and E. H. Hammond "Physical Elements of Geography," 4th ed., Copyright* 1957, *McGraw-Hill Book Company. Used with permission of McGraw-Hill Book Company.*)

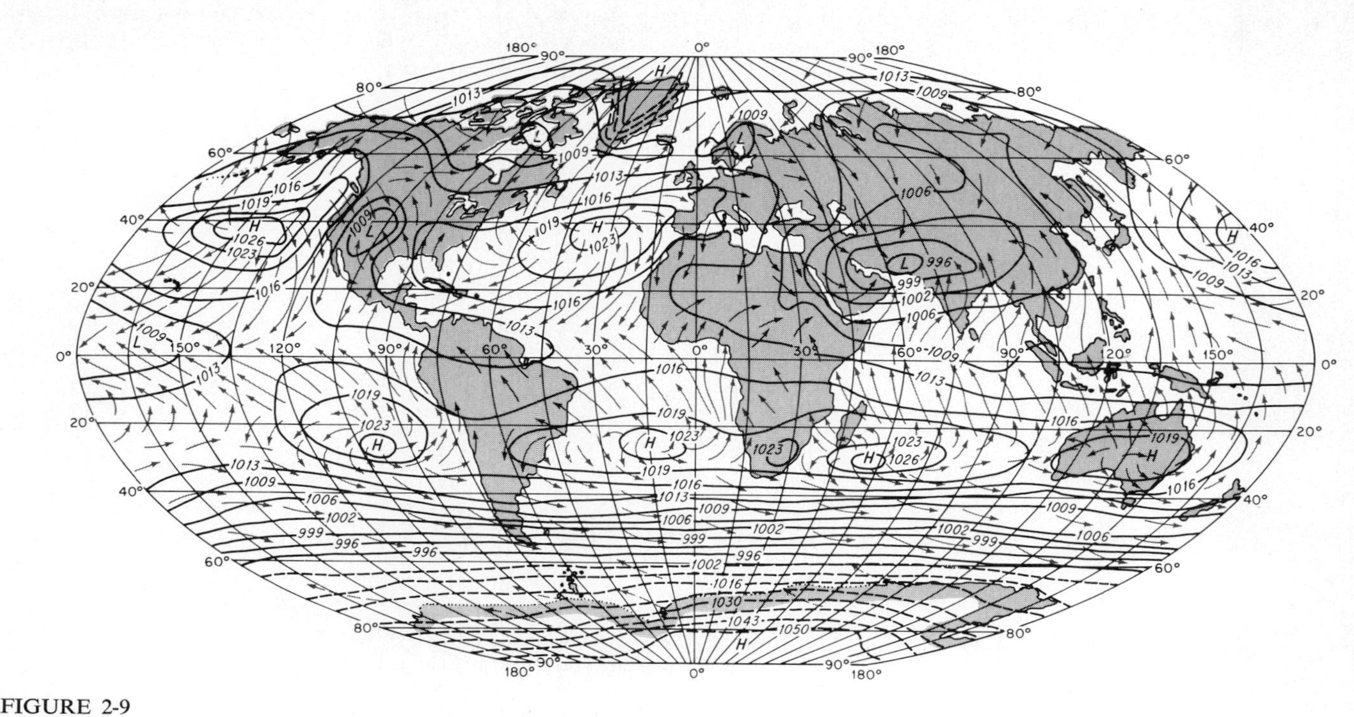

FIGURE 2-9
July mean sea-level pressure, in millibars, and prevailing winds. (*From V. C. Finch, G. T. Trewartha, A. H. Robinson, and E. H. Hammond, "Physical Elements of Geography," 4th ed., Copyright* 1957, *McGraw-Hill Book Company. Used with permission of McGraw-Hill Book Company.*)

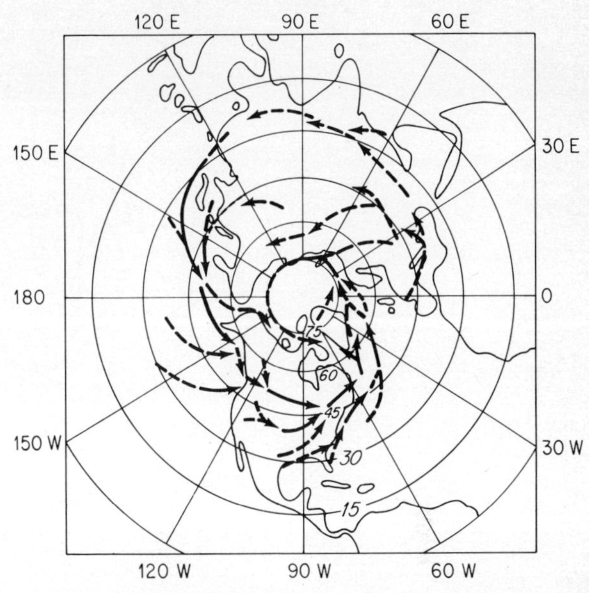

FIGURE 2-10
Principal tracks of Northern Hemisphere sea-level cyclones in January. Solid lines denote most frequent well-defined tracks; dashed lines, less frequent and less well-defined tracks. Locally preferred regions of genesis are indicated where tracks begin. Arrowheads end where cyclone frequency is a local minimum. (*From W. H. Klein, Principal Tracks and Mean Frequencies of Cyclones and Anticyclones in the Northern Hemisphere. U.S. Weather Bur. Res. Pap.* 40, 1957.)

and may develop into *hurricanes*, or *typhoons*, with winds exceeding 75 mi/h (33 m/s) over areas as large as 200 mi (300 km) in diameter. *Extratropical cyclones* usually form along the boundaries between warm and cold air masses. Such cyclones are usually larger than tropical cyclones and may produce precipitation over thousands of square miles. Figures 2-10 and 2-11 show average tracks of cyclones, or storm tracks, in January and July, respectively, in the Northern Hemisphere. An *anticyclone* is an area of relatively high pressure in which the winds tend to blow spirally outward in a clockwise direction in the Northern Hemisphere. Details on the general circulation and on the structure of cyclones and anticyclones can be found in meteorological textbooks.

2-10 Fronts

A *frontal surface* is the boundary between two adjacent air masses of different temperature and moisture content. Frontal "surfaces" are actually layers or zones of transition. Their thickness, however, is small with respect

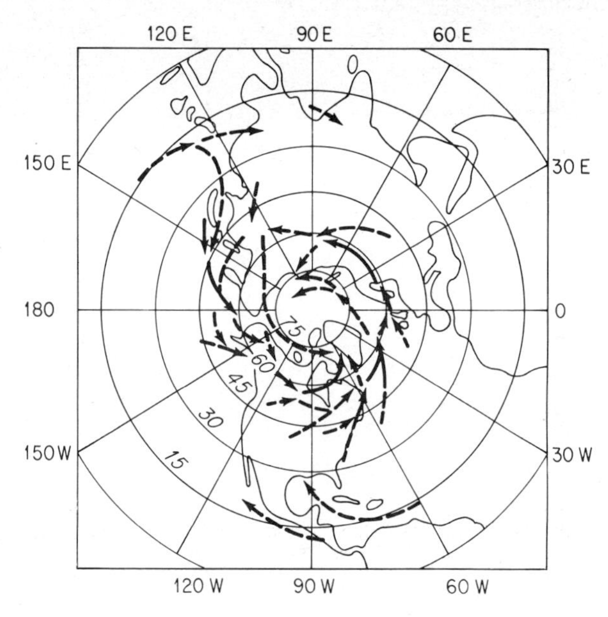

FIGURE 2-11
Principal tracks of Northern Hemisphere sea-level cyclones in July. (see legend of Fig. 2-10 for explanation). (*From W. H. Klein, Principal Tracks and Mean Frequencies of Cyclones and Anticyclones in the Northern Hemisphere, U.S. Weather Bur. Res. Pap.* 40, 1957.)

to dimensions of air masses. The line of intersection of a frontal surface with the earth is called a *surface front*. An *upper-air front* is formed by the intersection of two frontal surfaces aloft and hence marks the boundary between three air masses. If the air masses are moving so that warm air displaces colder air, the front is called a *warm front*; conversely, it is a *cold front* if cold air is displacing warmer air. If the front is not moving, it is called a *stationary front*.

The life history of a typical extratropical cyclone is shown in Fig. 2-12. For reasons not yet completely understood but with the jet stream apparently often a factor, a wave is generated on the boundary between two air masses (Fig. 2-12*B*). Under conditions of dynamic stability, this wave moves along the front with little change in form and with little or no precipitation. If the wave is unstable, and particularly with a jet stream aloft, it progresses through the successive stages of Fig. 2-12. In stage *C* the cyclone is deepening and has a well-formed warm sector. Warm air from the warm sector is forced upward along the warm-front surface, causing widespread precipita-

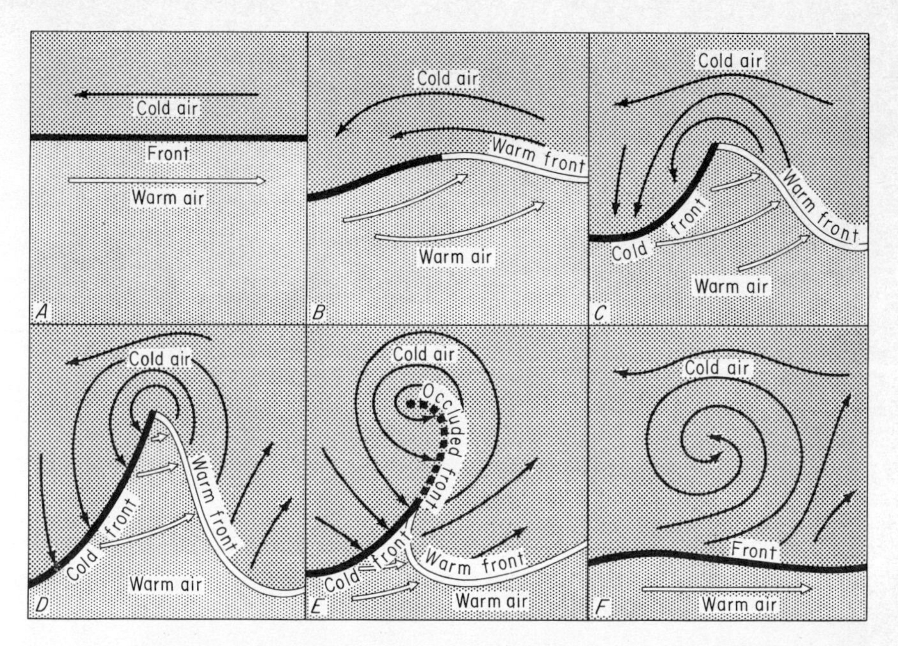

FIGURE 2-12
Life cycle of a Northern Hemisphere frontal cyclone: (*A*) surface front between cold and warm air; (*B*) wave beginning to form; (*C*) cyclonic circulation and wave have developed; (*D*) faster-moving cold front is overtaking retreating warm front and reducing warm sector; (*E*) warm sector has been eliminated, and cyclone is dissipating. (*U.S. National Weather Service.*)

tion ahead of the surface front. At the same time, the advancing wedge of cold air behind the cold front is lifting the warm air and causing convective showers behind the cold front. Showers also frequently occur in the warm sector.

Cold fronts move faster than warm fronts and usually overtake them (Fig. 2-12*D*, *E*). This process is called *occlusion*, and the resulting surface front is called an *occluded front*. Cloudiness associated with an occluding system is illustrated in Fig. 2-13, which also shows the position of the jet stream in relation to the cloudiness and surface position of the occluding system. As occlusion progresses, the warm sector is displaced farther from the cyclone center (Fig. 2-12*E*), which is eventually deprived of the warm air necessary to maintain its energy. Cold air replaces the warm, the center fills, and the occluded front is destroyed. A new cyclone center may form, however, at the peak of the remaining warm sector. The time from initial wave development to complete occlusion is usually of the order of 3 to 4 days.

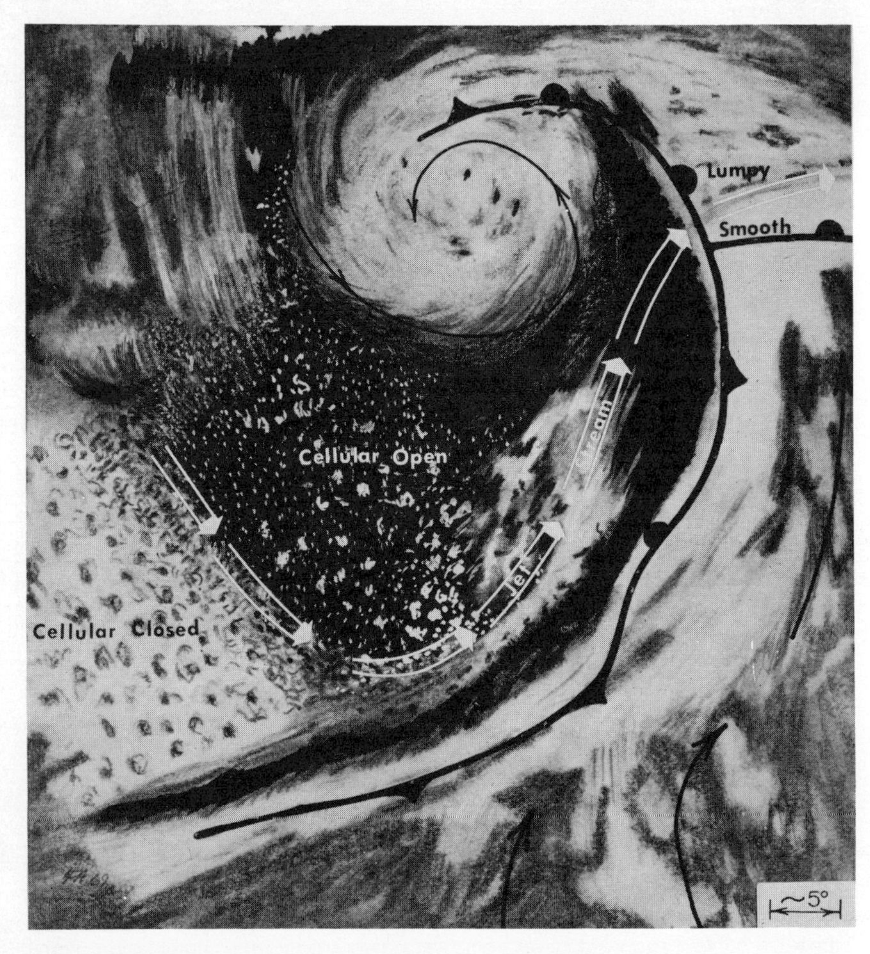

FIGURE 2-13
Idealized illustration showing relation of jet stream to occluding surface system. Note that the jet stream is over the boundary between open and closed cellular clouds and crosses the occluded front just above the junction of cold and warm fronts. (*U.S. National Environmental Satellite Center.*)

TEMPERATURE

2-11 Measurement of Temperature

In order to measure air temperature properly, thermometers must be placed where air circulation is relatively unobstructed, and yet they must be protected from the direct rays of the sun and from precipitation. In the United

FIGURE 2-14
Instrument shelter with maximum and minimum thermometers. (*U.S. National Weather Service.*)

States thermometers are placed in white, louvered, wooden *instrument shelters* (Fig. 2-14) through which the air can move readily. The shelter location must be typical of the area for which the measured temperatures are to be representative. Because of marked vertical temperature gradients just above the soil surface, all shelters should be about the same height above the ground for recorded temperatures to be comparable. In the United States shelters are set at about $4\frac{1}{2}$ ft (1.4 m) above the ground.

There are about 6000 stations in the United States for which official temperature records are compiled. Except for a few hundred stations equipped or staffed to obtain continuous or hourly temperatures, most make a daily observation consisting of the current, maximum, and minimum temperatures. The *minimum thermometer*, of the alcohol-in-glass type, has an index which remains at the lowest temperature occurring since its last setting. The *maximum thermometer* has a constriction near the bulb which prevents the mercury from returning to the bulb as the temperature falls and thus registers the highest temperature since its last setting. The *thermograph*, with either a bimetallic strip or a metal tube filled with alcohol or mercury for its thermometric element, makes an autographic record on a chart. Electrical-resistance thermometers, thermocouples, gas-bulb thermometers, and other types of instruments are used for special purposes. Electrical-resistance thermometers, for example, are widely used for measuring upper-air temperatures and in dew-cell hygrometers, which are described briefly in the section on humidity.

2-12 Terminology

A knowledge of terminology and methods of computation is required in order to avoid misuse of published temperature data. The terms *average*, *mean*, and *normal* are arithmetic means. The first two are used interchangeably, but the *normal* [6], generally used as a standard of comparison, is the average value for a particular date, month, season, or year over a specific 30-y period (1941 to 1970 as of 1974). Plans call for recomputing the 30-y normals every decade, dropping off the first 10 y and adding the most recent 10 y.

The *mean daily temperature* may be computed by several methods [7]. The most accurate practical method is to average hourly temperatures. Acceptably accurate results can be obtained by averaging 3- or 6-h observations, although the random error for an individual day with irregular variations may be of some importance, especially for 6-h observations. In some countries, climatological observations are made at hours (usually three per day: morning, noon, and evening) selected to permit computation of acceptable daily means by a formula which gives the mean as a linear function of the observed values with constants depending on observation times, time of year, and location.

In the United States the *mean daily temperature* is the average of the daily maximum and minimum temperatures. This yields a value usually less than a degree above the true daily average. Once-daily temperature observations are usually made about 7 A.M. or 5 P.M. Temperatures are published as of the date of the reading even though the maximum or minimum may have occurred on the preceding day. Mean temperatures computed from evening readings tend to be slightly higher than those from midnight readings. Morning readings yield mean temperatures with a negative bias, but the difference is less than that for evening readings. The maximum effect [8] on mean temperature of arbitrary changes in observation time varies with place and season, and may exceed 3 Fahrenheit (1.6 Celsius) degrees.

The *normal daily temperature* is the average daily mean temperature for a given date computed for a specific 30-y period. The *daily range* in temperature is the difference between the highest and lowest temperatures recorded on a particular day. The *mean monthly temperature* is the average of the mean monthly maximum and minimum temperatures. The *mean annual temperature* is the average of the monthly means for the year.

The *degree-day* is a departure of one degree for one day in the mean daily temperature from a specified base temperature. For snowmelt computations, the number of degree-days for a day is equal to the mean daily temperature minus the base temperature, all negative differences being taken as zero. The number of degree-days in a month or other time interval is the total of the daily values. Published degree-day values are for heating and cooling purposes and are based on departures below and above 65°F (18°C).

2-13 Lapse Rates

The *lapse rate*, or vertical temperature gradient, is the rate of change of temperature with height in the free atmosphere. The mean lapse rate is a decrease of about 0.7 Celsius degrees per 100 m (3.8 Fahrenheit degrees per 1000 ft) increase in height. The greatest variations in lapse rate are found in the layer of air just above the land surface. The earth radiates heat energy to space at a relatively constant rate which is a function of its absolute temperature, in kelvins.† Incoming radiation at night is less than the outgoing, and the temperature of the earth's surface and of the air immediately above it decreases. The surface cooling sometimes leads to an increase of temperature with altitude, or *temperature inversion*, in the surface layer. This condition usually occurs on still, clear nights because there is little turbulent mixing of air and because outgoing radiation is unhampered by clouds. Temperature inversions are also observed at higher levels when warm-air currents overrun colder air.

In the daytime there is a tendency for steep lapse rates because of the relatively high temperatures of the air near the ground. This daytime heating usually destroys a surface radiation inversion by early forenoon. As the heating continues, the lapse rate in the lower layers of the air steepens until it may reach the *dry-adiabatic lapse rate* (1 Celsius degree per 100 m, or 5.4 Fahrenheit degrees per 1000 ft), which is the rate of temperature change of unsaturated air resulting from expansion or compression as the air rises (lowering pressure) or descends (increasing pressure) without heat being added or removed.

Air having a dry-adiabatic lapse rate mixes readily, whereas a temperature inversion indicates a stable condition in which warm lighter air overlies cold denser air. Under optimum surface heating conditions, the air near the ground may be heated sufficiently for the lapse rate in the lowest layers to become *superadiabatic*, i.e., exceeding 1 Celsius degree per 100 m (5.4 Fahrenheit degrees per 1000 ft). This is an unstable condition since any parcel of air lifted dry-adiabatically remains warmer and lighter than the surrounding air and thus has a tendency to continue rising.

If a parcel of saturated air is lifted adiabatically, its temperature will decrease and its water vapor will condense, releasing latent heat of vaporization. This heat reduces the cooling rate of the ascending air. Hence, the *saturated-adiabatic lapse rate* is less than the dry-adiabatic. It varies inversely with the water-vapor content, and hence temperature, of the air. The average value for the lower layers at temperatures above freezing is roughly half the dry-adiabatic. At very low temperatures or high altitudes there is little

† In meteorology, absolute temperature is usually given in degrees Kelvin, called kelvins.

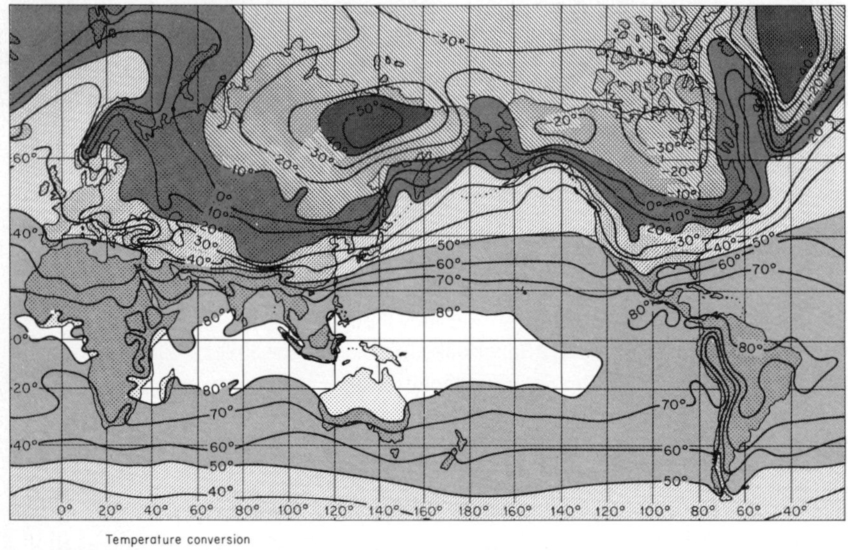

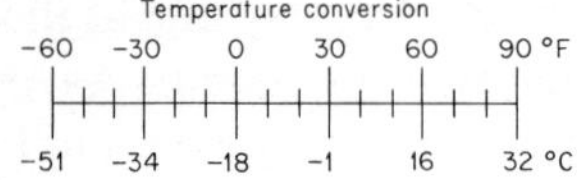

FIGURE 2-15
Average January temperature, in degrees Fahrenheit. (*From Arthur N. Strahler, "Introduction to Physical Geography," 2d ed., p. 64. Copyright © 1965, 1970, by John Wiley & Sons, Inc. Reprint by permission.*)

difference between the two lapse rates because of the very small amounts of water vapor available.

If the moisture in the rising air is precipitated as it is condensed, the temperature of the air will decrease at the *pseudo-adiabatic lapse rate*, which differs very little from the saturated-adiabatic. Actually, the process is not strictly adiabatic, as heat is carried away by the falling precipitation. A layer of saturated air having a saturated- or pseudo-adiabatic lapse rate is said to be in *neutral equilibrium*. If its lapse rate is less than the saturated- or pseudo-adiabatic, the air is stable; if greater, unstable.

2-14 Geographic Distribution of Temperature

In general, surface air temperature tends to be highest at low latitudes and to decrease poleward (Figs. 2-15 and 2-16). However, this trend is greatly distorted by the influence of land and water masses, topography, and vegetation. In the interior of large islands and continents, temperatures are higher in summer and lower in winter than on coasts at corresponding latitudes.

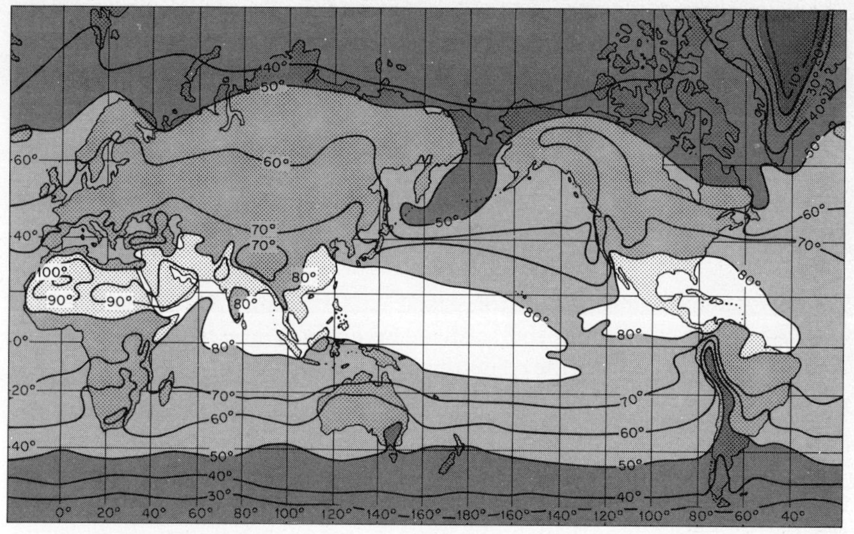

FIGURE 2-16
Average July temperature, in degrees Fahrenheit. (*From Arthur N. Strahler, "Introduction to Physical Geography," 2d ed., p. 66. Copyright* © 1965, 1970, *by John Wiley & Sons, Inc. Reprint by permission.*)

Temperatures at high elevations are colder than at low levels, and southern slopes have warmer temperatures than northern slopes. The average rate of decrease of surface air temperature with elevation is usually between $\frac{1}{2}$ and 1 Celsius degree per 100 m (3 and 5 Fahrenheit degrees per 1000 ft). Forested areas have higher daily minimum and lower daily maximum temperatures than barren areas. The mean temperature in a forested area may be 1 to 2 Celsius (2 to 4 Fahrenheit) degrees lower than in comparable open country, the difference being greater in summer.

The heat from a large city, which may roughly equal one-third of the solar radiation reaching it, produces local distortions in the temperature pattern so that temperatures recorded in cities may not represent the surrounding region. The mean annual temperature of cities averages about 1 Celsius degree (2 Fahrenheit degrees) higher than that of the surrounding region, most of the difference resulting from higher daily minima in the cities. Any comparison of city and country temperatures must allow for differences in exposure of thermometers. In cities the instrument shelters are sometimes located on roofs. On still, clear nights, when radiational cooling is particularly effective, the temperature on the ground may be as much as 8 Celsius (15 Fahrenheit) degrees colder than that at an elevation of 30 m (100 ft). A slight gradient in the opposite direction is observed on windy or cloudy nights. Daytime maxima tend to be lower at rooftop level than at the ground. In general, the average temperature from roof exposures is slightly lower than that on the ground.

2-15 Time Variations of Temperature

In continental regions the warmest and coldest points of the annual temperature cycle lag behind the solstices by about 1 month. In the United States, January is usually the coldest month and July the warmest. At oceanic stations the lag is nearer 2 months, and the temperature difference between the warmest and coldest months is much less.

The daily variation of temperature lags slightly behind the daily variation of solar radiation. The temperature begins to rise shortly after sunrise, reaches a peak 1 to 3 h (about $\frac{1}{2}$ h at oceanic stations) after the sun has reached its highest altitude, and falls through the night to a minimum about sunrise. The daily range of temperature is affected by the state of the sky. On cloudy days the maximum temperature is lower because of reduced insolation, and the minimum is higher because of reduced outgoing radiation. The daily range is also smaller over oceans.

HUMIDITY

2-16 Properties of Water Vapor

The process by which liquid water is converted into vapor is called *vaporization* or *evaporation*. Molecules of water having sufficient kinetic energy to overcome the attractive forces tending to hold them within the body of liquid water are projected through the water surface. Since kinetic energy increases and surface tension decreases as temperature rises, the rate of evaporation increases with temperature. Most atmospheric vapor is the product of evaporation from water surfaces. Molecules may leave a snow or ice surface in the same manner as they leave a liquid. The process whereby a solid is transformed directly to the vapor state, and vice versa, is called *sublimation*.

In any mixture of gases, each gas exerts a *partial pressure* independent of the other gases. The partial pressure exerted by water vapor is called *vapor pressure*. If all the water vapor in a closed container of moist air with an initial total pressure p were removed, the final pressure p' of the dry air alone would be less than p. The vapor pressure e would be the difference between the pressure of the moist air and that of the dry air, or $p - p'$.

Practically speaking, the maximum amount of water vapor that can exist in any given space is a function of temperature and is independent of the coexistence of other gases. When the maximum amount of water vapor for a given temperature is contained in a given space, the space is said to be *saturated*. The more common expression "the air is saturated" is not strictly correct. The pressure exerted by the vapor in a saturated space is called the *saturation vapor pressure*, which, for all practical purposes, is the maximum vapor pressure possible at a given temperature (Appendix Tables B-9 and B-10).

The process by which vapor changes to the liquid or solid state is called *condensation*. In a space in contact with a water surface, condensation and vaporization always go on simultaneously. If the space is not saturated, the rate of vaporization will exceed the rate of condensation, resulting in a net evaporation.† If the space is saturated, the rates of vaporization and condensation balance, provided that the water and air temperatures are the same.

Since the saturation vapor pressure over ice is less than that over water at the same temperature, the introduction of ice into a space saturated with respect to liquid water at the same or higher temperature will result in condensation of vapor on the ice.

Vaporization removes heat from the liquid being vaporized, while condensation adds heat. The *latent heat of vaporization* is the amount of heat absorbed by a unit mass of a substance, without change in temperature, while passing from the liquid to the vapor state. The change from vapor to the liquid state releases an equivalent amount of heat.

The heat of vaporization of water H_v in calories per gram varies with temperature but can be determined accurately up to 40°C (104°F) by

$$H_v = 597.3 - 0.564T \qquad (2\text{-}1)$$

where T is the temperature in degrees Celsius.

The *latent heat of fusion* for water is the amount of heat required to convert one gram of ice to liquid water at the same temperature. When one gram of liquid water at 0°C (32°F) freezes into ice at the same temperature, the latent heat of fusion (79.7 cal/g) is liberated.

The *latent heat of sublimation* for water is the amount of heat required to convert one gram of ice into vapor at the same temperature without passing through the intermediate liquid state. It is equal to the sum of the latent heat of vaporization and the latent heat of fusion. At 0°C (32°F) it is about 677 cal/g. Direct condensation of vapor into ice at the same temperature liberates an equivalent amount of heat.

† In hydrology, net evaporation is termed simply *evaporation*.

The *specific gravity* of water vapor is 0.622 times that of dry air at the same temperature and pressure. The density of water vapor ρ_v in grams per cubic centimeter is

$$\rho_v = 0.622 \frac{e}{R_g T} \tag{2-2}$$

where T is the absolute temperature in kelvins and R_g, the gas constant, equals 2.87×10^3 when the vapor pressure e is in millibars.†

The density of dry air ρ_d in grams per cubic centimeter is

$$\rho_d = \frac{p_d}{R_g T} \tag{2-3}$$

where p_d is the pressure in millibars.

The density of moist air is equal to the mass of water vapor plus the mass of dry air in a unit volume of the mixture. If p_a is the total pressure of the moist air, $p_a - e$ will be the partial pressure of the dry air alone. Adding Eqs. (2-2) and (2-3) and substituting $p_a - e$ for p_d gives

$$\rho_a = \frac{p_a}{R_g T}\left(1 - 0.378 \frac{e}{p_a}\right) \tag{2-4}$$

This equation shows that moist air is lighter than dry air.

2-17 Terminology

There are many expressions used for indicating the moisture content of the atmosphere. Each serves special purposes, and only those expressions common to hydrologic uses are discussed here. *Vapor pressure* e, usually expressed in millibars but sometimes in inches of mercury, is the pressure exerted by the vapor molecules. It is most commonly used in meteorology and hydrology to denote the partial pressure of the water vapor in the atmosphere. The *saturation vapor pressure* e_s is the maximum vapor pressure in saturated space and is a function of temperature alone. At any given temperature below the freezing point the saturation vapor pressure over liquid water is slightly greater than that over ice. The difference is at a maximum at about −12°C (10°F), but the ratio of the vapor pressures increases with decreasing temperature. Vapor pressure over water is generally used for most meteorological purposes regardless of temperature.

† The millibar is the standard unit of pressure in meteorology. It is equivalent to a force of 1000 dynes/cm², 0.0143 lb/in.², or 0.0295 inches of mercury (abbreviated in. Hg). Mean sea-level air pressure is 1013 millibars.

The computation of saturation vapor pressure is somewhat complicated, and its values are obtained generally from psychrometric tables such as those in Appendix B and in Smithsonian Meteorological Tables [9], which are based on the Goff-Gratch formula [10]. This formula yields values of saturation vapor pressure over water that are approximated to within 1 percent in the range of −50 to +55°C (−58 to +131°F) by the much simpler equation [11]

$$e_s \approx 33.8639[(0.00738T + 0.8072)^8 - 0.000019|1.8T + 48| + 0.001316] \tag{2-5}$$

where e_s is in millibars and T is in degrees Celsius.

It is sometimes necessary to convert vapor pressure over water to that over ice, or vice versa. Ratios for effecting these conversions are found in Table B-11. The following equation [12], derived for making the conversions on a computer, yields ratios to within 0.1 percent for the range 0 to −50°C (32 to −58°F):

$$\frac{e_{s,\,\text{ice}}}{e_{s,\,\text{water}}} \approx 1 + 0.00972T + 0.000042T^2 \tag{2-6}$$

where e_s is in millibars and T is in degrees Celsius.

The *dewpoint* T_d is the temperature at which space becomes saturated when air is cooled under constant pressure and with constant water-vapor content. It is the temperature having a saturation vapor pressure e_s equal to the existing vapor pressure e.

When the relative humidity is known, the dewpoint can be approximated to within 0.3 Celsius (0.5 Fahrenheit) degrees in the temperature range of −40 to 50°C (−40 to 122°F) by the following formula [13], which yields the dewpoint depression to be subtracted from the temperature to obtain the dewpoint

$$T - T_d \approx (14.55 + 0.114T)X + [(2.5 + 0.007T)X]^3 + (15.9 + 0.117T)X^{14} \tag{2-7}$$

where T is in degrees Celsius and X is the complement of the relative humidity f expressed as a decimal fraction, or $X = 1.00 - f/100$.

The *relative humidity* f is the percentage ratio of the actual to the saturation vapor pressure and is therefore a ratio of the amount of moisture in a given space to the amount the space could contain if saturated.

$$f = 100\,\frac{e}{e_s} \tag{2-8}$$

Relative humidity can also be computed directly from air temperature T and dewpoint T_d by an approximate formula [14] in convenient form for computer use:

$$f \approx \left(\frac{112 - 0.1T + T_d}{112 + 0.9T}\right)^8 \tag{2-9}$$

where temperatures are in degrees Celsius. The formula approximates relative humidity to within 1.2 percent in the meteorological range of temperatures and humidities and to within 0.6 percent in the range of -25 to 45°C (-13 to 113°F).

Psychrometric tables like those of Appendix B usually give dewpoint and relative humidity as a function of air temperature and *wet-bulb depression*, i.e., the difference between air and wet-bulb temperatures (Sec. 2-18).

The *specific humidity* q_h, usually expressed in grams per kilogram, is the mass of water vapor per unit mass of moist air:

$$q_h = 622 \frac{e}{p_a - 0.378e} \approx 622 \frac{e}{p_a} \tag{2-10}$$

where p_a is the total pressure of the air in millibars.

The *mixing ratio* w_r is the mass of water vapor per unit mass of perfectly dry air in a humid mixture. Usually expressed in grams per kilogram of dry air, it is given by

$$w_r = 622 \frac{e}{p_a - e} \tag{2-11}$$

The total amount of water vapor in a layer of air is often expressed as the depth of *precipitable water* W_p, in millimeters or inches, even though there is no natural process capable of precipitating the entire moisture content of the layer. The amount of precipitable water in any air column of considerable height is computed [15] by increments of pressure or height from surface and upper-air observations of temperature, humidity, and pressure. A convenient formula [16] for computing W_p in millimeters is

$$W_p = \sum 0.01\bar{q}_h \, \Delta p_a \tag{2-12}$$

where the pressure p_a is in millibars and $\bar{q}_h$, in grams per kilogram, is the average of the specific humidities at the top and bottom of each layer. Precipitable water may be approximated conveniently by nomograms [17] using dewpoints at specific pressure levels. Less accurate approximations, which may be considerably in error under certain conditions, can be made from surface dewpoints [18] or vapor pressure and assumed temperature and humidity lapse rates or relations based on observed precipitable water. Figure 2-17 gives the depth of precipitable water in a column of saturated air

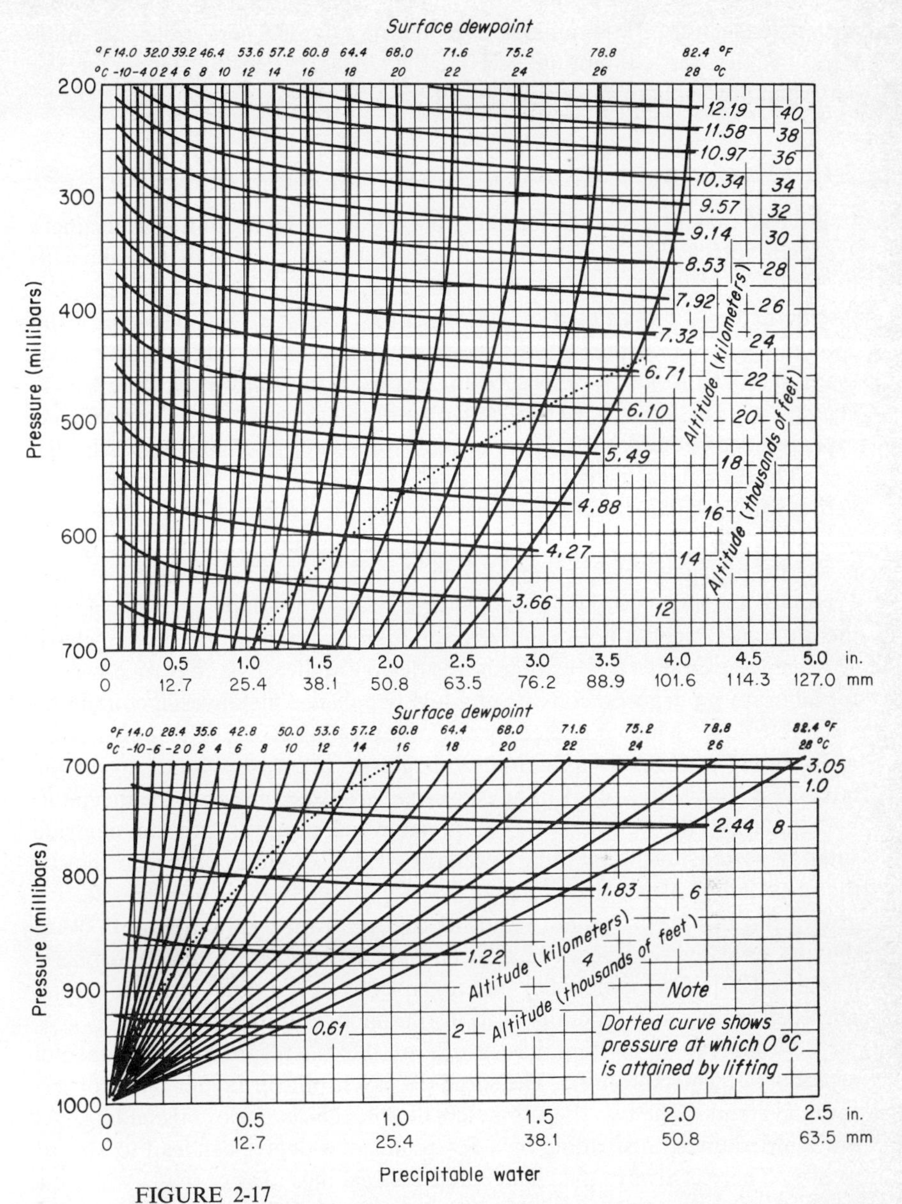

FIGURE 2-17
Depths of precipitable water in a column of air of any height above the 1000-millibar level as a function of the 1000-millibar dewpoint, assuming saturation and pseudo-adiabatic lapse rate. (*U.S. National Weather Service.*)

with its base at the 1000-millibar level and its top anywhere up to 200 millibars. Tables for computing precipitable water in various layers of the saturated atmosphere are available [19].

2-18 Measurement of Humidity

In general, measurements of humidity in the surface layers of the atmosphere are made with a psychrometer, which consists of two thermometers, one with its bulb covered by a jacket of clean muslin saturated with water. The thermometers are ventilated by whirling or by use of a fan. Because of the cooling effect of evaporation, the moistened, or *wet-bulb*, thermometer reads lower than the dry, the difference in degrees being known as the *wet-bulb depression*. The air and wet-bulb temperatures are used to obtain various expressions of humidity by reference to psychrometric tables (Appendix B).

The *hair hygrometer* makes use of the fact that the length of a hair varies with relative humidity. The changes are transmitted to a pointer indicating the relative humidity on a graduated scale. The *hair hygrograph* is a hair hygrometer operating a pen marking a trace on a chart. The *hygrothermograph*, combining the features of both the hair hygrograph and thermograph, records both relative humidity and temperature on one chart. A *dewpoint hygrometer*, which measures dewpoint directly and is used mostly for laboratory purposes, consists of a highly polished metal vessel containing a suitable liquid which is cooled by any of several methods. The temperature of the liquid at the time condensation begins to occur on the exterior of the metal vessel is the dewpoint. The *dew-cell hygrometer* measures the dewpoint by regulating the temperature of a saturated aqueous lithium chloride solution so that the water-vapor pressure of the solution is equal to that of the surrounding atmosphere. The *spectral hygrometer* measures the selective absorption of light in certain bands of the spectrum by water vapor. With the sun as a light source, it has been used to measure total atmospheric moisture [20]. Other humidity instruments have been developed for special purposes but are not generally used in routine operational activities.

Measurement of humidity is one of the least accurate instrumental procedures in meteorology. The standard psychrometer invites many observational errors. The two thermometers double the chance of misreading. At low temperatures, misreading by a few tenths of a degree can lead to absurd results. There is always the chance that the readings are not made when the wet-bulb thermometer is at its lowest temperature. In addition, there are errors with a positive bias resulting from insufficient ventilation, dirty or too thick muslin, and impure water.

Any instrument using a hair element is subject to appreciable error. The hair expands with increasing temperature, and its response to changes in humidity is very slow, the lag increasing with decreasing temperature until it becomes almost infinite at about −40°C (−40°F). This can lead to significant error in upper-air soundings, where large ranges in temperature and sharp variations of humidity with altitude are observed. Consequently, sounding instruments are now equipped with electrical hygrometers using various humidity elements. A carbon element [21] has been found most satisfactory of those tested so far.

At freezing temperatures there is some uncertainty whether the dewpoint or frost point is being measured by dewpoint hygrometers. The difference may lead to appreciable errors in computing relative humidity [22] and vapor pressure.

2-19 Geographic Distribution of Humidity

Atmospheric moisture tends to decrease with increasing latitude, but relative humidity, being an inverse function of temperature, tends to increase. Atmospheric moisture is greatest over oceans and decreases with distance inland. It also decreases with elevation and is greater over vegetation than over barren soil. The distribution of the average precipitable water over the Northern Hemisphere for 1958 is shown in Fig. 2-18. The distribution depicted is fairly representative of the mean annual pattern since there is generally relatively little year-to-year variation.

2-20 Time Variations in Humidity

Like temperature, atmospheric water vapor is at a minimum in winter and at a maximum in summer. In the Northern Hemisphere the driest months are January and February, and the most humid are July and August. In the middle and high latitudes average monthly precipitable water over continental interiors in the driest months is about half the mean annual; in the most humid months, about twice the mean annual. The seasonal variation is much less over oceans and coastal areas and is at a minimum over tropical seas. Unlike actual water vapor content, relative humidity is at a minimum in summer and at a maximum in winter.

The diurnal variation of atmospheric moisture is normally small, except where land and sea breezes bring air of differing characteristics. Near the ground surface, condensation of dew at night and reevaporation during the day may result in a minimum moisture content near sunrise and a maximum by noon. Relative humidity, of course, behaves in a manner opposite to that of temperature, being at a maximum in the early morning and at a minimum in the afternoon.

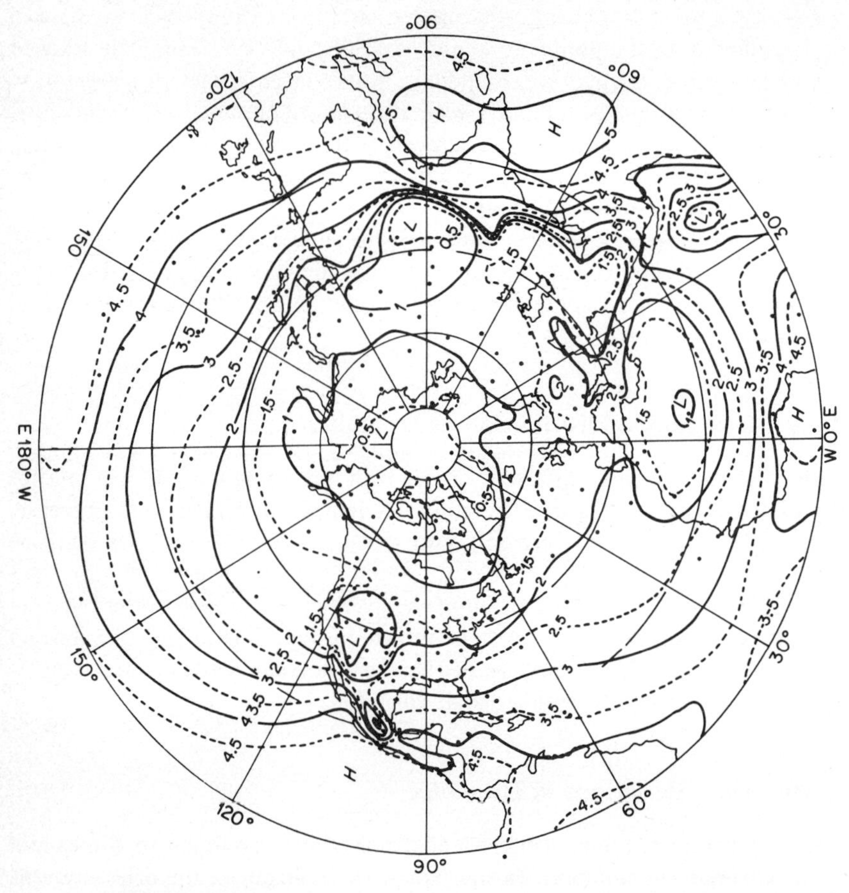

FIGURE 2-18
Average precipitable water, in centimeters, over the Northern Hemisphere for 1958. (*From V. P. Starr, J. P. Peixoto, and A. R. Crisi, Hemispheric Water Balance for the IGY, Tellus, vol.* 17, *no.* 14, *pp.* 463–472, 1965. *Used with permission of Svenska Geofysiska Foreningen.*)

WIND

Wind, which is air in motion, is a very influential factor in several hydrometeorological processes. Moisture and heat are readily transferred to and from air, which tends to adopt the thermal and moisture conditions of the surfaces with which it is in contact. Stagnant air in contact with a water surface eventually assumes the vapor pressure of the surface, so that no evaporation takes place. Similarly, stagnant air over a snow or ice surface eventually assumes the temperature and vapor pressure of the surface, so that melting by convection and condensation ceases. Consequently, wind exerts considerable influence in evaporative and snowmelt processes. It is also important in the production of precipitation, since it is only through sustained inflow of moist air into a storm that precipitation can be maintained.

2-21 Measurement of Wind

Wind has both speed and direction. The *wind direction* is the direction *from* which it is blowing. Direction is usually expressed in terms of 16 compass points (N, NNE, NE, ENE, etc.) for surface winds, and for winds aloft in degrees from north, measured clockwise. Wind speed is usually given in miles per hour, meters per second, or knots (1 m/s = 2.237 mi/h = 1.944 kn, and 1 kn = 1.151 mi/h = 0.514 m/s).

Wind speed is measured by instruments called *anemometers*, of which there are several types. The *three-* or *four-cup anemometer* with a vertical axis of rotation is most commonly used for official observations. It tends to register too high a mean speed in a variable wind because the cups accelerate faster than they lose speed. Vertical currents (turbulence) tend to rotate the cups and cause overregistration of horizontal speeds. Most cup anemometers will not record speeds below 1 or 2 mi/h because of starting friction. The *propeller anemometer* has a horizontal axis of rotation. *Pressure-tube anemometers*, of which the Dines is the best known, operate on the pitot-tube principle.

While wind speed varies greatly with height above the ground, no standard anemometer level has been adopted. Differences in wind speed resulting from differences in anemometer height, which may range anywhere from 30 to several hundred feet above the ground, usually exceed the errors from instrumental deficiencies. However, approximate adjustment can be made for differences in height [Eq. (2-14), (2-15), or (2-16)].

2-22 Geographic Variation of Wind

In winter there is a tendency for surface winds to blow from the colder interior of land masses toward the warmer oceans (Sec. 2-8). Conversely, in summer the winds tend to blow from the cooler bodies of water toward the warmer land. Similarly, diurnal land and sea breezes may result from temperature contrasts between land and water.

On mountain ridges and summits wind speeds at 10 m (30 ft) or more above the ground are higher than in the free air at corresponding elevations because of the convergence of the air forced by the orographic barriers. On lee slopes and in sheltered valleys wind speeds are light. Wind direction is greatly influenced by orientation of orographic barriers. With a weak pressure system, diurnal variation of wind direction may occur in mountain regions, the winds blowing upslope in the daytime and downslope at night.

Wind speeds are reduced and directions deflected in the lower layers of the atmosphere because of friction produced by trees, buildings, and other obstacles. These effects become negligible above about 600 m (2000 ft), and this lower layer is referred to as the *friction layer*. Over land the surface wind speed averages about 40 percent of that just above the friction layer, and at sea about 70 percent.

Because of its location in the general circulation, most of the conterminous United States has prevailing westerly winds. However, the winds are generally variable since most of the country is affected by migratory pressure systems, with winds circulating clockwise in the high-pressure areas and counterclockwise in low-pressure systems as they move across the country.

The variation of wind speed with height, or the *wind profile*, in the friction layer is usually expressed by one of two general relationships, namely, the logarithmic velocity profile or the power-law profile. In hydrology, the relationships are most often used to estimate the wind speed in the *surface boundary layer*, i.e., the thin layer of air between the ground surface and the anemometer level, usually about 10 m (30 ft) but often lower at special test sites or experimental stations. The common requirement is the wind speed above a snow or water surface for computations of snowmelt and evaporation.

One of the more common forms [23] of the logarithmic velocity profile for meteorological purposes is

$$\frac{\bar{v}}{v_*} = \frac{1}{k} \ln \frac{z}{z_0} \qquad z \geq z_0 \tag{2-13}$$

Table 2-2 REPRESENTATIVE VALUES OF ROUGHNESS LENGTH z_0 AND FRICTION VELOCITY v_* FOR NATURAL SURFACES

Neutral stability; values of v_* corresponding to mean velocity $\bar{v}$ of 5.0 m/s (11 mi/h) at 2 m (6½ ft)

Type of surface	z_0		v_*	
	cm	in.	cm/s	ft/s
Very smooth (mud flats, ice)	0.001	0.0004	16	0.5
Lawn, grass up to 1 cm (0.4 in.) high	0.1	0.4	26	0.9
Downland, thin grass up to 10 cm (4 in.) high	0.7	0.28	36	1.2
Thick grass, up to 10 cm (4 in.) high	2.3	0.91	45	1.5
Thin grass, up to 50 cm (20 in.) high	5	2.0	55	1.8
Thick grass, up to 50 cm (20 in.) high	9	3.5	63	2.1

SOURCE: From O. G. Sutton, "Micrometeorology" Copyright 1953, McGraw-Hill Book Company. Used with permission of McGraw-Hill Book Company.

where $\bar{v}$ is the mean wind speed for at least a few minutes at height z above the ground; k is the von Kármán constant, generally taken equal to 0.4; z_0 is the *roughness length*, which is a measure of surface roughness and presumably the height at which wind speed becomes zero and must therefore be less than z; and v_* is the *friction velocity*, which is equal to $|\tau/\rho|^{1/2}$, τ being the shearing stress, or *Reynolds stress*, and ρ the air density.

In meteorological investigations of the surface boundary layer, τ is generally considered independent of height, with the surface value assumed to apply throughout the layer, that is, $\tau = \tau_0$. Hence, the friction velocity v_* depends on the nature of the surface and the mean wind speed $\bar{v}$. It usually ranges [24] from about 3 to 12 percent of the mean wind speed $\bar{v}$, the lower values being associated with smooth surfaces. A rough assumption sometimes made in meteorological studies is that v_* is approximately equal to $\bar{v}/10$. Some measurements of roughness length z_0 and friction velocity v_* for a mean wind speed $\bar{v}$ of 5 m/s (11 mi/h) at 2 m ($6\frac{1}{2}$ ft) above the ground are presented in Table 2-2. A more detailed listing of z_0 for a wide range of surface roughness is given in Table 2-3.

A value of one-thirtieth the average height of surface irregularities is often taken as a fair estimate of z_0, but the data of Table 2-3 suggest that this simple approximation could be greatly in error under certain conditions, especially in the case of brush and trees. Table 2-3 shows that z_0 varies inversely with wind speed in the case of tall grasses, which tend to flatten out as speed increases [32]. In the case of a water surface, which roughens as wind speed increases, z_0 tends to be greater for higher wind speeds. However, it appears that there are discontinuities in this relationship, and, at least in the range of 2 to 10 m/s (4 to 22 mi/h) at 10 m (30 ft) above the surface, z_0 alternately increases and decreases as wind speed increases [33].

A convenient form [34] of the logarithmic velocity profile for relating mean wind speed $\bar{v}$ at some height z to the measured mean wind speed $\bar{v}_1$ at some standard height z_1 is

$$\frac{\bar{v}}{\bar{v}_1} = \frac{\ln\left(\dfrac{z}{z_0} + 1\right)}{\ln\left(\dfrac{z_1}{z_0} + 1\right)} \tag{2-14}$$

Another convenient form [35] of the logarithmic velocity profile for computing mean wind speed $\bar{v}_2$ at some intermediate height z_2 when mean wind speeds $\bar{v}_1$ and $\bar{v}_3$ at heights z_1 and z_3 are known is

$$\bar{v}_2 = \bar{v}_3 - (\bar{v}_3 - \bar{v}_1)\frac{\ln(z_3/z_2)}{\ln(z_3/z_1)} \tag{2-15}$$

Table 2-3 ROUGHNESS LENGTH z_0

	Wind speed at $z = 2$ m ($6\frac{1}{2}$ ft)		Roughness length z_0		
	m/s	mi/h	cm	in.	Ref.
Open water	2.1	4.7	0.001	0.0004	25
Smooth mud flats	...	...	0.001	0.0004	26
Smooth snow on short grass	...	...	0.005	0.002	26
Wet soil	1.8	4.0	0.02	0.008	25
Desert	...	...	0.03	0.012	26
Snow on prairie	...	...	0.10	0.04	26
Mown grass:					
1.5 cm (0.6 in.)	...	...	0.2	0.08	26
3.0 cm (1.2 in.)	...	...	0.7	0.28	26
4.5 cm (1.8 in.)	2	4.5	2.4	0.94	26
4.5 cm (1.8 in.)	6–8	13–18	1.7	0.67	26
Alfalfa:					
20–30 cm (8–12 in.)	1.9	4.3	1.4	0.55	25
30–40 cm (12–16 in.)	1.9	4.3	1.3	0.51	25
Long grass:					
60–70 cm (24–28 in.)	1.5	3.4	9.0	3.54	26
60–70 cm (24–28 in.)	3.5	7.8	6.1	2.40	26
60–70 cm (24–28 in.)	6.2	13.9	3.7	1.46	26
Maize:					
90 cm (35 in.)	...	...	2.0	0.79	27
170 cm (67 in.)	...	...	9.5	3.74	27
300 cm (118 in.)	...	...	22.0	8.66	27
Sugar cane:					
100 cm (39 in.)	...	...	4.0	1.57	28
200 cm (79 in.)	...	...	5.0	1.97	28
300 cm (118 in.)	...	...	7.0	2.76	28
400 cm (157 in.)	...	...	9.0	3.54	28
Brush, 135 cm (53 in.)	...	...	14.0	5.51	29
Orange orchard, 350 cm (138 in.)	...	...	50.0	19.7	29
Pine forest:					
5 m (16 ft)	...	...	65.0	25.6	30
27 m (89 ft)	...	...	300.0	118.1	31
Deciduous forest, 17 m (56 ft)	...	...	270.0	106.3	30

SOURCE: Adapted from P. S. Eagleson, "Dynamic Hydrology." Copyright 1970, McGraw-Hill Book Company. Used with permission of McGraw-Hill Book Company.

For most meteorological purposes the power-law profile is usually expressed as

$$\frac{\bar{v}}{\bar{v}_1} = \left(\frac{z}{z_1}\right)^k \tag{2-16}$$

with exponent k varying with surface roughness and atmospheric stability and usually ranging from 0.1 to 0.6 (Table 2-4) in the surface boundary layer [36, 37].

Comparative tests of the two profiles have yielded inconclusive results. The literature suggests that the logarithmic law should be the more representative of the two relationships, but this is not supported by tests except under certain conditions. The logarithmic law has been found generally more representative of the wind profile in the lowest 5 to 8 m (15 to 25 ft) above the ground when the atmospheric temperature lapse rate was adiabatic or near adiabatic [38, 39]. However, Johnson [36] found greater wind-speed increases with height over prairie grass and a snow surface than was indicated by the logarithmic law even under adiabatic conditions.

The power-law profile is considered by some investigators [40–42] to be more representative of the wind profile in the layer from several meters to about 100 m (300 ft) above the ground. Analyzing results of his own investigations, which were based on 1-h averaging periods, and those of other investigators, DeMarrais [42] concluded that for this range of elevation

Table 2-4 REPRESENTATIVE VALUES OF k FOR VARIOUS TEMPERATURE LAPSE RATES

Surface	Height range		Super-adiabatic	Neutral	Stable	Inversion	Ref.
	m	ft					
Meadows	10–70	33–230	0.25	0.27	. . .	0.61	43
Flat field	11–49	36–161	0.16	0.20	0.25	0.36	44
Grass field	8–120	26–394	0.14	0.17	0.27	0.32–0.77†	40‡
Airfield	9–27	30–89	0.09	0.08	0.18	. . .	45
Desert	6–61	20–200	0.15	0.18	0.22	. . .	46
Near wooded area	11–124	36–407	0.19	0.29	0.35	. . .	46

SOURCE: Adapted by permission from [35].

† $T_{400\,\text{ft}}-T_{5\,\text{ft}}$, F°	0–2	2–4	4–6	6–8	8–10	10–12
k	0.32	0.44	0.59	0.62	0.63	0.77

‡ Captive-balloon observations; all other measurements from towers.

exponent k increases with surface roughness and atmospheric stability except in the case of large superadiabatic lapse rates, when it increases with instability. Exponent k also varies with height of layer considered. Higher layers have lesser values of k when the lapse rate is superadiabatic, and they have greater values of k when the lapse rate is adiabatic or less. For adiabatic and superadiabatic lapse rates, k tends to range from 0.1 to 0.3, the variation being mainly in proportion to surface roughness. This is shown in Table 2-4, which shows also that for stable conditions k tends to range from 0.2 to 0.8, the higher values being associated with temperature inversions. A value of $\frac{1}{7}$ has been found applicable for a wide range of conditions in the layer 0 to 10 m (0 to 30 ft). Over a snow surface, under conditions favoring melting, a value of $\frac{1}{6}$ may be more appropriate.

Despite differences between the two types of relationships, when z_1/z_0 and k in the logarithmic and power-law profiles, respectively, are determined from observed winds at two levels in the surface boundary layer, say at 1 and 10 m (3 and 30 ft), differences between wind speeds indicated by the two profiles are usually within the limits of accuracy of wind-measuring devices.

2-23 Time Variation of Winds

Wind speeds are highest and most variable in winter, whereas middle and late summer is the calmest period of the year. In winter westerly winds prevail over the United States up to at least 6 km (20,000 ft), except near the Gulf of Mexico, where there is a tendency for southeasterly winds up to about $1\frac{1}{2}$ km (5000 ft). In summer, while there is still a tendency for westerly winds, there is generally more variation of direction with altitude. In the plains west of the Mississippi River there is a tendency for southerly winds up to $1\frac{1}{2}$ km (5000 ft), and on the Pacific Coast the winds at the lower altitudes are frequently from the northwest.

The diurnal variation of wind is significant only near the ground and is most pronounced during the summer. Surface-wind speed is usually at a minimum about sunrise and increases to a maximum in early afternoon. At about 300 m (1000 ft) above the ground, the maximum occurs at night and the minimum in the daytime.

REFERENCES

1. E. R. Anderson, Water-Loss Investigations: Lake Hefner Studies, Technical Report, *U.S. Geol. Surv. Prof. Pap.* 269, pp. 71–119, 1954.
2. V. A. Myers, Infrared Radiation from Air to Underlying Surface, *ESSA Weather Bur. Tech. Note* 44-HYDRO-1, 1966.

3. E. A. Anderson and D. R. Baker, Estimating Incident Terrestrial Radiation under All Atmospheric Conditions, *Water Resour. Res.*, vol. 3, no. 4, pp. 975–988, 1967.
4. K. R. Cooley and S. B. Idso, A Comparison of Energy Balance Methods for Estimating Atmospheric Thermal Radiation, *Water Resour. Res.*, vol. 7, pp. 39–45, February 1971.
5. W. Smith and R. J. Younkin, An Operationally Useful Relationship between the Polar Jet Stream and Heavy Precipitation, *Mon. Weather Rev.*, vol. 100, pp. 434–440, June 1972.
6. World Meteorological Organization, A Note on Climatological Normals, *Tech. Note* 84, Geneva, 1967.
7. World Meteorological Organization, Guide to Climatological Practices, WMO no. 100, *Tech. Pap.* 44, pp. V.9–V.11, Geneva, 1960.
8. J. M. Mitchell, Jr., Effect of Changing Observation Time on Mean Temperature, *Bull. Am. Meteorol. Soc.,* vol. 39, pp. 83–89, February 1958.
9. R. J. List (ed.), "Smithsonian Meteorological Tables," 6th ed., pp. 350–373, Smithsonian Institution, Washington, 1966.
10. J. A. Goff and S. Gratch, Low-Pressure Properties of Water from −160 to 212°F, *Trans. Am. Soc. Heat. Vent. Eng.*, vol. 52, pap. 1286, pp. 95–122, 1946. The formula presented in this paper was adopted as standard in 1947 by Resolution 964 of the Twelfth Conference of Directors of the International Meteorological Organization.
11. J. F. Bosen, A Formula for Approximation of the Saturation Vapor Pressure over Water, *Mon. Weather Rev.*, vol. 88, p. 275, August 1960.
12. J. F. Bosen, Formula for Approximation of the Ratio of the Saturation Vapor Pressure over Ice to That over Water at the Same Temperature, *Mon. Weather Rev.*, vol. 92, p. 508, November 1961.
13. Developed by J. F. Bosen but never published.
14. J. F. Bosen, An Approximation Formula to Compute Relative Humidity from Dry Bulb and Dew Point Temperatures, *Mon. Weather Rev.*, vol. 86, p. 486, December 1958.
15. L. P. Harrison, Calculation of Precipitable Water, *ESSA Tech. Mem. WBTM TDL* 33, June 1970.
16. S. Solot, Computation of Depth of Precipitable Water in a Column of Air, *Mon. Weather Rev.*, vol. 67, pp. 100–103, April 1939.
17. K. R. Peterson, A Precipitable Water Nomogram, *Bull. Am. Meteorol. Soc.,* vol. 42, pp. 119–121, February 1961.
18. W. L. Smith, Note on the Relationship between Total Precipitable Water and Surface Dew Point, *J. Appl. Meteorol.*, vol. 5, pp. 726–727, October 1966. Comments by L. Berkofsky, *J. Appl. Meteorol.*, vol. 6, pp. 959–961, October 1967, and F. K. Schwarz, *J. Appl. Meteorol.*, vol. 7, pp. 509–510, June 1968.
19. For English units see Tables of Precipitable Water, *U.S. Weather Bur. Tech.*

Pap. 14, 1951. For metric units see W. O. Eihle, R. J. Powers, and R. A. Clark, Meteorological Tables for Determination of Precipitable Water, Temperatures, and Pressures Aloft for a Saturated Pseudoadiabatic Atmosphere—in the Metric System, *Tex. A & M Univ. Water Resour. Inst. Rep*. 16, December 1968.

20. N. B. Foster, D. T. Volz, and L. W. Foskett, A Spectral Hygrometer for Measuring Total Precipitable Water, in A. Wexler (ed.), "Humidity and Moisture: Measurement and Control in Science and Industry," vol. 1, pp. 455–464, Reinhold, New York, 1965.
21. S. L. Stine, Carbon Humidity Elements: Manufacture, Performance and Theory, in A. Wexler (ed.), "Humidity and Moisture: Measurement and Control of Science and Industry," vol. 1, pp. 316–330, Reinhold, New York, 1965.
22. H. S. Appleman, Relative Humidity Errors Resulting from Ambiguous Dew-Point Hygrometer Readings, *J. Appl. Meteorol*., vol. 3, pp. 113–115, February 1964.
23. O. G. Sutton, "Micrometeorology," pp. 78–85, 229–241, McGraw-Hill, New York, 1953.
24. See [23], p. 233.
25. C. H. M. Van Bavel, Potential Evapotranspiration, *Water Resour. Res*., vol. 2, pp. 455–467, 1966.
26. C. H. B. Priestly, "Turbulent Transfer in the Lower Atmosphere," p. 21, University of Chicago Press, Chicago, 1959.
27. Z. Uchijima and J. L. Wright, An Experimental Study of Air in the Corn-Plant Air Layer, *Bull. Natl. Inst. Agric. Sci.* (*Jap.*), ser. A, pp. 19–66, 1964.
28. Hawaiian Sugar Planters' Report, 1959.
29. C. B. Tanner and W. L. Pelton, Potential Evapotranspiration Estimates by the Approximate Energy Balance Method of Penman, *J. Geophys. Res*., vol. 65, pp. 3391–3413, October 1960.
30. J. L. Rauner, O Teplovom Balanse Listvennogo Lesav Zimniy Period, *Isv. Akad. Nauk. USSR, Ser. Geogr*., no. 4, 1965.
31. S. Tajchman, Energie und Wasserhaushalt verschiedener Pflanzenbestande bei München, Ph.D. thesis, *Univ. Muench. Meteorol. Inst. Wiss. Mitt*., no. 12, 1967.
32. G. Szeicz, G. Endrodi, and S. Tajchman, Aerodynamic and Surface Factors in Evaporation, *Water Resour. Res*., vol. 5, pp. 380–394, April 1969.
33. K. W. Ruggles, The Vertical Mean Wind Profile over the Ocean for Light to Moderate Winds, *J. Appl. Meteorol*., vol. 9, pp. 389–395, June 1970.
34. See [23], p. 84.
35. G. A. DeMarrais, Wind-Speed Profiles at Brookhaven National Laboratory, *J. Meteorol*., vol. 16, pp. 181–190, April 1959.
36. O. Johnson, An Examination of the Vertical Wind Profile in the Lowest Layers of the Atmosphere, *J. Meteorol*., vol. 16, pp. 144–148, April 1959.
37. R. Geiger, "The Climate near the Ground," pp. 116–117, Harvard University Press, Cambridge, Mass., 1965.

38. A. C. Best, Transfer of Heat and Momentum in the Lowest Layers of the Atmosphere, *Meteorol. Off. Geophys. Mem.*, vol. 7, no. 65, 1935.
39. C. W. Thornthwaite and P. Kaser, Wind-Gradient Observations, *Trans. Am. Geophys. Union*, vol. 24, pp. 166–182, 1943.
40. R. Frost, The Velocity Profile in the Lowest 400 Feet, *Meteorol. Mag.*, vol. 76, pp. 14–18, 1947.
41. R. Frost, Atmospheric Turbulence, *Q. J. R. Meteorol. Soc.* ,vol. 74, pp. 316–338, 1948.
42. See [35], p. 189.
43. E. Frankenberger and W. Rudloff, Windmessungen an den Quickborner Funkmasten 1947–1948, *Meteorologisches Amt für Nordwestdeutschland Zentralamt für die Britische Zone,* Hamburg, 1949.
44. P. O. Huss and D. J. Portman, Study of Natural Wind and Computation of the Austausch Turbulence Constant, *Daniel Guggenheim Airship Inst. Rep.* 156, 1949.
45. N. G. Stewart, H. J. Gale, and R. N. Crooks, The Atmospheric Diffusion of Gases Discharged from the Chimney of the Harwell Pile, *At. Energy Res. Establ.*, HP/R 1452, Harwell, England, 1954.
46. See [35], p. 187.

BIBLIOGRAPHY

General Circulation

LORENZ, E. N.: The Nature and Theory of the General Circulation of the Atmosphere, WMO no. 216, *World Meteorol. Organ. Tech. Pap.* 115, Geneva, 1967.

MANABE, S.: The Atmospheric Circulation and the Hydrology of the Earth's Surface, *Mon. Weather Rev.*, vol. 97, pp. 739–774, November 1969.

PALMEN, E.: Evaluation of the Atmospheric Moisture Transport for Hydrological Purposes, *Rep. WMO/IHD Proj.* 1, World Meteorological Organization, Geneva, 1967.

——— and NEWTON, C. W.: "Atmospheric Circulation Systems," International Geophysics. Series, vol. 13, Academic, New York, 1969.

PEIXOTO, J. P.: Atmospheric Vapor Flux Computations for Hydrological Purposes, *Rep. WMO/IHD Proj., no.* 20, *Publ.* 357, World Meteorological Organization, Geneva, 1974.

RASMUSSON, E.: Atmospheric Water Vapor Transport and the Water Balance of North America, *Mon. Weather Rev.*, vol. 95, pp. 403–426, July 1967.

———: A Study of the Hydrology of Eastern North America Using Atmospheric Vapor Flux Data, *Mon. Weather Rev.*, vol. 99, pp. 119–135, February 1971.

General Climatology

CRITCHFIELD, H. J.: "General Climatology," 2d ed., Prentice-Hall, Englewood Cliffs, N.J., 1966.

LANDSBERG, H.: "Physical Climatology," 2d ed., Pennsylvania State University, University Park, Pa., 1958.

SELLERS, W. D.: "Physical Climatology," University of Chicago Press, Chicago, 1965.

TREWARTHA, G. T.: "Introduction to Climate," 4th ed., McGraw-Hill, New York, 1968.

General and Jet-Stream Meteorology

BLAIR, T. A., and FITE, R. C.: "Weather Elements," 5th ed., Prentice-Hall, Englewood Cliffs, N.J., 1958.

BYERS, H. R.: "General Meteorology," 3d ed., McGraw-Hill, New York, 1959.

DONN, W. L.: "Meteorology," 3d ed., McGraw-Hill, New York, 1965.

HUSCHKE, R. E. (ed.): "Glossary of Meteorology," American Meteorological Society, Boston, 1959.

PETTERSSEN, S.: "Introduction to Meteorology," 3d ed., McGraw-Hill, New York, 1969.

REITER, E. R.: "Jet-Stream Meteorology," University of Chicago Press, Chicago, 1963.

RIEHL, H.: "Introduction to the Atmosphere," 2d ed., McGraw-Hill, New York, 1972.

TAYLOR, G. F.: "Elementary Meteorology," Prentice-Hall, Englewood Cliffs, N.J., 1963.

Meteorological Instruments and Observations

GREAT BRITAIN METEOROLOGICAL OFFICE: "Handbook of Meteorological Instruments," 2 vols., London, 1961.

MIDDLETON, W. E. K.: "Invention of the Meteorological Instruments," Johns Hopkins, Baltimore, 1969.

——— and SPILHAUS, A. F.: "Meteorological Instruments," 3d ed., University of Toronto Press, Toronto, 1953.

STERZAT, M.S.: "Instruments and Methods for Meteorological Observations," *Proc. All-Am. Meteorol. Conf.*, vol. 9, 1966; available from U.S. Dept. of Commerce National Technical Information Center, Springfield, Va.

U.S. NATIONAL WEATHER SERVICE: Substation Observations, *Obs. Handb.*, no. 2, 1970; rev. December 1972.

WORLD METEOROLOGICAL ORGANIZATION: Guide to Meteorological Instrument and Observing Practices, WMO no. 8, *Tech. Pap.* 3, Geneva, 1969.

Solar and Terrestrial Radiation

BUDYKO, M. I.: "The Heat Balance of the Earth's Surface," trans. N. A. Stepanova, U.S. Dept. of Commerce, 1958.

CANADIAN METEOROLOGICAL SERVICE: "Manual of Radiation Instruments and Observation," Toronto, 1968.

DRUMMOND, A. J. (ed.): "Precision Radiometry," Advances in Geophysics, vol. 14, Academic, New York, 1969.

ELSASSER, W. M.: An Atmospheric Radiation Chart and Its Use, *Q. J. Suppl. R. Meteorol. Soc.*, vol. 66, pp. 41–56, 1940.

FRITZ, S., and MACDONALD, T. H.: Average Solar Radiation in the United States, *Heat. Vent.*, vol. 46, pp. 61–64, July 1949.

HAND, I. F.: Weekly Mean Values of Daily Total Solar and Sky Radiation, *U.S. Weather Bur. Tech. Pap.* 11, 1949.

KONDRATYEV, K. YA.: "Radiation in the Atmosphere," International Geophysics Series, vol. 12, Academic, New York, 1969.

———: "Radiation Processes in the Atmosphere," World Meteorological Organization, Geneva, 1972.

LATIMER, J. R.: Radiation Measurement, *Int. Field Year Great Lakes Tech. Man. Ser.*, no. 2, Information Canada, Ottawa, 1972.

ROBINSON, N.: "Solar Radiation," Elsevier, New York, 1966.

U.S. WEATHER BUREAU: "Manual of Radiation Observations," 1962.

WORLD METEOROLOGICAL ORGANIZATION: Preciseness of Pyrrheliometric Measurements, WMO no. 209, *Tech. Pap.* 109, Geneva, 1966.

Urban Effects on Temperature

BORNSTEIN, R. D.: Observations of the Heat Island Effect of New York City, *J. Appl. Meteorol.*, vol. 7, pp. 575–582, August 1968.

DEMARRAIS, G. A.: Vertical Temperature Difference Observed over an Urban Area, *Bull. Am. Meteorol. Soc.*, vol. 42, pp. 548–554, August 1961.

DUCKWORTH, F. S., and SANDBERG, J. S.: The Effect of Cities upon Horizontal and Vertical Temperature Gradients, *Bull. Am. Meteorol. Soc.*, vol. 35, pp. 198–207, May 1954.

MOFFITT, B. J.: The Effects of Urbanization on Mean Temperatures at Kew Observatory, *Weather*, vol. 27, pp. 121–129, March 1972.

MUNN, R. E., HIRT, M. S., and FINDLAY, B. F.: A Climatological Study of the Urban Temperature Anomaly in the Lakeshore Environment at Toronto, *J. Appl. Meteorol.*, vol. 8, pp. 411–422, June 1969.

MYRUP, L. O.: A Numerical Model of the Urban Heat Island, *J. Appl. Meteorol.*, vol. 8, pp. 908–918, December 1969.

Wind Profiles

BERNSTEIN, A. B.: An Examination of Three Wind Profile Hypotheses, *J. Appl. Meteorol.*, vol. 5, pp. 217–219, April 1966.

BRADLEY, E. F.: A Micrometeorological Study of Velocity Profiles and Surface Drag in the Region Modified by a Change in Surface Roughness, *Q. J. R. Meteorol. Soc.*, vol. 94, pp. 361–379, 1968.

DEACON, E. L.: Vertical Profiles of Mean Wind in the Surface Layers of the Atmosphere, *Meteorol. Off. Geophys. Mem.* 91, 1953.

MCVEHIL, G. E.: Wind and Temperature Profiles near the Ground in Stable Stratification, *Q. J. R. Meteorol. Soc.*, vol. 90, pp. 136–146, 1964.

PANOFSKY, H. A., BLACKADAR, A. K., and MCVEHIL, G. E.: The Diabatic Wind Profile, *Q. J. R. Meteorol. Soc.*, vol. 86, pp. 390–398, 1960.
SWINBANK, W. C.: The Exponential Wind Profile, *Q. J. R. Meteorol. Soc.*, vol. 90, pp. 119–135, 1964.
TAYLOR, P. A.: On Wind and Shear Stress Profiles above a Change in Surface Roughness, *Q. J. R. Meteorol. Soc.*, vol. 95, pp. 77–91, 1969.
WEBB, E. K.: Profile Relationships: The Log Linear Range and Extension to Strong Stability, *Q. J. R. Meteorol. Soc.*, vol. 96, pp. 67–90, 1970.

UNITED STATES DATA SOURCES

The main sources of data on temperature, humidity, and wind are the monthly bulletins entitled *Climatological Data* published by the Environmental Data Service of the National Oceanic and Atmospheric Administration (NOAA). Daily values of direct and diffuse solar radiation are published for about 80 stations in the monthly *Climatological Data National Summary*. The *Climatic Summary of the United States* summarizes monthly and annual data from the beginning of record through 1960. *Monthly Weather Review* contains maps summarizing the weather of the previous month for the country. Summaries of normals may be found in the following:

Daily Normals of Temperature and Heating and Cooling Degree Days, *Climatography of the United States*, no. 84, NOAA Environmental Data Service, September 1973.
Monthly Normals of Temperature, Precipitation, and Heating and Cooling Degree Days, *Climatography of the United States*, no. 81 (by states), NOAA Environmental Data Service, August 1973.
"Climatic Atlas of the United States," ESSA Environmental Data Service, 1968. (This publication presents summaries of wind, temperature, precipitation, humidity, evaporation, and solar radiation on maps.)

Other summaries may be found in:

Climates of the States, *Climatography of the United States*, no. 60-1–60-52 (by states and other areas), U.S. Weather Bureau.
Decennial Census of the United States Climate: Summary of Hourly Observations 1951–1960, *Climatography of the United States*, no. 82-1–82-48 (by cities), U.S. Weather Bureau. (This publication gives monthly and annual frequencies of occurrence of various values of temperature, precipitation, relative humidity, wind speed and direction, and cloudiness for over 250 cities.)

In cooperation with the World Meteorological Organization, the Environmental Data Service also publishes *Monthly Climatic Data for the World*, which contains monthly precipitation amounts and averages of pressure, temperature, and relative humidity for about 1000 stations distributed throughout the world. It also includes monthly averages of upper-air temperature and dewpoint for about 250 stations.

Information on all climatological data published by the Environmental Data Service is contained in its Selective Guide to Climatic Data Sources, *Key Meteorol. Rec. Docum.*, no. 4.11, 1969.

PROBLEMS

2-1 Show why the theoretical eastward velocity (constant angular momentum) of air at rest relative to earth's surface at the equator would be 1560 mi/h if the parcel were displaced to 60°N latitude.

2-2 How many degree-days above 32°F are there in a day with a minimum temperature of 26°F and a maximum of 48°F?

2-3 A parcel of moist air at 60°F initially at 2000 ft, mean sea level, is forced to pass over a mountain ridge at 8000 ft, mean sea level, and then descends to its original elevation. Assuming that a lift of 2000 ft produces saturation and precipitation and that the average pseudo-adiabatic lapse rate is one-half the dry-adiabatic, what is the final temperature of the parcel?

2-4 What is the heat of vaporization, in calories per gram, for water at (*a*) 15°C and (*b*) 77°F?

2-5 What is the density, in kilograms per cubic meter, of (*a*) dry air at 30°C and a pressure of 900 millibars and (*b*) moist air with relative humidity of 70 percent at the same temperature and pressure?

2-6 Assuming dry- and wet-bulb temperatures of 80 and 62°F, respectively, and using the psychrometric tables of Appendix B, determine (*a*) dewpoint temperature, (*b*) relative humidity, (*c*) saturation vapor pressure, and (*d*) actual vapor pressure, in millibars.

2-7 A radio sounding in saturated atmosphere shows temperatures of 16.0, 11.6, and 6.2°C at the 900-, 800-, and 700-millibar levels, respectively. Compute the precipitable water, in millimeters, in the layer between 900 and 700 millibars, and compare the result with that obtained from Fig. 2-17. (The temperature of 16.0°C at the 900-millibar level reduces pseudo-adiabatically to 20.0°C at 1000 millibars.)

2-8 A formula for estimating potential evapotranspiration from a lawn requires wind speed at 2 m above the surface. What is the estimated speed at this level if the roughness length is 1.0 cm and an anemometer 10 m above the ground indicates a mean wind speed of 5.0 m/s?

2-9 A pilot-balloon observation shows wind speed of 40 kn at 300 m above the ground. What is the estimated speed, in miles per hour, 30 ft above the ground indicated by the power-law profile with values of (*a*) $\frac{1}{7}$ and (*b*) $\frac{1}{5}$ for exponent k?

2-10 The saturation vapor pressure over water at 10°F is 2.40 millibars. What is it over ice at the same temperature?

2-11 Compute the weight, in kilograms, of 1 m^3 of dry air at a temperature of (*a*) 0°C and a pressure of 1000 millibars and (*b*) 20°C at the same pressure.

2-12 How many calories are required to evaporate 1 gal (U.S.) of water at 70°F? How many pounds of ice at 14°F would the same amount of heat melt? (Specific heat of ice = 0.5.)

2-13 Anemometers mounted on a tower at 2 and 16 m above the ground indicate mean wind speeds of 2.5 and 5.0 m/s, respectively.

(*a*) What is the roughness length, in centimeters?

(*b*) Using the roughness length computed in part (*a*), determine the wind speed at 5 m.

(*c*) What does Eq. (2-15) yield for the wind speed at 5 m?

2-14 How many calories per square foot are required (*a*) to melt a 1-ft layer of ice with a specific gravity of 0.90 at 20°F and (*b*) to evaporate the resulting meltwater without raising its temperature above 32°F? (Use 0.5 for specific heat of ice.)

2-15 Anemometers at 10 and 100 m on a tower record average wind speeds of 5.0 and 10.0 m/s, respectively. Compute average speeds at 30 and 60 m using (*a*) Eq. (2-15) and (*b*) Eq. (2-16).

2-16 What is the relative humidity if air temperature and dewpoint are (*a*) 20 and 10°C and (*b*) 40 and 4°F?

2-17 What is the dewpoint for air temperature and relative humidity of (*a*) 15°C and 49 percent and (*b*) 25°F and 24 percent?

2-18 Convert Eq. (2-9) so it can be used for the Fahrenheit temperature scale.

2-19 Given a mean wind speed of 2.0 m/s at 2 m above the ground and a roughness length of 0.5 cm, compute (*a*) the friction velocity, in centimeters per second, and (*b*) the wind speed, in meters per second, at 0.5 m.

3
PRECIPITATION

Hydrologists have long known that only about one-fourth the total precipitation that falls on continental areas is returned to the seas by direct runoff and underground flow. Hence, it was generally believed that continental evaporation constituted the principal source of moisture for continental precipitation. Many ideas for increasing precipitation were based on the premise (now known to be erroneous) that more precipitation would result from increased atmospheric moisture through local evaporation. Impounding streamflow in lakes and ponds and selecting farm crops with high transpiration rates were among the methods suggested. That such methods are ineffective is demonstrated by the Caspian Sea. Although this sea has an area of about 438,000 km^2 (169,000 mi^2), which is larger than California, and its annual evaporation is roughly estimated [1] to be about 500 to 600 billion cubic meters (400 to 500 million acre-feet), the annual precipitation along its shores is generally less than 250 mm (10 in.).

It is now recognized that evaporation from ocean surfaces is the chief source of moisture for precipitation, and probably no more than about 10 percent of continental precipitation can be attributed to continental evaporation. Nearness to oceans, however, does not necessarily lead to adequate precipitation, as evidenced by the many desert islands. The location of a region with respect to the general circulation system, latitude, and distance to a moisture source is primarily responsible for its climate. Orographic barriers often exert more influence on the climate of a region than nearness to a moisture source does. These climatic and geographic factors determine the amount of atmospheric moisture over a region, the frequency and kind of precipitation-producing storms passing over it, and thus its precipitation.

3-1 Formation of Precipitation

Moisture is always present in the atmosphere, even on cloudless days. For precipitation to occur, some mechanism is required to cool the air sufficiently to bring it to or near saturation. The large-scale cooling needed for significant amounts of precipitation is achieved by lifting the air, and this is accomplished by convective or convergence systems resulting from unequal radiative heating or cooling of the earth's surface and atmosphere or by orographic barriers. Saturation, however, does not necessarily lead to precipitation.

Condensation and freezing nuclei Assuming that the air is saturated or nearly so, formation of fog or cloud droplets or ice crystals generally requires the presence of *condensation* or *freezing nuclei* on which the droplets or crystals form. These nuclei are small particles of various substances, not necessarily hygroscopic, usually ranging in size from about 0.1 to 10 μm in diameter. Those with diameters less than 3 μm are within the size range of aerosols and might remain airborne indefinitely except for precipitation fallouts. Condensation nuclei usually consist of products of combustion, oxides of nitrogen, and salt particles. The last are most effective and may result in condensation with relative humidity as low as 75 percent.

Freezing nuclei usually consist of clay minerals, kaolin perhaps being the most common [2]. Artificial freezing nuclei most frequently used in weather modification (cloud seeding) are carbon dioxide (dry ice) and silver iodide. Freezing nuclei serve only to nucleate the liquid phase and thus initiate the growth of ice crystals. Pure water droplets may remain in the liquid state to temperatures as low as -40°C (-40°F), and it is only in the presence of such supercooled droplets that freezing nuclei are active. Carbon dioxide, CO_2, can produce ice crystals in a supercooled cloud at any temperature up to about 0°C (32°F); silver iodide, AgI, has a nucleating threshold of about -4°C (25°F).

Growth of water droplets and ice crystals Upon nucleation the droplet or ice crystal grows to visible size in a fraction of a second through diffusion of water vapor to it, but growth thereafter is very slow. Diffusion by itself leads only to fog or cloud elements generally smaller than 10 μm in diameter, with some reaching 50 μm. Since condensation tends to enlarge droplets or ice crystals at about the same rate, differences in size result mainly from differences in size of the nuclei on which they are formed. While cloud elements tend to settle, the average element weighs so little that only a slight upward motion of the air is needed to support it. Most droplets in nonprecipitating stratus have diameters under 10 μm, and an upward current of less than 0.5 cm/s (0.2 in./s) is sufficient to keep them from falling. Ice crystals of equivalent weight, because of their shape and larger size, can be supported by even lower velocities.

Upward velocities under and in clouds often greatly exceed those required to support cloud elements. Hence, for precipitation to occur, cloud elements must increase in size until their falling speeds exceed the ascensional rate of the air. The cloud element must also be large enough to penetrate the unsaturated air below the cloud base without evaporating completely before reaching the ground. A drop falling out of a cloud base at 1 km in air of 90 percent relative humidity rising at 10 cm/s (0.3 ft/s) would require [3] a diameter of about 440 μm in order to reach the ground with a diameter of 200 μm, which is often considered to mark the boundary between cloud droplet size and precipitation size.

Some growth of cloud elements through diffusion is to be expected because of differences in vapor pressure resulting from differences in size and temperature. Electric charges appear to have some effect on the growth of cloud elements also. Diffusion is most effective when ice crystals and liquid-water droplets are both present in a cloud. Saturation vapor pressure over ice is less than that over liquid water at the same temperature, and this results in a net transport of moisture from liquid droplets to ice crystals. Growth is particularly rapid when liquid droplets greatly outnumber ice crystals, and this is usually the case. The process is considered important in causing heavy precipitation, but since heavy rainfall can occur from *warm clouds*, with above-freezing temperatures throughout, other factors must sometimes be involved.

Collision and coalescence of cloud and precipitation elements (*accretion*) are considered the most important factors leading to significant precipitation. Collisions between cloud and precipitation particles arise mostly from differences in falling speeds as a result of differences in size. Heavier particles fall faster (or ascend more slowly in rising currents) than smaller particles do. Particles that collide usually coalesce to form larger particles, and the process may be repeated a number of times. It has been estimated [4] that in a typical heavy rain seven collisions occur for every kilometer of fall.

Raindrops may grow as large as 6 mm (0.25 in.) in diameter. Maximum falling speeds, or *terminal velocities* (Table 3-1), tend to level off as the drops approach maximum size because of increasing air resistance due to flattening. For large diameters, the deformation may be sufficient to break up the drops before they can attain terminal velocity.

Single ice crystals may reach the ground, but usually a number of them collide and coalesce to form a cluster and fall as *snowflakes*. Coalescence is most effective when temperatures are near freezing. The fall velocity of snow and ice crystals is usually no more than about 1 m/s (3 ft/s) thus providing considerable time for growth by diffusion. At temperatures near freezing, rime is formed. Since riming enhances both coalescence and inequalities in falling speeds, it increases the probability of collision; it is then that the largest snowflakes are observed.

Maximum liquid-water content of clouds Speeds of ascending air in vigorous convective systems usually far exceed terminal velocities (Table 3-1) of the largest raindrops. Radar observations have indicated ascensional rates as high as 40 to 50 m/s (90 to 110 mi/h) in cumulonimbus. Aircraft observa-

Table 3-1 TERMINAL VELOCITY OF WATER-DROPS IN STILL AIR
Pressure 1013.3 millibars, temperature 20°C, relative humidity 50 percent

Drop diameter		Terminal velocity	
mm	in.	cm/s	ft/s
0.5	0.02	206	6.8
1.0	0.04	403	13.2
1.5	0.06	541	17.7
2.0	0.08	649	21.3
3.0	0.12	806	26.4
4.0	0.16	883	29.0
5.0	0.20	909	29.8
5.5	0.22	915	30.0
5.8	0.23	917	30.1

SOURCE: From R. Gunn and G. D. Kinzer, The Terminal Velocity of Fall for Water Droplets in Stagnant Air, *J. Meteorol.*, vol. 6, pp. 243–248, August 1949. Used with permission of American Meteorological Society.

tions indicate a horizontal diameter of about 1.5 km (5000 ft) for updrafts in a thunderstorm cell, which usually ranges from about 6 to 10 km (4 to 6 mi) in diameter. Downdrafts are about the same size as updrafts. Ascensional rates in updrafts are not uniform, however, and high-speed vertical jets in updrafts may have diameters of no more than a few hundred meters.

The stronger updrafts prevent even the largest raindrops from falling and carry all precipitation elements to the upper portions of the clouds to produce an accumulation of liquid water far exceeding that of the ordinary cloud elements. Theory and radar observations suggest [5] that the height of the accumulation depends on updraft speed. Eventually, the accumulated water is precipitated as a result of the weakening of the updraft or by horizontal displacement from the supporting updraft to a weaker one or, as often happens, to a downdraft, which possibly may be initiated [6] by the mass of accumulated water. When suddenly precipitated in a downdraft, the resulting torrential downpour lasts but a few minutes, and point rainfall is usually less than 100 mm (4 in.). In a thunderstorm there may be several such downpours, or bursts, from a number of cells, and total point rainfall in 1 h may exceed 200 mm (8 in.).

In nonprecipitating cumulus congestus the maximum concentration of liquid water may be near 4 g/m^3, but the mean for the cloud might be only half this value. Mean values for the cloud as a whole apparently have little significance in relation to natural precipitation. The maximum liquid-water content of nonprecipitating clouds usually ranges from about 0.5 g/m^3 in thin stratus to about 4 g/m^3 in thick cumulus, but values exceeding 30 g/m^3 have been measured [7]. Clouds having concentrations of 4 g/m^3 or more usually produce precipitation that reaches the ground. Rainfall rates tend to be correlated with maximum liquid-water content. For heavy rains it appears that the rainfall rate increases about 25 mm/h (1 in./h) for each gram per cubic meter.

3-2 Forms of Precipitation

Any product of condensation of atmospheric water vapor formed in the free air or at earth's surface is a *hydrometeor*. Since the hydrologist is primarily interested in precipitation, only those hydrometeors referring to falling moisture are here defined. Among those hydrometeors not included are damp haze, fog, ice fog, drifting snow, blowing snow, and frost.

Drizzle, sometimes called *mist*, consists of tiny liquid water droplets, usually with diameters between 0.1 and 0.5 mm (0.004 and 0.02 in.), with such slow settling rates that they occasionally appear to float. Drizzle usually falls from low stratus and rarely exceeds 1 mm/h (0.04 in./h).

Rain consists of liquid water drops mostly larger than 0.5 mm (0.02 in.)

in diameter. *Rainfall* usually refers to amounts of liquid precipitation. In the United States rain is reported in three intensities:

Light For rates of fall up to 0.10 in./h (2.5 mm/h) inclusive
Moderate From 0.11 to 0.30 in./h (2.8 to 7.6 mm/h)
Heavy Over 0.30 in./h (7.6 mm/h)

In the United States also, rainfall is labeled *excessive precipitation* when amounts R, in hundredths of an inch, for durations t from 5 to 180 min equal or exceed those indicated by $R = t + 20$. Tabulations of excessive-precipitation data have been compiled and published since 1896, but caution should be exercised in their use because of changes in the threshold formulas and methods of compilation during the period of record.

Glaze is the ice coating, generally clear and smooth but usually containing some air pockets, formed on exposed surfaces by the freezing of supercooled water deposited by rain or drizzle. Its specific gravity may be as high as 0.8 to 0.9.

Rime is a white, opaque, granular deposit of ice composed essentially of ice granules more or less separated by trapped air and formed by rapid freezing of supercooled water drops impinging on exposed objects. Its specific gravity may be as low as 0.2 to 0.3.

Snow is composed of white or translucent ice crystals, chiefly in complex, branched hexagonal form, often mixed with simple crystals and often agglomerated into *snowflakes*, which may reach several inches in diameter. The density of freshly fallen snow varies greatly; 125 to 500 mm (5 to 20 in.) of snow is generally required to equal 25 mm (1 in.) of liquid water. The average density (specific gravity) is often assumed to be 0.1.

Snow pellets, also called *soft hail*, *graupel*, or *tapioca snow*, consist of white, opaque, rounded or conical ice particles of snowlike structure and about 2 to 5 mm (0.1 to 0.2 in.) in diameter. Snow pellets are crisp and easily crushed. They rebound upon striking hard surfaces and often break up.

Hail is precipitation in the form of balls or irregular lumps of ice, produced in convective clouds, mostly cumulonimbus. *Hailstones* may be spheroidal, conical, or irregular in shape, and range from about 5 to over 125 mm (0.2 to over 5 in.) in diameter. They are generally composed of alternating layers of glaze and rime, and their specific gravity is about 0.8. The largest hailstone of record in the United States fell at Coffeyville, Kansas, Sept. 3, 1970. It measured 44 cm (17.5 in.) in circumference, and weighed 766 g (1.67 lb).

Ice pellets are composed of transparent or translucent ice. They may be spherical, irregular, or (sometimes) conical and are usually less than 5 mm (0.2 in.) in diameter. Ice pellets usually bounce when hitting hard surfaces and make a sound upon impact. Two basically different types of precipitation, known in the United States as sleet and small hail, are included in ice pellets:

Sleet, or *grains of ice* Generally transparent, globular, solid grains of ice formed by the freezing of raindrops or refreezing of largely melted ice crystals falling through a layer of subfreezing air near the earth's surface

Small hail Composed of generally translucent particles consisting of snow pellets encased in a thin layer of ice. The ice cover may form either by accretion of liquid droplets upon the snow pellet or by melting and refreezing of the surface of the snow pellet.

3-3 Types of Precipitation

Precipitation is often typed according to the factor mainly responsible for lifting the air to effect the large-scale cooling required for significant amounts of precipitation. Thus, *cyclonic precipitation* results from the lifting of air converging into a low-pressure area, or cyclone. Cyclonic precipitation may be classified further as frontal or nonfrontal. *Frontal precipitation* results from the lifting of warm air on one side of a frontal surface over colder, denser air on the other side. *Warm-front precipitation* is formed in the warm air advancing upward over a colder air mass. The rate of ascent is relatively slow since the average slope of the frontal surface is usually between 1/100 and 1/300. Precipitation may extend 300 to 500 km (200 to 300 mi) ahead of the surface front and is generally light to moderate and nearly continuous until passage of the front. *Cold-front precipitation*, on the other hand, is of showery nature and is formed in the warm air forced upward by an advancing mass of cold air, the leading edge of which is the surface cold front. Cold fronts move faster than warm fronts, and the frontal surfaces, with slopes usually averaging 1/50 to 1/150, are steeper. Consequently, the warm air is forced upward more rapidly than by the warm front, and precipitation rates are generally much higher. Heaviest amounts and intensities occur near the surface front. *Nonfrontal precipitation* is precipitation unrelated to fronts.

Convective precipitation is caused by the rising of warmer, lighter air in colder, denser surroundings. The difference in temperature may result from unequal heating at the surface, unequal cooling at the top of the air layer, or mechanical lifting when the air is forced to pass over a denser, colder air mass or over a mountain barrier. Convective precipitation is spotty, and its intensity may range from light showers to cloudbursts. *Orographic precipitation* results from mechanical lifting over mountain barriers. In rugged terrain the orographic influence is so marked that storm precipitation patterns tend to resemble that of mean annual precipitation.

In nature, the effects of these various types of cooling are often interrelated, and the resulting precipitation cannot be identified as being of any one type.

3-4 Artificially Induced Precipitation

Weather modification, sometimes referred to as *weather control*, is the general term for efforts to alter artificially the natural meteorological phenomena of the atmosphere. Attempts to increase or decrease precipitation, suppress hail and lightning, mitigate hurricanes, dissipate fog, prevent frost, alter radiation balance, etc., are all included under weather modification. *Cloud modification*, or *cloud seeding*, is one type of weather modification, and usually has as its goal either dissipation of the cloud or stimulation of precipitation.

The demonstration, in 1946, that solid carbon dioxide (dry ice) can cause precipitation in a cloud† containing supercooled water droplets gave new impetus to cloud seeding. Interest stimulated by this discovery soon led to further discoveries that certain salts, notably silver iodide, can also induce precipitation. Both dry ice and silver iodide, the two most commonly used seeding agents, act as freezing nuclei in supercooled clouds. Seeding clouds with dry ice requires delivery into the cloud by aircraft, balloons, or rockets. Silver iodide, which is most effective when heated to vaporization, may be delivered into the cloud either by airborne or ground-based generators but has the disadvantage that its effectiveness is reduced by exposure to sunlight, the number of effective particles decreasing by a factor of about 10 for every hour of exposure. Nevertheless, the low operation cost of ground-based generators has made this method the most commonly used for augmentation of precipitation.

The nucleating threshold temperature of silver iodide, however, is −4°C (25°F), and for cloud-base temperatures warmer than freezing, it makes a difference whether the silver iodide is released from ground-based or airborne generators. In such situations, the silver iodide released from a ground-based generator may become embedded in liquid droplets, and crystallization will not occur until the temperature has been lowered by convective lifting to between −10 and −15°C (14 and 5°F). Thus, unless there is enough convective lifting to provide the necessary cooling, seeding from a ground-based generator is ineffective. On the other hand, silver iodide introduced into the cloud at levels where temperatures are −5°C (23°F) or colder will result in immediate crystallization of pure liquid droplets upon contact.

The effectiveness of cloud seeding depends on many factors such as

† Limited inconclusive experiments with solid CO_2 were made in Holland in 1930 by August W. Veraart.

height of cloud base and top, cloud temperature, difference between density inside the cloud and that outside (*buoyancy*), updraft velocity distribution, amount and concentration of liquid water in the cloud, number and distribution of natural freezing or condensation nuclei, the number of artificial nuclei added, and where they are introduced into the cloud. Tests of the effectiveness of cloud seeding are now made where results can be observed over relatively dense raingage networks and by radar. Precipitation over the seeded area, called the *target area*, is usually compared with that over carefully selected, nearby unseeded areas, or *control areas*.

There are two general approaches to cloud seeding. The *static approach* consists of introducing a certain quantity of artificial freezing nuclei, say one nucleus per liter (61 $in.^3$) of cloud air, to produce ice crystals which, through diffusion and accretion, eventually grow into precipitation particles. The *dynamic approach* involves massive seeding, say 100 to 1000 nuclei per liter of cloud air, of cumulus clouds to stimulate, through release of heat of fusion, the buoyancy forces and circulations that sustain them. With favorable temperature and moisture distributions, massive seeding of supercooled cumulus clouds with tops at heights below the −40°C (−40°F) level may raise the cloud tops 3 to 5 km (10,000 to 16,000 ft) within a few minutes so that they extend well above the −40°C (−40°F) level, where spontaneous natural crystallization can occur, thus increasing the probability of natural precipitation.

The dynamic approach is particularly useful in lower latitudes, where tops of cumuli are often below the −25°C (−13°F) level. Models have been developed [8] that can predict cloud growth after seeding. The difference between the predicted maximum cloud-top heights after seeding and the actual height before seeding, called *seedability*, is highly correlated with rainfall augmentation expected from seeding. Rainfall decreases are associated with seedabilities less than about 0.8 km (2600 ft), but appreciable increases are found [9] with seedabilities of over 3 km (10,000 ft). Tests over Florida indicated that seeded clouds are generally larger and longer-lasting than unseeded clouds and produce about twice as much rainfall.

Stratus clouds offer poor prospects for effective seeding with the purpose of increasing rainfall. The moisture content is relatively low, and atmospheric conditions attending such clouds are too stable. Seeding may result in weak turbulence which distributes the ice nuclei so that the cloud may be dissipated with little or no precipitation if the cloud temperature is below the nucleating threshold temperature of the seeding agent used.

Orographic clouds generally have a much lower moisture content than large summer cumuli, but they have the advantage that their formation is continuous so long as there is a moist air flow directed upslope. Seeding of orographic clouds under favorable conditions has been fairly effective. When cloud-top temperatures are from −12 to −20°C (10 to −4°F), when the water content is relatively high, and when there are few natural freezing nuclei (which are relatively ineffective at these temperatures), seeding may increase snowfall by a factor of 2 to 3. On the other hand, snowfall may be reduced appreciably by seeding when cloud-top temperatures are −24°C (−11°F) or colder. In seeding experiments in the Colorado mountains (1960–1965) average daily precipitation for 120 seeded days was 6 to 11 percent greater than that for 131 unseeded days, but there was a high probability that the difference might have occurred by chance alone. The same experiments, however, indicated that statistically significant increases occurred [10] when wind speeds at the 500 millibar level were 22 to 28 m/s (49 to 63 mi/h). Later experiments indicated that the dominant effect of seeding the warmer cloud systems is the increase in the number of snowfall hours [11]. A smaller beneficial effect is the increase in precipitation rates during natural snowfall hours.

Seeding orographic clouds over Kings River Basin in California for a 10-y period (1955–1964) is reported [12] to have resulted in an average increase in streamflow of 6 percent, with increases for individual years ranging from 2 to 68 percent. Other seeding experiments involving orographic clouds are reported [13] to have produced precipitation increases of over 100 percent for individual periods when seeding was done only under the most favorable temperature and wind conditions.

Because of heavy damage caused by hailstorms, much attention has been given to cloud seeding for suppressing hail. The basic idea is to introduce so many freezing nuclei into the cloud that only very small ice particles will form. Since the ice particles compete for the supercooled droplets, the larger the number of particles, the smaller the average size of the hailstones. Both silver iodide and lead iodide have been widely used for hail suppression. The effectiveness of cloud seeding in suppressing hail apparently depends on storm characteristics and the method and rate at which the seeding agent is delivered into the cloud. Soviet scientists have experimented with silver iodide and lead iodide delivered to likely hail clouds by artillery shells and rockets and claim effective results [14]. A summary [15] of the results of several hail-suppression projects in scattered parts of the world during 1953–1967 suggests that seeding rates less than 1000 g/h per storm may increase hailfall but seeding rates of 2000 to 3000 g/h per storm can reduce major hailfalls.

Seeding with freezing nuclei is ineffective in warm clouds because of the absence of supercooled droplets. In such clouds, the only way raindrops can be formed is through collision and coalescence, which cannot occur until some cloud droplets have reached a radius of about 20 μm. This growth of cloud droplets can be initiated or stimulated either by introducing into the cloud large condensation nuclei or droplets of the required size or larger. Experiments have been conducted using sodium chloride and water drops sprayed from aircraft, apparently with some success [16, 17]. The better results of the relatively few experiments were obtained when the seeding was done in the cloud base.

There is little question that seeding can increase precipitation from a cloud under favorable conditions. The effectiveness of localized seeding in increasing precipitation over large areas is difficult to evaluate. While massive seeding of cumulus clouds may produce appreciable rainfall per cloud, it is important to know whether this artificially induced rainfall represents a net increase, a decrease, or merely a redistribution of the rainfall over the area encompassing the cloud. The effect of individual seedings of many supercooled cumuli over hundreds of square kilometers has yet to be determined.

Attempts to seed precipitating cloud systems of extratropical cyclones over flat or hilly terrain have failed to produce any statistically significant increases. The processes producing steady precipitation over large areas may be so efficient in utilizing the total cloud condensate that any possible increases from seeding would be small [18].

Evaluation of cloud-seeding effectiveness is difficult [19] because increases of 10 to 20 percent are hard to detect against natural variability. In reviewing cloud-seeding results, one finds that until about 1960 reports on seeding experiments were highly conflicting. Some differences undoubtedly arose from indiscriminate seeding without regard to cloud characteristics. Some resulted from inadequate controls for proper evaluation. Also, some reports were probably biased in favor of the rainmakers, some of whom may well have been more interested in the monetary than in the scientific aspects of cloud seeding. Much confusion has been eliminated by (1) increased knowledge of cloud physics, (2) more stringent, statistically designed controls, and (3) more and better radar and rain-gage network observations. The effects of prolonged cloud seeding on the climate, hydrology, ecology, and economy of a region are largely unknown. A discussion of these effects [20, 21] and the legal problems [22] arising from cloud seeding is outside the scope of this text.

MEASUREMENT OF PRECIPITATION

A variety of instruments and techniques have been developed for gathering information on various phases of precipitation. Instruments for measuring amount and intensity of precipitation are the most important. Other instruments include devices for measuring raindrop-size distribution and for

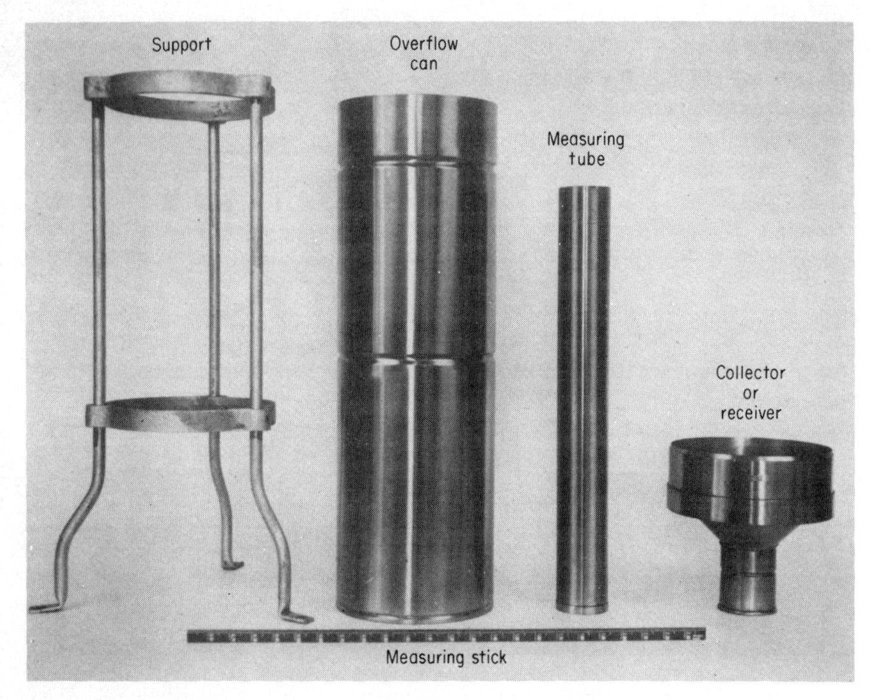

FIGURE 3-1
Standard 8-in. precipitation gage. (*U.S. National Weather Service.*)

determining the time of beginning and ending of precipitation. All forms of precipitation are measured on the basis of the vertical depth of water that would accumulate on a level surface if the precipitation remained where it fell. In the United States precipitation is measured in inches and hundredths. In the metric system precipitation is measured in millimeters and tenths.

3-5 Precipitation Gages

Any open receptacle with vertical sides is a convenient rain gage, but because of varying wind and splash effects the measurements are not comparable unless the receptacles are of the same size [23] and shape [24] and similarly exposed. The standard gage [25] (Fig. 3-1) of the U.S. National Weather Service has a collector (receiver) of 8-in. (20.3-cm) diameter. Rain passes from the collector into a cylindrical measuring tube inside the overflow can. The measuring tube has a cross-sectional area one-tenth that of the collector so that a 0.1-in. rainfall will fill the tube to a depth of 1 in. With a measuring stick marked in tenths of an inch, rainfall can be measured to the nearest 0.01 in. (0.25 mm). The collector and tube are removed when snow is

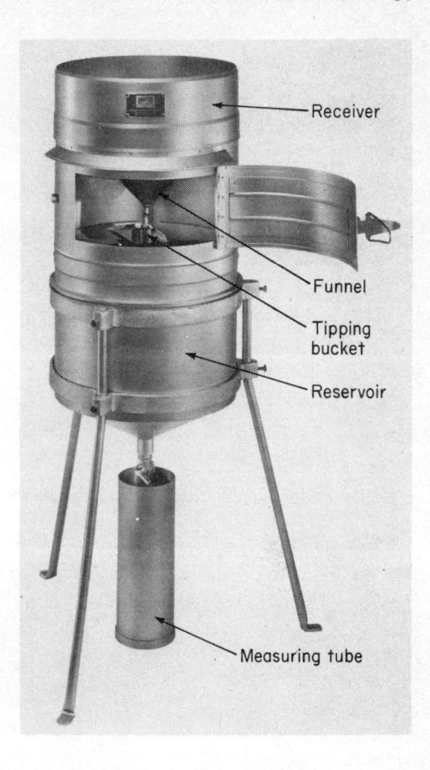

FIGURE 3-2
Tipping-bucket rain gage. (*Belfort Instrument Co.*)

expected. The snow caught in the outer container, or overflow can, is melted, poured into the measuring tube, and measured.

The three types of recording gages in common use are the tipping-bucket gage, the weighing gage, and the float gage. In the *tipping-bucket gage* (Fig. 3-2) the water caught in the collector is funneled into a two-compartment bucket: 0.01 in. (or 0.1 mm) of rain will fill one compartment and overbalance the bucket so that it tips, emptying into a reservoir and moving the second compartment into place beneath the funnel. As the bucket is tipped, it actuates an electric circuit, causing a pen to mark on a revolving drum. This type of gage is not suitable for measuring snow without heating the collector.

The *weighing-type gage* (Fig. 3-3) weighs the rain or snow which falls into a bucket set on a platform of a spring or lever balance. The increasing weight of the bucket and its contents is recorded on a chart. The record thus shows the accumulation of precipitation.

There are several types of *float recording gages*. In most, the rise of the float with increasing catch of rainfall is recorded on a chart. Some gages must be emptied manually, while others are emptied automatically by self-starting siphons. In most gages the float is placed in the receiver, but in some the receiver rests in a bath of oil or mercury and the float measures the

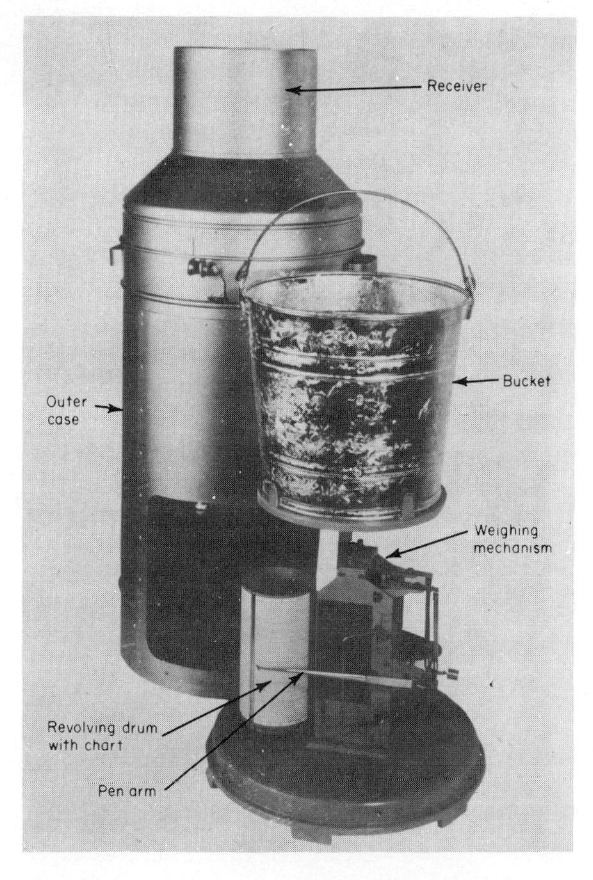

FIGURE 3-3
Weighing-type precipitation gage, 12-in. dual traverse. (*Belfort Instrument Co.*)

rise of the oil or mercury displaced by the increasing weight of the receiver as the rainfall catch accumulates. Floats may be damaged if the rainfall catch freezes.

Most gages record by a pen trace on a chart. The *punched-tape recorder* punches the amount of precipitation accumulated in the collector on a tape in digital code, which later can be run through a translator for adapting to computer evaluation of the record.

Storage gages are used in remote regions where frequent servicing is impracticable. Weighing-type storage gages operate from 1 to 2 months

FIGURE 3-4
Tower-mounted storage gage with Alter shield. (*U.S. National Weather Service.*)

without servicing, and some nonrecording storage gages are designed to operate for an entire season without attention. Storage gages located in heavy-snowfall areas should have collectors in the form of an inverted frustrum of a cone to prevent wet snow from clinging to the inside walls and clogging the orifice. The orifice should be above the maximum snow depth expected (Fig. 3-4). Storage gages are customarily charged with calcium chloride or other antifreeze solution to liquefy the snow and prevent damage to the gage. Interim measurements of the gage catch are made by stick or tape, while the initial charge and final measurement of the seasonal catch are usually made by weighing. Losses due to evaporation of the gage contents are practically eliminated by a thin film of oil [26].

Precipitation measurements are subject to various errors, most being individually small but with a general tendency to yield measurements that are too low. Except for mistakes in reading the scale of the gage, observational errors are usually small but cumulative; e.g., light amounts may be neglected, and extended immersion of the measuring stick may result in water creeping up the stick. Errors in scale reading, although large, are usually random and compensating. Instrumental errors may be quite large and are cumulative. The water displaced by the measuring stick increases the reading about 1 percent. Dents in the collector rim may change its receiving area. It is estimated that 0.01 in. (0.25 mm) of each rain measured with a gage initially dry is required to moisten the funnel and inside surfaces. This loss could easily amount to 1 in./y (25 mm/y) in some areas. Another loss results from raindrop splash from the collector.

In rainfall of 5 to 6 in./h (125 to 150 mm/h) the bucket of a tipping-bucket gage tips every 6 to 7 s. About 0.3 s is required to complete the tip, during which some water is still pouring into the already filled compartment. The recorded rate may be 5 percent too low [27]. However, the water, which is all caught in the gage reservoir, is measured independently of the recorder count, and the difference is prorated through the period of excessive rainfall.

Frictional effects in the weighing mechanism of weighing-type gages, in float guides in float-type gages, or in the recording-pen linkages of both types can result in inaccurate indications of rainfall rates. In self-emptying float-type gages the siphoning takes at least a few seconds, and rain falling into the receiver during the siphoning period is recorded inaccurately. Another source of error is that the rainfall amounts siphoned out are not always the same for all emptying cycles.

Of all the errors the most serious is the deficiency of measurements due to wind. The vertical acceleration of air forced upward over a gage imparts an upward acceleration to precipitation about to enter and results in a deficient catch. The deficiency is greater for small raindrops than for large and is thus greater for light than for heavy rain. The deficiency is greater for snow than for rain and larger for "dry" than for "wet" snow; hence it is inversely related to air temperature [28]. Reliable evaluation of wind errors is difficult because of problems involved in determining the true, or actual, precipitation reaching the ground. Attempts at assessing wind errors usually consist of comparing gage measurements with weight changes in nearby lysimeters or with changes in lake levels, or merely comparing measurements of shielded and unshielded gages. Figure 3-5 shows average deficiencies in rain and snow measurements. The curve shown for rain is actually an average of separate curves for shielded and unshielded gages which showed a spread of only 3 percent at 10 mi/h (4.5 m/s) and about 5 percent at 20 mi/h (8.9 m/s), the curve for unshielded gages showing the greater deficiencies.

Various types of shields [29] have been used, but the Alter shield (Fig. 3-4) has been adopted as standard in the United States. Its open construction provides less opportunity than solid shields for snow buildup, and the flexible design allows wind movement to help keep the shield free from accumulated snow. Artificial windshields, including snow fences, cannot overcome the effects of inherently poor gage exposure. Also, the higher the gage, the greater the wind error. Roof installations and windswept slopes should be avoided. The best site is on level ground with bushes or trees serving as a windbreak, provided that they are not so close that they reduce the gage

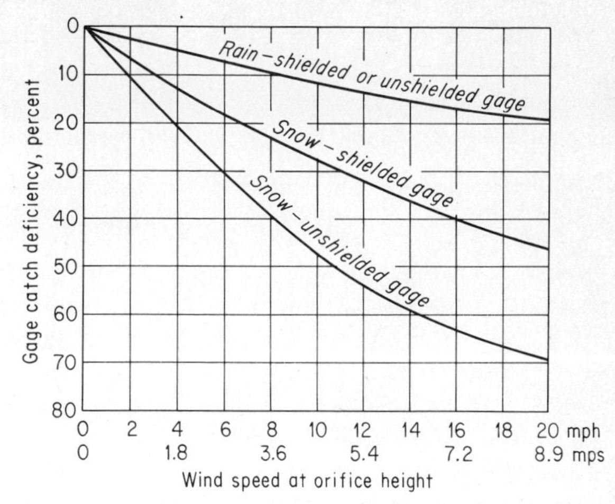

FIGURE 3-5
Effect of wind speed on the catch of precipitation gages. (*From L. W. Larson and E. L. Peck, Accuracy of Precipitation Measurements for Hydrologic Modeling, U.S. National Weather Service, in process.*)

catch [30, 31]. Trees or other obstacles serving as a windbreak should subtend angles of 20 to 30° from the gage orifice, with none greater than 45°, and should fairly well surround the gage to provide protection from all directions. An ideal site would be a clearing in a coniferous forest.

Various attempts have been made to estimate true, or actual, precipitation from gage measurements. One is based on the premise that there is a relationship between the ratio of the unshielded-gage catch P_{UG} to the actual precipitation P_A and the ratio of the unshielded-gage catch P_{UG} to the shielded-gage catch P_{SG}, or

$$\ln \frac{P_{UG}}{P_A} = b \ln \frac{P_{UG}}{P_{SG}} \qquad (3\text{-}1)$$

where b is a calibration coefficient, which depends on the type of gage. The model [32] is presumably independent of wind and form and type of precipitation.

When rain is falling vertically, a gage inclined 10° from the vertical will catch 1.5 percent less than it should. If a gage on level ground is inclined slightly toward the wind, it will catch more than the true amount. Some investigators [33–35] feel that gages should be perpendicular to land slopes. However, the area of a basin is its projection on a horizontal plane, and measurements from tilted gages must be reduced by multiplying by the cosine

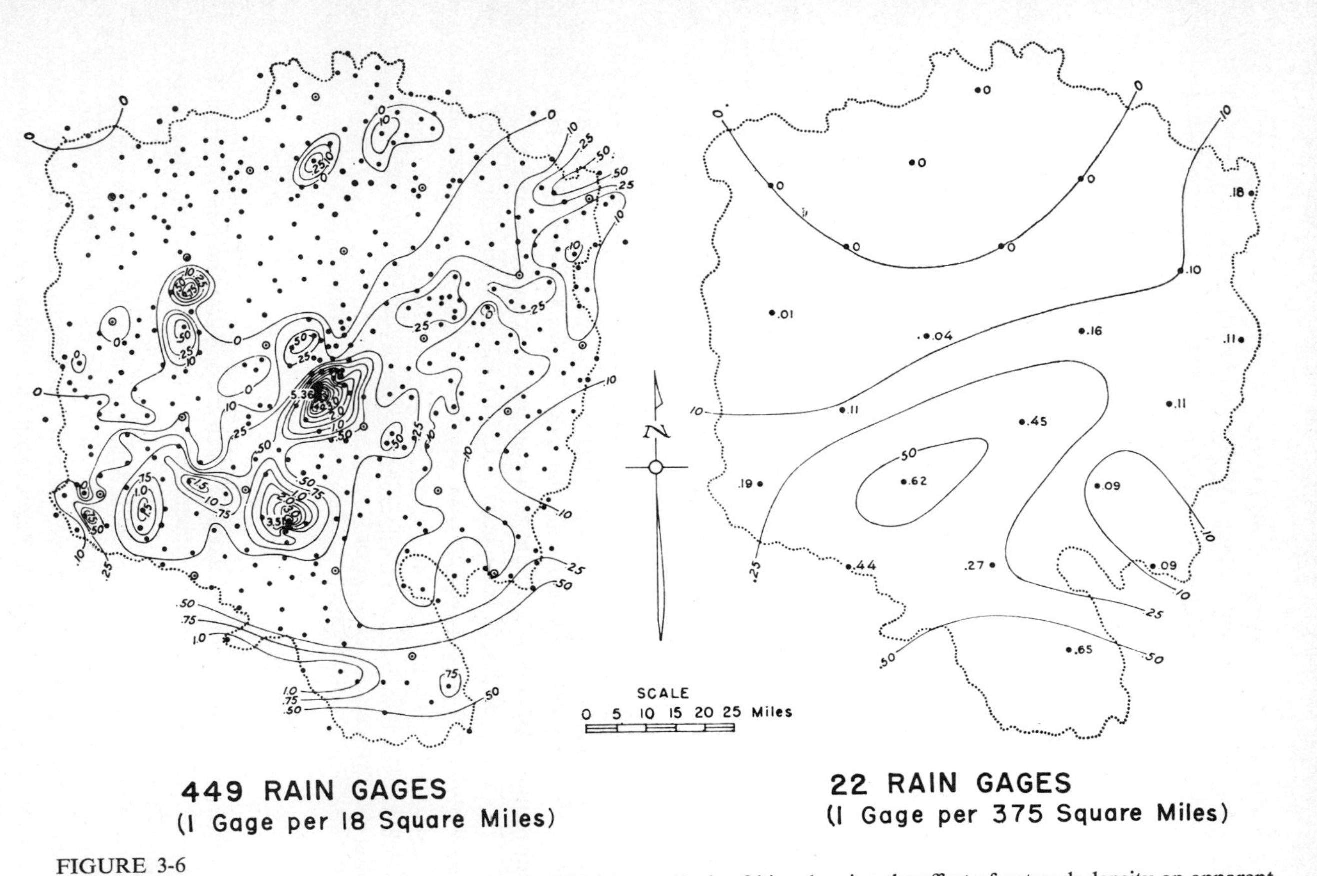

FIGURE 3-6
Isohyetal maps of the storm of Aug. 3, 1939, in the Muskingum Basin, Ohio, showing the effect of network density on apparent storm pattern. (*U.S. National Weather Service.*)

of the angle of inclination. Considering the variability of slope, aspect, and wind direction, it is virtually impossible to install a network of tilted gages for general purposes. Practically speaking, no gage has been designed which will give reliable measurements on steep slopes experiencing high winds, and such sites should be avoided. All official precipitation records in the United States are obtained from vertical gages.

3-6 The Precipitation-Gage Network

The uses for which precipitation data are intended should determine network density. A relatively sparse network of stations would suffice for studies of large general storms of for determining annual averages over large areas of level terrain. A very dense network is required to determine the rainfall pattern in thunderstorms. The probability that a storm center will be recorded by a gage varies with network density (Fig. 3-6). A network should be planned to yield a representative picture of the areal distribution of precipitation. There should be no concentration of gages in heavy-rainfall areas at the expense of dry areas, or vice versa. Unfortunately, the cost of installing and maintaining a network and accessibility of the gage site to an observer are always important considerations. The U.S. National Weather Service networks consist of about 3500 recording and 11,000 nonrecording gages of various types at about 13,000 stations, or one station per 230 mi^2 (600 km^2) on an average.

The error of rainfall averages computed from networks of various densities has been investigated [36]. Figure 3-7, based on an analysis [37] of storm rainfall in the Muskingum Basin, Ohio, shows the standard error of rainfall averages as a function of network density and area. In general, sampling errors, in terms of depth, tend to increase with increasing areal mean precipitation and to decrease with increasing network density, duration of precipitation, and size of area. Thus, a particular network would yield greater average errors for storm precipitation than for monthly or seasonal precipitation. Average errors also tend to be greater for summer than for winter precipitation because of the generally greater spatial variability of summer rainfall. Summer-storm rainfall may require a network density 2 to 3 times that required for winter storms in order to maintain equivalent degrees of accuracy.

The following minimum densities of precipitation networks have been recommended [38] for general hydrometeorological purposes:

1 For flat regions of temperate, mediterranean, and tropical zones, 600 to 900 km^2 (230 to 350 mi^2) per station

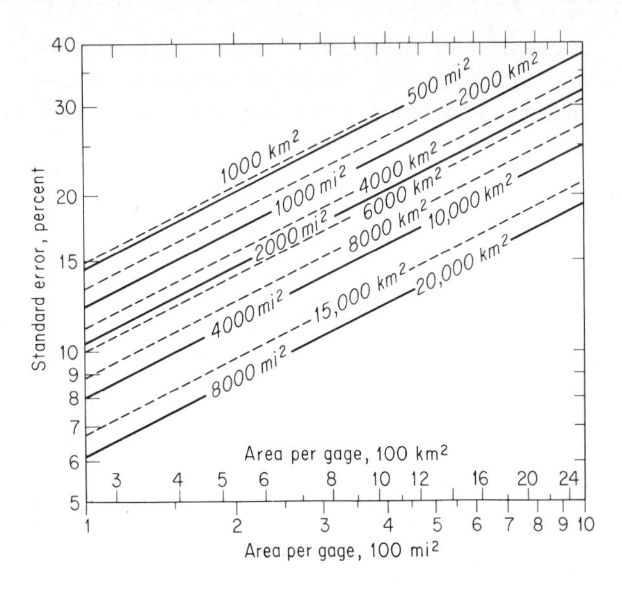

FIGURE 3-7
Standard error of storm precipitation averages as a function of network density and area for the Muskingum Basin. (*U.S. National Weather Service.*)

2 For mountainous regions of temperate, mediterranean, and tropical zones, 100 to 250 km^2 (40 to 100 mi^2) per station

3 For small mountainous islands with irregular precipitation, 25 km^2 (10 mi^2) per station

4 For arid and polar zones, 1500 to 10,000 km^2 (600 to 4000 mi^2) per station

Information on the difference between calculated and true watershed precipitation is of climatological interest but does not answer the hydrologic question: What error results in estimation of streamflow because of imperfect precipitation gaging? The answer depends on the climatic characteristics of precipitation in the area (including the effects of watershed topography), hydrologic characteristics of the watershed, the streamflow characteristics being estimated, and possibly the method used to estimate streamflow.

If only one gage is used as an index to watershed precipitation and if storm patterns are randomly distributed over the watershed, the observed precipitation will have a greater variance than the true average precipitation; i.e., the gage may on occasion record amounts much greater or much less than the basin average. Streamflow estimated from this index will display a greater variance than observed streamflow. On the other hand, over a

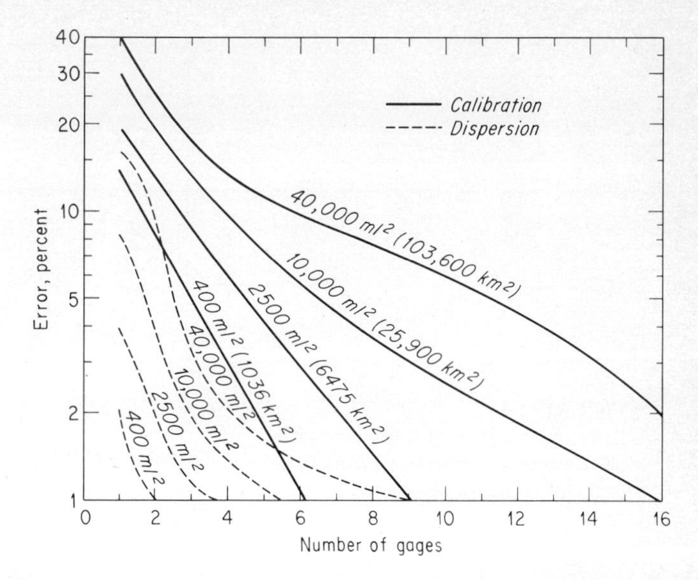

FIGURE 3-8
Error of simulation of annual volume of streamflow because of imperfect precipitation gaging. (*Adapted from* [40].)

sufficient period of time the gage should indicate average precipitation close to the watershed mean, and thus the mean estimated streamflow should be reasonably accurate.

While not enough work has been done to yield general results, the problem of minimum precipitation-network density for various hydrologic purposes has been investigated. Eagleson [39], for example, found that two properly located gages might be adequate for the determination of long-term basin average precipitation. Johanson [40] explored the errors involved in simulating streamflow from a dense gage network in central Illinois and found that the number of gages is more important than the density. Figure 3-8 indicates the calibration and dispersion errors to be expected when simulation methods (Chap. 10) are used to estimate annual streamflow. In the calibration phase one attempts to match each event in the historic record with an event in the simulated record. The calibration error is the ratio (percent) of the standard error of estimate to the standard deviation of the observed flows. When simulation is used to estimate flows for planning or design, the main concern is in accurately reproducing the mean and dispersion of the flow series. The dispersion error is expressed as a percentage of the coefficient of variation of the historic record. Figure 3-8 shows that the calibration error is much the larger and that neither error is closely related to network density.

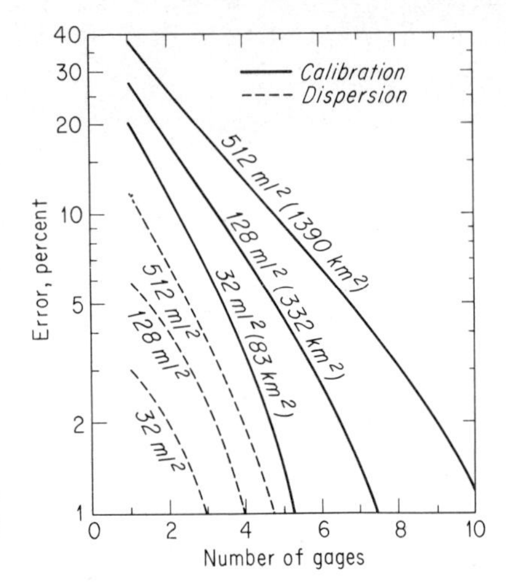

FIGURE 3-9
Error of simulation of storm direct runoff because of imperfect precipitation gaging. (*Adapted from* [40].)

Figure 3-9 presents similar data for simulation of storm direct runoff under summer thunderstorm conditions. Errors are larger but show a similar trend. Johanson's data precluded extending this case to larger areas. His sample period was too short to yield conclusive results for flood peaks, but the evidence indicates that errors are somewhat larger than for direct runoff, as might be expected.

3-7 Radar Measurement of Precipitation

A radar transmits a pulse of electromagnetic energy as a beam in a direction determined by a movable antenna. The beam width and shape are determined by the antenna size and configuration. The radiated wave, which travels at the speed of light, is partially reflected by cloud or precipitation particles and returns to the radar, where it is received by the same antenna.

Energy returned to the radar is called the *target signal*, the amount is termed *returned power*, and its display on the radarscope is called an *echo*. The brightness of an echo, or *echo intensity*, is an indication of the magnitude of returned power, which in turn is a measure of the *radar reflectivity* of the hydrometeors. The reflectivity of a group of hydrometeors depends on such factors as (1) drop-size distribution, (2) number of particles per unit volume, (3) physical state, i.e., solid or liquid, (4) shape of the individual elements, and (5) if asymmetrical, their aspect with respect to the radar. Generally speaking, the more intense the precipitation, the greater the reflectivity.

The time interval between emission of the pulse and appearance of the echo on the radarscope is a measure of the distance, or *range*, of the target from the radar. Direction of the target from the radar is determined by the orientation of the antenna at the time the target signal is received. Both range and direction are displayed in their proper perspectives by the locations of the echoes on the radarscope. The areal extent of an entire storm can be obtained by rotating the antenna. Nodding the antenna in a vertical plane yields information on precipitation structure and height.

Loss of radar energy due to passage through precipitation is called *attenuation*. Part of the loss results from scattering and part from absorption. The larger the rain and snow diameter-to-wavelength ratios, the greater the attenuation. For a given particle diameter, the shorter the wavelength, the greater the attenuation. Thus, at short wavelengths the total energy may be greatly diminished or completely dissipated by a relatively short penetration into a storm. Even the lightest precipitation intensities will seriously attenuate radar energy of wavelengths less than 1 cm. Wavelengths less than 5 cm are considered [41] unsuitable for measuring precipitation. These short wavelengths, however, are useful for defining very light rainfall within short distances, drizzle, and cloud forms. Wavelengths of 10 cm or more are generally recommended for precipitation measurement.

The *average returned power* $\bar{P}_r$ is a measure of the radar reflectivity of all particles at range r intercepting the radiated beam, or

$$\bar{P}_r = \frac{C}{r^2} \sum d^6 \qquad (3\text{-}2)$$

where C is dependent on wavelength, beam shape and width, pulse length, transmitted power, antenna gain, and refractive index of the target, and d is the diameter of the individual particles.

Many studies have been made of the relationship of raindrop-size distribution to rainfall intensity, and radar measurement of precipitation is based on empirical relationships between $\sum d^6$, usually represented by Z, and rainfall rate R in the form

$$Z = aR^b \qquad (3\text{-}3)$$

Values of a and b can be obtained by direct measurement of raindrop-size distribution or comparison of radar and rain-gage measurements.

The main obstacle to the accurate determination of the Z-R relationship arises from the fact that radar measures precipitation in the atmosphere while gages measure it at the ground. In order to avoid interference from hills, trees, buildings, etc., i.e., *ground clutter*, the radar beam is directed upward at an angle of $\frac{1}{2}$ to 1° above the horizontal, and the height of the beam thus increases with distance. The magnitude of the discrepancy between radar and gage measurements varies with the angle of the beam elevation, beam width, and range. Another factor leading to error is evaporation of precipitation before reaching the ground, which happens fairly frequently in arid areas. Also, strong winds may carry precipitation away from beneath the producing cloud. Solid forms of precipitation, especially if covered with liquid water, and discontinuities in the vertical distribution of precipitation in the cloud affect radar reflectivity and thus are also sources of error.

With Z in mm^6/m^3 and R in mm/h, values of a and b have been found to vary from about 15 to 1100 and from 1.2 to 3.2, respectively, for rainfall. While errors in the determination of the relationship are undoubtedly responsible for part of the large variation of a and b values, differences in climate and storm types are apparently important factors [42]. Also, some tests involved precipitation in the cloud rather than at the ground. It is generally agreed that values of 200 and 1.6 for a and b, respectively, yield the most reliable results on the average, and the Z-R relationship using these values is sometimes referred to as the standard. Also well known is the so-called *Miami relationship*, which uses a and b values of 300 and 1.4, respectively. Very few studies of the Z-R relationship have been made for solid precipitation, but values of about 2000 and 2.0 for a and b, respectively, appear applicable for snowflake aggregates [43].

The hydrologist is usually more interested in the volume of precipitation than in instantaneous rates, and much attention has been given to the development of procedures for integrating radar-indicated intensities with respect to time and area. The first method developed [44] involved the use of a camera to take a continuous exposure of the radarscope for a specific period. A photodensitometer was then used to measure the echo-image intensity, or film density, from the film negatives. The density was related to rain-gage measurements, and curves were drawn showing the relation of rainfall amount to film density and range. The relation permitted crude estimation of amounts over ungaged areas. This photographic method, however, is slow and difficult to standardize and is becoming obsolete.

In a method developed in the mid-1960s, special equipment automatically measures electronically the returned power for incremental areas within the beam and converts it to equivalent rainfall rates, which are then integrated with respect to time. The totals for any duration are displayed on a computer printout in grid form on which isohyets, or contours of equal precipitation, can be drawn on the basis of both radar and gage measurements.

Because of factors that affect returned power, the accuracy of radar measurements of precipitation varies with duration, area, storm type, and range. Numerous comparisons suggest that radar measurements of rainfall are within one-half to twice the gage measurements within a 60-nmi (110-km) range (where nmi is the abbreviation for nautical miles), with larger deviations for longer distances. The areal extent of rainfall may be depicted reliably by radar for ranges up to about 125 nmi (230 km). Since measurements by ordinary gage networks may be appreciably in error (Fig. 3-7) as a result of inadequate sampling, and since radar can detect and estimate precipitation between gages in a network of ordinary density, conjunctive use of radar and gage network should yield more accurate measurements than can be obtained from either one alone [45].

3-8 Satellite Estimates of Precipitation

Studies of water balance on a global scale require information on precipitation over areas where gage networks are inadequate or nonexistent, as over oceans. It has been suggested [46] that observational data from meteorological satellites might be used for estimating rainfall amounts for 1 month or longer. The chief problem is that satellites cannot measure rainfall directly, and solution requires evaluation of a rainfall coefficient on the basis of the amount and type of clouds and the probability of rainfall and likely rainfall intensity associated with each cloud type. These factors are necessarily based on data over land surfaces, and acceptability of the approach as a whole depends on precipitation processes over seas resembling those over land.

Attempts have been made to estimate precipitation for short as well as long durations by relating or calibrating satellite photographs with radar measurements so that the relation or calibration could then be applied to satellite photographs of remote regions to estimate precipitation intensity, frequency, and extent. As of 1974, such attempts had been generally unsuccessful. A major problem is that satellite photographs often do not reveal precipitation-producing clouds because of overlying cloud layers. Future developments in instrumentation and techniques may lead eventually to reasonably accurate satellite estimates of precipitation. Satellite-borne microwave radiometers, which can be used to calculate the liquid-water content of clouds, may provide the ultimate answer to precipitation measurement from space [47].

INTERPRETATION OF PRECIPITATION DATA

In order to avoid erroneous conclusions it is important to give the proper interpretation to precipitation data, which often cannot be accepted at face value. For example, a mean annual precipitation value for a station may have little significance if the gage site has been changed significantly during the period for which the average is computed. Also, there are several ways of computing average precipitation over an area, each of which may give a different answer.

3-9 Estimating Missing Precipitation Data

Many precipitation stations have short breaks in their records because of absences of the observer or because of instrumental failures. It is often necessary to estimate this missing record. In the procedure [48] used by the U.S. Environmental Data Service, precipitation amounts are estimated from observations at three stations as close to and as evenly spaced around the station with the missing record as possible. If the normal annual precipitation at each of the index stations is within 10 percent of that for the station with the missing record, a simple arithmetic average of the precipitation at the index stations provides the estimated amount.

If the normal annual precipitation at any of the index stations differs from that at the station in question by more than 10 percent, the *normal-ratio method* is used. In this method, the amounts at the index stations are weighted by the ratios of the normal-annual-precipitation values. That is, precipitation P_X at station X is

$$P_X = \frac{1}{3}\left(\frac{N_X}{N_A} P_A + \frac{N_X}{N_B} P_B + \frac{N_X}{N_C} P_C\right) \tag{3-4}$$

in which N is the normal annual precipitation.

Another method, used by the U.S. National Weather Service [49] for river forecasting, estimates precipitation at a point as the weighted average of that at four stations, one in each of the quadrants delineated by north-south and east-west lines through the point. Each station is the nearest in its quadrant to the point for which precipitation is being estimated. The weight applicable to each station is equal to the reciprocal of the square of the distance between the point and the station. Multiplying the precipitation for the storm (or other period) at each station by its weighting factor, adding the four weighted amounts, and dividing by the sum of the weights yields the estimated precipitation for the point. If one or more quadrants contain no precipitation stations, as might be the case for a point in a coastal area, then the estimation involves only the remaining quadrants.

The method can be used to estimate precipitation for a regular network station which failed to report or for points in a grid over a basin or other area so as to permit depth-area studies. The use of a grid also makes possible the machine plotting of isohyets or the direct estimation of average basin precipitation by arithmetic averaging of precipitation estimated for the grid points lying within the basin. A shortcoming of the method is that it can never yield a point estimate greater than the largest amount observed or less than the smallest. In mountainous regions precipitation values should therefore be expressed in percent of normal, as in Eq. (3-4).

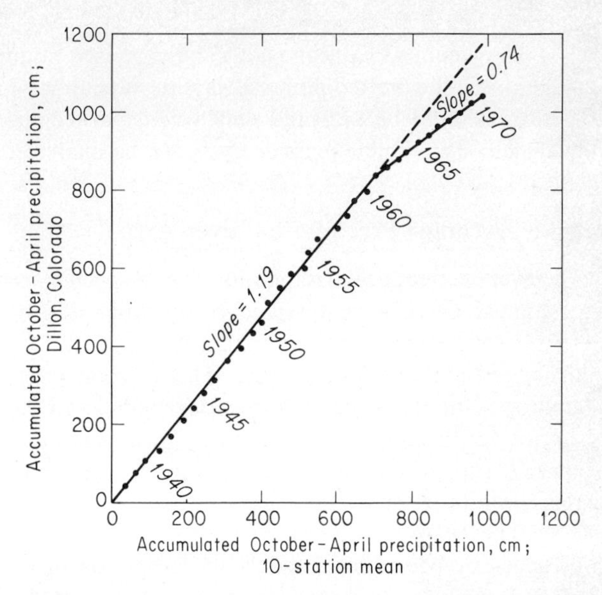

FIGURE 3-10
Adjustment of precipitation data for Dillon, Colorado, by double-mass curve.

3-10 Double-Mass Analysis

Changes in gage location, exposure, instrumentation, or observational procedure may cause a relative change in the precipitation catch. Frequently these changes are not disclosed in the published records. Current U.S. Environmental Data Service practice calls for a new station identification whenever the gage location is changed by as much as 5 mi (8 km) and/or 100 ft (30 m) in elevation.

Double-mass analysis [50] tests the consistency of the record at a station by comparing its accumulated annual or seasonal precipitation with the concurrent accumulated values of mean precipitation for a group of surrounding stations. In Fig. 3-10, for example, a change in slope about 1961 indicates a change in the precipitation regime at Dillon, Colorado. A change due to meteorological causes would not cause a change in slope, as all base stations would be similarly affected. The station history for Dillon discloses a change in gage location in June 1961. To make the record prior to 1961 comparable with that for the more recent location, it should be adjusted by the ratio of the slopes of the two segments of the double-mass curve (0.74/1.19). The consistency of the record for each of the base stations should be tested, and those showing inconsistent records should be dropped before other stations are tested or adjusted.

Considerable caution should be exercised in applying the double-mass technique. The plotted points always deviate about a mean line, and changes in slope should be accepted only when marked or substantiated by other evidence. The double-mass analysis can be made on a computer [51].

3-11 Average Precipitation over Area

The average depth of precipitation over a specific area, on a storm, seasonal, or annual basis, is required in many types of hydrologic problems. The simplest method of obtaining the average depth is to average arithmetically the gaged amounts in the area. This method yields good estimates in flat country if the gages are uniformly distributed and the individual gage catches do not vary widely from the mean. These limitations can be partially overcome if topographic influences and areal representativity are considered in the selection of gage sites [52].

The *Thiessen method* [53] attempts to allow for nonuniform distribution of gages by providing a weighting factor for each gage. The stations are plotted on a map, and connecting lines are drawn (Fig. 3-11*b*). Perpendicular bisectors of these connecting lines form polygons around each station. The sides of each polygon are the boundaries of the effective area assumed for the station. The area of each polygon is determined by planimetry and is expressed as a percentage of the total area. Weighted average rainfall for the total area is computed by multiplying the precipitation at each station by its assigned percentage of area and totaling. The results are usually more accurate than those obtained by simple arithmetical averaging. The greatest limitation of the Thiessen method is its inflexibility, a new Thiessen diagram being required every time there is a change in the gage network. Also, the method does not allow for orographic influences. Actually, the method simply assumes linear variation of precipitation between stations and assigns each segment of area to the nearest station.

The most accurate method of averaging precipitation over an area is the *isohyetal method*. Station locations and amounts are plotted on a suitable map, and contours of equal precipitation (*isohyets*) are then drawn (Fig. 3-11*c*). The average precipitation for an area is computed by weighting the average precipitation between successive isohyets (usually taken as the average of the two isohyetal values) by the area between isohyets, totaling these products, and dividing by the total area.

The isohyetal method permits the use and interpretation of all available data and is well adapted to display and discussion. In constructing an isohyetal map the analyst can make full use of his knowledge of orographic effects and storm morphology, and in this case the final map should represent a more realistic precipitation pattern than could be obtained from the gaged

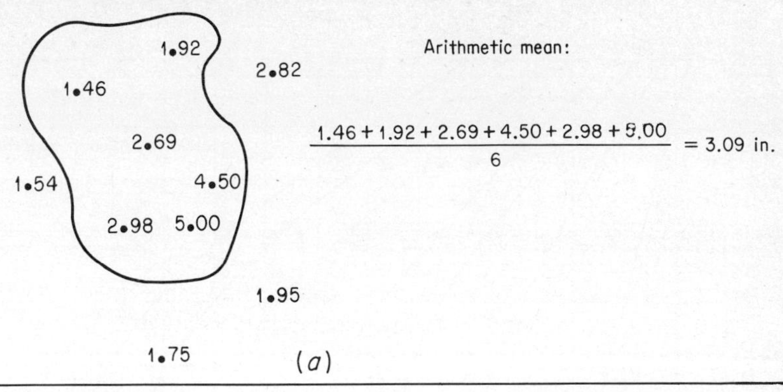

Arithmetic mean:

$$\frac{1.46 + 1.92 + 2.69 + 4.50 + 2.98 + 5.00}{6} = 3.09 \text{ in.}$$

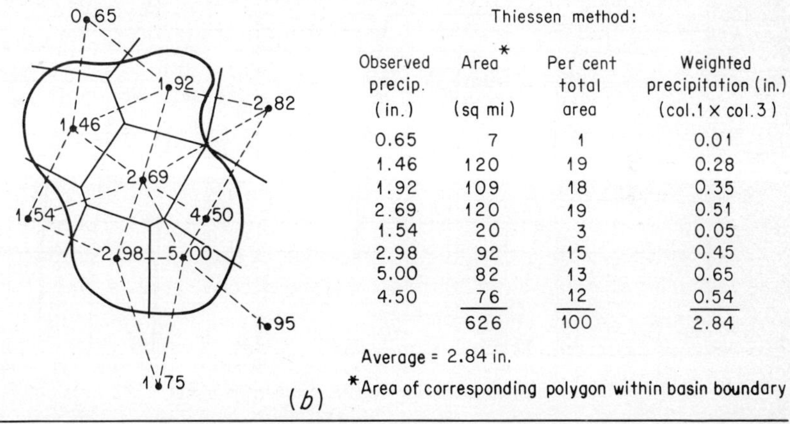

Thiessen method:

Observed precip. (in.)	Area* (sq mi)	Per cent total area	Weighted precipitation (in.) (col.1 × col.3)
0.65	7	1	0.01
1.46	120	19	0.28
1.92	109	18	0.35
2.69	120	19	0.51
1.54	20	3	0.05
2.98	92	15	0.45
5.00	82	13	0.65
4.50	76	12	0.54
	626	100	2.84

Average = 2.84 in.

*Area of corresponding polygon within basin boundary

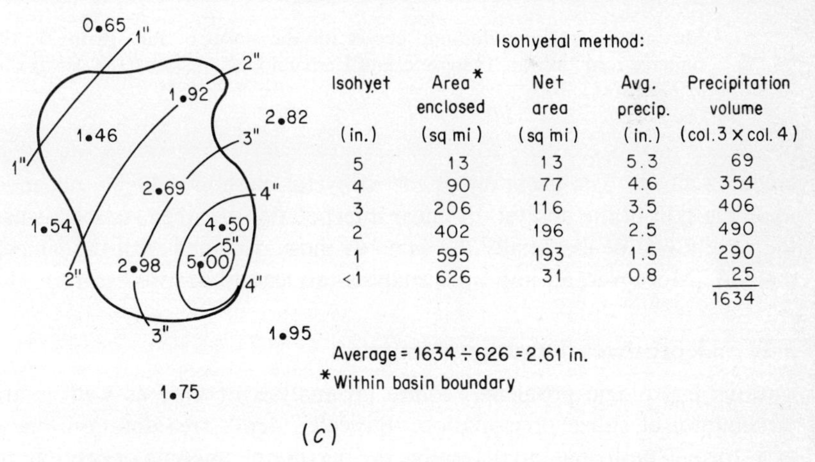

Isohyetal method:

Isohyet (in.)	Area* enclosed (sq mi)	Net area (sq mi)	Avg. precip. (in.)	Precipitation volume (col.3 × col. 4)
5	13	13	5.3	69
4	90	77	4.6	354
3	206	116	3.5	406
2	402	196	2.5	490
1	595	193	1.5	290
<1	626	31	0.8	25
				1634

Average = 1634 ÷ 626 = 2.61 in.

*Within basin boundary

FIGURE 3-11
Areal averaging of precipitation by (*a*) arithmetic method, (*b*) Thiessen method, and (*c*) isohyetal method.

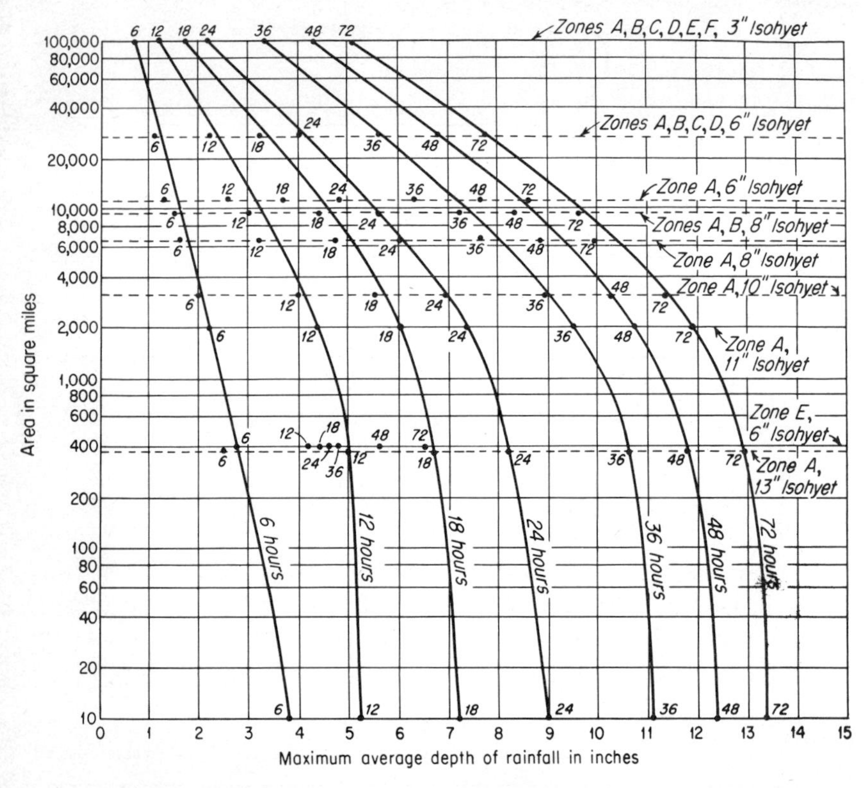

FIGURE 3-12
Maximum depth-area-duration curves for the storm of Jan. 18 to 21, 1935, centered near Bolivar, Tennessee, and Hernando, Mississippi. (*U.S. Army Corps of Engineers.*)

amounts alone. The accuracy of the isohyetal method is highly dependent upon the skill of the analyst. If linear interpolation between stations is used, the results will be essentially the same as those obtained with the Thiessen method. Moreover, an improper analysis can lead to serious error.

3-12 Depth-Area-Duration Analysis

Various hydrologic problems require an analysis of time as well as areal distribution of storm precipitation. Basically, depth-area-duration analysis of a storm is performed to determine the maximum amounts of precipitation within various durations over areas of various sizes. The method [54] discussed here is somewhat arbitrary but has been standardized so that results are comparable. The procedure has been computerized [55].

For a storm with a single major center, the isohyets are taken as boundaries of individual areas. The average storm precipitation within each isohyet is computed as described in Sec. 3-11. The storm total is distributed through successive increments of time (usually 6 h) in accordance with the distribution recorded at nearby stations [56]. When this has been done for each isohyet, data are available showing the time distribution of average rainfall over areas of various sizes. From these data, the maximum rainfall for various durations (6, 12, 18 h, etc.) can be selected for each size of area. These maxima are plotted (Fig. 3-12), and an enveloping depth-area curve is drawn for each duration. Storms with multiple centers are divided into zones for analysis.

VARIATIONS IN PRECIPITATION

3-13 Geographic Variations

In general, precipitation is heaviest near the equator and decreases with increasing latitude. However, the irregularity and orientation of the isohyets on the mean-annual-precipitation maps of the world (Fig. 3-13) and the United States (Fig. 3-14) indicate that the geographic distribution of precipitation depends on more effective factors than distance from the equator.

The main source of moisture for precipitation is evaporation from the surface of large bodies of water. Therefore, precipitation tends to be heavier near coastlines, as indicated by the isohyets of Figs. 3-13 and 3-14. Distortions in the isohyets reflect orographic effects.

Since lifting of air masses accounts for almost all precipitation, amounts and frequency are generally greater on the windward side of mountain barriers. Conversely, since downslope motion of air results in decreased relative humidity, the lee sides of barriers usually experience relatively light precipitation. However, the continued rise of air immediately downwind from the ridge and the slanting fall of the precipitation produce heavy amounts on the lee slopes near the crest.

The variation of precipitation with elevation and other topographic factors has been investigated [57] with somewhat varying conclusions. Perhaps the most detailed study of orographic influences on precipitation is that by Spreen [58], who correlated mean seasonal precipitation with elevation, slope, orientation, and exposure for western Colorado. While elevation alone accounted for only 30 percent of the variation in precipitation, the four parameters together accounted for 85 percent. Relations of this type are very useful for constructing isohyetal maps in rugged areas having sparse data.

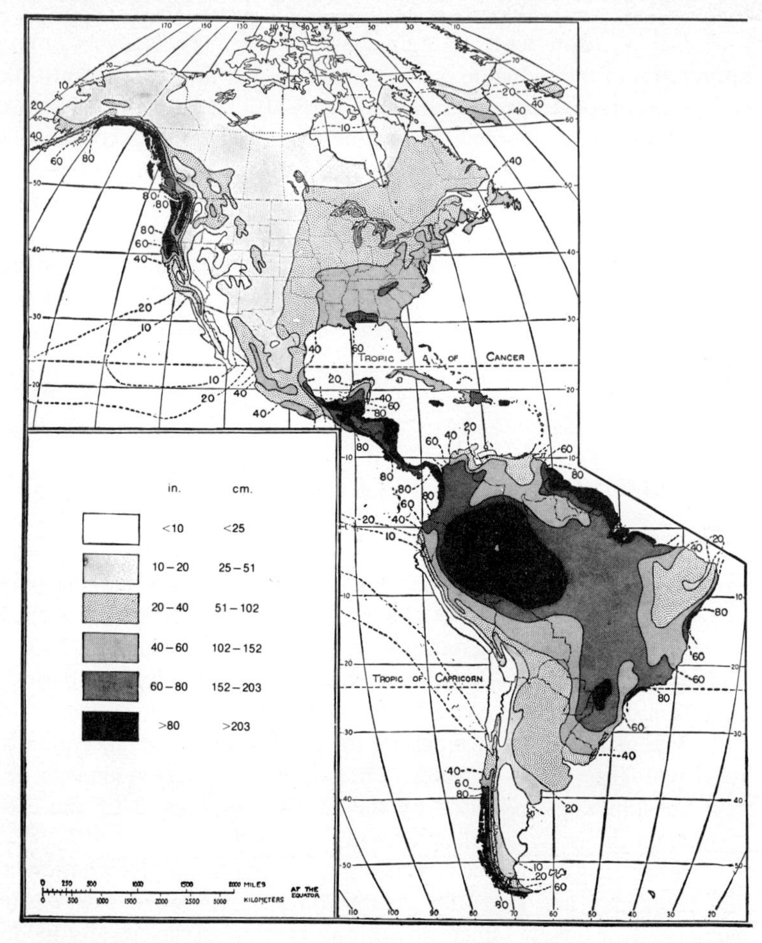

FIGURE 3-13
World distribution of mean annual precipitation, in inches. (*From G. T. Trewartha, "Introduction to Climate"*, 3*d ed. Copyright* 1954, *McGraw-Hill Book Company. Used with permission of McGraw-Hill Book Company.*)

The development of a relation like Spreen's is a tedious and lengthy procedure. In a mountainous area having relatively homogeneous geographic features, elevation generally accounts for a large proportion of the variation in normal annual precipitation. Some investigators [59, 60] develop only the precipitation-elevation relation for such an area. Differences between observed station precipitation and precipitation estimated by the relation are determined and plotted on a map, and isanomalous lines are then drawn. Precipitation for any ungaged point can then be obtained by using the elevation of the point and the precipitation-elevation relation to make a

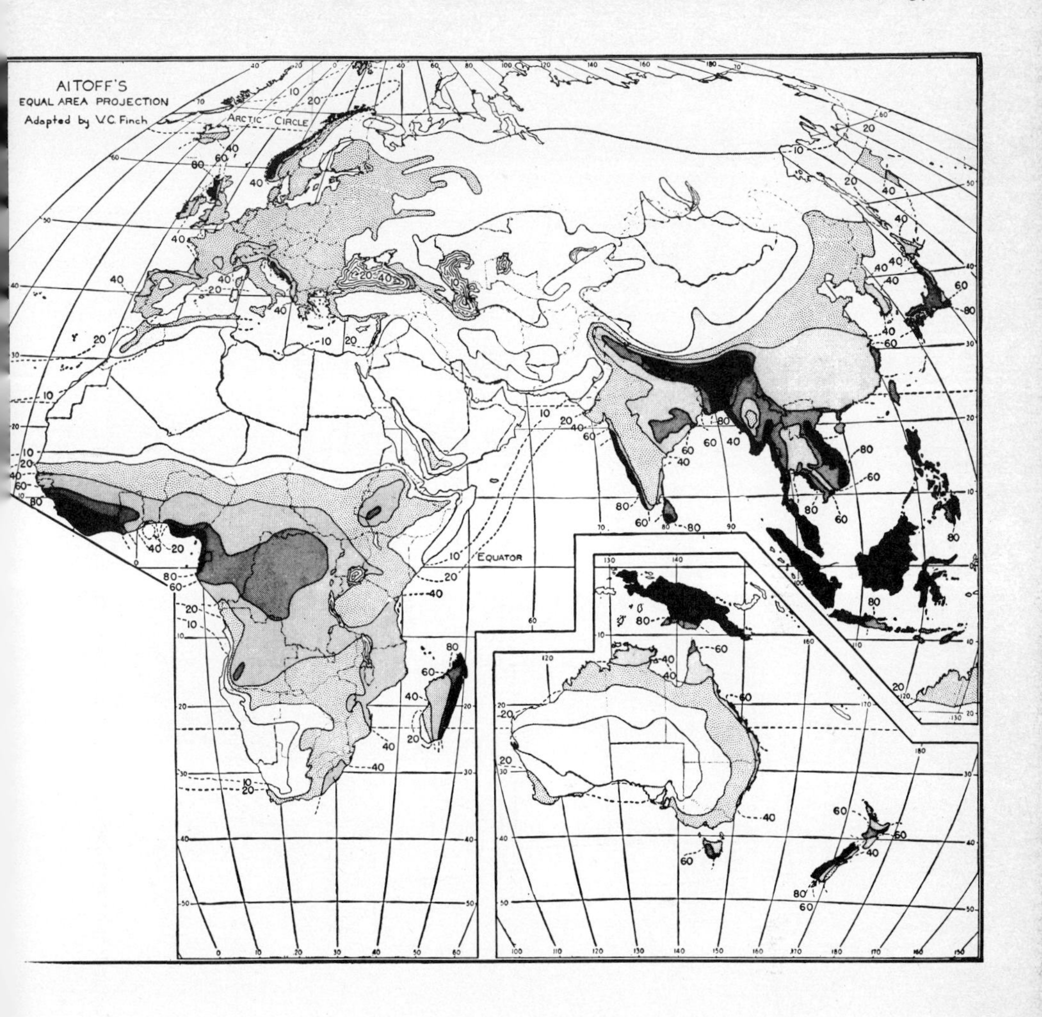

preliminary estimate and then adjusting the estimate as indicated by the anomaly map.

3-14 Time Variations

While portions of a precipitation record may suggest an increasing or decreasing trend, there is usually a tendency to return to the mean; abnormally wet periods tend to be balanced by dry periods. The regularity of these fluctuations has been repeatedly investigated. More than 100 apparent cycles,

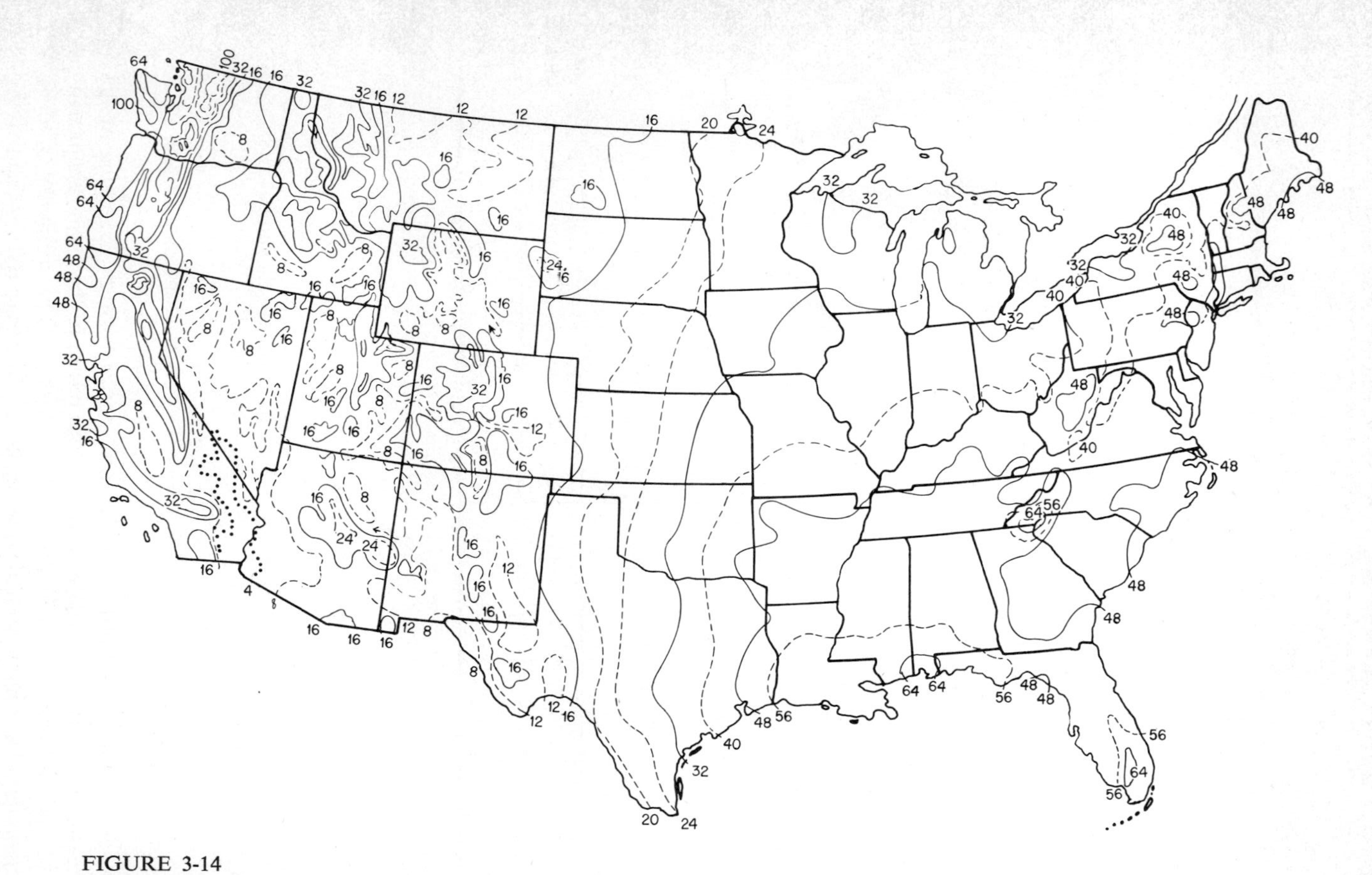

FIGURE 3-14
Mean annual precipitation in the United States, in inches. (*U.S. Environmental Data Service.*)

ranging in period from 1 to 744 y, have been reported [61]. The bibliography lists several reports on attempts to detect these variations. However, with the exception of diurnal and seasonal variations, no persistent regular cycles of any appreciable magnitude have been conclusively demonstrated [62].

The seasonal distribution of precipitation varies widely within the United States. Figure 3-15 shows typical seasonal distribution graphs for selected stations.

3-15 Record Rainfalls

Table 3-2 lists the world's greatest observed point rainfalls. Values of Table 3-2 plotted on logarithmic paper define an enveloping curve closely approximating a straight line (Fig. 3-16).

Table 3-2 WORLD'S GREATEST OBSERVED POINT RAINFALLS

Duration	Depth		Location	Date
	in.	mm		
1 min	1.50	38	Barot, Guadeloupe	Nov. 26, 1970
8 min	4.96	126	Füssen, Bavaria	May 25, 1920
15 min	7.80	198	Plumb Point, Jamaica	May 12, 1916
20 min	8.10	206	Curtea-de-Argeș, Roumania	July 7, 1889
42 min	12.00	305	Holt, Mo.	June 22, 1947
2 h 10 min	19.00	483	Rockport, W.Va.	July 18, 1889
2 h 45 min	22.00	559	D'Hanis, Tex. (17 mi NNW)	May 31, 1935
4 h 30 min	30.8+	782+	Smethport, Pa.	July 18, 1942
9 h	42.79	1,087	Belouve, Réunion	Feb. 28, 1964
12 h	52.76	1,340	Belouve, Réunion	Feb. 28–29, 1964
18 h 30 min	66.49	1,689	Belouve, Réunion	Feb. 28–29, 1964
24 h	73.62	1,870	Cilaos, Réunion	Mar. 15–16, 1952
2 d	98.42	2,500	Cilaos, Réunion	Mar. 15-17, 1952
3 d	127.56	3,240	Cilaos, Réunion	Mar. 15–18, 1952
4 d	137.95	3,504	Cilaos, Réunion	Mar. 14–18, 1952
5 d	151.73	3,854	Cilaos, Réunion	Mar. 13–18, 1952
6 d	159.65	4,055	Cilaos, Réunion	Mar. 13–19, 1952
7 d	161.81	4,110	Cilaos, Réunion	Mar. 12–19, 1952
8 d	162.59	4,130	Cilaos, Réunion	Mar. 11–19, 1952
15 d	188.88	4,798	Cherrapunji, India	June 24–July 8, 1931
31 d	366.14	9,300	Cherrapunji, India	July 1861
2 mo	502.63	12,767	Cherrapunji, India	June–July 1861
3 mo	644.44	16,369	Cherrapunji, India	May–July 1861
4 mo	737.70	18,738	Cherrapunji, India	Apr.–July 1861
5 mo	803.62	20,412	Cherrapunji, India	Apr.–Aug. 1861
6 mo	884.03	22,454	Cherrapunji, India	Apr.–Sept. 1861
11 mo	905.12	22,990	Cherrapunji, India	Jan.–Nov. 1861
1 y	1041.78	26,461	Cherrapunji, India	Aug. 1860–July 1861
2 y	1605.05	40,768	Cherrapunji, India	1860–1861

FIGURE 3-15
Normal monthly distribution of precipitation in the United States, in inches. (*U.S. Environmental Data Service.*)

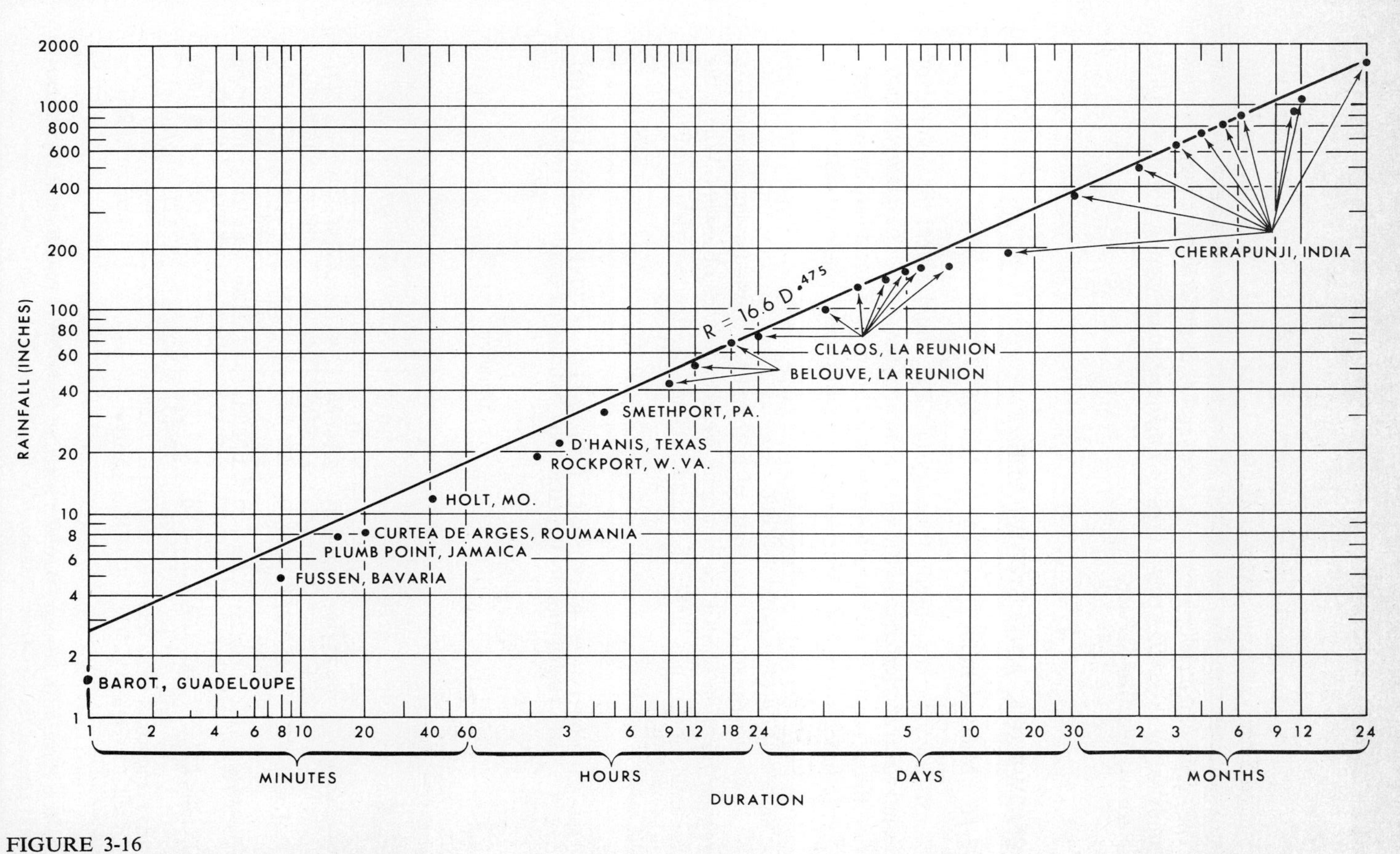

FIGURE 3-16
World's greatest observed point rainfalls.

The maximum rainfalls of record for durations up to 24 h at five major United States cities are given in Table 3-3. Table 3-4 lists maximum depth-area-duration data for the United States and the storms producing them. They represent enveloping values for over 400 of the country's major storms analyzed by the U.S. Army Corps of Engineers in cooperation with the U.S. National Weather Service.

SNOWPACK AND SNOWFALL

3-16 Measurement

Snow is usually composed of ice crystals and liquid water; the amount of liquid water is referred to as the *water content* of the snow. The *quality of snow*, i.e., the percentage by weight which is ice, can be determined by a calorimetric process [63, 64]. Most evaluations of quality made to date indicate values of 90 percent or more, but values as low as 50 percent have been obtained at times of rapid melting.

Measurement of the depth of accumulated snow on the ground is a regular function of all U.S. National Weather Service observers. Where the accumulation is not large, the measurements are made with a yardstick or

Table 3-3 MAXIMUM RECORDED RAINFALLS AT FIVE MAJOR UNITED STATES CITIES, IN INCHES AND (MILLIMETERS)

	Duration					
	Minutes			Hours		
Station	5	15	30	1	6	24
New York, N.Y.	0.75 (19) 8/12/26	1.63 (41) 7/10/05	2.34 (59) 8/12/26	2.97 (75) 8/26/47	4.44 (113) 10/1/13	9.55 (243) 10/8/03
St. Louis, Mo.	0.88 (22) 3/5/97	1.39 (35) 8/8/23	2.56 (65) 8/8/23	3.47 (88) 7/23/33	5.82 (148) 7/9/42	8.78 (223) 8/15/46
New Orleans, La.	1.00 (25) 2/5/55	1.90 (48) 4/25/53	3.18 (81) 4/25/53	4.71 (120) 4/25/53	8.62 (219) 9/6/29	14.01 (356) 4/15/27
Denver, Colo.	0.91 (23) 7/14/12	1.57 (40) 7/25/65	1.99 (51) 7/25/65	2.20 (56) 8/23/21	2.91 (74) 8/23/21	6.53 (166) 5/21/76
San Francisco, Calif.	0.33 (8) 11/25/26	0.65 (17) 11/4/18	0.83 (21) 3/4/12	1.07 (27) 3/4/12	2.31 (59) 1/14/56	4.67 (119) 1/29/81

SOURCE: From A. H. Jennings, Maximum Recorded United States Point Rainfall for 5 minutes to 24 hours at 296 First-Order Stations, *U.S. Weather Bur. Tech. Pap.* 2, rev., 1963.

Table 3-4 MAXIMUM OBSERVED DEPTH-AREA-DURATION DATA FOR THE UNITED STATES
Average rainfall in inches and (millimeters)

Area mi² (km²)	Duration, h						
	6	12	18	24	36	48	72
10	24.7[a]	29.8[b]	36.3[c]	38.7[c]	41.8[c]	43.1[c]	45.2[c]
(26)	(627)	(757)	(922)	(983)	(1062)	(1095)	(1148)
100	19.6[b]	26.3[c]	32.5[c]	35.2[c]	37.9[c]	38.9[c]	40.6[c]
(259)	(498)	(668)	(826)	(894)	(963)	(988)	(1031)
200	17.9[b]	25.6[c]	31.4[c]	34.2[c]	36.7[c]	37.7[c]	39.2[c]
(518)	(455)	(650)	(798)	(869)	(932)	(958)	(996)
500	15.4[b]	24.6[c]	29.7[c]	32.7[c]	35.0[c]	36.0[c]	37.3[c]
(1,295)	(391)	(625)	(754)	(831)	(889)	(914)	(947)
1,000	13.4[b]	22.6[c]	27.4[c]	30.2[c]	32.9[c]	33.7[c]	34.9[c]
(2,590)	(340)	(574)	(696)	(767)	(836)	(856)	(886)
2,000	11.2[b]	17.7[c]	22.5[c]	24.8[c]	27.3[c]	28.4[c]	29.7[c]
(5,180)	(284)	(450)	(572)	(630)	(693)	(721)	(754)
5,000	8.1[bd]	11.1[b]	14.1[b]	15.5[c]	18.7[e]	20.7[e]	24.4[e]
(12,950)	(206)	(282)	(358)	(394)	(475)	(526)	(620)
10,000	5.7[d]	7.9[f]	10.1[g]	12.1[g]	15.1[e]	17.4[e]	21.3[e]
(25,900)	(145)	(201)	(257)	(307)	(384)	(442)	(541)
20,000	4.0[d]	6.0[f]	7.9[g]	9.6[g]	11.6[e]	13.8[e]	17.6[e]
(51,800)	(102)	(152)	(201)	(244)	(295)	(351)	(447)
50,000	2.5[gh]	4.2[i]	5.3[g]	6.3[g]	7.9[g]	9.9[l]	13.2[l]
(129,500)	(64)	(107)	(135)	(160)	(201)	(251)	(335)
100,000	1.7[h]	2.5[hk]	3.5[g]	4.3[g]	5.6[g]	6.6[j]	8.9[j]
(259,000)	(43)	(64)	(89)	(109)	(142)	(168)	(226)

SOURCE: Compiled from U.S. Army Corps of Engineers, Storm Rainfall in the United States.

Storm	Date	Location of center
a	July 17–18, 1942	Smethport, Pa.
b	Sept. 8–10, 1921	Thrall, Tex.
c	Sept. 3–7, 1950	Yankeetown, Fla.
d	June 27–July 4, 1936	Bebe, Tex.
e	June 27–July 1, 1899	Hearne, Tex.
f	Apr. 12–16, 1927	Jefferson Parish, La.
g	Mar. 13–15, 1929	Elba, Ala.
h	May 22–26, 1908	Chattanooga, Okla.
i	Apr. 15–18, 1900	Eutaw, Ala.
j	July 5–10, 1916	Bonifay, Fla.
k	Nov. 19–22, 1934	Millry, Ala.
l	Sept. 19–24, 1967	Sombreretillo, Mex.

rain-gage measuring stick. In regions where large accumulations are the rule, permanent *snow stakes*, graduated in inches, are normally used. *Aerial snow-depth markers*, a type of stake adapted for visual reading from low-flying aircraft, are used in some remote areas. All stakes for measuring snow depth should be installed where they will be least affected by blowing or drifting snow [65].

The hydrologist is usually more interested in the water equivalent of the snowpack than in its depth. The *water equivalent* of the snowpack, i.e., the depth of water that would result from melting, depends on snow density as well as on depth. *Snow density*, the ratio between the volume of meltwater from a sample of snow and the initial volume of the sample, has been observed to vary from 0.004 for freshly fallen snow at high latitudes to 0.91 for compacted snow in glaciers. An average density of 0.10 for freshly fallen snow is often assumed. In regions of heavy snow accumulation, densities of 0.4 to 0.6 are common by the time the spring thaw begins.

Measurements of water equivalent are usually made by sampling with a snow tube (Fig. 3-17). The tube is driven vertically into the snowpack, sections of tubing being added as required. The cutting edge on the leading section is designed to penetrate ice layers when the tube is rotated. When the bottom of the snowpack is reached, the snow depth is determined from the graduations on the tube. The tube and its contents are then withdrawn and weighed to determine the water equivalent.

Because of the highly variable snowpack resulting from drifting and nonuniform melting, a number of measurements are made along an established line, or *snow course* [66]. Courses are selected at sites free from extensive wind effects and lateral drainage of meltwater. Even then the measurement for each course is considered to be only an index to areal snowpack.

Snow tubes tend to overmeasure water equivalent by 7 to 12 percent, the error increasing with snow density [67]. Chief disadvantages of snow surveys are that they are costly and disturb snow at the sampling sites so that natural changes in the snowpack at any particular site cannot be determined reliably.

The water equivalent of the snowpack can also be measured by *pressure pillows*. The pillows are made of thin butyl rubber, and various sizes from 5 to 12 ft (1.5 to 3.7 m) in diameter have been tested. Accuracy increases with size [68]. The pillows are filled with a mixture of water and antifreeze. As the snow accumulates on the pillow, the internal pressure increases. The weight of the snow resting on the pillow is then determined by measuring the pressure with a manometer or pressure transducer.

The chief advantage claimed for the pressure-pillow snow gage is that it is not subject to the wind errors of the usual type of precipitation gage so

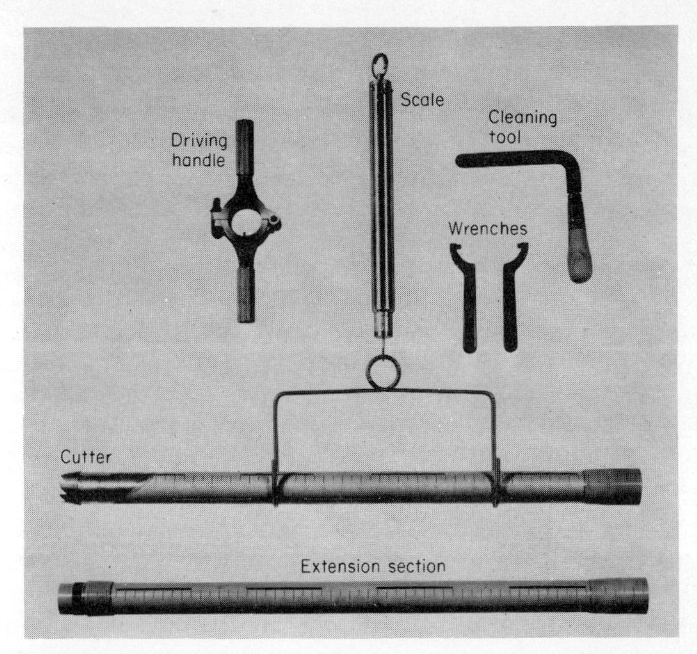

FIGURE 3-17
Mt. Rose snow sampler. (*Leupold & Stevens, Inc.*)

that the snow accumulating on it is more likely to be representative of the true snow cover on the ground. The disadvantages are that ice in the snow cover may form a bridge over the pillow, thus leading to false readings, and that the pillows are vulnerable to puncturing by hunters and animals. Also, unexplained diurnal variations in pressure readings may be large enough to cause appreciable error in daily melt measurements [69, 70].

Nuclear-radiation snow gages have been developed [71, 72] for measuring water equivalent of the snowpack. Measurement depends on attenuation of the radiation beam by the snow. Early models had the radioisotope source at the ground and detector overhead, but positions were reversed in later models to minimize temperature effects on the detector. One such gage [73] has the source about 15 ft (5 m) above the detector at ground level. Nuclear-radiation gages are expensive, and few are in use. Also, changes in the physical state of the snow affect measurement accuracy.

Special types of radioactive snow gages have been developed [74, 75] to determine the variation of the water equivalent, or density, with depth in the snowpack. These gages, known generally as *profiling radioactive snow gages* or *twin-probe snow-density gages*, consist of a gamma photon source and detector moved synchronously through vertical tubes about 2 ft (60 cm) apart in the snowpack. The gages are used for studying variations with time of the internal structure of the snowpack.

Areal measurements of the snow-cover water equivalent can be made by aircraft. Natural gamma emissions from the soil are attenuated by the snow cover. The degree of attenuation is related to the mass of water in the snow. Gamma spectral and total counting rates are collected and recorded by an airborne system using sodium iodide scintillation crystals [76]. Corrections are made for soil moisture, background radiation, altitude, and air density. Comparison of results with extensive ground measurements indicates that gamma spectral data yield areal values of water equivalent accurate within 0.2 to 0.5 in. (5 to 12 mm). Natural gamma radiation for measuring water equivalent has been used for many years in the Soviet Union, where depths of up to 12 in. (300 mm) have been measured successfully [77].

The extent of snow cover can be mapped from satellite photographs [78, 79] with an accuracy of ±20 mi (30 km) in flat terrain,† which still provides more detailed mapping than can be obtained from the conventional snow-reporting station network. Accuracy of ±1 mi (2 km) is possible in mountainous regions, where the snowline is readily identifiable with elevation contours. Qualitative estimates of snow depth can also be obtained from the photographs. In nonforested areas, snow brightness, or reflectivity, increases with depth up to about 4 to 6 in. (10 to 15 cm), there being no further increase for greater depths. Brightness estimates from satellite photography must be used cautiously, however, since anisotropy of snow seems to be a major factor affecting its appearance in photographs. No discernible relationship between reflectivity and snow depth is observed in forested areas.

Differentiation between snow and cloud is a major problem, but cloudiness can be determined from weather reports and charts. Since a cloud layer precludes photographs of snow cover, satellite observations cannot be made on a regular schedule. Indications of little or no cloudiness can be gained from topographic or pattern recognition of such features as lakes, dendritic stream patterns in mountains, dark forested areas, etc. Composite minimum brightness (CMB) charts [80] have proved useful in minimizing the cloudiness problem. These charts are computer-composited for a period of 5 to 10 d using only the minimum brightness observed for each array point during the compositing period. Cloudiness is thus retained in the CMB chart only when present in a given area every day of the period. Since cloudiness is usually transitory, the bright areas in the CMB chart generally represent the relatively stationary boundaries of the snow and ice fields.

† This degree of accuracy is expected to improve with the launching of planned satellites with better resolution. See D. R. Baker, M. Deutsch, and N. L. Durocher, Use of Earth Satellites for Hydrology, in *Sci. Pap. Tech. Conf. Hydrol. Meteorol. Serv., World Meteorol. Org., Geneva, 28 Sept.–6 Oct., 1970.*

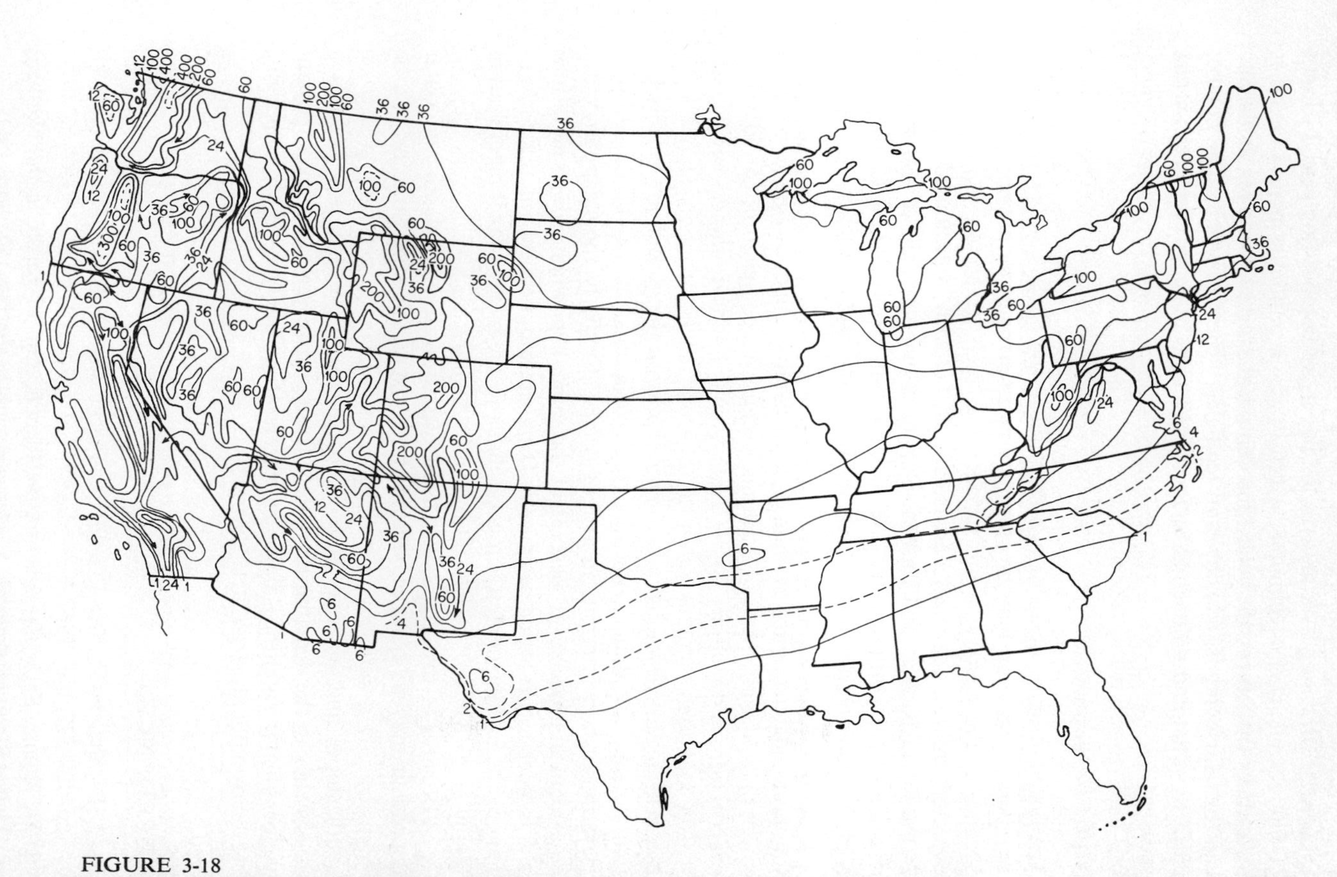

FIGURE 3-18
Mean annual snowfall in the United States, in inches. (*U.S. Environmental Data Service.*)

Cloudiness persisting over an area for the entire compositing period cannot be filtered out by the technique, which is why a minimum period of 5 d is advisable. Brightness for such time-composited areas, however, is usually less than for composited snow and ice fields. The CMB technique is most effective in flat and unforested regions and less so in heavy coniferous-forest areas.

3-17 Variations

The distribution of mean annual snowfall in the United States is shown in Fig. 3-18. This map may be considerably in error in mountainous regions because of the paucity of measurements at high elevations. As should be expected, there is a gradual increase of snowfall with latitude and elevation. In the Sierra Nevada and Cascade Range, annual snowfalls of 400 in. (1000 cm) are not uncommon. In general, maximum annual snowfall occurs at slightly higher elevations than maximum annual precipitation.

Maps of mean depth of snow on the ground or mean water equivalent for specific dates are not available. Snow depths would, of course, be much lower than annual snowfall amounts because of compaction of old snow, evaporation, and melting. Snow depth usually builds up rapidly early in the season and then remains relatively constant as compaction of old snow compensates for new falls. Maximum depths on the ground are usually less than one-half the annual snowfall at high elevations and are still less at lower elevations, where intermittent melting occurs. Similarly, the maximum water equivalent approaches 10 percent of the annual snowfall at high elevations and is much less at lower elevations. The mean water equivalent at the beginning of the melting season is a good index to the mean annual precipitation [81] in areas which experience little winter melting.

Because of drifting, considerable variation in snow depth and water equivalent may be observed within short distances. This variation can be intensified by differences in melting rates, which are generally greater on south slopes and in areas without forest cover. Nevertheless, there is a consistency in the distribution pattern from year to year as many important factors affecting melting—slope, aspect, elevation, forest cover, wind currents, etc.—normally undergo little change.

REFERENCES

1. J. E. McDonald, The Evaporation-Precipitation Fallacy, *Weather*, vol. 17, pp. 168–177, May 1962.
2. M. Kumai, Snow Crystals and the Identification of Nuclei in the Northern United States of America, *J. Meteorol.*, vol. 18, pp.139–150, April 1961.

3. B. J. Mason, Precipitation of Rain and Drizzle by Coalescence in Stratiform Clouds, *Q. J. R. Meteorol. Soc.*, vol. 78, pp. 377–386, July 1952.
4. R. Gunn, Collision Characteristics of Freely Falling Water Drops, *Science*, vol. 150, no. 3697, pp. 695–791, November 5, 1965.
5. P. M. Hamilton, Vertical Profiles of Total Precipitation in Shower Situations, *Q. J. R. Meteorol. Soc.*, vol. 92, pp. 346–362, April 1966.
6. P. Das, Role of Condensed Water in the Life Cycle of a Convective Cloud, *J. Atmos. Sci.*, vol. 21, pp. 404–418, July 1964.
7. G. P. Roys and F. Kessler, Measurement by Aircraft of Condensed Water in Great Plains Thunderstorms, *ESSA Tech. Note* 49-NSSP-19, July 1966.
8. J. Simpson and V. Wiggert, Models of Precipitating Cumulus Towers, *Mon. Weather Rev.*, vol. 97, pp. 471–489, July 1969.
9. W. L. Woodley, Rainfall Enhancement by Dynamic Cloud Modification, *Science*, vol. 170, no. 3954, pp. 127–132, Oct. 9, 1970.
10. P. W. Mielke, Jr., L. O. Grant, and C. F. Chappell, Elevation and Spatial Variation Effects of Wintertime Orographic Cloud Seeding, *J. Appl. Meteorol.*, vol. 9, pp. 476–488, June 1970.
11. C. F. Chappell, L. O. Grant, and P. W. Mielke, Jr., Cloud Seeding Effects on Precipitation Intensity and Duration of Wintertime Orographic Clouds, *J. Appl. Meteorol.*, vol. 10, pp. 1006–1010, October 1971.
12. T. J. Henderson, Cloud Seeding on the Kings River Watershed, Atmospherics Inc., Fresno, Calif., 1965.
13. 2d *Natl. Conf. Weather Modif. Am. Meteorol. Soc., April 6–9, 1970, Santa Barbara, Calif.*, pp. 34–90.
14. G. K. Sulakvelidze, N. Sh. Bibilashvili, and V. F. Lapcheva, "Formation of Precipitation and Modification of Hail Processes," Israel Program for Scientific Translation, pp. 155–199, 1967; available from U.S. Dept. of Commerce National Technical Information Center, Springfield, Va.
15. R. A. Schleusener, Hailfall Damage Suppression by Cloud Seeding: A Review of Evidence, *J. Appl. Meteorol.*, vol. 7, pp. 1004–1011, December 1968.
16. R. G. Fleagle (ed.), "Weather Modification: Science and Public Policy," University of Washington Press, Seattle, pp. 31–32, 1969.
17. K. R. Biswas and A. S. Dennis, Formation of a Rain Shower by Salt Seeding, *J. Appl. Meteorol.*, vol. 10, pp. 780–784, August 1971.
18. R. Wexler, Efficiency of Natural Rain, *Am. Geophys. Union Geophys. Monog.* 5, pp. 158–163, 1960.
19. J. Neyman, E. L. Scott, and M. A. Wells, Statistics in Meteorology, *Rev. Int. Stat. Inst.*, vol. 37, pp. 119–148, 1969.
20. See [16], pp. 87–142.
21. W. R. D. Sewell, Human Dimensions of Weather Modification, *Univ. Chicago Dept. Geogr. Res. Pap.* 105, 1966.
22. H. J. Taubenfeld (ed.), "Controlling the Weather: A Study of Law and Regu-

latory Procedures," Cambridge University Press, Cambridge, and Dunellen, New York, 1970.

23. F. A. Huff, Comparison between Standard and Small Orifice Raingages, *Trans. Am. Geophys. Union*, vol. 36, pp. 689–694, August 1955.
24. D. M. A. Jones, Effect of Housing Shape on the Catch of Recording Gages, *Mon. Weather Rev.*, vol. 97, pp. 604–606, August 1969.
25. U.S. National Weather Service, Substation Observations, *Obs. Handb.*, no. 2, pp. 17–37, 1970, rev. December 1972.
26. R. W. Beausoleil and K. W. Davis, Final Report of the Fischer and Porter Precipitation Gage Installation, *U.S. Natl. Weather Serv. Syst. Dev. Off. Rep.* 5, pp. 15–23, June 1970.
27. D. A. Parsons, Calibration of a Weather Bureau Tipping-Bucket Gage, *Mon. Weather Rev.*, vol. 69, p. 205, July 1941.
28. L. R. Struzer, Method of Measuring the Correct Value of Solid Atmospheric Precipitation, *Sov. Hydrol. Sel. Pap.*, no. 6, pp. 560–565, 1969.
29. See Windshields in Bibliography.
30. M. J. Brown and E. L. Peck, Reliability of Precipitation Measurements as Related to Exposure, *J. Appl. Meteorol.*, vol. 1, pp. 203–207, June 1962.
31. L. L. Weiss, Securing More Nearly True Precipitation Measurements, *J. Hydraul. Div. ASCE*, vol. 89, pp. 11–18, March 1963.
32. W. R. Hamon, Computing Actual Precipitation, in Distribution of Precipitation in Mountainous Areas, WMO/OMM no. 326, pp. 159–173, World Meteorological Organization, Geneva, 1973.
33. E. L. Hamilton, Rainfall Sampling on Rugged Terrain, *U.S. Dept. Agric. Tech. Bull.* 1096, 1954.
34. E. Hoeck, Report of the Committee on the Measurement of Precipitation, *Trans. IUGG Assoc. Sci. Hydrol.*, vol. 3, pp. 81–93, 1952.
35. L. Serra, The Correct Measurements of Precipitation, *Houille Blanche*, spec. no. A, pp. 152–158, March–April 1953.
36. See Network Density and Estimates of Areal Precipitation and Runoff in Bibliography.
37. Thunderstorm Rainfall, *Hydrometeorol. Rep.* 5, U.S. Weather Bureau in cooperation with Corps of Engineers, pp. 234–259, 1947.
38. World Meteorological Organization, Guide to Hydrometeorological Practices, 2d ed., WMO no. 168, *Tech. Pap.* 82, pp. III-8–III-11, Geneva, 1970.
39. P. S. Eagleson, Optimum Density of Rainfall Networks, *Water Resour. Res.*, vol. 3, no. 4, pp. 1021–1033, 1967.
40. R. C. Johanson, Precipitation Network Requirements for Streamflow Estimation, *Stanford Univ. Dept. Civ. Eng. Tech. Rep.* 147, August 1971.
41. E. Kessler and K. E. Wilk, The Radar Measurement of Precipitation for Hydrological Purposes, *Rep. WMO/IHD Proj.*, no. 5, World Meteorological Organization, Geneva, 1968.

42. G. E. Stout and E. A. Mueller, Survey of Relationships between Rainfall Rate and Radar Reflectivity in the Measurement of Precipitation, *J. Appl. Meteorol.*, vol. 7, pp. 465–474, June 1968.
43. P. E. Carlson and J. S. Marshall, Measurement of Snowfall by Radar, *J. Appl. Meteorol.*, vol. 11, pp. 494–500, April 1972.
44. H. W. Hiser, H. V. Senn, and L. F. Conover, Rainfall Measurement by Radar Using Photographic Integration Techniques, *Trans. Am. Geophys. Union*, vol. 39, pp. 1043–1047, December 1958.
45. J. W. Wilson, Integration of Radar and Raingage Data for Improved Rainfall Measurements, *J. Appl. Meteorol.*, vol. 9, pp. 489–497, June 1970.
46. E. C. Barrett, The Estimation of Monthly Rainfall from Satellite Data, *Mon. Weather Rev.*, vol. 98, pp. 322–327, April 1970.
47. W. L. Woodley and B. Sancho, A First Step toward Rainfall Estimation from Satellite Photographs, *Weather*, vol. 26, pp. 279–289, June 1971.
48. J. L. H. Paulhus and M. A. Kohler, Interpolation of Missing Precipitation Records, *Mon. Weather Rev.*, vol. 80, pp. 129–133, August 1952.
49. U.S. National Weather Service, *Natl. Weather Serv. River Forecast Syst. Forecast Proc.*, *NOAA Tech. Memo. NWS HYDRO* 14, pp. 3.1–3.14, December 1972.
50. M. A. Kohler, Double-Mass Analysis for Testing the Consistency of Records and for Making Required Adjustments, *Bull. Am. Meteorol. Soc.*, vol. 30, pp. 188–189, May 1949.
51. R. Singh, Double-Mass Analysis on the Computer, *J. Hydraul. Div. ASCE*, vol. 94, pp. 139–142, January 1968.
52. H. C. Wilm, A. Z. Nelson, and H. C. Storey, An Analysis of Precipitation Measurements on Mountain Watersheds, *Mon. Weather Rev.*, vol. 67, pp. 163–172, May 1939.
53. A. H. Thiessen, Precipitation for Large Areas, *Mon. Weather Rev.*, vol. 39, pp. 1082–1084, July 1911.
54. World Meteorological Organization, Manual for Depth-Area-Duration Analysis of Storm Precipitation, WMO no. 237, *Tech. Pap.* 129, pp. 1–31, Geneva, 1969.
55. See [54], pp. 32–56.
56. A. L. Shands and G. N. Brancato, Applied Meteorology: Mass Curves of Rainfall, *U.S. Weather Bur. Hydrometeorol. Tech. Pap.* 4, 1947.
57. See Spatial and Time Distribution of Precipitation in Bibliography.
58. W. C. Spreen, Determination of the Effect of Topography upon Precipitation, *Trans. Am. Geophys. Union*, vol. 28, pp. 285–290, April 1947.
59. D. R. Dawdy and W. B. Langbein, Mapping Mean Areal Precipitation, *Bull. Int. Assoc. Sci. Hydrol.*, vol. 5, pp. 16–23, September 1960.
60. E. L. Peck and M. J. Brown, An Approach to the Development of Isohyetal Maps for Mountainous Areas, *J. Geophys. Res.*, vol. 67, pp. 681–694, February 1962.

61. N. Shaw, "Manual of Meteorology," 2d ed., vol. 2, pp. 320–325, Cambridge University Press, London, 1942.
62. J. M. Mitchell, Jr., A Critical Appraisal of Periodicities in Climate, *Iowa State Univ., Cent. Agric. Econom. Devel. Rep.* 20, pp. 189–227, 1964.
63. M. Bernard and W. T. Wilson, A New Technique for the Determination of the Heat Necessary to Melt Snow, *Trans. Am. Geophys. Union*, vol. 22, pt. 1, pp. 178–181, 1941.
64. U. Radok, S. K. Stevens, and K. L. Sutherland, On the Calorimetric Determination of Snow Quality, *Int. Assoc. Sci. Hydrol. Pub.* 54, pp. 132–135, 1961.
65. R. W. Miller, Aerial Snow Depth Marker Configuration and Installation Considerations, *West. Snow Conf. Proc. 1962*, pp. 1–5.
66. U.S. Soil Conservation Service, Snow Survey and Water Supply Forecasting, "SCS National Engineering Handbook," sec. 22, April 1972.
67. R. A. Work, H. J. Stockwell, and R. T. Beaumont, Accuracy of Field Snow Surveys, *U.S. Army Corps Eng. Cold Regions Res. Eng. Lab. Tech. Rep.* 163, 1965.
68. R. T. Beaumont, Mt. Hood Pressure Pillow Snow Gage, *J. Appl. Meteorol.*, vol. 4, pp. 626–631, October 1965.
69. V. E. Penton and A. C. Robertson, Experience with the Pressure Pillow as a Snow Measuring Device, *Water Resour. Res.*, vol. 3, no. 2, pp. 405–408, 1967.
70. A. Tollan, Experience with Snow Pillows in Norway, *Bull. Int. Assoc. Sci. Hydrol.*, vol. 15, pp. 113–120, June 1970.
71. R. W. Gerdel, B. L. Hanson, and W. C. Cassidy, The Use of Radioisotopes for the Measurement of the Water Equivalent of a Snowpack, *Trans. Am. Geophys. Union*, vol. 31, pp. 449–453, June 1950.
72. C. C. Warnick and V. E. Penton, New Methods of Measuring Water Equivalent of Snow Pack for Automatic Recording at Remote Mountain Locations, *J. Hydrol.*, vol. 13, pp. 201–215, 1971.
73. G. A. McKean, A Nuclear Radiation Snow Gage, *Univ. Idaho Eng. Exp. Sta. Bull.* 13, August 1967.
74. J. L. Smith, W. D. Willen, and M. S. Owens, Isotope Snow Gage for Determining Hydrologic Characteristics of Snowpacks, in Isotope Techniques in the Hydrologic Cycle, *Am. Geophys. Union Geophys. Monogr.* 11, 1967.
75. J. L. Smith, The Profiling Radioactive Snow Gage, *Trans. Isotopic Snow Gage Infor. Meet., Sun Valley, Idaho, Oct. 28, 1940,* Idaho Nuclear Energy Commission and U.S. Soil Conservation Service,
76. E. L. Peck, V. C. Bissell, E. B. Jones, and D. L. Burge, Evaluation of Snow Water Equivalent by Airborne Measurement of Passive Terrestrial Gamma Radiation, *Water Resour. Res.*, vol. 7, pp. 1151–1159, October 1971.
77. N. V. Zotimov, Investigation of a Method of Measuring Snow Storage by Using the Gamma Radiation of the Earth, *Sov. Hydrol.: Sel. Pap.*, no. 3, pp. 254–265, 1968.

78. J. C. Barnes and C. J. Bowley, Snow Cover Distribution as Mapped from Satellite Photography, *Water Resour. Res.*, vol. 4, pp. 257–272, April 1968.
79. R. W. Popham, Satellite Applications to Snow Hydrology, *Rep. WMO/IHD Proj.*, no. 7, World Meteorological Organization, Geneva, 1968.
80. E. P. McClain and D. R. Baker, Experimental Large Scale Snow and Ice Mapping with Composite Brightness Charts, *ESSA Tech. Mem., NESCTM* 12, September 1969.
81. J. L. H. Paulhus, C. E. Erickson, and J. T. Riedel, Estimation of Mean Annual Precipitation from Snow Survey Data, *Trans. Am. Geophys. Union*, vol. 33, pp. 763–767, October 1952.

BIBLIOGRAPHY

Artificial Induction of Precipitation

BATTAN, L. J.: "Cloud Physics and Cloud Seeding," Anchor, Garden City, N.Y., 1962.

———: Silver-Iodide Seeding and Precipitation Initiation in Convective Clouds, *J. Appl. Meteorol.*, vol. 6, pp. 317–322, April 1967.

HOUGHTON, H. G.: On Precipitation Mechanisms and Their Artificial Modification, *J. Appl. Meteorol.*, vol. 7, pp. 851–859, October 1968.

LUMB, A. M., and LINSLEY, R. K.: Hydrologic Consequences of Rainfall Augmentation, *J. Hydraul. Div. ASCE*, vol. 97, pp. 1065–1080, July 1971.

NATIONAL RESEARCH COUNCIL: "Weather and Climate Modification: Problems and Progress," National Academy of Sciences, Washington, D.C., 1973.

NEIBURGER, M.: Artificial Modification of Clouds and Precipitation, *World Meteorol. Org. Tech. Note* 105, Geneva, 1970.

TRIBUS, M.: Physical View of Cloud Seeding, *Science*, vol. 168, no. 3298, pp. 201–211, Apr. 10, 1970.

Depth-Area Relations

COURT, A.: Area-Depth Rainfall Formulas, *J. Geophys. Res.*, vol. 66, pp. 1823–1832, June 1961.

STOUT, G. E., and HUFF, F. A.: Studies of Severe Rainstorms in Illinois, *J. Hydraul. Div. ASCE*, vol. 88, pp. 129–146, July 1962.

Measurement of Precipitation

BATTAN, L. J.: "Radar Observation of the Atmosphere," University of Chicago Press, Chicago, 1973.

BISWAS, A. K.: Development of Rain Gages, *J. Irrig. Drain. Div. ASCE*, vol. 93, *Proc. Pap.* 5416, September 1967.

BROWN, M. J., and PECK, E. L.: Reliability of Precipitation Measurements as Related to Exposure, *J. Appl. Meteorol.*, vol. 1, pp. 203–207, June 1962.

CATANEO, R.: A Method for Estimating Rainfall Rate—Radar Reflectivity Relationships, *J. Appl. Meteorol.*, vol. 8, pp. 815–819, October 1969.

HARROLD, T. W.: Radar Measurement of Rainfall, *Weather*, vol. 21, pp. 247–249, July 1966.

ISRAELSEN, C. E.: Reliability of Can-Type Precipitation Gage Measurements, *Utah State Univ., Utah Water Resour. Lab. Tech. Rep.* 2, July 1967.

KESSLER, E., and WILK, K. E.: Radar Measurement of Precipitation for Hydrological Purposes, *Rep. WMO/IHD Proj.*, no. 5, World Meteorological Organization, Geneva, 1968.

KURTYKA, J. C.: "Precipitation Measurements Study," State Water Survey Division, Urbana, Ill., 1953.

LARSON, L. W.: Approaches to Measuring "True" Snowfall, paper presented at *29th East. Snow Conf., Oswego, N.Y., Feb. 3–4, 1972.*

MARTIN, D. W., and SCHERER, W. D.: Review of Satellite Rainfall Estimation Methods, *Bull. Am. Meteorol. Soc.*, vol. 54, pp. 661–674, 1973.

MIDDLETON, W. E. K., and SPILHAUS, A. F.: "Meteorological Instruments," 3d ed., pp. 118–131, University of Toronto Press, Toronto, 1953.

RODDA, J. C.: Annotated Bibliography on Precipitation Measurement Instruments, *Rep. WMO/IHD Proj.*, no. 17, WMO no. 343, World Meteorological Organization, Geneva, 1973.

STOUT, G. E., and MUELLER, E. A.: Survey of Relationship between Rainfall Rate and Radar Reflectivity in the Measurement of Precipitation, *J. Appl. Meteorol.*, vol. 7, pp. 465–474, June 1968.

UNDERHILL, H. W.: Rainfall Recorders: A Comparison of Different Types, *Bull. Int. Assoc. Sci. Hydrol.*, vol. 10, pp. 50–55, September 1965.

U.S. WEATHER BUREAU: Excessive Precipitation Techniques, *Key Meteorol. Rec. Doc.*, no. 3.081, 1958.

———: History of Weather Bureau Precipitation Measurements, *Key Meteorol. Rec. Doc.*, no. 3.082, 1963.

Network Density and Estimates of Areal Precipitation and Runoff

ALVAREZ, F., and HENRY, W. K.: Rain Gage Spacing and Reported Rainfall, *Bull. Int. Assoc. Sci. Hydrol.*, vol. 15, pp. 97–107, March 1970.

AMOROCHO, J., BRANDSTETTER, A., and MORGAN, D.: The Effects of Density of Recording Rain Gage Networks on the Description of Precipitation Patterns, *Int. Assoc. Sci. Hydrol. Publ.* 78, pp. 189–202, 1968.

HERSHFIELD, D. M.: On the Spacing of Raingages, *Int. Assoc. Sci. Hydrol. Publ.* 67, pp. 72–81, 1965.

HUFF, F. A.: Sampling Errors in Measurement of Mean Precipitation, *J. Appl. Meteorol.*, vol. 9, pp. 35–44, February 1970.

NĚMEC, J.: A Contribution to the Design of Recording Gage Network, *Bull. Int. Assoc. Sci. Hydrol.*, vol. 10, pp. 70–73, June 1965.

NORDENSON, T. J.: Preparation of Co-ordinated Precipitation, Runoff and Evaporation Maps, *Rep. WMO/IHD Proj.*, no. 6, World Meteorological Organization, Geneva, 1967.

RAINBIRD, A. F.: Methods of Estimating Areal Average Precipitation, *Rep. WMO/IHD Proj.*, no. 3, World Meteorological Organization, Geneva, 1967.

RODDA, J. C.: Hydrological Network Design: Needs, Problems and Approaches, *Rep. WMO/IHD Proj.*, no. 12, World Meteorological Organization, Geneva, 1969.

SCHICKEDANZ, P. T.: Application of Dense Rainfall Data to Regions of Sparse Data Coverage, State Water Survey, Urbana, Ill., October 1971.

STEPHENSON, P. M.: Objective Assessment of Adequate Numbers of Raingauges for Estimating Areal Rainfall Depths, *Int. Assoc. Sci. Hydrol. Publ.* 78, pp. 252–264, 1968.

WORLD METEOROLOGICAL ORGANIZATION, Hydrological Network Design Practices, *WMO Publ.* 324, *Geneva*, 1972.

Precipitation Cycles

BRIER, G. W., SHAPIRO, R., and MACDONALD, N. J.: A Search for Rainfall Calendricities, *J. Atmos. Sci.*, vol. 20, pp. 529–532, November 1963.

LAMBOR, J.: "Hydrologic Forecasting Methods," Scientific Publications Foreign Cooperation Center of the Central Institute for Scientific, Technical, and Economic Information, Warsaw, pp. 314–336, 1962; available from U.S. Dept. of Commerce National Technical Information Center, Springfield, Va.

NAMIAS, J.: Long-Range Weather Forecasting History, Current Status and Outlook, *Bull. Am. Meteorol. Soc.*, vol. 49, pp. 438–470, May 1968.

Precipitation Forms and Formation

BYERS, H. R.: "Elements of Cloud Physics," University of Chicago Press, Chicago, 1965.

FLETCHER, N. H.: "The Physics of Rain Clouds," Cambridge University Press, London, 1962.

HARDY, K. R.: The Development of Raindrop-Size Distributions and Implications Related to the Physics of Precipitation, *J. Atmos. Sci.*, vol. 20, pp. 299–312. July 1963.

MASON, B. J.: "The Physics of Clouds," 2d ed., Oxford University Press, London, 1971.

Spatial and Time Distribution of Precipitation

ATKINSON, B. W.: The Effect of an Urban Area on the Precipitation from a Moving Thunderstorm, *J. Appl. Meteorol.*, vol. 10, pp. 47–55, February 1971.

CHANGNON, S. A., JR.: Recent Studies of Urban Effects on Precipitation in the United States, *Bull. Am. Meteorol. Soc.*, vol. 50, pp. 411–421, June 1969.

FEDOROV, S. F., and BUROV, A. S.: Influence of the Forest on Precipitation, *Sov. Hydrol.: Sel. Pap.*, no. 3, pp. 217–227, 1967.

HOVIND, E. L.: Precipitation Distribution around a Windy Mountain Peak, *J. Geophys. Res.*, vol. 70, pp. 3271–3278, July 15, 1965.

HUFF, F. A.: Spatial Distribution of Rainfall Rates, *Water Resour. Res.*, vol. 6, pp. 254–260, February 1970.

———: Time Distribution Characteristics of Rainfall Rates, *Water Resour. Res.*, vol. 6, pp. 447–454, April 1970.

——— and CHANGNON, S. A., JR.: Precipitation Modification by Major Urban Areas, *Bull. Am. Meteorol. Soc.*, vol. 54, pp. 1220–1233, December 1973.

OGDEN, T. L.: The Effect on Rainfall of a Large Steelworks, *J. Appl. Meteorol.*, vol. 8, pp. 585–591, August 1969.

POTTER, J. G.: Changes in Seasonal Snowfall in Cities, *Can. Geogr.*, vol. 5, no, 1, pp. 37–42, 1961.

SCHERMERHORN, V. P.: Relations between Topography and Annual Precipitation in Western Oregon and Washington, *Water Resour. Res.*, vol. 3, pp. 707–711, 3d quarter, 1967.

WILLIAMS, G. C.: An Occurrence of Lake Snow: One of the Direct Effects of Lake Michigan on the Climate of the Chicago Area, *Mon. Weather Rev.*, vol. 9, pp. 465–467, September 1963.

Windshields

RECHARD, P. A., and LARSON, L. W.: Snow Fence Shielding of Precipitation Gages, *J. Hydraul. Div. ASCE*, vol. 97, pp. 1427–1440, September 1971.

WARNICK, C. C.: Experiments with Wind Shields for Precipitation Gages, *Trans. Am. Geophys. Union*, vol. 34, pp. 379–388, June 1953.

WEISS, L. L.: Relative Catches of Snow in Shielded and Unshielded Gages, *Mon. Weather Rev.*, vol. 89 pp. 397–400, October 1961.

UNITED STATES DATA SOURCES

The main source of daily precipitation data is *Climatological Data* published monthly through 1965 by the Weather Bureau and then by the Environmental Data Service. Hourly rainfall intensities are found in *Hydrologic Bulletin* (1940–1948), *Climatological Data* (1948–1951), and since 1951 in *Hourly Precipitation Data.* Weekly and monthly maps of precipitation are found in *Weekly Weather and Crop Bulletin* and *Monthly Weather Review.* Monthly and annual data are summarized from beginning of record through 1950 in *Climatic Summary for the United States* and from 1951 through 1960 in *Climatography of the United States*, no. 86. Monthly data for about 1000 stations scattered throughout the world are published in *Monthly Climatic Data for the World*, which is sponsored by the World Meteorological Organization in cooperation with the Environmental Data Service.

The *Water Bulletin* of the International Boundary and Water Commission contains data for the Rio Grande Basin, and *Precipitation in the Tennessee River Basin*, published by the Tennessee Valley Authority, summarizes data in its service area. Many states and local groups publish data summaries which contain information on precipitation.

Results of snow surveys in California are published by the State Department of Water Resources in *Water Conditions in California*. Elsewhere in the West, snow-survey data are published by the U.S. Soil Conservation Service in the *Federal-State-Private Cooperative Snow Surveys and Water-Supply Outlook*. In the East snow-survey data are published by the U.S. National Weather Service in *Snow Cover Surveys by Eastern Snow Conference*.

Special summaries and data on precipitation normals, extremes, and frequencies may be found in:

"Climatological Data, National Summary," U.S. Environmental Data Service.

"Climatic Atlas of the United States," U.S. Environmental Data Service.

Monthly Normals of Temperature, Precipitation, and Heating and Cooling Degree Days, *Climatography of the United States*, no. 81 (by states), NOAA Environmental Data Service, August 1973.

HERSHFIELD, D. M.: Rainfall-Frequency Atlas of the United States for Durations from 30 Minutes to 24 Hours and Return Periods from 1 to 100 Years, *U.S. Weather Bur. Tech. Pap.* 40, 1961. (Similar data for durations up to 10 d are also available in the *Weather Bureau Technical Paper* series for the United States, including Alaska and Hawaii, Puerto Rico, and Virgin Islands.)

JENNINGS, A. H.: Maximum Recorded United States Point Rainfall for 5 Minutes to 24 Hours at 296 First-Order Stations, *U.S. Weather Bur. Tech. Pap.* 2, rev. 1963.

JORGENSEN, D. L., KLEIN, W. H., and ROBERTS, C. F.: Conditional Probabilities of Precipitation Amounts in the Conterminous United States, *ESSA Tech. Mem. WBTM TDL* 18, March 1969.

LUDLUM, D. M.: Extremes of Snowfall in the United States, *Weatherwise*, vol. 15, pp. 246–253, December 1962.

MILLER, J. F., and FREDERICK, R. H.: Normal Monthly Number of Days with Precipitation of 0.5, 1.0, 2.0, and 4.0 Inches or More in the Conterminous United States, *ESSA Tech. Pap.* 57, 1966.

———, ———, and TRACEY, R. J.: "Precipitation-Frequency Atlas for the Conterminous Western United States (by States)," *NOAA Atlas* 2 (11 vols.), 1974.

THOM, H. C. S.: Probabilities of One-Inch Snowfall Thresholds for the United States, *Mon. Weather Rev.*, vol. 85, pp. 269–271, August 1957.

———: Distribution of Maximum Annual Water Equivalent of Snow on the Ground, *Mon. Weather Rev.*, vol. 94, pp. 265–271, April 1966.

Detailed descriptions of most of the above and additional sources are given in Selective Guide to Climatic Data Sources, *Key Meteorolog. Rec. Doc.* 4.11, U.S. Environmental Data Service, 1969.

PROBLEMS

3-1 Assuming rain falling vertically, express the catch of a gage inclined 15° from the vertical as a percentage of the catch for the same gage installed vertically.

3-2 Precipitation station *X* was inoperative for part of a month during which a storm occurred. The respective storm totals at three surrounding stations, *A*, *B*, and *C*, were 107, 89, and 122 mm. The normal annual precipitation amounts at stations *X*, *A*, *B*, and *C* are, respectively, 978, 1120, 935, and 1200 mm. Estimate the storm precipitation for station *X*.

3-3 The annual precipitation at station *X* and the average annual precipitation at 15 surrounding stations are as shown in the following table.

(*a*) Determine the consistency of the record at station *X*.

(*b*) In what year is a change in regime indicated?

(*c*) Compute the mean annual precipitation for station *X* for the entire 34-y period without adjustment.

(*d*) Repeat part (*c*) for station *X* at its 1971 site with the data adjusted for the change in regime.

	Annual precipitation, in.			Annual precipitation, in.	
Year	Sta. *X*	15-station average	Year	Sta. *X*	15-station average
1938	13.4	13.9	1955	13.3	9.6
1939	10.7	9.9	1956	16.3	10.2
1940	10.9	10.1	1957	22.7	15.9
1941	12.0	13.7	1958	13.9	10.9
1942	13.3	13.1	1959	14.7	10.2
1943	14.6	13.2	1960	14.0	10.3
1944	9.0	10.9	1961	11.4	10.2
1945	11.8	11.4	1962	13.8	11.8
1946	9.7	10.2	1963	10.0	9.2
1947	15.4	13.9	1964	10.5	10.2
1948	12.5	13.0	1965	16.7	14.0
1949	11.5	13.1	1966	9.3	8.4
1950	10.9	9.2	1967	18.4	11.5
1951	13.9	10.9	1968	14.1	9.1
1952	14.1	13.2	1969	19.8	13.0
1953	10.4	10.0	1970	17.1	13.1
1954	7.9	8.8	1971	16.0	10.7

3-4 The average annual precipitation for the four subbasins constituting a large river basin is 28.9, 33.4, 44.2, and 39.7 in. The areas are 360, 275, 420, and 650 mi^2, respectively. What is the average annual precipitation for the basin as a whole?

3-5 Construct maximum depth-area-duration curves for the United States from the data of Table 3-4. Tabulate maximum values, in inches, for areas of 50, 7500, and 30,000 mi^2 for durations of 6, 12, 18, 24, 36, 48, and 72 h.

3-6 Plot the 15-station average precipitation of Prob. 3-3 as a time series. Also plot 5-y moving averages and accumulated annual departures from the 34-y mean. Are there any apparent cycles or time trends? Discuss.

3-7 A shielded and an unshielded gage at a station indicate a storm rainfall of 110 and 100 mm, respectively. Estimate the true rainfall. Assume $b = 1.8$.

3-8 Develop an equation for the 4000-mi^2 (10,000-km^2) line of Fig. 3-7 showing the percentage standard error (SE) as a function of average area A per gage (*a*) in square miles and (*b*) in square kilometers.

3-9 With a rainfall rate of 100 mm/h, what is the value of Z, in mm^6/m^3, for a Z-R relationship having a and b values of (*a*) 200 and 1.6 and (*b*) 300 and 1.4, respectively?

3-10 What is the value of Z, in mm^6/m^3, for a Z-R relationship with a and b values of 200 and 1.6, respectively, if the rainfall rate is (*a*) 25 mm/h and (*b*) 50 mm/h?

3-11 A calorimetric procedure for evaluating quality Q_t of melting snow consists of inserting into a thermos bottle or jug containing a known weight W_1 of warm water a sample of snow of known weight W_2. Write an equation for evaluating Q_t as a percentage when the initial and final temperatures of the water are, respectively, T_1 and T_2, in degrees Celsius, and the calorimeter constant is k.

3-12 Using Fig. 3-7, develop a formula for approximating the percentage standard error (SE) as a function of area A, in square miles, when the average precipitation network density is one gage per 100 mi^2.

3-13 Compute the mean annual precipitation for some river basin selected by your instructor. Use the arithmetic average, Thiessen network, and isohyetal map, and compare the three values. Which do you feel is the most accurate? How consistent are the answers determined by each method among members of your class?

3-14 Using the formula developed in Prob. 3-11, determine the snow quality when 100 g of snow inserted into a thermos bottle containing 300 g of water at 25°C lowers the temperature to 5°C. Assume a calorimeter constant of 50 g.

4

STREAMFLOW

Most data used by hydrologists serve other purposes in meteorology, climatology, or other earth sciences. Streamflow data are gathered primarily by hydrologists for hydrologic studies. To the engineering hydrologist, streamflow is the dependent variable in most studies, since engineering hydrology is concerned mainly with estimating rates or volumes of flow or the changes in these values resulting from man-made causes.

WATER STAGE

4-1 Manual Gages

River stage is the elevation above some arbitrary zero datum of the water surface at a station. The datum is sometimes taken as mean sea level but more often is slightly below the point of zero flow in the stream. Because it is difficult to make a direct, continuous measurement of the rate of flow in a stream but relatively simple to obtain a continuous record of stage, the

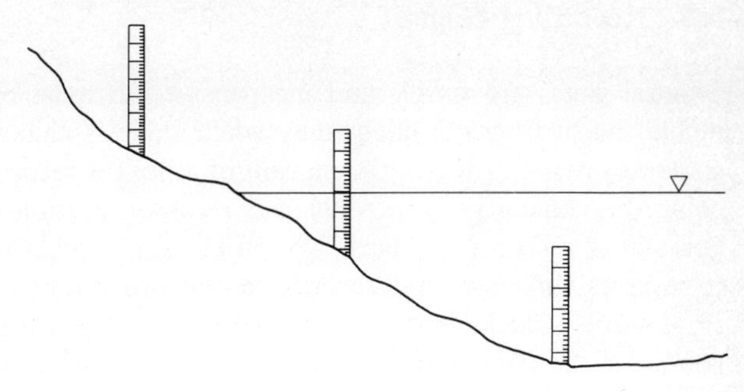

FIGURE 4-1
A sectional staff gage.

primary field data gathered at a streamflow measurement station are river stage [1].

The simplest way to measure river stage is by means of a *staff gage*, a scale set so that a portion of it is immersed in the water at all times. The gage may consist of a single vertical scale attached to a bridge pier, piling, wharf, or other structure that extends into the low-water channel of the stream. If no suitable structure exists in a location accessible at all stages, a *sectional staff gage* (Fig. 4–1) may be used. Short sections of staff are mounted on available structures or on specially constructed supports in such a way that one section is always accessible. An alternative to the sectional staff is an *inclined staff gage* which is placed on the slope of the stream bank and graduated so that the scale reads directly in vertical depth.

The gage scale may be painted on an existing structure or on a special gage board. Painted scales are usually graduated in feet and tenths or in centimeters. Markings are often in the pattern used on stadia rods for visibility. Enameled metal sections are available where particularly accurate stage data are desired. If a stream carries a large amount of fine sediment or industrial waste, scale markings may be quickly obliterated. In this case, a serrated edge on the staff or raised marking symbols may be helpful.

In another type of manual gage a weight is lowered from a bridge or other overhead structure until it reaches the water surface. By subtracting the length of line paid out from the elevation of a fixed reference point on the structure, the water-surface elevation can be determined. The *wire-weight gage* has a drum with a circumference such that each revolution unwinds 1 ft of wire. A counter records the number of revolutions of the drum while a fixed reference point indicates hundredths of feet on a scale around the circumference.

4-2 Recording Gages

Manual gages are simple and inexpensive but must be read frequently to define the hydrograph adequately when stage is changing rapidly. Water-stage recorders, in which the motion of a float is recorded on a chart, overcome this difficulty. In a *continuous recorder*, motion of the float moves a pen across a long strip chart. When the pen reaches the edge of the chart it reverses direction and records in the other direction across the chart (Fig. 4-2*a*). Clocks may be weight-driven and will run as long as there is room for the clock weight to drop. Electric clocks that can operate for a year on a battery are also used.

A short-term recorder usually has a chart wrapped around a drum rotated by the float while the pen is driven at constant speed parallel to its axis. The circumference of the drum represents any selected change in stage. Larger changes are recorded beginning again at the bottom of the chart (Fig. 4-2*b*).

Digital computers offer considerable advantage for processing large quantities of data, and recorders which punch the stage at fixed intervals (usually 15 min) on paper tape are also used. The tape (Fig. 4-2*c*) can be read, verified, and converted to streamflow by electronic equipment. Since punched tapes do not offer ready visual inspection of the stage record for observing stage or detecting errors in the stage recording, chart recorders are preferred in some instances and both chart and tape equipment may be used where visual reading is required.

Float-type water-stage recorders are generally installed in a shelter house and stilling well (Fig. 4-3). The stilling well serves to protect the float and counterweight cables from floating debris and (if the intakes are properly designed) suppress fluctuations from surface waves in the stream. Generally two or more intake pipes are placed from the stilling well into the stream so that at least one will admit water at all times. When the well is attached to a bridge pier and water enters through the open bottom of the well, an inverted cone may be placed over the bottom of the well to reduce the size of the opening and suppress the effect of surface waves. The open-bottom type of stilling well has the advantage of being less likely to fill with sediment. If a stilling well with a closed bottom is installed on a stream with a high sediment load, it is necessary to provide for removal of the sediment which accumulates in the well. It is customary to install staff gages inside and outside the well to check the performance of the recorder.

Recent developments in recording gages include a recorder which uses a tubular float and requires only a small-diameter pipe as a stilling well, and bubbler gages [2], which record the pressure required to maintain a small flow of gas from an orifice submerged in the stream. The aim is to eliminate

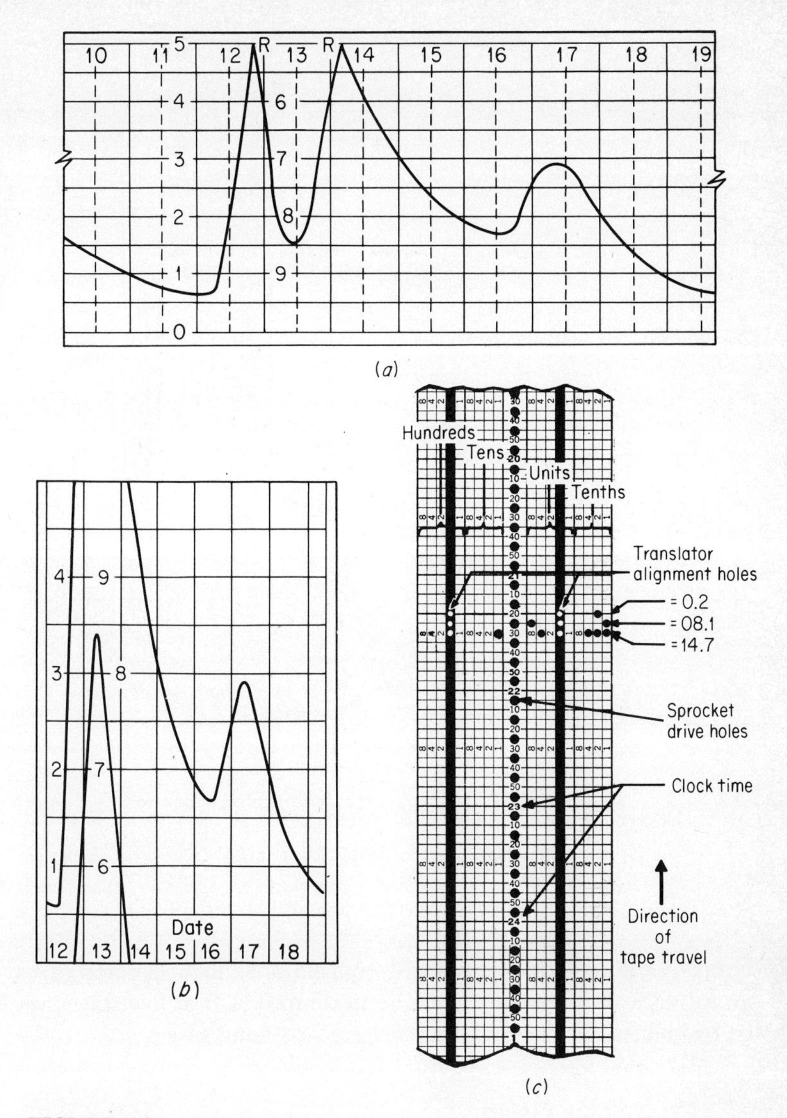

FIGURE 4-2
Recording-gage charts: (*a*) continuous-strip chart, (*b*) weekly chart, and (*c*) punched tape. [*Part* (*c*) *Fischer & Porter Co.*]

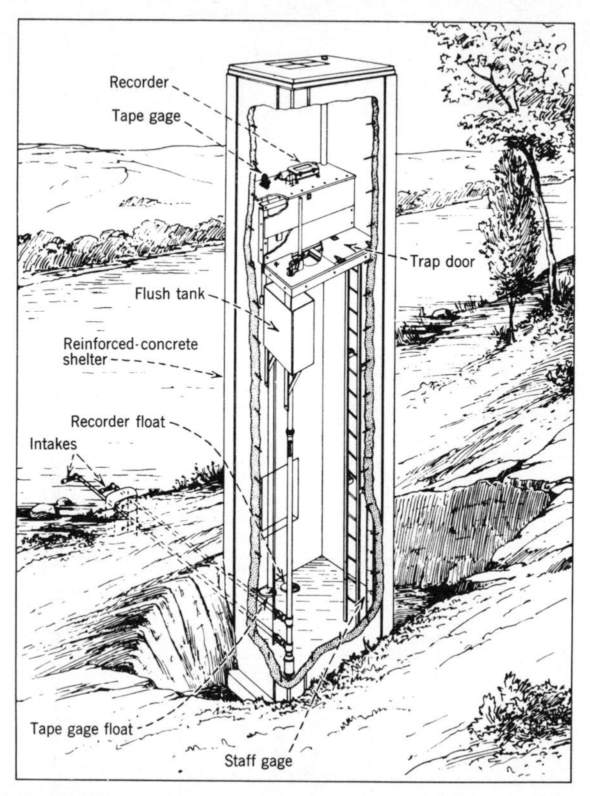

FIGURE 4-3
A typical water-stage recorder installation. (*U.S. Geological Survey.*)

the costly stilling well required with conventional float-operated gages. It has proved difficult to develop gages accurate to 0.01 ft at low stages yet having the ruggedness and range required to record flood stages.

4-3 Crest-Stage Gages

Crest gages provide low-cost, supplementary records of crest stages at locations where recorders are not justified and where manually read staff gages are inadequate. A variety of such gages have been devised, including small floats which rise with stage but are restrained at the maximum level [3, 4] and water-soluble paints [5] on bridge piers where they are protected from rain and can indicate a definite high-water mark. The gage used by the U.S. Geological Survey consists of a length of pipe (Fig. 4-4) containing a

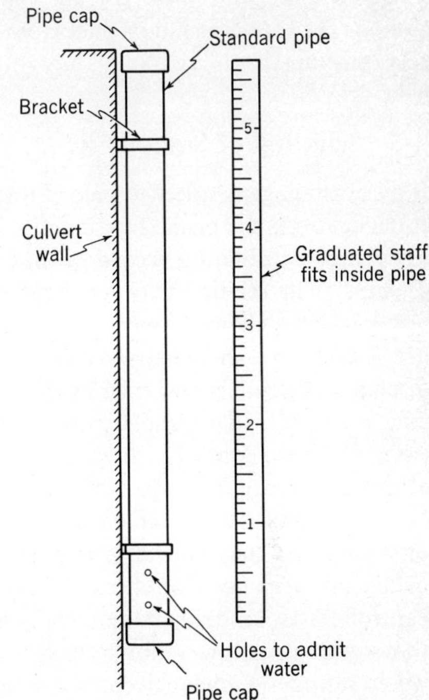

FIGURE 4-4
Crest-stage gage used by the U.S. Geological Survey.

graduated stick and a small amount of ground cork [6]. The cork floats as the water rises, and some adheres to the stick at the highest level reached by the water. The stick can be removed, the crest reading recorded, the cork wiped off, and the stick replaced ready for the next rise.

4-4 Miscellaneous Stage Gages

Water- or mercury-filled manometers are often used to indicate reservoir water levels or to actuate recording devices. Remote recorders in which a system of selsyn motors is used to transmit water-level information from streamside to a recorder at a distance are available, as are numerous remote transmitting telephonic or radio gages. These latter gages use a coding device which converts stage to a signal transmitted as a series of impulses which can be counted, a change in frequency of oscillation which can be measured, or the time interval required for a sensing element to move from a zero point to the water surface at constant speed. Such remote recording devices are used primarily for flood forecasting or reservoir operation. Cost usually precludes their use for routine data collection. Use of earth satellites as relay stations for transmission of data from remote stations may eliminate surface relay stations.

4-5 Selection of Station Site

If a stream gage is solely to record water level for flood warning or as an aid to navigation, the prime factor in its location is accessibility. If the gage is to be used to obtain a record of discharge, the location should be carefully selected. The relation between stage and discharge is controlled by physical features of the channel downstream from the gage. When controlling features are situated in a short length of channel, a *section control* exists. If the stage-discharge relation is governed by the slope, size, and roughness of the channel over a considerable distance, the station is under *channel control.* In many cases no single control is effective at all stages, but a complex of controlling elements function as stage varies.

The ideal low-water control is a section control consisting of rapids or a riffle. If this control is in rock, it will be reasonably permanent and once calibrated need be checked only infrequently. Where no such natural control exists, an artificial control consisting of a low concrete weir, sometimes with a shallow V notch, may be constructed to maintain a stable low-water rating. A channel control is more likely to change with time as a result of scour or deposition of sediment, and more frequent flow measurements are required to maintain an accurate stage-discharge relation.

Rapids may also be effective controls at high flows if the slope of the stream is steep, but where slopes are flat, the section control is likely to be submerged and ineffective at high flows. High-water controls are more likely to be channel controls although in some cases the contraction at a bridge or the effect of a dam may control at high stages. It is advisable to avoid locations where varying backwater from a dam, an intersecting stream, or tidal action occurs. These situations require special ratings (Sec. 4-9) which are usually less accurate.

DISCHARGE

4-6 Current Meters

The stage record is transformed to a discharge record by calibration. Since the control rarely has a regular shape for which discharge can be computed, calibration is accomplished by relating field measurements of discharge with the simultaneous river stage (Secs. 4-7 to 4-10).

The most common current meter in the United States, the *Price meter* (Fig. 4-5), consists of six conical cups rotating about a vertical axis [7, 8].

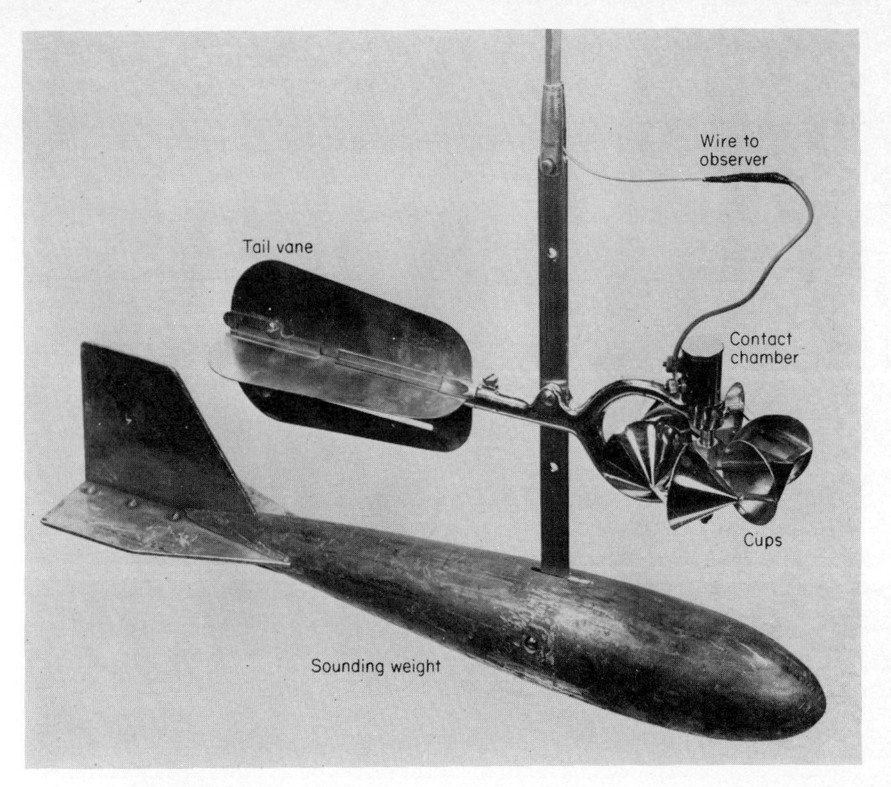

FIGURE 4-5
Price current meter and 30-lb C-type sounding weight. (*U.S. Geological Survey.*)

Electric contacts driven by the cups close a circuit through a battery and the wire of the supporting cable to cause a click for each revolution (or each fifth revolution) in headphones worn by the operator. For measurements in deep water, the meter is suspended from a cable. Tail vanes keep the meter facing into the current, and a heavy weight keeps the meter cable as nearly vertical as possible. Special cranes are available to support the meter over a bridge rail, to simplify handling the heavy weights, and to permit measuring the length of cable paid out. In shallow water the meter is mounted on a rod, and the observer wades the stream. A *pygmy* Price meter has been used for measuring flow at very shallow depths.

Propeller-type current meters employ as rotating element a propeller turning about a horizontal axis (Fig. 4-6). The contacting mechanism of a propeller meter is similar to that of a Price meter, and similar suspensions are used. The vertical-axis meter has an important advantage in that the bearings supporting the shaft can be enclosed in inverted cups which trap air

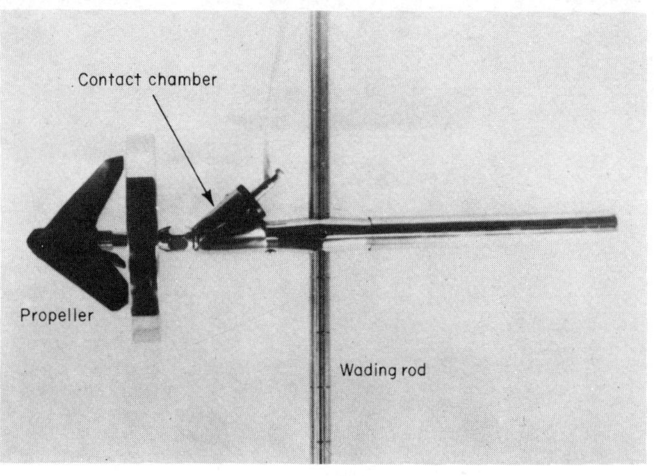

FIGURE 4-6
Propeller-type current meter. (*U.S. Geological Survey.*)

and prevent entrance of sediment-laden water. The bearings of propeller meters cannot be so protected and are exposed to damage by abrasion. On the other hand, vertical currents or upstream velocity components will rotate the cups of a vertical-axis meter in the same direction as downstream currents. A Price meter moved vertically in still water will indicate a positive velocity. Hence, it tends to overestimate stream velocity. If the measuring section is well chosen with the current flow nearly parallel to the channel axis and with a minimum of turbulence, the error is probably not more than 2 percent [9].

The relation between revolutions per second N of the meter cups and water velocity v is given by an equation of the form

$$v = a + bN \qquad (4\text{-}1)$$

where a is the starting velocity or velocity required to overcome mechanical friction. Some differences in these constants must be expected as a result of manufacturing variations and effects of wear. Consequently each meter should be individually calibrated [10] by mounting the meter on a carriage which moves it through still water. The carriage may run on rails along a straight channel or may rotate about a central pivot in a circular basin. The speed of the carriage is determined by the time required to travel a known distance. With several runs at various speeds it is possible to plot a curve showing the relation between meter contacts per unit time and water speed.

4-7 Current-Meter Measurements

A discharge measurement [11] requires determination of sufficient point velocities to permit computation of an average velocity in the stream. Cross-sectional area multiplied by average velocity gives the total discharge. The number of velocity determinations must be limited to those which can be made within a reasonable time, especially if stage is changing rapidly, since it is desirable to complete the measurement with a minimum change in stage.

The practical procedure involves dividing the stream into a number of vertical sections (Fig. 4-7). No section should include more than about 10 percent of the total flow; thus 20 to 30 vertical sections are typical, depending on the width of the stream. Velocity varies approximately as a parabola (Fig. 4-8) from zero at the channel bottom to a maximum at (or near) the surface. On the basis of many field tests, the variation for most channels is such that the average of the velocities at two-tenths and eight-tenths depth below the surface equals the mean velocity in the vertical. The velocity at six-tenths depth below the surface closely approximates the mean in the vertical. The adequacy of these assumptions for a particular stream can be tested by making a large number of velocity determinations in the vertical.

Sample current-meter field notes are shown in Fig. 4-9. Determination of the mean velocity in a vertical is as follows:

1 Measure total depth of water by sounding with the meter cable.

2 Raise meter to eight-tenths depth and measure velocity by starting a stopwatch on an impulse from the meter and stopping it on another impulse about 45 s later. The number of impulses counted (taking the first as zero) and the elapsed time permit calculation of velocity from the meter calibration.

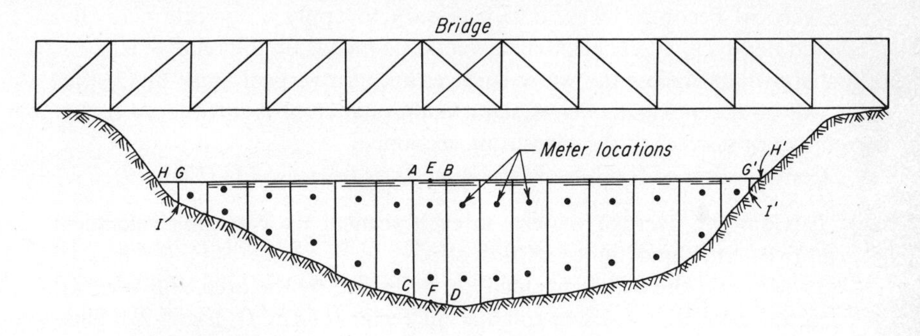

FIGURE 4-7
The procedure for current-meter measurement.

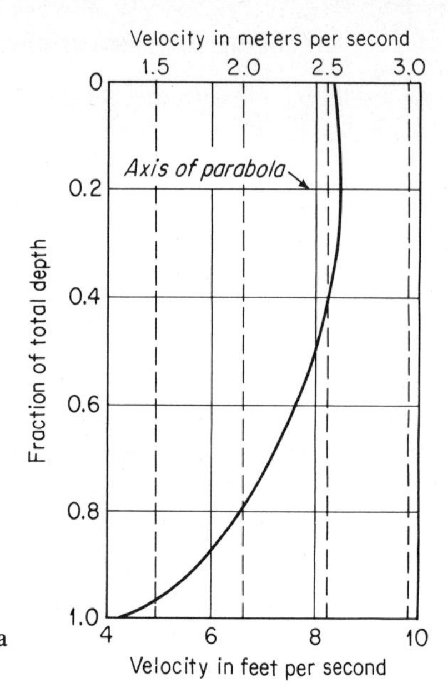

FIGURE 4-8
A typical vertical velocity profile in a stream.

3 Raise meter to two-tenths depth and repeat step *2*.

In shallow water near the shore a single velocity determination at six-tenths depth may be used.

If velocities are high, the meter and weight will not hang vertically below the point of suspension but will be carried downstream by the current (Fig. 4-10). Under these conditions the length of line paid out is greater than the true vertical depth, and the meter is higher than indicated. Heavy weights are used to minimize this effect, but if the angle between the line and a vertical becomes large, it is necessary to apply a correction to the measured depths [12]. The actual correction depends on the relative lengths of line above and below the water surface, but at a vertical angle of 12° the error will be about 2 percent. A slight additional error is introduced if the current is not normal to the measuring section.

Computation of total discharge is made as follows [13] (Fig. 4-9):

1 Compute average velocity in each vertical by averaging velocities at two-tenths and eight-tenths depths.
2 Multiply the average velocity in a vertical by the area of a vertical section extending halfway to adjacent verticals (*ABCD*, Fig. 4-7). This area is taken as the measured depth at the vertical (*EF*) times the width of the section (*AB*).

9-275-F
Jan. 1952

UNITED STATES
DEPARTMENT OF THE INTERIOR
GEOLOGICAL SURVEY
WATER RESOURCES DIVISION

DISCHARGE MEASUREMENT NOTES

Meas. No. 270
Comp. by TJB
Checked by

BIG CREEK NR DOGWOOD, VA.

Date SEP 25, 19 62 Party T. J. Buchanan
Width 70 Area 143 Vel. 0.53 G. H. 1.93 Disch. 76.4
Method 0.6 and 0.2+0.8 No. secs. 28 G. H. change −0.02 in 1 hrs. Susp. Rod
Method coef. 1 Hor. angle coef. Varies Susp. coef. 1 Meter No. 3684

GAGE READINGS

Time		Recorder	Inside	Outside
1310		1.94	1.94	1.92
1330	Start		(1.94)	
1430	FINISH		(1.92)	
1445		1.92	1.92	1.90
Weighted M. G. H.			1.93	
G. H. correction				
Correct M. G. H.			—	—

Date rated 2-16-62 Used rating for rod ✓ susp. Meter — ft. above bottom of wt. Tags checked
Spin before meas. 2:45 after Free
Meas. plots % diff. from rating
Wading, cable, ice, boat, upstr., downstr., side bridge 25 feet, mile, above, below gage, and
Check-bar, chain found —
changed to at
Correct —
Levels obtained —

Measurement rated excellent (2%), (good (5%)), fair (8%), poor (over 8%), based on following conditions: Cross section Sand and gravel, fairly even
Flow Good distribution Weather Cloudy
Other Greatest flow in center Air 67 °F@ 1435
Gage Water 54 °F@ 1435
Record removed Yes Intake flushed
Observer Talked with
Control Clear
Remarks

G. H. of zero flow 1.92 − 1.61 = 0.3 ft. ± 0.1 ft

16—58354-3

Big Cr. River at— Dogwood, Va.

Angle coefficient	Dist. from initial point	Width	Depth	Observation depth	Revolutions	Time in seconds	Velocity: At point	Velocity: Mean in vertical	Adjusted for hor. angle or	Area	Discharge
	1	1.5	0							0	0
	4	3	0.95	.6	5	40			.31	2.8	.87
	7	3	1.4	.6	7	43			.40	4.2	1.68
	10	3	2.0	.6	7	52			.33	6.0	1.98
	13	3	2.1	.6	7	40			.42	6.3	2.65
	16	3	2.3	.6	10	55			.44	6.9	3.04
	19	3	2.25	.6	10	53			.46	6.8	3.13
	22	2.5	2.2	.6	10	48			.50	5.5	2.75
	24	2	2.5	.2	15	50	.70		.59	5.0	2.95
				.8	10	50	.48				
	26	2	2.8	.2	15	45	.78		.64	5.6	3.58
				.8	10	48	.50				
	28	2	3.0	.2	15	43	.82		.67	6.0	4.02
				.8	10	46	.52				
	30	2	2.95	.2	20	52	.90		.74	5.9	4.37
				.8	10	40	.54				
	32	2	3.1	.2	20	50	.93		.75	6.2	4.65
				.8	10	42	.57				
	34	2	3.2	.2	20	48	.97		.74	6.4	4.74
				.8	10	46	.52				
	36	2	3.05	.2	20	52	.90		.77	6.1	4.70
				.8	15	55	.64				
	38	2	3.1	.2	15	42	.83		.68	6.2	4.22
				.8	10	46	.52				

FIGURE 4-9
Sample current-meter field notes used by the U.S. Geological Survey.

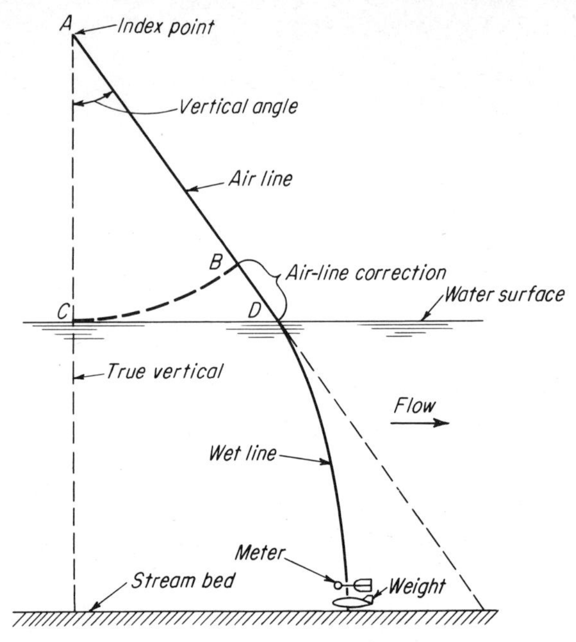

FIGURE 4-10
Position of sounding line in swift water.

3 Add the increments of discharge in the several verticals. Incremental discharge in the shore section (*GHI*, Fig. 4-7) is taken as zero.

Access to individual verticals of a section may be obtained by wading if the water is shallow. At high stages the meter must be lowered from an overhead support. Where possible, bridges are used as the measuring section, if the bridge is normal to the stream axis and the current essentially parallel to the stream axis. The measuring section need not be at the same location as the control section. However, the distance between sections should be short enough to ensure that the intervening inflow is not large. Where no existing bridge is suitable, a special cableway may be used. The hydrographer rides in a small car suspended beneath the cable and lowers the meter through an opening in the floor of the car. Where neither bridge nor cableway is practical, measurements may be made from a boat. This is far less satisfactory because of the difficulty of maintaining position during a measurement and because either vertical or horizontal motion of the boat results in a positive velocity indication by a Price meter. In many countries cableways are used to carry the meter over the stream with the observer remaining on the shore. This precludes horizontal angle corrections.

4-8 Chemical Gaging

Current-meter gaging is difficult and sometimes impossible in boulder-strewn mountain torrents, very small streams, etc. Here chemical gaging may prove useful. Common salt, fluorescein dye [14], a radioactive material, or any easily measurable material not present in the stream and not likely to be lost by chemical combination with materials in the stream may be used. The tracer can be used to measure mean velocity through a reach by measuring the time between injection of a pulse and its arrival at a downstream point [15].

In the dilution method [16] a tracer of concentration c_t is injected into the stream at rate q_t. At a downstream point samples are taken, and after an equilibrium concentration c_e is reached, the discharge q is

$$q = \left(\frac{c_t}{c_e} - 1\right) q_t \tag{4-2}$$

Complete mixing of the tracer in the flow and accurate determination of the initial and final concentrations are essential.

4-9 Stage-Discharge Relations

Periodic meter measurements of flow and simultaneous stage observations provide data for a calibration curve called a *rating curve* or *stage-discharge relation* [17]. For most stations a simple plot of stage versus discharge (Fig. 4-11) is satisfactory. Such a curve is approximately parabolic but may show some irregularities if the control changes between low and high flows or if the cross section is irregular.

The adequacy of a rating curve is measured by the scatter of the measured flows about the mean line. If the control is reasonably permanent and the slope of the energy grade line at the station reasonably constant for all occurrences of a given stage, a simple rating will suffice.

If the control is altered by scour or deposition, frequent meter measurements are necessary. Under conditions of *shifting control*, discharge is usually estimated by noting the difference between the stage at the time of a discharge measurement and the stage on the mean rating curve which shows the same discharge. This difference is applied as a correction to all stages before entering the rating. If the correction changes between measurements, a linear variation with time is usually assumed.

Individual measurements may deviate from the mean stage-discharge relation as a result of differences in water-surface slope at the control. Since velocity head is usually small, the slopes of the water surface and of the energy grade line are nearly equal. Differences in slope may be caused

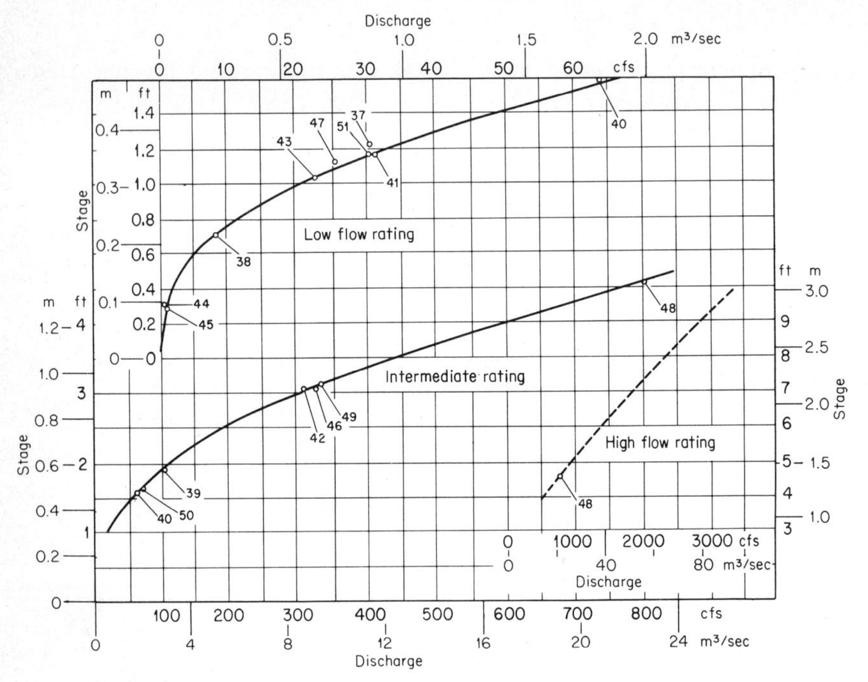

FIGURE 4-11
A simple stage-discharge relation.

by varying backwater as a result of an obstruction downstream or high stages in an intersecting stream. If either of these factors is present, the rating curve must include slope as a parameter [18, 19]. The basic approach is represented by

$$\frac{q}{q_0} = \left(\frac{s}{s_0}\right)^m = \left(\frac{F}{F_0}\right)^k \tag{4-3}$$

The equation states that discharge q is proportional to a power of water-surface slope s. From fluid mechanics the exponent m would be expected to be $\frac{1}{2}$. The fall F is the difference in water-surface elevation between two fixed sections and is usually measured by two conventional river gages. There is no assurance that the water-surface profile between these gages is a straight line, i.e., that $F/L = s$. Consequently the exponent k need not be $\frac{1}{2}$ and must be determined empirically.

A *slope-stage-discharge relation* requires a base gage and an auxiliary gage. The gages should be far enough apart for F to be at least 1 ft (30 cm) to minimize the effect of observational errors. The fall F applicable to each

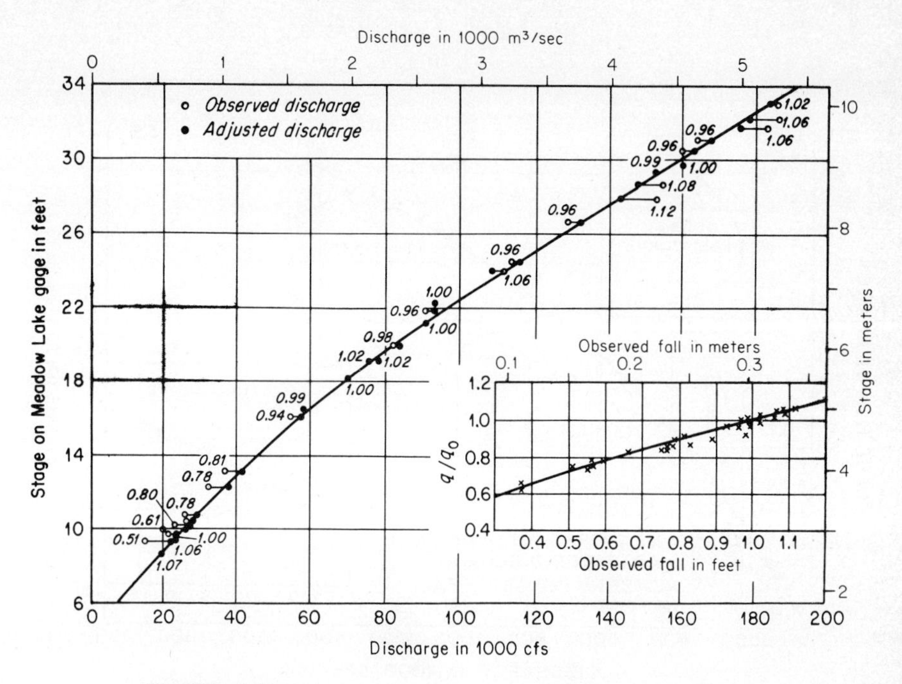

FIGURE 4-12
A slope-stage-discharge rating curve for the Tennessee River at Chattanooga, Tennessee. (*U.S. Geological Survey.*)

discharge measurement is determined, and if the observed falls do not vary greatly, an average value F_0 is selected. All measurements with values of $F \approx F_0$ are plotted as a simple stage-discharge relation, and a curve is fitted (Fig. 4-12). This is the q_0 curve representing the discharge when $F = F_0$. If $F \neq F_0$, the ratio F/F_0 is plotted against q/q_0 on an auxiliary chart. The discharge at any time can be computed by calculating the ratio F/F_0 and selecting a value of q/q_0 from the auxiliary curve. A value of q_0 corresponding to the existing stage is taken from the q_0 curve and multiplied by q/q_0 to give q. If the auxiliary curve plots as a straight line on logarithmic paper, the slope of the line is k in Eq. (4-3). The rating just described is known as a *constant-fall rating*, since the adopted mean fall F_0 is constant.

In some cases the range of F is large, and its variation is related to stage. In this case a *normal-fall rating* may be used. This rating is similar to the constant-fall rating except that the normal fall F_n which is used in place of F_0 is taken as a function of stage and defined by a second auxiliary curve (Fig. 4-13). To compute discharge with a normal-fall rating, the actual

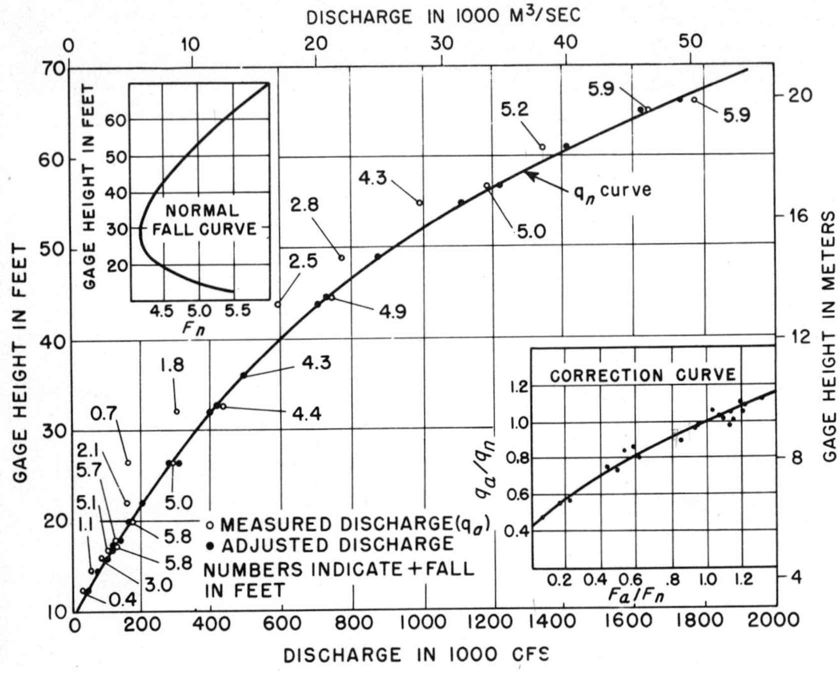

FIGURE 4-13
Normal-fall rating curve for the Ohio River at Metropolis, Illinois. (*U.S. Geological Survey.*)

fall F is computed, a value of F_n is read from the curve, and the ratio F/F_n is calculated. The ratio q/q_n can then be determined and multiplied by q_n from the rating curve to give q.

Figure 4-14 shows the profile of a flood wave as it moves past a station. The slope is equal to $s_b + s_r$, where s_b is the slope of the channel bottom (or the slope of the water surface in uniform flow) and $s_r = dg/u\ dt$. Here dg/dt is the slope of the flood wave expressed as a rate of change of stage with time and u is the celerity of movement of the flood wave. Since s_r is the variable factor in the slope, a rating can be constructed by plotting stage against discharge for those measurements for which dg/dt is near zero. If u is assumed to be constant, a correction curve can then be established by plotting q/q_c versus dg/dt, where q_c is the discharge corresponding to the existing stage when $dg/dt = 0$. Theoretically, dg/dt is the tangent to the water-stage recorder trace, but in practice it is convenient to use Δg, the change in stage during a finite time period, usually 1 h but sometimes much longer.

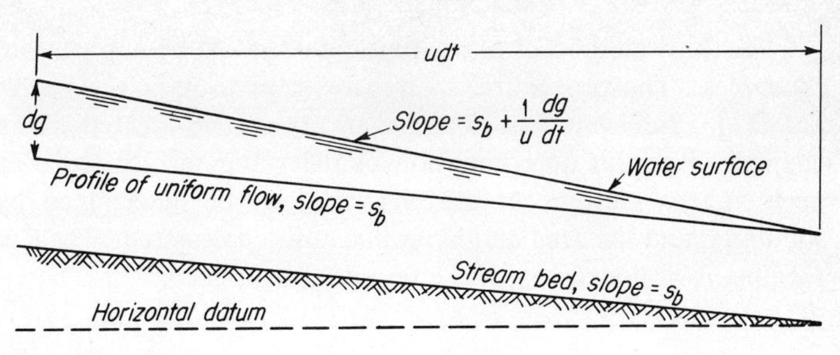

FIGURE 4-14
Profile of a flood wave.

4-10 Extension of Rating Curves

There is no completely satisfactory method for extrapolating a rating curve beyond the highest measured discharge. It is often assumed that the equation of the rating curve is

$$q = k(g - a)^b \tag{4-4}$$

where a, b, and k are constants for the station. This is the equation of a parabola in which a is the distance between the zero elevation of the gage and the elevation of zero flow. If the correct value of a is determined by trial, the equation will plot as a straight line on logarithmic paper and is easily extended. However, the procedure includes the assumption of a parabolic rating and cannot account for any marked change in the hydraulic geometry of the stream at high flows.

Another method [20, 21] of extending rating curves is based on the Chézy formula

$$q = AC\sqrt{Rs} \tag{4-5}$$

where C is a roughness coefficient, s is the slope of the energy line, A is the cross-sectional area, and R is the hydraulic radius. If $C\sqrt{s}$ is assumed to be constant for the station and D, the mean depth, is substituted for R,

$$q = kA\sqrt{D} \tag{4-6}$$

Known values of q and $A\sqrt{D}$ are plotted on a graph and usually define something close to a straight line which can readily be extended. Values of $A\sqrt{D}$ for stages above the existing rating can be obtained by field measurement and used with the extended curve for estimates of q. An abrupt discontinuity may be observed at bankfull stage.

A third method of estimating high flows is by application of hydraulic principles. The procedure is often referred to as a *slope-area computation* [22, 23]. Sufficient high-water marks must be located along a reach of channel to permit determination of the water-surface slope at the time of peak. Cross sections of the channel may be determined by leveling or sounding, and the area and hydraulic radius calculated. The Chézy-Manning formula is ordinarily used to compute discharge:

$$q = \frac{1.49}{n} AR^{2/3}s^{1/2} \tag{4-7}$$

The main source of error in applying Eq. (4-7) is in estimating the roughness coefficient n (Appendix Table B-4). Since q is inversely proportional to n and the average value of n for natural streams is about 0.035, an error of 0.001 in n represents about 3 percent in discharge. Considerable doubt may exist whether the cross section measured after the flood is the section which existed at the time of peak. A stream often scours during rising stages and redeposits material during falling stages (Fig. 4-15). Under the most favorable conditions, an error of 10 percent may be expected in a slope-area estimate of flow. The coefficient in Eq. (4-7) applies for dimensions in feet and becomes 1.0 when metric units (meters and cubic meters per second) are used.

4-11 Effect of Ice on Streamflow

When ice covers a stream, a new friction surface is formed and the stream becomes a closed conduit with lower discharge because of the decreased hydraulic radius. The underside of the ice sheet may be extremely rough if ice cakes are tilted helter-skelter and then frozen together. Movement of the water under the ice gradually develops a smooth surface. If the stage falls, leaving the ice as a bridge across the stream, the stage-discharge characteristics return to those of a free stream.

In turbulent streams the first ice to form is *frazil ice*, small crystals suspended in the turbulent flow. Frazil ice collecting on rocks on the streambed is called *anchor ice* and may cause a small increase in stage. If the turbulence is not sufficient to keep the frazil ice mixed in the stream, it rises to the surface to form sheet ice. Until a complete ice sheet is formed, small variations in the stage-discharge relation must be expected from time to time.

When ice conditions exist, it is necessary to make periodic measurements [24] through holes in the ice and to interpolate the discharge between these measurements in any manner which seems reasonable. The meter must be moved rapidly from hole to hole and should be kept in the water at all times (except when being moved) to prevent freezing. Fortunately, if the

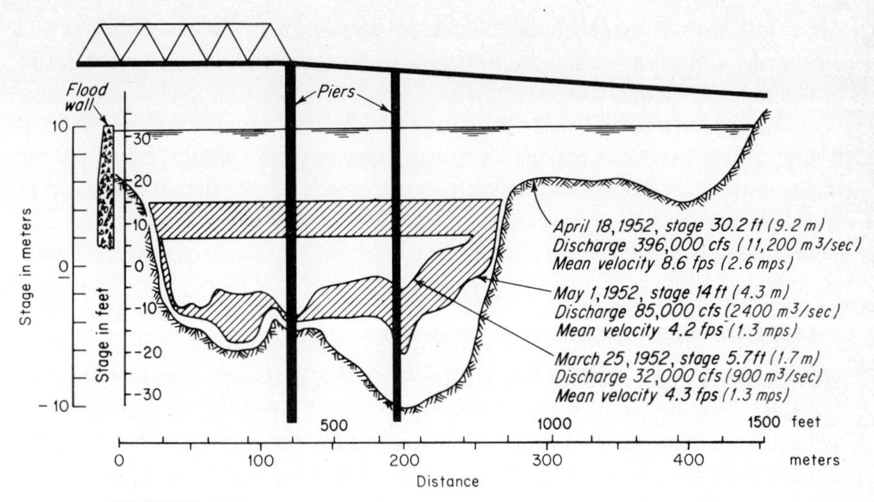

FIGURE 4-15
Cross sections of the Missouri River at Omaha, Nebraska, showing the progressive scour and filling during the passage of a flood wave.

stream is solidly frozen over, the flow is usually small since there will be little snowmelt or other source of runoff within the tributary area.

4-12 Other Methods of Obtaining Streamflow Data

On large streams the discharge at dams can be determined from calibration of the spillway, sluiceway, and turbine gates. If a careful record of gate and turbine operation is maintained, the discharge can be computed. Some progress has been made in generalized rating curves based on the geometry of the gate structure [25].

On small streams, flow measurements may be made with weirs or flumes [26]. These devices are commonly rated on the basis of laboratory calibration although the rating may be checked in place with current meters. For small streams, a combination of a V-notch weir for low flows and a venturi flume for high flows may be necessary to assure accuracy. Large weirs are usually unsatisfactory because deposition of sediment in the upstream pool changes the discharge characteristics.

Highway culverts have frequently been suggested as flow-measuring devices, and in many instances this is feasible. However, the hydraulics of culverts is quite complex [27]. In flat terrain, care must be taken to avoid culverts affected by backwater unless both headwater and tailwater stages are measured. On steep slopes it is necessary to establish whether the culvert flows full with control in the barrel or partially full with control at the entrance. Temporary changes in discharge capacity may occur as the result of sediment or debris deposits.

Rough measurements of discharge can be made by timing the speed of floats. A surface float travels with a velocity which is about 1.2 times the mean velocity. Objects floating with a greater submerged depth will travel at speeds closer to the mean velocity in the section. A float extending from the surface to mid-depth travels with a velocity about 1.1 times the mean velocity.

4-13 Planning a Streamflow Network

How many gaging stations should be installed, and where should they be located? This decision of *network design* must be faced by the manager of a streamflow network. The design of a streamflow network is both a problem of statistical sampling over area and a matter of sampling the locations where data are most likely to be needed.

It is convenient to recognize three types of stations. *Operational stations* are required for streamflow forecasting, project operation, water allocation, etc. They are located as required for the purposes they serve and are operated as long as the purposes exist. *Special stations* are installed to secure data for a project investigation, special studies, or research. Their location is determined by the special need, and they are operated until the study is completed. *Basic data stations* are operated to obtain data for future use. The time and nature of this future use are usually unknown when the station is established. Planning for such stations is the main problem of streamflow-network design.

The decision problem is how best to utilize a limited budget for capital costs and operating expenses of a network [28]. Since there is no rational evaluation of the value of the data, it seems logical to attempt to define a minimum size of drainage area for which reasonable estimates of streamflow should be available. This minimum will be determined by the level of regional development, the hydrologic characteristics of the region, and the probable information needs (culvert design, irrigation diversion, etc.). The minimum area might range from 10 mi^2 (26 km^2) in a well-developed region to 100 mi^2 (260 km^2) or more in developing areas.

Every major stream in the region should be gaged near its mouth and a number of tributaries as well. If it is obvious which streams will be developed, they are the ones which should be gaged. This includes those streams which may pose present or future flood problems as well as those streams which may be used for power, irrigation, etc. The network should sample watersheds of all sizes larger than the specified minimum and should also sample the range of hydrologic and geologic characteristics in the region. It will be a practical impossibility to gage all streams at every site where data may be needed. A major function of the engineering hydrologist is to estimate flow at ungaged locations. His task is made easier and the results more reliable by a network which samples the regional characteristics effectively.

Because hydrologic design information is most often expressed in terms of probability (Chaps. 11 and 12), there is a considerable value in long flow records provided they are internally homogeneous, i.e., represent essentially similar basin characteristics throughout the entire period of record. Man's activities are now so widespread that few streams remain unaffected. *Bench-mark stations* [29] should be maintained permanently on all streams that are substantially unaffected by man. Operational stations will in most cases be relatively permanent. Basic data stations, however, can be discontinued as soon as data are sufficient for a synthetic record to be derived if needed.

If the budget for data is severely constrained, there may be considerable advantage in terminating a data station after say 10 y and moving the equipment to another site to enlarge the sample of gaged watersheds. If this policy is adopted, a suitable method for deriving a synthetic record should be available, and all the necessary data for its use should be collected. Precipitation data generally provide the best base for deriving synthetic records (Chaps. 7, 8, 10), and hence a precipitation-gage network must be installed at the same time as the streamflow station and maintained after the station is discontinued. The decision is an economic one represented by

$$kF_P + O_P + pS < kF_q + O_q \qquad (4\text{-}8)$$

where F is first cost, O is operating cost, the subscripts P and q refer to precipitation and streamflow, respectively, k denotes a capital recovery factor, S is the cost of synthesizing the record if needed, and p is the probability that synthesis will be needed. Since the precipitation stations will probably be needed for other purposes in any case and both p and S are difficult to estimate, an exact solution is impossible. A decision based on judgment may be sufficiently accurate. Since the cost of synthetic record S is almost certain to increase as drainage area increases, it is better to continue stations on large drainage areas more or less permanently and move the stations on small drainage areas. Reliability of the synthesis must also be considered, since a synthesis of low quality is less valuable than a record of good quality.

INTERPRETATION OF STREAMFLOW DATA

4-14 Units

The basic flow unit in the United States is the *cubic foot per second* (ft^3/s), also called *second-foot* and abbreviated *cusec*, or *cfs*. Countries using the metric system usually express flow in *cubic meters per second* (m^3/s). Volume of flow may be expressed in cubic feet or cubic meters, but since this leads to very large numbers, larger volume units are commonly used. In the United States the *second-foot-day* (sfd), or the volume of water discharged in twenty-four hours, with a flow of one cubic foot per second is widely used; 1 sfd is $24 \times 60 \times 60 = 86{,}400$ ft^3. The average flow in cubic feet per second for any 24-h period is the volume of flow in second-foot-days. The other common unit of volume is the *acre-foot*, the volume of water required to cover one acre to a depth of one foot. Hence, 1 acre-ft contains 43,560 ft^3 and equals 0.504 sfd. Within an error of 1 percent, 1 acre-ft equals $\frac{1}{2}$ sfd. In some cases it is convenient to use second-foot-hours ($\frac{1}{24}$ sfd) as a volume unit. In countries using the metric system, volume is usually expressed in some multiple of cubic meters, commonly, cubic hectometers; 1 hm^3 = 10^6 m^3 = 409 sfd (Appendix Table B-1).

Less common units of flow are the cubic foot per second per square mile (csm), the inch, and the miner's inch. Cubic feet per second per square mile (cubic meters per second per square kilometer) is a convenient unit for comparing rates of flow on streams with differing tributary area and is the flow in cubic feet per second divided by the drainage area in square miles. The *inch*, the amount of water required to cover the drainage area to a depth of one inch, is a useful unit for comparing streamflow with the precipitation which caused it. The inch is a unit of volume only when associated with a specific drainage area. In the metric system the millimeter or centimeter is used as the unit of runoff depth and as a unit of volume when associated with a specific drainage area. The *miner's inch* is an old unit used in the western United States and defined as the rate of discharge through an orifice one inch square under a specified head. By statute the miner's inch has been defined between 0.020 and 0.028 ft^3/s, depending on the state (Appendix Table B-3). The unit is no longer used except where old water rights are specified in terms of miner's inches.

It is desirable to treat annual streamflow data in such a way that the flood season is not divided between successive years. Various *water years* have been used for special purposes; the U.S. Geological Survey uses the period Oct. 1 to Sept. 30 for data publication, which is the usual water year in the United States.

4-15 Hydrographs

A *hydrograph* is a graph of stage or discharge versus time. Many different methods of plotting are used, depending on the purpose of the chart. Monthly and annual mean or total flow is used to display the record of past runoff at a station. The characteristics of a particular flood, however, usually cannot be shown successfully by plotting average flow for periods longer than 1 d.

Preferably, flood hydrographs are plotted by computing instantaneous flow values from the water-stage recorder chart. A sufficient number of values should be plotted to indicate adequately all significant changes in the slope of the hydrograph. The plotting scales will vary with the problem—cubic feet per second and minutes on the smallest basins to thousands of cubic feet per second and hours or days on very large basins. The shape of the hydrograph is determined by the scales used, and in any particular study it is good practice to use the same scales for all floods on a given basin.

4-16 Mean Daily Flows

Streamflow data are usually published in the form of mean daily flows (Fig. 4-16). This is the average discharge rate in cubic feet per second for the period from midnight to midnight. Since 1945, the U.S. Geological Survey has included the magnitude and time of occurrence of all significant flood peaks. On large streams this form of publication is quite satisfactory, but on small streams it leaves something to be desired. The picture presented by mean daily flows depends on the chance relation between time of storm occurrence and fixed clock hours (Fig. 4-17). On large streams the maximum instantaneous flow may be only slightly higher than the maximum mean daily flow. On small streams the maximum instantaneous flow is usually much greater than the highest mean daily flow. Wherever possible, the hydrologist should secure copies of the recorder charts and work with a hydrograph of instantaneous flow when dealing with small basins. If this is not feasible, Fig. 4-18 may be used as a guide to estimate the peak flow and time of peak. The figure is an average relationship [30] developed from data for stations all over the United States and cannot be expected to give exact results. On very small basins, where a complete rise and fall of a storm hydrograph may take only a fraction of a day, use of the recorder charts is essential if one wishes to study rates of flow.

4-17 Adjustment of Streamflow Data

Published streamflow data should be carefully reviewed and adjusted for errors resulting from instrumental and observational deficiencies until they are as accurate a presentation of the flow as it is possible to make. For a number of reasons, the published flow may not represent the data actually required by the analyst. The location of the station may have changed during the period of record with a resultant change in drainage area and hence volume and rate of flow. In this case an adjustment of the record is possible by use of the double-mass curve (Sec. 3-10). The base for the double-mass curve may be either flow at one or more gaging stations that have not been

POTOMAC RIVER BASIN

01619500 Antietam Creek near Sharpsburg, Md.

LOCATION.--Lat 39°27'01", long 77°43'52", Washington County, on left bank 400 ft downstream from Burnside Bridge, 1 mile southeast of Sharpsburg, and 4 miles upstream from mouth.

DRAINAGE AREA.--281 sq mi.

PERIOD OF RECORD.--June 1897 to September 1905. August 1928 to current year. Monthly discharge only for some periods, published in WSP 1302.

GAGE.--Water-stage recorder. Concrete control since Mar. 29, 1934. Datum of gage is 311.00 ft above mean sea level, adjustment of 1912. June 24, 1897, to Aug. 25, 1905, nonrecording gage a few hundred feet downstream from Middle Bridge, 1.2 miles upstream at datum about 12 feet higher. Aug. 21, 1928, to July 13, 1933, nonrecording gage at Burnside Bridge at present datum.

AVERAGE DISCHARGE.--48 years (1897-1903, 1904-5, 1930-1971), 257 cfs (12.42 inches per year), adjusted for inflow since 1930.

EXTREMES.--Current year: Maximum discharge, 2,670 cfs Feb. 13 (gage height, 7.15 ft); minimum, 122 cfs part of each day Sept. 8-11 (gage height, 2.45 ft).
Period of record: Maximum discharge, 12,600 cfs July 20, 1956 (gage height, 16.73 ft), from rating curve extended above 4,300 cfs on basis of contracted-opening measurement of peak flow; minimum discharge, 9.4 cfs Nov. 22, 1957, result of regulation caused by construction work above station; minimum daily, 37 cfs Jan. 30, 1966.

REMARKS.--Records good. Some diurnal fluctuation caused by powerplant above station. Since 1928, records include pumpage from Potomac River for municipal supply of Hagerstown. This water later enters Antietam Creek above station as sewage. Records of chemical analyses and water temperatures for the water year 1971 are published in Part 2 of this report.

REVISIONS (WATER YEARS).--WSP 192: 1897-1905. WSP 726: Drainage area. WSP 1432: 1929-31(M), 1933, 1935(M), 1937(M), 1949(M), 1952(M).

DISCHARGE, IN CUBIC FEET PER SECOND, WATER YEAR OCTOBER 1970 TO SEPTEMBER 1971

DAY	OCT	NOV	DEC	JAN	FEB	MAR	APR	MAY	JUN	JUL	AUG	SEP
1	146	206	228	320	260	876	392	262	391	216	219	136
2	144	166	211	300	240	909	388	269	327	208	372	136
3	141	289	205	300	230	790	389	297	344	200	241	134
4	135	426	208	391	230	910	370	271	400	188	348	132
5	134	256	202	864	290	722	360	256	379	183	341	128
6	137	208	192	760	392	724	359	351	374	182	243	126
7	135	187	185	600	365	784	453	489	387	183	192	127
8	134	174	186	520	350	912	428	473	467	177	176	125
9	132	166	183	500	480	714	382	479	377	173	168	125
10	132	170	183	480	398	676	365	408	339	170	165	133
11	131	232	181	456	340	657	354	373	319	171	189	127
12	130	264	201	439	382	628	346	354	310	195	192	198
13	130	313	226	426	1,160	629	343	479	302	181	169	178
14	129	283	201	421	1,590	623	338	588	302	177	156	178
15	164	294	191	432	691	585	326	471	506	173	151	150
16	144	336	191	380	582	576	322	483	434	176	153	135
17	132	280	352	340	567	542	320	495	352	174	150	157
18	127	260	388	320	593	514	319	438	321	167	149	151
19	125	248	301	310	640	528	306	421	298	172	151	144
20	126	269	278	300	694	645	301	398	285	173	154	152
21	150	377	267	310	816	564	301	420	275	165	146	341
22	292	306	376	321	992	526	297	386	270	160	146	216
23	212	272	504	340	1,630	514	290	349	259	157	143	163
24	154	266	661	340	1,360	494	284	331	251	157	139	151
25	142	247	545	327	1,060	474	276	328	242	203	136	145
26	137	237	495	320	939	462	274	326	235	171	134	146
27	135	230	453	300	1,040	450	272	301	227	165	167	155
28	130	225	416	280	1,000	437	273	291	219	158	207	156
29	129	219	386	260	------	433	284	292	222	158	162	145
30	142	222	360	300	------	422	274	376	219	173	143	141
31	202	------	340	290	------	404	------	496	------	192	139	------
TOTAL	4,531	7,628	9,296	12,247	19,311	18,924	9,986	11,951	9,633	5,498	5,741	4,631
MEAN	146	254	300	395	690	607	333	386	321	177	185	154
MAX	292	426	661	864	1,630	876	453	588	506	216	372	341
MIN	125	166	181	260	230	404	272	256	219	157	134	125
(†)	-13.1	-11.0	-6.9	-5.7	-6.0	-5.5	-6.5	-5.5	-6.8	-10.8	-11.3	-13.0
MEAN ‡	133	243	293	389	684	602	326	380	314	166	174	141
CFSM ‡	.47	.86	1.04	1.38	2.43	2.14	1.16	1.35	1.12	.59	.62	.50
IN ‡	.54	.96	1.20	1.59	2.53	2.47	1.29	1.56	1.25	.68	.72	.56

CAL YR 1970 TOTAL 124,639 MEAN 341 MAX 1,810 MIN 110 MEAN‡ 333 CFSM‡ 1.19 IN‡ 16.15
WTR YR 1971 TOTAL 119,277 MEAN 327 MAX 1,630 MIN 125 MEAN‡ 318 CFSM‡ 1.13 IN‡ 15.34

PEAK DISCHARGE (BASE, 1,500 CFS)

DATE	TIME	G. H.	DISCHARGE	DATE	TIME	G. H.	DISCHARGE
2-13	2115	7.15	2,670	2-23	1800	5.67	1,680

† Pumpage, in cubic feet per second, from Potomac River for municipal supply of Hagerstown.
‡ Adjusted for pumpage.

FIGURE 4-16
Sample page from U.S. Geological Survey *Water Resources Data*.

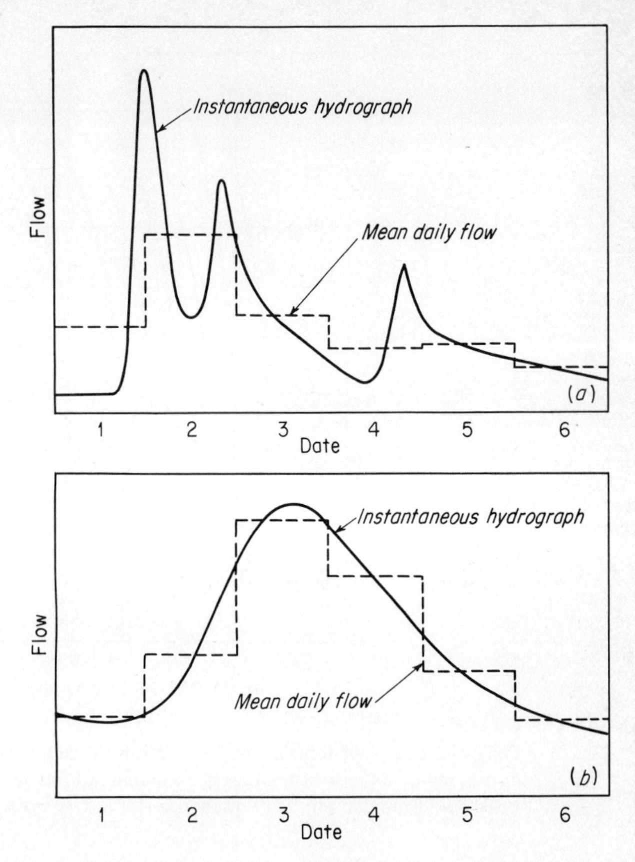

FIGURE 4-17
The relationship between instantaneous and mean daily flows (upper and lower charts typical of small and large streams, respectively).

moved or average precipitation at a number of stations in the area. The double-mass-curve method implies a relationship of the form $q = kP$, which may not be correct if precipitation is used as a base. A more effective procedure is to develop a relationship between precipitation and runoff and make a double-mass curve of observed streamflow versus runoff as estimated from this relation (Sec. 8-12).

Man-made works—storage reservoirs, diversions, levees, etc.—cause changes in either total flow volume or rate of flow or both. An analysis of the effects on the record at a given station requires a careful search to determine the number and size of reservoirs, the number and quantity of

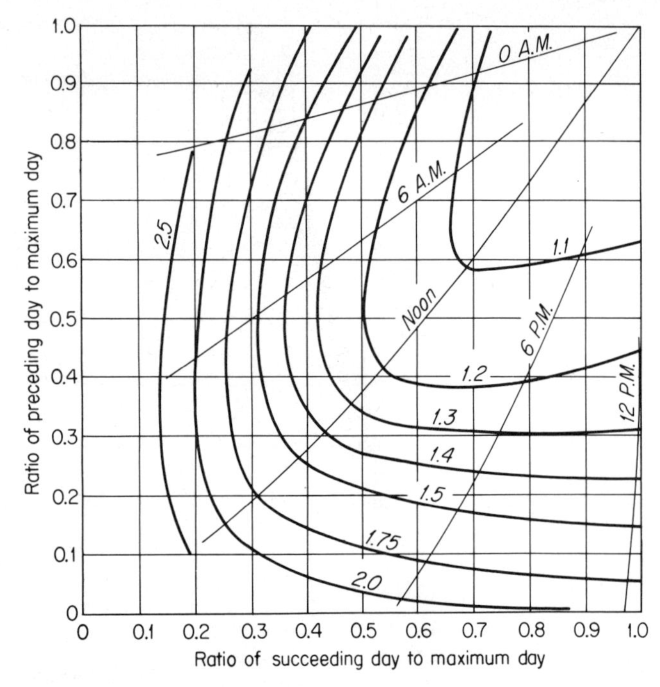

FIGURE 4-18
Peak discharge and time of crest in relation to daily mean discharge. Labels on curves indicate clock time of crest on maximum day and ratio of instantaneous peak flow to maximum daily discharge. (*U.S. Geological Survey.*)

diversions, and the date of their construction. Many small diversions may be unmeasured, and estimates of the flow diverted must be based on electric-power consumption of pumps, capacity of pump equipment, duration of pumping, or conduit capacities for gravity diversions. Diversions for irrigation may be estimated from known irrigable or irrigated acreage and estimated unit water requirements (Chap. 5). The adjustment of the stream-flow record for the effect of reservoirs or diversions on flow volume requires the addition of the net change in storage and/or the total diversion to the reported total flow. It may be necessary to consider also channel losses and losses by evaporation from the reservoirs.

Adjustment of short-period or instantaneous flow rates for the effect of storage or diversion is a much more complex problem. Levees, channel improvement, and similar works may also affect flow rate. In some instances pumping of groundwater has markedly reduced low flows, as has the construction of stock ponds [31]. Correction for the effect of storage or diversion on flow rates is made by adding the rate of change of storage or the diversion rate to the observed flows. In addition, it may be necessary to use storage-routing techniques (Chap. 9) to correct for the effect of channel storage between the reservoir or diversion point and the gaging station. Channel improvement and levee work change flow through their effects on the channel storage of the stream. Unless it is possible to establish "before" and "after" correlations with some station outside the influence of the channel works, corrections must be made entirely by storage-routing methods.

Land-use changes, urbanization, deforestation, or reforestation affect streamflow and cause apparent shifts in the flow record. Unless the timing and areal extent of such changes are well documented, correction of the record is almost impossible. Even with good documentation of changes, the adjustment process is complex. The most direct solution may be the use of simulation (Chap. 10) to reconstruct a flow series from rainfall and other meteorological factors using parameters appropriate for current or expected future conditions.

4-18 Mean Annual Runoff

Figure 4-19 is a map of mean annual runoff in the United States. In addition to the problems of representing any highly variable element by isopleths drawn on the basis of limited data, there are other problems peculiar to streamflow. The record of flow at a gaging station represents the integrated runoff for the entire basin above the station. In many areas the production of runoff is not uniform over the basin. The average annual runoff of the Missouri River at Omaha is 10,370,000 sfd (25,400 hm^3), or 1.20 in. (30 mm) over the 322,800-mi^2 (836,000-km^2) drainage area. The headwater areas produce much heavier runoff [Yellowstone River at Corwin Springs = 15.6 in. (396 mm)], whereas other large areas must average below the mean for Omaha. In addition, storage and diversion further complicate the picture. Figure 4-19 was prepared by the Geological Survey [32] using all available information and represents the most reliable map available for the United States. Such a map is intended to present a general picture of geographical variations in runoff. It cannot possibly show fine detail and should be used only for general information and preliminary studies. It should not be used as a source of information for a specific design problem. Such information can be developed more reliably through use of actual streamflow data and the analytical techniques discussed in the chapters which follow. It is helpful to coordinate precipitation maps and evaporation maps with the runoff map [33].

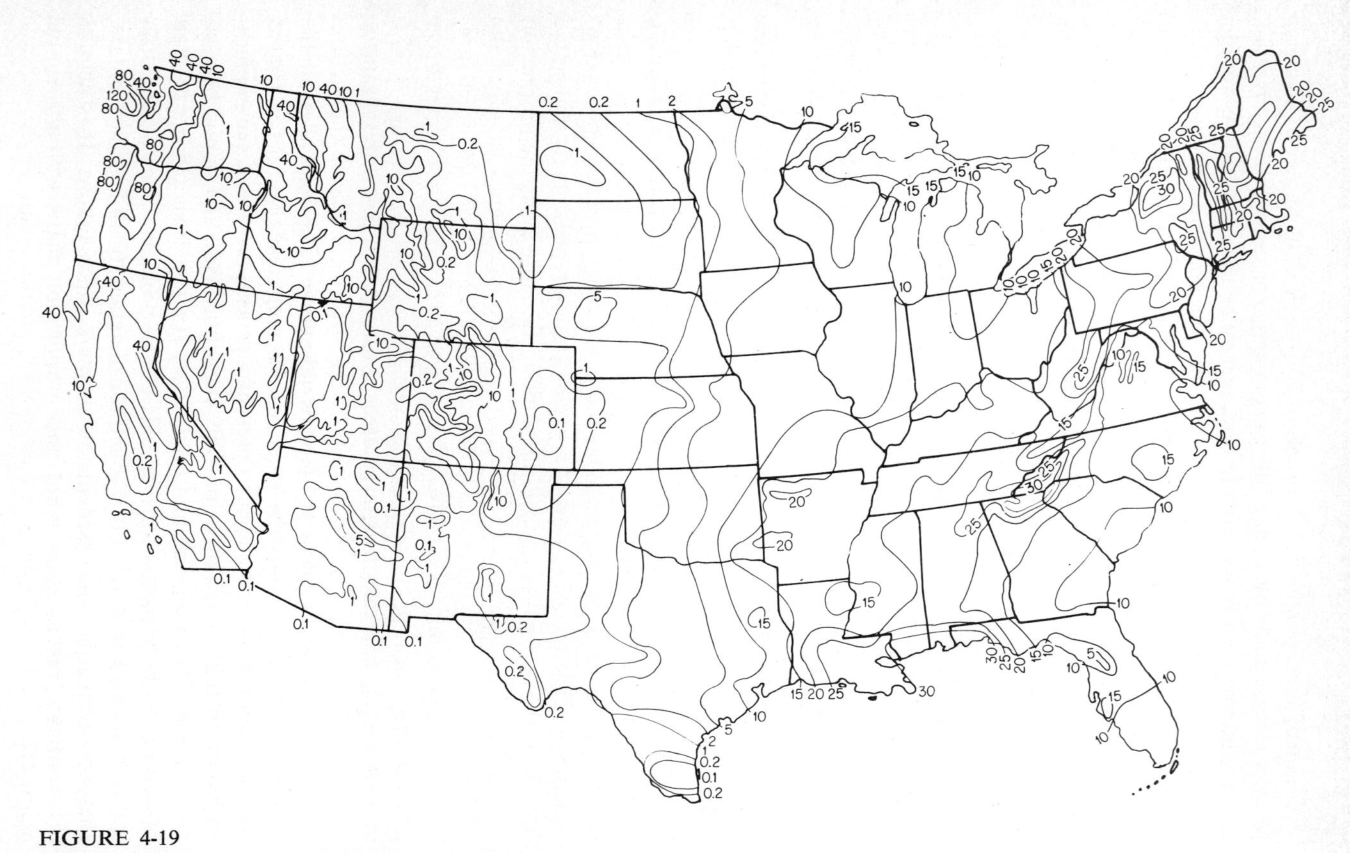

FIGURE 4-19
Average annual runoff (inches) in the United States. (*U.S. Geological Survey.*)

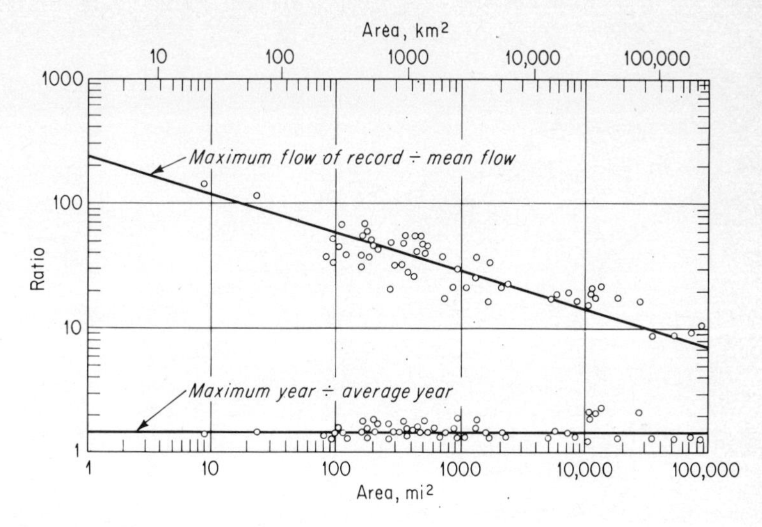

FIGURE 4-20
Ratios of maximum annual runoff to average annual runoff and maximum flow of record to mean flow for stations in the upper Ohio River basin.

4-19 Streamflow Variations

The normal or average values of runoff serve an important purpose, but they do not disclose all the pertinent information concerning the hydrology of an area. Especially significant are variations of streamflow about this normal. These variations include the following:

1 Variations in total runoff from year to year
2 Variations of daily rates of runoff throughout the year
3 Seasonal variations in runoff

To some extent these variations are regional characteristics, but size of the drainage area is invariably a factor. Figure 4-20 shows the ratio of the maximum annual runoff to the mean annual runoff as a function of drainage area for a number of stations in the upper Ohio River basin having at least 20 y of record. On the same chart is a curve showing the ratio of the maximum flood of record to the average flow. In the first instance, little effect of area is evident, but in the second case substantially higher ratios are associated with the smaller basins. The relationships are not perfect, for basin shape, geology, and climatic exposure are also significant factors. The map of Fig. 4-21 indicates the variation of these two ratios over the United States. To minimize the effect of area, ratios are given for stations having

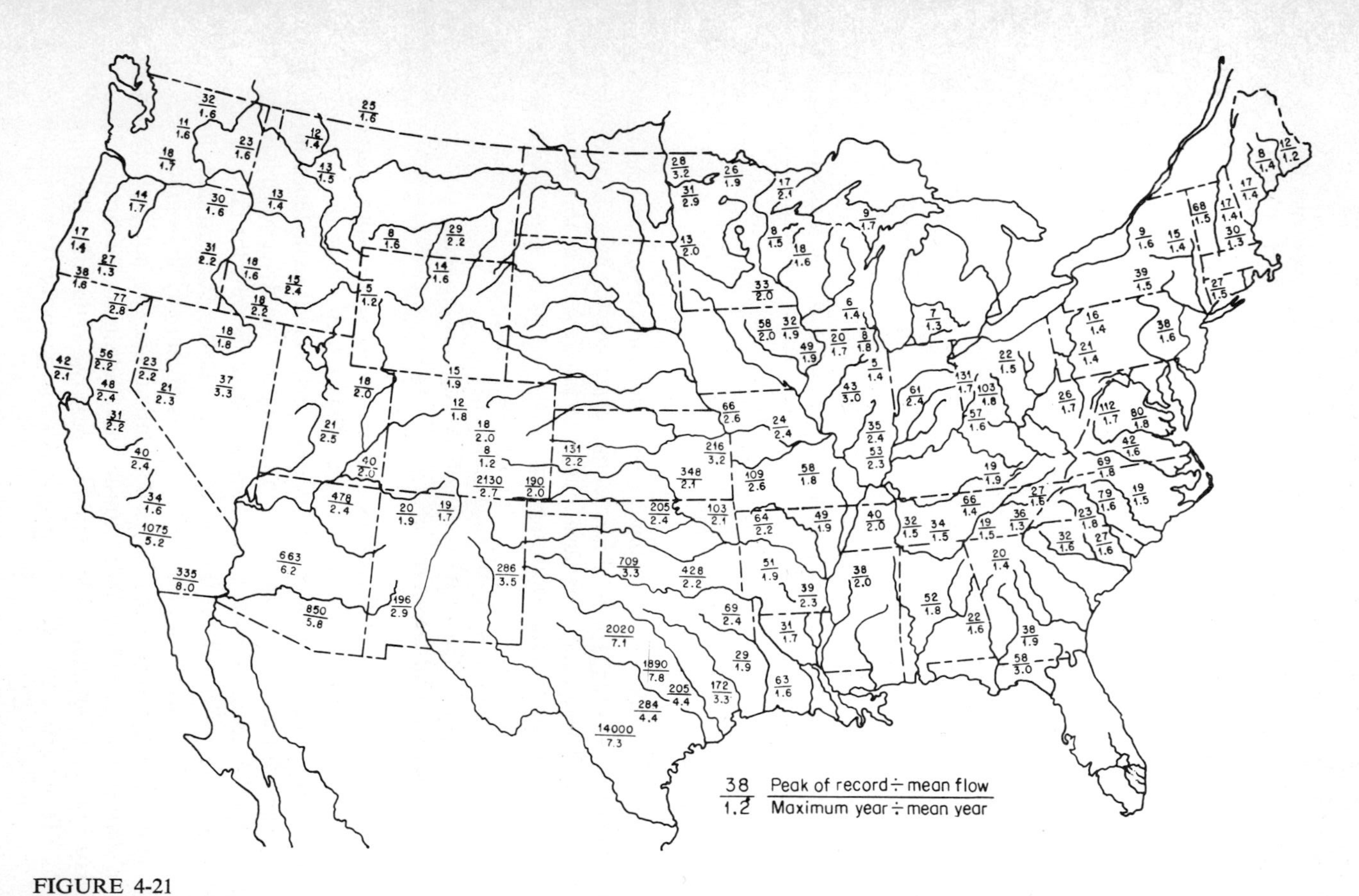

FIGURE 4-21
Ratios of maximum annual flow and maximum peak flow to mean flow for selected United States stations with drainage areas between 1000 and 2000 mi² (2590 and 5180 km²).

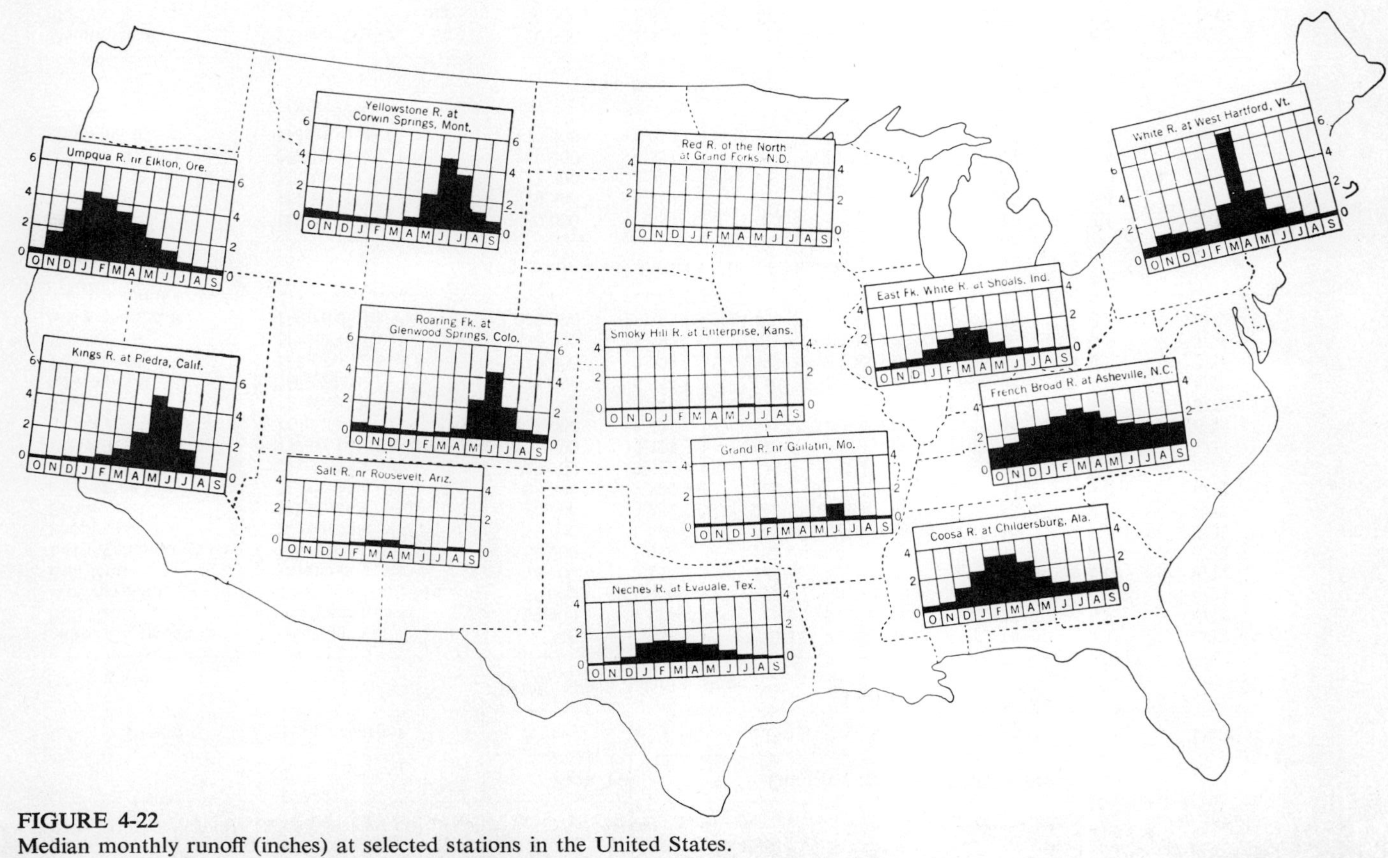

FIGURE 4-22
Median monthly runoff (inches) at selected stations in the United States.

Table 4-1 SELECTED RECORD PEAK FLOWS IN THE UNITED STATES

Stream	Station	Peak flow		Drainage area		Flow per unit area		Date
		ft³/s	m³/s	mi²	km²	ft³/s-mi²	m³/s-km²	
North Atlantic Slope								
Nibbs Cr. tributary	†Amelia, Va.	740	21	0.35	0.91	2,114	23.1	6/72
Bull Run	†Catharpin, Va.	39,400	1,116	25.8	66.8	1,527	16.7	6/72
Western Run	‡Western Run, Md.	38,000	1,076	59.8	154.9	635	6.9	6/72
Bull Run	†Manassas, Va.	76,100	2,155	148	383	514	5.6	6/72
East Mahantango Cr.	†Dalmatia, Pa.	69,900	1,979	162	420	431	4.7	6/72
Esopus Cr.	‡Colbrook, N.Y.	59,600	1,688	192	497	310	3.4	3/51
Patapsco R.	‡Hollofield, Md.	80,600	2,282	285	738	283	3.1	6/72
Conestoga Cr.	‡Lancaster, Pa.	88,300	2,500	324	839	273	3.0	6/72
Loyalsock Cr.	‡Loyalsock, Pa.	88,700	2,512	443	1,147	200	2.2	6/72
White R.	‡West Hartford, Vt.	120,000	3,398	690	1,787	174	1.9	11/27
Tioga R.	‡Lindley, N.Y.	128,000	3,625	771	1,997	166	1.8	6/72
Tioga R.	†Erwins, N.Y.	190,000	5,380	1,377	3,566	138	1.5	6/72
Chemung R.	†Big Flats, N.Y.	235,000	6,654	2,150	5,568	109	1.2	6/72
Potomac R.	‡Hancock, Md.	340,000	9,628	4,073	10,549	83	0.9	3/36
Potomac R.	‡Point of Rocks, Md.	480,000	13,592	9,651	24,996	50	0.5	3/36
Susquehanna R.	‡Harrisburg, Pa.	1,020,000	28,883	24,100	62,419	42	0.5	6/72
South Atlantic and Eastern Gulf of Mexico								
Morgan Cr.	†Chapel Hill, N.C.	30,000	850	27	70	1,110	12.1	8/24
Linville R.	†Nebo, N.C.	39,500	1,119	67	174	590	6.4	8/40
Catawba R.	‡Catawba, N.C.	71,400	2,022	171	443	416	4.6	8/40
Yadkin R.	‡Wilkesboro, N.C.	160,000	4,531	493	1,277	324	3.5	8/40
Catawba R.	†Marion, N.C.	177,000	5,012	1,535	3,976	115	1.3	8/40
Ohio River Basin								
Wautauga R.	‡Sugar Grove, N.C.	50,800	1,438	91	236	559	6.1	8/40
Buffalo R.	†Flat Woods, Tenn.	90,000	2,549	447	1,158	202	2.2	2/48

New R.	†Galax, Va.	141,000	3,993	1,131	2,929	125	1.4	8/40
Caney Fork	†Rock Island, Tenn.	210,000	5,947	1,678	4,346	125	1.4	3/29
Great Miami R.	‡Dayton, Ohio	250,000	7,079	2,511	6,503	100	1.1	3/13
Great Miami R.	‡Hamilton, Ohio	352,000	9,968	3,630	9,402	97	1.1	3/13
Ohio R.	‡Sewickly, Ohio	574,000	16,250	19,500	50,500	29	0.3	3/36
Ohio R.	‡Owensboro, Ohio	1,210,000	34,260	97,200	251,750	12	0.1	1/37
Ohio R.	‡Metropolis, Ill.	1,850,000	52,390	203,000	525,770	9	0.1	2/37
St. Lawrence Basin								
Quigg Hollow Bk.	†Andover, N.Y.	9,200	260	4.2	11.0	2,170	23.7	6/72
Trumansburg Cr.	‡Trumansburg, N.Y.	17,800	504	11.5	29.8	1,550	16.9	7/35
East Cr.	‡Rutland, Vt.	36,500	1,034	51	132	719	7.8	6/47
Winooski R.	†Essex Junction, Vt.	113,000	3,200	1,044	2,704	108	1.2	11/27
Auglaize R.	†Defiance, Ohio	120,000	3,398	2,318	6,004	52	0.6	3/13
Hudson Bay and Upper Mississippi River Basin								
East Fork, Galena R.	‡Council Hill, Ill.	16,600	470	20	52	830	9.0	4/47
Farm Cr.	‡East Peoria, Ill.	22,000	623	61	158	361	3.9	4/47
Platte R.	†Rockville, Wis.	43,500	1,232	149	386	313	3.2	7/50
Big Eau Pleine R.	†Stratford, Wis.	41,000	1,161	224	580	183	2.0	9/38
Cuivre R.	†Troy, Mo.	120,000	3,398	903	2,339	133	1.5	10/41
Sangamon R.	‡Oakford, Ill.	123,000	3,483	5,120	13,260	24	0.3	5/43
Des Moines R.	†Tracy, Iowa	155,000	4,389	12,479	32,320	12	0.1	6/47
Mississippi R.	‡Clinton, Iowa	250,000	7,079	85,600	221,700	3	0.03	6/1880
Mississippi R.	‡Keokuk, Iowa	360,000	10,190	119,000	308,200	3	0.03	6/1851
Missouri River Basin								
Estes Cr.	†Nemo, S.D.	6,620	187	6.1	15.8	1,080	11.8	6/72
E. Fork, Fishing R.	†Excelsior Springs, Mo.	12,000	340	20	52	600	6.5	7/51
Battle Cr.	†Keystone, S.D.	26,200	742	66	171	397	4.4	6/72
Big Bull Cr.	†Hillsdale, Kans.	45,200	1,280	147	381	308	3.4	7/51
Little Nemaha R.	†Syracuse, Nebr.	225,000	6,371	212	549	1,061	11.6	5/50
Little Nemaha R.	‡Auburn, Nebr.	164,000	4,644	793	2,054	207	2.3	5/50
Marais des Cygnes R.	‡Ottawa, Kans.	142,000	4,021	1,250	3,237	114	1.2	7/51

† Near. ‡ At.

Table 4-1 (*continued*)

Stream	Station	Peak flow		Drainage area		Flow per unit area		Date
		ft³/s	m³/s	mi²	km²	ft³/s-mi²	m³/s-km²	
Lower Mississippi River Basin								
Green Acre Br.	†Rolla, Mo.	1,900	54	0.6	1.6	3,062	33.8	6/50
Sallisaw Cr.	†Sallisaw, Okla.	110,000	3,115	182	471	604	6.6	4/45
Neosho R.	‡Council Grove, Kans.	121,000	3,426	250	647	484	5.3	7/51
Little Missouri R.	†Murfreesboro, Ark.	120,000	3,398	380	984	316	3.4	3/45
Neosho R.	‡Strawn, Kans.	400,000	11,330	2,933	7,596	136	1.5	7/51
Neosho R.	‡Parsons, Kans.	410,000	11,610	4,905	12,704	84	0.9	7/51
Western Gulf of Mexico								
E. Fork, James R.	‡Old Noxville, Tex.	105,000	2,970	61	158	1,730	18.8	7/32
Johnson Cr.	†Ingram, Tex.	138,000	3,908	114	295	1,211	13.2	7/32
West Nueces R.	†Brackettville, Tex.	550,000	15,570	700	1,813	786	8.6	6/35
Nueces R.	§Uvalde, Tex.	616,000	17,440	1,947	5,043	317	3.5	6/35
Colorado River and Great Basin								
Skyrocket Cr.	‡Ouray, Colo.	2,000	57	1	2.6	2,000	21.9	7/23
Whitewater R.	‡Whitewater, Calif.	42,000	1,189	57	148	730	8.0	3/38
Deep Cr.	†Hesperia, Calif.	46,600	1,320	136	352	343	3.8	3/38
Mojave R.	†Victorville, Calif.	70,600	1,999	514	1,331	137	1.5	3/38
San Pedro R.	‡Charleston, Ariz.	98,000	2,775	1,219	3,157	80	0.9	9/26
Bill Williams R.	‡Planet, Ariz.	200,000	5,663	5,140	13,312	39	0.4	2/1891
Humboldt R.	†Imlay, Nev.	6,080	172	157,000	406,630	0.04	0.0004	5/52

Pacific Slope Basins in California								
Cucamonga Cr.	†Upland, Calif.	14,100	399	10	26	1,410	15.3	1/69
Tujunga R.	†Sunland, Calif.	50,000	1,416	106	275	471	5.1	3/38
S. Fork, Eel R.	†Miranda, Calif.	199,000	5,635	537	1,391	371	4.0	12/64
Smith R.	†Crescent City, Calif.	228,000	6,456	609	1,577	374	4.1	12/64
Eel R.	‡Scotia, Calif.	752,000	21,290	3,113	8,063	242	2.6	12/64
Pacific Slope Basins in Washington and Upper Columbia Basin								
N. Fork, Skykomish R.	†Index, Wash.	27,000	765	57	148	474	5.2	11/34
Elwha R.	†Port Angeles, Wash.	41,600	1,178	269	697	155	1.7	11/1897
Skykomish R.	†Gold Bar, Wash.	88,700	2,512	535	1,386	166	1.8	12/33
Snake River Basin								
E. Fork, Wallowa R.	†Joseph, Ore.	450	13	10	26	45	0.5	7/37
Lake Fork, Payette R.	†McCall, Idaho	2,600	74	49	127	53	0.6	6/48
Clearwater R.	†Peck, Idaho	118,000	3,341	8,040	20,820	15	0.16	6/64
Pacific Slope Basins in Oregon and Lower Columbia River Basin								
Willow Cr.	†Heppner, Oreg.	36,000	1,019	87	225	401	4.5	7/03
Lewis R.	‡Ariel, Wash.	129,000	3,653	731	1,893	177	1.9	12/33
Santiam R.	‡Jefferson, Oreg.	202,000	5,720	1,790	4,636	113	1.2	11/21
Willamette R.	‡Albany, Oreg.	340,000	9,628	4,840	12,540	70	0.8	12/1861
Willamette R.	‡Salem, Oreg.	500,000	14,160	7,280	18,860	69	0.8	12/1861
Columbia R.	†The Dalles, Oreg.	1,240,000	35,110	237,000	613,800	5	0.06	6/1894
Hawaii								
Kalihi Stream	†Honolulu	12,400	351	2.6	6.7	4,769	52.4	11/30
N. Fork, Wailua R.	†Kapaa	53,200	1,506	18.7	48.4	2,845	31.1	11/55
S. Fork, Wailua R.	†Lihue	87,300	2,472	22.4	58	3,897	42.6	4/63
Waimea R.	†Waimea	37,100	1,051	58	150	640	7.0	2/49
Wailuku R.	‡Piihonua	63,400	1,795	125	324	507	5.5	8/40

† Near. ‡ At. § Below.

drainage areas between 1000 and 2000 mi^2 (2600 to 5200 km^2). High ratios are generally located in arid regions with low normal annual runoff.

Median monthly runoff for selected stations is shown in Fig. 4-22 (p. 141). Here again there are significant regional differences. A marked summer dry season is characteristic of the Pacific Southwest and is an important factor in the need for irrigation in that area. In contrast, the low-flow season in the northern tier of states is during the winter months, when precipitation is largely in the form of snow. In the eastern and southeastern states runoff is more uniformly distributed throughout the year as a result of a more uniform precipitation distribution (Fig. 3-15).

Table 4-1, which begins on page 142, summarizes some of the largest flood peaks in the various parts of the United States.

REFERENCES

1. T. J. Buchanan and W. P. Somers, Stage Measurement at Gaging Stations, *U.S. Geol. Surv. Tech. Water Resour. Inv.*, bk. 3, chap. A7, 1968.
2. E. G. Barron, New instruments for Surface-Water Investigations, in Selected Techniques for Water-Resources Investigations, *U.S. Geol. Surv. Water-Supply Pap.* 1692-Z, pp. Z-4–Z-8, 1963.
3. M. H. Collet, Crest-Stage Meter for Measuring Static Heads, *Civ. Eng.*, vol. 12, p. 396, 1942.
4. J. C. Stevens, Device for Measuring Static Heads, *Civ. Eng.*, vol. 12, p. 461, 1942.
5. F. J. Doran, High Water Gaging, *Civ. Eng.*, vol. 12, pp. 103–104, 1942.
6. G. E. Ferguson, Gage to Measure Crest Stages of Streams, *Civ. Eng.*, vol. 12, pp. 570–571, 1942.
7. For some history of the development of cup-type meters see A. H. Frazier, Daniel Farrand Henry's Cup Type "Telegraphic" Current Meter, *Tech. Cult.*, vol. 5, pp. 541–565, 1964.
8. A. H. Frazier, William Gunn Price and the Price Current Meters, *U.S. Natl. Mus. Bull.*, vol. 252, pp. 37–68, 1967.
9. M. P. O'Brien and R. G. Folsom, Notes on the Design of Current Meters, *Trans. Geophys. Union*, vol. 29, pp. 243–250, April 1948.
10. G. F. Smoot and C. E. Novak, Calibration and Maintenance of Vertical-Axis Type Current Meters, *U.S. Geol. Surv. Tech. Water Resour. Inv.*, bk. 3, chap. A8, 1968.
11. T. J. Buchanan and W. P. Somers, Discharge Measurements of Gaging Stations, *U.S. Geol. Surv. Tech. Water Resour. Inv.*, bk. 3, chap. A8, 1969.
12. D. M. Corbett and others, Stream-gaging Procedure, *U.S. Geol. Surv. Water-Supply Pap.* 888, pp. 43–51, 1945.

13. K. B. Young, "A Comparative Study of Methods for Computation of Discharge," U.S. Geological Survey, February 1950.
14. J. F. Wilson, Fluorometric Procedures for Dye Tracing, *U.S. Geol. Surv. Tech. Water Resour. Inv.*, bk. 3, chap. A11, 1968.
15. C. M. Allen and E. A. Taylor, The Salt Velocity Method of Water Measurement, *Trans. ASME*, vol. 45, p. 285, 1923.
16. H. Addison, "Applied Hydraulics," 4th ed., pp. 583–584, Wiley, New York, 1954.
17. R. W. Carter and J. Davidian, Discharge Ratings at Gaging Stations, *U.S. Geol. Surv. Surface-Water Tech.*, bk. 1, chap. 12, 1965.
18. M. C. Boyer, Determining Discharge at Gaging Stations Affected by Variable Slope, *Civ. Eng.*, vol. 9, p. 556, 1939.
19. W. D. Mitchell, Stage-Fall-Discharge Relations for Steady Flow in Prismatic Channels, *U.S. Geol. Surv. Water-Supply Pap.* 1164, 1954.
20. J. C. Stevens, A Method of Estimating Stream Discharge from a Limited Number of Gagings, *Eng. News*, pp. 52–53, July 18, 1907.
21. W. T. Sittner, Extension of Rating Curves by Field Surveys. *J. Hydraul. Div., ASCE*, vol. 89, pp. 1–9, March 1963.
22. M. A. Benson, Measurement of Peak Discharge by Indirect Methods, *World Meteorol. Org. Tech. Note* 90, Geneva, 1968.
23. T. Dalrymple and M. A. Benson, Measurement of Peak Discharge by the Slope-Area Method, *U.S. Geol. Surv. Tech. Water-Resour. Inv.*, bk. 3, chap. A2, 1967.
24. W. G. Hoyt, The Effect of Ice on Stream Flow, *U.S. Geol. Surv. Water-Supply Pap.* 337, pp. 24–30, 1913.
25. J. N. Bradley, Rating Curves for Flow Over Drum Gates, *Trans. ASCE*, vol. 119, pp. 403–431, 1954.
26. U.S. Bureau of Reclamation, "Water Measurement Manual," 1953.
27. C. L. Bodhaine, Measurement of Peak Discharge of Culverts by Indirect Measurements, *U.S. Geol. Surv. Tech. Water-Resour. Inv.*, bk. 3, chap. A3, 1968.
28. M. A. Kohler, Design of Hydrological Networks, *World Meteorological Organization Tech. Note*, no. 25, 1958.
29. W. B. Langbein, Hydrological Bench Marks, *Rep. WMO/IHD Proj.*, no. 8, World Meteorological Organization, Geneva, 1968.
30. W. B. Langbein, Peak Discharge from Daily Records, *U.S. Geol. Surv. Water-Resour. Bull.*, p. 145, Aug. 10, 1944.
31. R. C. Culler and H. V. Peterson, Effect of Stock Reservoirs on Runoff in the Cheyenne River Basin above Angostura Dam, *U.S. Geol. Surv. Circ.* 223, 1953.
32. W. B. Langbein and others, Annual Runoff in the United States, *U.S. Geol. Surv. Circ.* 52, June 1949.
33. T. J. Nordenson, Preparation of Coordinated Maps of Precipitation, Runoff, and Evaporation, *Rep. WMO/IHD Proj.*, no. 6, World Meteorological Organization, Geneva, 1968.

BIBLIOGRAPHY

BENSON, M. A., and DALRYMPLE, T.: General Field and Office Procedures for Indirect Measurements, *U.S. Geol. Surv. Tech. Water-Resour. Inv.*, bk. 3, chap. A1, 1967.

BYIKOV, V. D., and VASILEV, A. V.: "Gidrometria," in Russian (Hydrometry), Gidromeoizdat, Leningrad, 1965.

CARTER, R. W., and DAVIDIAN, J.: General Procedure for Gaging Streams, *U.S. Geol. Surv. Tech. Water-Resour. Inv.*, bk. 3, chap. A6, 1968.

KING, H. W., and BRATER, E. F.: "Handbook of Hydraulics," 5th ed., McGraw-Hill, New York, 1963.

SCHAEFER, H.: Report on Hydrometry, *Hydrol. Sci. Bull.*, vol. 17, pp. 145–166, July 1972.

WORLD METEOROLOGICAL ORGANIZATION: "Guide to Hydrological Practices," 3d ed., Geneva, 1974.

———: Casebook on Hydrological Network Design Practices, *WMO Publ.* 432, Geneva, 1972.

———: Technical Regulations Vol. III, Operational Hydrology, *WMO Publ.* 49, Geneva, 1971.

——— and ECONOMIC COMMISSION FOR ASIA AND THE FAR EAST: Standards for Methods and Records of Hydrologic Measurements, *ECAFE Flood Control Ser.*, no. 6, United Nations, New York, 1945.

UNITED STATES DATA SOURCES

The main sources of streamflow data are the publications of the U.S. Geological Survey and *Daily River Stages*, published by the National Weather Service. More detailed information may be obtained by writing to either of these agencies or their field offices, located in principal cities throughout the country.

The U.S. Corps of Engineers, U.S. Bureau of Reclamation, and U.S. Soil Conservation Service make occasional observations of flow. Information from these sources can usually be obtained from the nearest field office. Many states have agencies which make or cooperate in streamflow measurements.

PROBLEMS

4-1 Compute the streamflow for the measurement data below. Take the meter rating from Eq. (4-1) with $a = 0.1$ and $b = 2.2$ for v in ft/s. Note that the rule against more than 10 percent of the flow in any vertical section is violated in this problem to reduce computation.

Distance from bank, ft	Depth, ft	Meter depth, ft	Revolu- tions	Time, s
2	1	0.6	10	50
4	3.5	2.8	22	55
		0.7	35	52
6	5.2	4.2	28	53
		1.0	40	58
9	6.3	5.0	32	58
		1.3	45	60
11	4.4	3.5	28	45
		0.9	33	46
13	2.2	1.3	22	50
15	0.8	0.5	12	49
17	0			

4-2 The table that follows gives discharge, base stage, and stage at an auxiliary gage 2000 ft downstream. Develop a slope-stage-discharge relationship from these data. Compute the average error of the rating, using the tabulated data. What is the estimated discharge for base and auxiliary stages of 25.00 and 24.20 ft, respectively?

Base stage, ft	Discharge, ft^3/s	Auxiliary stage, ft	Base stage, ft	Discharge, ft^3/s	Auxiliary stage, ft
14.02	2,400	13.00	26.40	55,000	25.70
23.80	29,600	23.25	22.20	74,200	21.00
17.70	21,200	16.60	16.20	9,550	15.30
24.60	85,500	23.55	21.10	43,500	20.13
20.40	28,200	19.55	25.60	84,000	24.60
17.00	7,400	16.40	23.20	93,500	21.95
18.65	34,000	17.50			

4-3 Given below are data for a station rating curve. Extend the relation and estimate the flow at a stage of 14.5 ft by both the logarithmic and $A\sqrt{D}$ methods.

Stage, ft	Area, ft^2	Depth, ft	Discharge, ft^3/s	Stage, ft	Area, ft^2	Depth, ft	Discharge, ft^3/s
1.72	263	1.5	1,020	2.50	674	1.8	2,700
3.47	1,200	2.1	4,900	4.02	1,570	2.8	6,600
4.26	1,790	3.2	7,700	5.08	2,150	3.9	9,450
5.61	2,380	4.6	10,700	5.98	2,910	4.9	13,100
6.70	3,280	5.2	15,100	6.83	3,420	5.4	16,100
7.80	3,960	5.7	19,000	8.75	4,820	6.0	24,100
9.21	5,000	6.1	25,000	9.90	5,250	6.5	27,300
14.50	8,200	9.0					

4-4 What volume is represented by 1.43 in. of runoff from a basin of 254 mi^2? Give answer in cubic feet, second-foot-days, and acre-feet.

4-5 Given below are the daily mean flows in cubic feet per second at a gaging station for a period of 5 d. What is the mean flow rate for the period in cubic feet per second? What is the total discharge during the period in second-foot-days? Acre-feet? If the drainage area is 756 mi^2, what is the runoff volume in inches?

Day	1	2	3	4	5
Flow, ft^3/s	700	4800	3100	2020	1310

4-6 Using the data of Prob. 4-5 and the relationship of Fig. 4-18, estimate the peak flow and the time of peak.

4-7 Obtain a copy of a *U.S. Geological Survey Water-Supply Paper*, and for some stream in your area determine the maximum flow of record in cubic feet per second and in cubic feet per second per square mile. Find the average annual runoff in acre-feet, second-foot-days, and inches.

4-8 For some stream selected by your instructor determine the mean flow for each month on the basis of 10 y of record. What percent of the total annual flow occurs each month? Compare these percentages with the monthly distribution of precipitation. Can you explain the differences?

4-9 For three basins in your area find the ratio of the maximum peak flow of record to the average daily flow rate. Is there an apparent relationship with drainage area? Can you explain the differences?

4-10 Compare the data in Table 4-1 for the several regions of the country. What regional differences do you see? Look for such things as date of floods, relative magnitude of flood peaks, apparent areal extent of floods, etc. Can you explain the differences?

4-11 What volume of runoff in cubic meters is represented by a depth of 37 mm on a basin of 600 km^2? How many hectares can be irrigated with this volume if 60 cm of water is required for irrigation?

4-12 Taking the flow data of Prob. 4-5 in cubic meters per second, what is the total discharge in cubic meters? If the tributary area is 100,000 km^2, what is the equivalent runoff depth in millimeters?

5

EVAPORATION AND TRANSPIRATION

This chapter discusses that phase of the hydrologic cycle in which precipitation reaching the earth's surface is returned to the atmosphere as vapor. Of the precipitation falling earthward, a portion evaporates before reaching the ground. Since the hydrologist normally measures precipitation just a few feet above the ground, evaporation from raindrops is of no practical concern, except in the interpretation of radar reflectivity as a measure of precipitation (Sec. 3-7). Likewise, evaporation from the oceans lies beyond his scope of direct interest. Precipitation caught on vegetation (interception) eventually evaporates, and the quantity of water actually reaching the soil surface is correspondingly reduced below that observed in a precipitation gage. Other evaporative mechanisms to be considered are transpiration by plants and evaporation from soil, snow, and free-water surfaces (lakes, reservoirs, streams, and depressions).

Anticipated evaporation is a decisive element in design of reservoirs to be constructed in arid regions. Ten reservoirs the size of Lake Mead would evaporate virtually the entire flow of the Colorado River in a normal year, and normal evaporation from Lake Mead alone is equivalent to almost one-third of the minimum annual inflow to the reservoir. Evaporation and transpiration are indicative of changes in the moisture deficiency of a basin and, in this capacity, are sometimes used to estimate storm runoff in the preparation of river forecasts. Estimates of these factors are also used in determining water-supply requirements of proposed irrigation projects.

EVAPORATION

Although there is always continuous exchange of water molecules to and from the atmosphere, the hydrologic definition of *evaporation* is restricted to the net rate of vapor transport to the atmosphere. This change in state requires an exchange of approximately 600 cal for each gram of water evaporated (Sec. 2-16). If the temperature of the surface is to be maintained, these large quantities of heat must be supplied by radiation and conduction from the overlying air or at the expense of energy stored below the surface.

5-1 Factors Controlling the Evaporation Process

Rates of evaporation vary, depending on meteorological factors and the nature of the evaporating surface. Much of the ensuing discussion of meteorological factors is couched in terms of evaporation from a free-water surface. The extent to which the material is applicable to other surfaces of interest in hydrology is believed to be self-evident.

Meteorological factors If natural evaporation is viewed as an energy-exchange process, it can be demonstrated that radiation is by far the most important single factor and that the term *solar evaporation* is basically applicable. On the other hand, theory and wind-tunnel experiments have shown that the rate of evaporation from water of specified temperature is proportional to wind speed and is highly dependent on the vapor pressure of the overlying air. How are these two conclusions to be reconciled? In essence, it can be said that water temperature is not independent of wind speed and vapor pressure. If radiation exchange and all other meteorological elements were to remain constant over a shallow lake for an appreciable time, the water temperature and the evaporation rate would become constant. If the wind speed were then suddenly doubled, the evaporation rate would also double momentarily. This increased rate of evaporation would immediately begin to extract heat from the water at a more rapid rate than it could be replaced by radiation and conduction. The water temperature would approach a new, lower equilibrium value, and evaporation would diminish accordingly. On a long-term basis, a change of 10 percent in wind speed will change evaporation only 1 to 3 percent, depending on other meteorological factors. In deep lakes with capacity for considerable heat storage, sudden changes in wind and humidity have longer-lasting effects; heat into or from storage assists in balancing the energy demands. Thus by utilizing stored energy excessive evaporation during a dry, windy week can reduce evaporation which would otherwise occur in subsequent weeks.

The relative effect of controlling meteorological factors is difficult to evaluate at best, and any conclusions must be qualified in terms of the time period considered. Evaluation of the relative importance of meteorological factors without an understanding of the subsequent discussion on energy-budget and mass-transfer equations would be placing the cart before the horse. It can be stated, however, that the rate of evaporation is influenced by solar radiation, air temperature, vapor pressure, wind, and possibly atmospheric pressure. Since solar radiation is an important factor, evaporation varies with latitude, season, time of day, and sky condition.

Nature of evaporating surface All surfaces exposed to precipitation, such as vegetation, buildings, and paved streets, are potentially evaporation surfaces. Since the rate of evaporation during rainy periods is small, the quantity of storm rainfall disposed of in this manner is essentially limited to that required to saturate the surface. Although this evaporation is appreciable on an annual basis, it is seldom evaluated separately but is considered part of overall evaporation and transpiration.

The rate of evaporation from a saturated soil surface is approximately the same as that from an adjacent water surface of the same temperature. As the soil begins to dry, evaporation decreases and its temperature rises to maintain the energy balance. Eventually, evaporation virtually ceases since there is no effective mechanism for transporting water from appreciable depths. Thus the rate of evaporation from soil surfaces is limited by the availability of water, or *evaporation opportunity*.

Evaporation from snow and ice constitutes a special problem since the melting point lies within the range of temperatures normally experienced. Evaporation can occur only when the vapor pressure of the air is less than that of the snow surface [Eq. (5-11)], i.e., only when the dewpoint is lower than the temperature of the snow. The vapor pressure of the surface film on melting snow is 6.11 millibars, and this value represents the maximum possible air-to-snow vapor-pressure difference under evaporating conditions. The maximum rate of evaporation from snow is only about one-fourth that from a water surface at 80°F (26.7°C) when the dewpoint is 45°F (7.2°C) under the same wind conditions. These and other considerations lead to the conclusion that with temperatures much above freezing, the rate of snowmelt must exceed evaporation unless a large part of the area consists of exposed, wet soil [1]. The impression that a chinook, or foehn, is conducive to excessive evaporation from the snow cover is fallacious unless the wind is strong enough to result in blowing snow. The dewpoint increases downslope, and evaporation from the snow must cease when the dewpoint rises to the freezing point. Reasonable assumptions yield an upper limit of about 0.2 in. (5 mm) water equivalent per day for evaporation from a snow surface. There is evidence [2] that considerable quantities of snow can disappear during periods when the wind is strong enough to whip the snow into the overrunning air.

Effects of water quality The effect of salinity, or dissolved solids, is brought about by the reduced vapor pressure of the solution. The vapor pressure of seawater (35,000 ppm dissolved salts) is about 2 percent less than that of pure water at the same temperature. The reduction in evaporation is less than that indicated by the change in vapor pressure [Eq. (5-11)] because with reduced evaporation there is an increase in water temperature which partially offsets the vapor-pressure reduction [3, 4]. Even for seawater, the reduction in evaporation is never in excess of a few percent (over an extended period of time), so that salinity effects can be neglected in the estimation of reservoir evaporation. Any foreign material which tends to seal the water surface or change its vapor pressure or albedo (Sec. 5-8) will affect the evaporation.

5-2 Water-Budget Determinations of Reservoir Evaporation

The direct measurement of evaporation under field conditions is not feasible, at least in the sense that one is able to measure river stage, rainfall, etc. As a consequence, a variety of techniques have been derived for determining or estimating vapor transport from water surfaces. The most obvious approach involves the maintenance of a water budget. Assuming that storage S, surface inflow I, surface outflow O, subsurface seepage O_g, and precipitation P can be measured, evaporation E can be computed from

$$E = (S_1 - S_2) + I + P - O - O_g \tag{5-1}$$

This approach is simple in theory, but application rarely produces reliable results since all errors in measuring outflow, inflow, and change in storage are reflected directly in the computed evaporation.

Of those factors required, seepage is usually the most difficult to evaluate since it must be estimated indirectly from measurements of groundwater levels, permeability, etc. If seepage approaches or exceeds evaporation, reliable evaporation determinations by this method are usually not possible. Under some circumstances, both seepage and evaporation can be evaluated by simultaneous solution of Eqs. (5-1) and (5-11) for periods of insignificant inflow and outflow [5]. The water budget can then be applied on a continuing basis through application of a stage-seepage relation.

The determination of rainfall generally does not represent a major obstacle, provided the average of on-shore measurements is representative of the reservoir. Difficulties in this respect may be expected when the surrounding topography is of high relief and for very large lakes which modify local weather. Errors in the measurement of snowfall can be large during periods of high wind. In addition to the usual deficiency of gage catch, a small reservoir can trap considerable quantities of blowing snow.

Water-stage recorders are sufficiently precise for determining the storage changes provided that the stage-area relationship is accurately established. Variations in bank storage are sometimes an important source of error in monthly computations but can usually be neglected in estimates of annual evaporation. Similarly, expansion or contraction of stored water with large temperature changes can introduce appreciable errors. At Lake Hefner, Oklahoma [6], corrections as large as 0.4 in. (10 mm) per month were required for changes in density.

The relative effect of errors in the surface inflow and outflow terms varies considerably from lake to lake, depending upon the extent of ungaged areas, the reliability of the rating curves, and the relative magnitude of flows with respect to evaporation. Determinations of streamflow to within 5 percent are normally considered excellent, and corresponding evaporation errors may be expected in an off-channel reservoir without appreciable outflow. If the quantity of water passing through a reservoir is large in comparison with evaporation losses, water-budget results are of questionable accuracy.

Under somewhat idealized conditions, it was found that daily evaporation from Lake Hefner, Oklahoma, could be reliably computed from a water budget; results were considered to be within 5 percent one-third of the time and within 10 percent two-thirds of the time. It should be emphasized that Lake Hefner was selected, after a survey of more than 100 lakes and reservoirs [7], as one of the three or four best meeting water-budget requirements. The requirements are not nearly so stringent for estimates of annual or mean annual reservoir evaporation, and satisfactory estimates have been made for many reservoirs.

5-3 Energy-Budget Determinations of Reservoir Evaporation

The energy-budget approach, like the water budget, employs a continuity equation and solves for evaporation as the residual required to maintain a balance. Although the continuity equation in this case is one of energy, an approximate water budget is required as well, since inflow, outflow, and storage of water represent energy values which must be considered in conjunction with the respective temperatures [8]. Application of the energy budget has been attempted by numerous investigators, with cases selected to minimize the effect of terms that could not be evaluated. However, the Lake Hefner experiment is believed to constitute the first test of the method with adequate control. The energy-budget approach is receiving increasing application to special studies but is not likely to be used on a broad-scale, continuing basis until instrumentation is improved.

The energy budget for a lake or reservoir may be expressed as

$$Q_n - Q_h - Q_e = Q_\theta - Q_v \tag{5-2}$$

where Q_n is the net (all-wave) radiation absorbed by the water body, Q_h sensible-heat transfer (conduction) to the atmosphere, Q_e energy used for evaporation, Q_θ the increase in energy stored in the water body, and Q_v energy advected† into the water body, all in calories per square centimeter. When we let H_v represent the latent heat of vaporization and R the ratio of heat loss by conduction to heat loss by evaporation (*Bowen ratio*), Eq. (5-2) becomes

$$E = \frac{Q_n + Q_v - Q_\theta}{\rho H_v (1 + R)} \tag{5-3}$$

where E is the evaporation in centimeters and ρ is the density of water. The Bowen ratio [9] can be computed from the equation

$$R = 0.61 \frac{T_0 - T_a}{e_0 - e_a} \frac{p}{1{,}000} \tag{5-4}$$

where p is the atmospheric pressure, T_a the temperature of the air, e_a the vapor pressure of the air, T_0 the water-surface temperature, and e_0 the saturation vapor pressure corresponding to T_0; all temperatures and pressures are in degrees Celsius and millibars.

Sensible-heat transfer cannot readily be observed or computed, and the Bowen ratio was conceived as a means of eliminating this term from the energy-budget equation. The validity of the constant in Eq. (5-4) has been the subject of much discussion [10]. Bowen found limiting values of 0.58 and 0.66, depending on the stability of the atmosphere, and concluded that 0.61 is applicable under normal atmospheric conditions. Using an independent approach, Pritchard [92] derived values of 0.57 and 0.66 for smooth and rough surfaces, respectively. At Lake Hefner, Oklahoma, monthly values of the ratio [computed from Eq. (5.4)] were found to vary from −0.32 in February to 0.25 in November, while the annual value was −0.03. It is obvious that one need not be concerned over variations in the constant of Eq. (5-4) for annual computations. If the correct value is assumed to have been one of the limits determined by Bowen, the extreme error in monthly evaporation at Lake Hefner would be only about 4 percent.

† Net energy content of inflowing and outflowing water is termed *advected energy*.

In the application of Eq. (5-3), it is important that the net radiation exchange be accurately evaluated. The net radiation can be expressed in terms of its five components

$$Q_n = Q_s - Q_r + Q_a - Q_{ar} - Q_0 \qquad (5\text{-}5)$$

where Q_s is sun and sky shortwave radiation incident upon the water surface, Q_r reflected shortwave radiation, Q_a incident atmospheric longwave radiation, Q_{ar} reflected longwave radiation, and Q_0 emitted longwave radiation. Most routine observations from established networks provide measurements of incident shortwave radiation only, and prior to the development of longwave and all-wave radiometers, radiation exchange was necessarily estimated from empirical relations [11, 12]. Radiometers can be designed to measure either the total incoming or net radiation (Sec. 2-4). Net radiometers must be exposed over the water at one or more points constituting a representative surface temperature. Because of the difficulties of maintaining observations over a lake, net radiometers are sometimes exposed over a tank of water. Assuming that the emissivity ε and reflectivity are the same for the water in the tank and lake, the incident minus reflected all-wave radiation for the tank $\hat{Q}_{ir}$ and the adjacent lake Q_{ir} can be obtained from the net radiation $\hat{Q}_n$ and absolute temperature $\hat{T}_0$ of the tank water surface:

$$\hat{Q}_{ir} = Q_{ir} = \hat{Q}_n + \varepsilon\sigma(\hat{T}_0)^4 \qquad (5\text{-}6)$$

where σ is the Stefan-Boltzmann constant (11.71×10^{-8} cal/cm^{2}K^{4}d) and ε may be taken as 0.97. The net radiation for the lake can then be obtained from

$$Q_n = Q_{ir} - \varepsilon\sigma(T_0)^4 \qquad (5\text{-}7)$$

or

$$Q_n = \hat{Q}_n + \varepsilon\sigma(\hat{T}_0 - T_0)^4 \qquad (5\text{-}8)$$

Another promising approach to the determination of net radiation involves the application of the energy budget to an insulated evaporation pan (*radiation integrator*) [13, 14]. The assumption is again made that Q_{ir} is the same for the pan and an adjacent lake, and values of this term are computed from Eqs. (5-2) and (5-6).

The energy budget has also been applied to small lakes using onshore measurements of incoming shortwave Q_s and total hemispherical radiation $Q_s + Q_a$. In this case, Q_{ir} is computed assuming average reflectivities ($Q_r = 0.05Q_s$, and $Q_{ar} = 0.03Q_a$).

The energy advection and storage terms ($Q_v - Q_\theta$) of Eq. (5-3) are computed from an approximate water budget and temperatures of the respective water volumes. Equation (5-1) can be written

$$S_2 - S_1 = I + P - O - O_g - E \qquad (5\text{-}9)$$

Variations in density are neglected, and all terms are expressed in cubic centimeters. The energy content per gram of water is the product of its specific heat and temperature. Assuming unity for the values of density and specific heat gives

$$Q_v - Q_\theta = \frac{1}{A}(IT_I + PT_P - OT_O - O_gT_g - ET_E + S_1T_1 - S_2T_2) \tag{5-10}$$

where T_I, T_P, ... are the Celsius temperatures of the respective volumes of water and the surface area A of the lake is introduced to convert energies into units of calories per square centimeter. Equation (5-9) should be balanced before solving Eq. (5-10), although approximate values of the individual terms will suffice. The temperature of precipitation can be taken as the wet-bulb temperature, seepage temperature as that of the water in the lowest levels of the lake, and T_E as the lake-surface temperature. Advected energy and change in energy storage tend to balance for most lakes, particularly over long periods of time, and are frequently assumed to cancel when considering annual or mean annual evaporation.

5-4 Aerodynamic Determination of Reservoir Evaporation

The theoretical development of turbulent-transport equations has followed two basic approaches: the discontinuous, or mixing-length, concept introduced by Prandtl and Schmidt, and the continuous-mixing concept of Taylor. An extensive physical and mathematical review [15] of the two approaches was prepared in advance of the Lake Hefner experiment, and a number of the equations were tested at Lake Hefner and Lake Mead [16]. Equations derived by Sverdrup and Sutton gave good results at Lake Hefner but were considered inadequate when applied to Lake Mead.

Numerous empirical formulas [17] have been derived which express evaporation as a function of atmospheric elements and which parallel the turbulent-transport approach in some respects. Many such equations are of the Dalton type [18] and may be written in the form

$$E = (e_0 - e_a)(a + bv) \tag{5-11}$$

where e_0 and e_a are vapor pressures of the water surface and at some fixed height in the overrunning air, respectively, and v is the wind speed (also at some fixed height). The saturation vapor pressure at the temperature of the air e_s is sometimes used instead of e_0.

Several empirical equations have been derived from the data collected at Lake Hefner:

$$E = 0.00304(e_0 - e_2)v_4 \qquad e_2 \text{ and } v_4 \text{ over lake} \tag{5-12}$$

$$E = 0.00241(e_0 - e_8)v_8 \qquad e_8 \text{ and } v_8 \text{ over lake} \tag{5-13}$$

$$E = 0.00270(e_0 - e_2)v_4 \qquad e_2 \text{ upwind and } v_4 \text{ over lake} \tag{5-14}$$

where E is lake evaporation in inches per day, vapor pressures are in inches of mercury, wind is in miles per day, and numerical subscripts designate heights above the surface in meters. With vapor pressures in millibars and wind speeds in meters per second, the constants in Eqs. (5-12) to (5-14) become 0.122, 0.097, and 0.109, respectively, for evaporation in millimeters per day. Equation (5-12) was found to yield excellent results for Lake Mead, and there is good reason to believe that it is of general applicability [19]. Equation (5-13) yielded a satisfactory value of annual evaporation for Lake Mead, but there was a seasonal bias which appeared to be correlated with atmospheric stability.

Vapor pressure of the air increases downwind across an open-water surface; hence turbulent-transport concepts have led to the belief that point evaporation decreases downwind. Sutton [20] concluded that average depth of evaporation from a circular water surface is proportional to the -0.11 power of its diameter under adiabatic conditions, and this functional relation has been verified in wind-tunnel experiments [21]. The theory assumes that water temperature and wind are unchanging downwind, and this condition tends to prevail in a wind tunnel, where solar radiation is not a factor. Observations show that wind speed increases downwind from the upwind edge of a lake, however, and a consideration of energy conservation requires that any immediate reduction in evaporation rate brought about by decreased vapor-pressure gradient be followed by an increase in water temperature. Although experimental data are insufficient to determine the magnitude of the *size effect*, Eq. (5-12) can be applied to lakes ranging up to several hundred square miles without appreciable error in this respect if all observations are well centered in the lake.

Using studies of many reservoirs up to 29,000 acres (120 km^2) in area, Harbeck [22] found that the data fit Eq. (5-11) such that $a = 0$ and b is given by

$$b = 0.00014A^{-0.05} \tag{5-15}$$

for E is in inches per day, A is the area of the reservoir in acres, v_2 is in miles per day, vapor pressures are in millibars,and e_a is measured in the unmodified air (upwind). The coefficient in Eq. (5-15) becomes 0.29 for E in millimeters per day, A in square meters, v_2 in meters per second, and vapor pressures in millibars. It should be stressed that Harbeck's results do not demonstrate a size effect in actual evaporation or in Eq. (5-12). The relation between e_2 over the water and e_a is a function of lake size, and this would assure a size effect in Eq. (5-14) if Eq. (5-12) were entirely independent of lake area.

5-5 Combination Methods of Estimating Reservoir Evaporation

The simultaneous solution of aerodynamic and water-balance equations to estimate evaporation and seepage is suggested in Sec. 5-2. A similar approach can be used to eliminate the need for water-surface temperature observations by simultaneous solution of aerodynamic and energy-budget equations [23–26]. Assuming a thin free-water surface (without heat storage or conduction from below), Penman derived the equation

$$E = \frac{1}{\Delta + \gamma}(Q_n \Delta + \gamma E_a) \qquad (5\text{-}16)$$

where Δ is the slope of the saturation-vapor-pressure versus temperature curve at the air temperature T_a, E_a is the evaporation given by Eq. (5-11) assuming the water-surface temperature $T_0 = T_a$, Q_n is the net radiation exchange expressed in the same units as E, and γ is the psychrometric constant in the Bowen ratio equation [Eq. (5-4)]:

$$R = \gamma \frac{T_0 - T_a}{e_0 - e_a} \qquad (5\text{-}17)$$

Figure 5-1 is based on an application [27] of the Penman equation to the Class A pan (Fig. 5-4) modified [28] to reflect differences in γ for the pan and an extended free-water surface and assuming that the class A pan coefficient is 0.70 when pan-water temperature is the same as air temperature (Sec. 5-6). Figure 5-1 provides a convenient means of estimating reservoir evaporation when net advection is not appreciable; otherwise, adjustment should be made using Eq. (5-21).

Equation (5-16) assumes that Q_n is representative of exchange at the water surface; E_a is based on an aerodynamic equation which yields correct values of E when used with observed vapor pressure of the water surface; and Δ at T_a is a good approximation of its average value between T_a and T_0. Net radiation must be measured over the water surface or estimated so as to reflect the water temperature (Sec. 5-3). While Penman's derivation takes into account the effect of any difference between T_0 and T_a on evaporation and convective heat transfer, his method of estimating Q_n assumes that the two temperatures are equal. This assumption can result in an appreciable overestimate of evaporation under calm, humid conditions and a corresponding underestimate for dry, windy conditions.

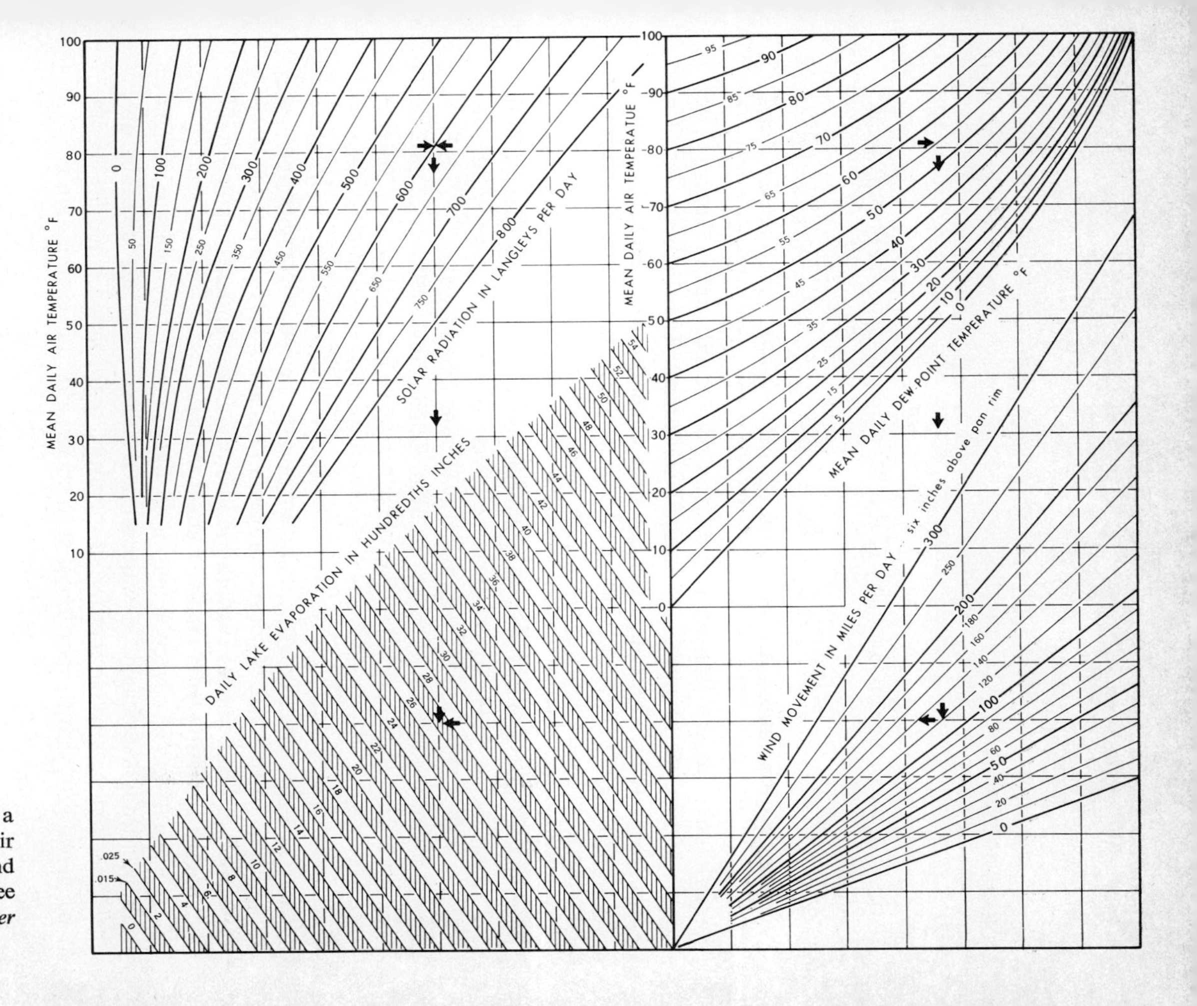

FIGURE 5-1
Shallow-lake evaporation as a function of solar radiation, air temperature, dewpoint, and wind movement. For metric units see Fig. C-1. (*U.S. National Weather Service.*)

By using Eq. (5-7), T_0 can be eliminated from the radiation term in Penman's derivation [29], yielding

$$E = \frac{(Q_{ir} - \varepsilon\sigma T_a^4)\,\Delta + E_a\,[\gamma + 4\varepsilon\sigma T_a^3/f(v)]}{\Delta + [\gamma + 4\varepsilon\sigma T_a^3/f(v)]} \qquad (5\text{-}18)$$

where T_a is in kelvins and $f(v)$ is the wind function in the aerodynamic equation

$$E_a = (e_s - e_2)(0.181 + 0.00236v_4) \qquad (5\text{-}19)$$

where all observations are over land (inches per day, miles per day, and inches of mercury). The constants in Eq. (5-19) become 0.136 and 0.095 in metric units (millimeters per day, meters per second, and millibars). The units of Eqs. (5-18) and (5-19) must be consistent. The similarity between Eqs. (5-16) and (5-18) is apparent: the term appearing parenthetically with γ corrects for the difference between emitted longwave radiation at T_0 and T_a. Methods of obtaining values of Q_{ir} are discussed in Sec. 5-3.

Equation (5-18) assumes that net radiation is entirely dissipated through evaporation and sensible-heat exchange with the atmosphere, and it is therefore necessary to consider the effects of heat exchange within the water body as they are related to energy storage and advection.

It can be shown that the effects of advected energy (inflow and outflow of water) are unimportant except when flows are large relative to the rate of evaporation. Even then, the inflow and outflow temperatures must be appreciably different. The effects of changes in energy storage, on the other hand, may be relatively large, depending on the period of computation and the depth of the lake. Energy storage can be safely neglected in the computation of mean annual evaporation in all cases and in the computation of annual evaporation from relatively shallow lakes.

The effects of advection and/or changes in energy storage are brought about through a change in water-surface temperature and are thus distributed among sensible transfer, evaporation, and emitted radiation. It can be shown [30] that the portion of such energy affecting evaporation is

$$\alpha = \frac{\Delta}{\Delta + \gamma + 4\varepsilon\sigma T_0^3/f(v)} \qquad (5\text{-}20)$$

The ratio α is analogous to the Bowen ratio. Although the wind function in Eq. (5-20) should be from an over-water aerodynamic equation, that appearing in Eq. (5-19) can be used without appreciable error. Figure 5-2 provides a convenient solution to Eq. (5-20). Two sets of curves are shown to reflect the effects of pressure (elevation) on γ.

When it becomes necessary to take advection and energy storage into account, the adjusted lake evaporation E_L can be computed from

$$E_L = E + \alpha(Q_v - Q_\theta) \qquad (5\text{-}21)$$

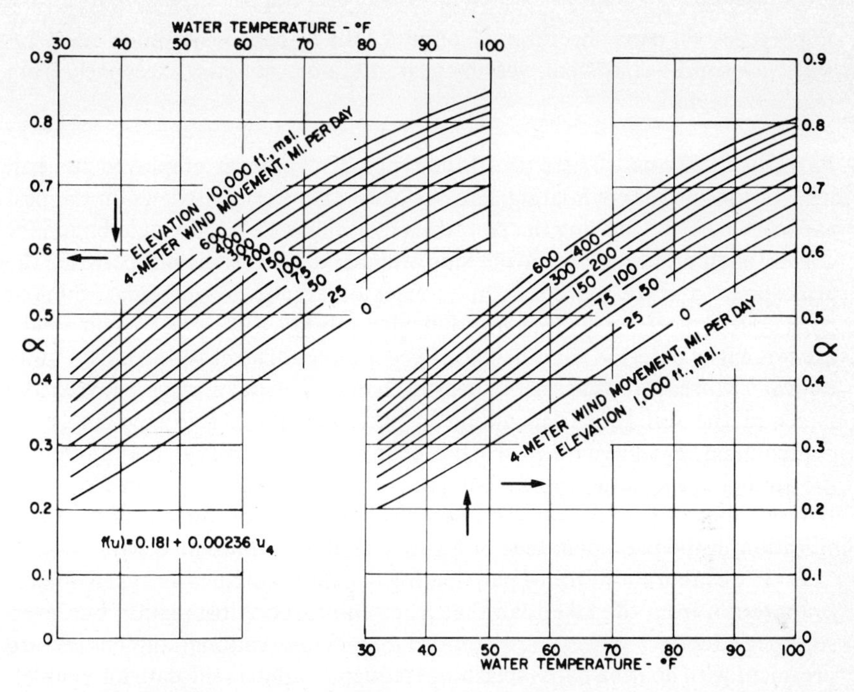

FIGURE 5-2
Variation of α with water temperature, wind movement, and elevation. For metric units, see Fig. C-2. (*From* [26].)

where E is evaporation computed from Eq. (5-18) and the *net advection*, $Q_v - Q_\theta$ [Eq. (5-10)], is expressed in equivalent units of evaporation. Thus, while Eq. (5-18) provides generalized estimates of evaporation from a very shallow lake, it may be necessary to use Eq. (5-21) when comparing the results with evaporation from specific lakes. Approximate adjustments for net advection may also be feasible when considering reservoir design. Annual net advection is usually small and may generally be neglected. Average net advection into Lake Mead is equivalent to about 12 in. (30 cm) of evaporation per year, however, as a result of low-level outflow which is colder than the inflow.

5-6 Estimation of Reservoir Evaporation from Pan Evaporation and Related Meteorological Data

The pan is undoubtedly the most widely used evaporation instrument today, and its application in hydrologic design and operation is of long standing. Although criticism of the pan may be justified on theoretical grounds, for some types of pans the ratio of annual lake-to-pan evaporation (*pan coefficient*) is quite consistent, year by year, and does not vary excessively from region to region.

Pan observations There are three types of exposures employed for pan installations—sunken, floating, and surface—and divergent views on the best exposure persist. Burying the pan tends to eliminate objectionable boundary effects, such as radiation on the side walls and heat exchange between the atmosphere and the pan itself, but creates observational problems. Sunken pans collect more trash; they are difficult to install, clean, and repair; leaks are not easily detected; and height of vegetation adjacent to the pan is quite critical. Moreover, appreciable heat exchange does take place between the pan and the soil [31], depending on such factors as soil type, moisture content, and vegetative cover. Heat exchange with the soil can readily change the annual evaporation from a 2-m pan 10 percent and a 5-m pan 7 percent. It is therefore obvious that such heat exchange will produce large climatic variation in the pan coefficient of a small, sunken, uninsulated pan.

The evaporation from a pan floating in a lake more nearly approximates evaporation from the lake than that from an on-shore installation but, even so, the boundary effects are appreciable. Observational difficulties are prevalent with floating pans (splashing frequently renders the data unreliable), and installation and operational expense is excessive. Since relatively few such installations are now in existence, floating pans are not considered in the subsequent discussion.

Pans exposed above ground experience greater evaporation than sunken pans, primarily because of the radiant energy intercepted by the side walls, and heat exchange through the pan produces unrealistic effects which must be taken into account. Both deficiencies can be minimized by insulating the pan. The principal advantages of surface exposure are economy and ease of installation, operation, and maintenance.

Of the various sunken pans used, only three have gained prominence in the United States: the Young screened pan, the Colorado pan, and the Bureau of Plant Industry (BPI) pan. The Young pan [32] is 2 ft (61 cm) in diameter and 3 ft (91.5 cm) deep and is covered with $\frac{1}{4}$-in.-mesh (6-mm) hardware cloth. The screen modifies the pan coefficient to near unity, on an average, but the small size of the pan leads to an unstable coefficient and the overall effect of screening may be adverse. The Colorado pan is 3 ft (91.5 cm) square and 18 in. (46 cm) deep. The BPI pan, 6 ft (183 cm) in diameter by 2 ft (61 cm) deep, provides by far the best index to lake evaporation because of its size.

The standard Weather Bureau Class A pan is the most widely used evaporation pan in the United States; in 1974 records were published for

FIGURE 5-3
Class A evaporation station: 1 = instrument shelter, 2 = evaporation pan, 3 = anemometer, 4 = standard 8-in. precipitation gage, 5 = weighing-type recording precipitation gage. (*U.S. National Weather Service.*)

about 450 stations. It is of unpainted galvanized iron 4 ft (122 cm) in diameter by 10 in. (25.4 cm) deep and is exposed on a wood frame to let air circulate beneath the pan (Fig. 5-3). It is filled to a depth of 8 in. (20 cm), and instructions [33] require that it be refilled when the depth has fallen to 7 in. (18 cm). Water-surface level is measured daily with a hook gage in a stilling well, and evaporation is computed as the difference between observed levels, adjusted for any precipitation measured in a standard rain gage. Alternatively, water is added each day to bring the level up to a fixed point in the stilling well. This method assures proper water level at all times.

Many other types of pans are in use in different parts of the world, and the need for international standardization has long been recognized by the World Meteorological Organization. Pending standardization, many intercomparisons of the various types of pans have been made [34] which show that pan-to-pan ratios display appreciable geographic (climatic) variation. The two most widely used pans are the Class A and the GGI-3000 [35]. The latter pan is circular, 3000 cm^2 in area (61.8 cm or 24.3 in. in diameter). The depth is 60 cm (23.6 in.) at the wall and somewhat greater at the center. It is fabricated of galvanized sheet iron. The pan and a precipitation gage of similar dimensions are both sunk into the ground.

The value of a pan as an index to lake evaporation must depend on energy-exchange considerations rather than aerodynamic similarities. As a "calorimeter," there is obviously much to be gained by insulating the pan, and if this is done, it seems that the disadvantages of a sunken exposure should be avoided. Experiments with an insulated pan of approximately the same dimensions as the GGI-3000 are extremely promising [14]. The geographic variation of the pan coefficient for such a pan is appreciably less than for any small pan in general use (see Table 5-2 on page 169).

In some localities, it is necessary to screen evaporation pans to eliminate loss of water due to birds and animals drinking from the pan. It should be realized that screening changes the pan coefficient: tests in Kenya showed that Class A pan evaporation is reduced as much as 14 percent by screening with wire mesh [34].

Pan evaporation and meteorological factors Many attempts have been made to derive reliable relations between pan evaporation and meteorological factors [36]. Obvious purposes to be served by such relations are as follows:

1 To increase our knowledge of evaporation.
2 To estimate missing pan records (pans are not operated during winter in areas where ice cover would occur much of the time, and records for days with snow or heavy rain are frequently fallacious).
3 To estimate data for stations at which pan observations are not made.
4 To test the reliability and representativeness of observed data.
5 To aid the study of lake-pan relations.

Some of the relations developed involve the substitution of air temperature for water temperature with a resultant seasonal and geographic bias.

The Penman approach [Eq. (5-16)] was applied to the composite record from a number of stations over the United States to derive (Appendix A) the coaxial relation [28] shown in Fig. 5-4. To compute pan evaporation, E_a is first determined from the upper left-hand chart by entering with values of air temperature, dewpoint, and wind. Equation (5-16) is solved by entering the other three curve families of Fig. 5-4 with air temperature, solar radiation, E_a, and air temperature. Although the correlation was based on daily data, experience has shown that only minor errors result when monthly evaporation, i.e., mean daily value for the month, is computed from monthly averages of the daily values of the parameters.

There are only a limited number of locations where all the data required for application of Fig. 5-4 are available. In 1974 solar radiation was observed at only about 85 stations in the United States. There are reasonably reliable means of estimating this factor [37, 38], however, and the other required

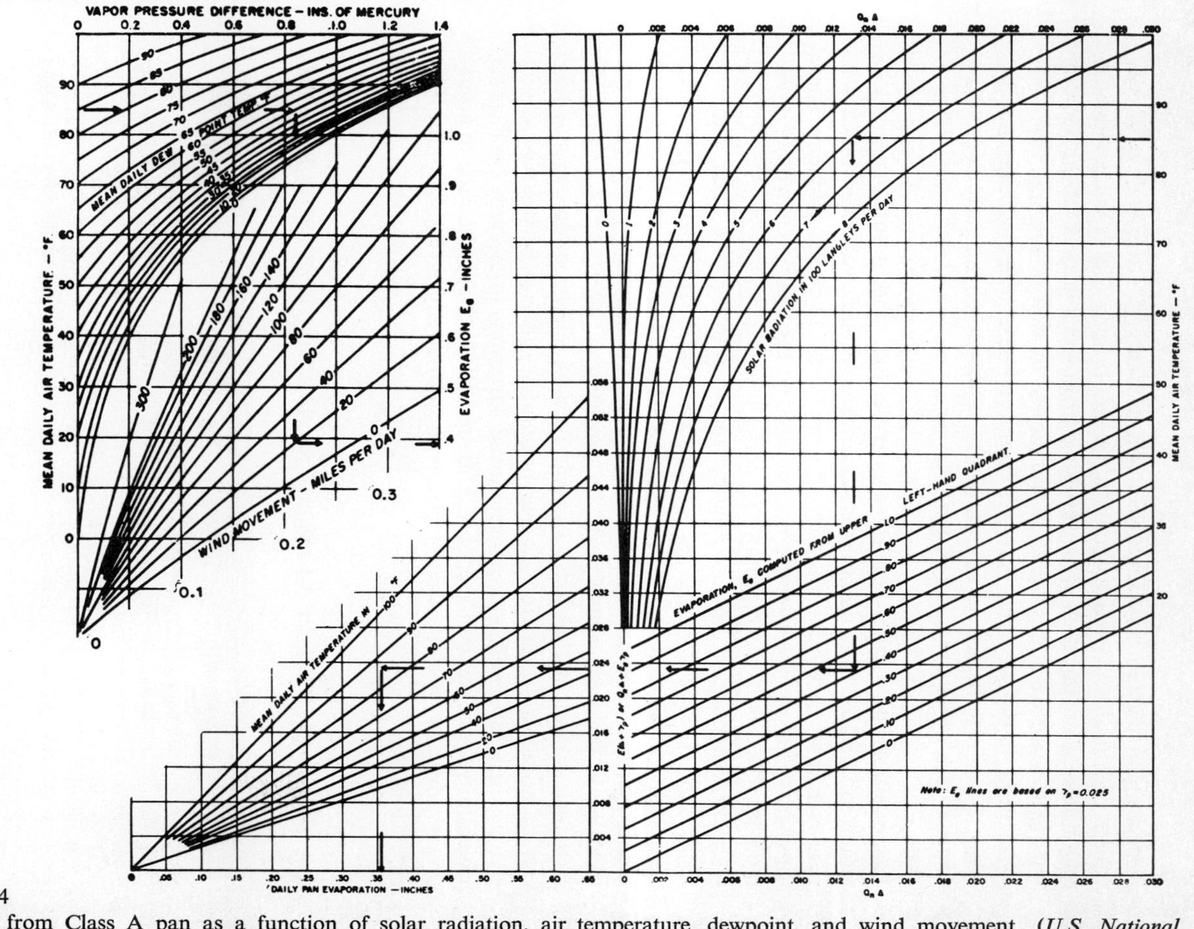

FIGURE 5-4
Evaporation from Class A pan as a function of solar radiation, air temperature, dewpoint, and wind movement. (*U.S. National Weather Service.*)

elements can usually be estimated accurately enough to provide adequate values of monthly or normal monthly evaporation. Table 5-1 shows the magnitude of evaporation error resulting from error in each of the various factors, for selected meteorological situations. In view of the high degree of correlation found to exist between computed and observed evaporation, Table 5-1 is also indicative of the relative influence of the various elements under typical meteorological situations.

Pan coefficients Water-budget, energy-budget, and aerodynamic techniques can be used to estimate evaporation from existing reservoirs and lakes. However, these methods are not directly applicable to design problems, since water-temperature observations are required in their use. The combination methods (Sec. 5-5) are beginning to come into use, but most estimates of reservoir evaporation, both for design and operation, have been made by applying a coefficient (Table 5-2) to observed or derived pan evaporation. Although too few determinations have been made to appraise the approach accurately, assuming an annual class A pan coefficient of 0.70 for the lakes included in the table would result in a maximum difference of 12 percent. Part-year coefficients are more variable because energy storage in the lake can be appreciably different at the beginning and end of the period and changes in heat storage cause pronounced variations in monthly coefficients which must be taken into account. Equation (5-21) should be used to adjust for net advection (in this case, E is pan evaporation reduced by an appropriate coefficient).

Table 5-1 VARIATION OF CLASS A PAN EVAPORATION WITH RELATED FACTORS

Case no.	T_a, °F	% error/deg change in T_a	T_d, °F	% error/deg change in T_d	Q_s, Ly/d	% error/% change in Q_s	v, mi/d	% error/% change in v	E, in./d
1	91	1.8	41	0.4	700	0.7	50	0.2	0.51
2	91	2.3	63	0.8	700	0.8	50	0.1	0.46
3	84	3.8	75	2.7	600	0.9	50	0.1	0.28
4	66	6.0	55	4.0	300	0.6	50	0.2	0.12
5	45	6.3	28	2.9	250	0.3	50	0.3	0.09
6	91	1.8	41	0.3	700	0.6	100	0.3	0.60
7	91	2.3	63	1.0	700	0.7	100	0.2	0.52
8	84	4.4	75	3.1	600	0.8	100	0.2	0.31
9	66	6.2	55	4.6	300	0.5	100	0.4	0.15
10	45	6.2	28	3.1	250	0.3	100	0.5	0.11

Effects of advected energy on pan evaporation Observations demonstrate that the sensible-heat transfer through the pan can be appreciable and may flow in either direction, depending upon water and air temperatures. Since heat transfer through the bottom of the reservoir is essentially zero, pan data require adjustment.

Table 5-2 SUMMARY OF PAN COEFFICIENTS
See text for description of pans

Location	Years of record	Period	Surface exposure		Sunken pans			
			Class A	X-3[a]	BPI	Young	Colo.	GGI-3000
Davis Calif.[b]	1966–69	Ann.	0.72	0.71	...	...	...	0.94
Denver, Colo.[c]	1915–16	Ann.	0.67					
	1916	June–Oct.	...	...	0.94			
Felt Lake, Calif.	1955	Ann.	0.77	...	0.91	0.99	0.85	
Ft. Collins, Colo.	1926–28	Apr.–Nov.	0.70	...	...	...	0.79	
Fullerton, Calif.[c]	1936–39	Ann.	0.77	...	0.94	0.98	0.89	
Lake Colorado City, Tex.	1954–55	Ann.	0.72					
Lake Elsinore, Calif.	1939–41	Ann.	0.77	...	...	0.98		
Lake Hefner, Okla.	1950–51	Ann.	0.69	...	0.91	0.91	0.83	
Lake Mead, Ariz.–Nev.[b]	1966–69	Ann.	0.66[d]	0.73[d]	...	...	...	0.71[d]
Lake Okeechobee, Fla.	1940–46	Ann.	0.81	...	...	...	0.98	
Red Bluff Res., Tex.	1939–47	Ann.	0.68					
Salton Sea, Calif.	1967–69	Ann.	0.64	0.64				
Silver Hill, Md.[e]	1955–60	Apr.–Nov.	0.74	...	1.05	...	0.97	
Sterling, Va.[b]	1965–68	Apr.–Nov.	0.69[f]	0.71[f]	...	...	...	1.11[f]
India (Poona)[b]	1965–68	Ann.	0.69	...	...	...	...	0.78
Israel (Lod Airport)[c]	1954–60	Ann.	0.74					
Sudan (Khartoum)[c]	1960–61	Ann.	0.65					
U.K. (London)	1956–62	Ann.	0.70					
U.S.S.R. (Dubovka)[b]	1957–59	May–Oct.	0.64	...	...	...	...	0.91
	1962–67	May–Oct.	0.64	...	...	...	...	0.84
U.S.S.R. (Valdai)[b]	1949–53	May–Sept.	0.82	...	...	...	...	0.93
	1958–63	May–Sept.	0.67	...	...	...	...	0.98

SOURCE: References 13, pp. 127–148; 14; 16; 17; 28; 31; 32; and 34; also unpublished material.
[a] Insulated; dimensions approximately the same as for the GGI-3000 pan.
[b] Assuming that evaporation from a tank 5 m (16.5 ft) in diameter is equivalent to lake evaporation.
[c] Assuming that evaporation from a tank 12 ft (3.65 m) in diameter is equivalent to lake evaporation.
[d] Correction for heat flow from soil into 5-m tank would reduce coefficient at least 5 percent.
[e] Assuming that evaporation (adjusted for heat flow to the soil) from a tank 15 ft (4.57 m) in diameter is equivalent to lake evaporation.
[f] Correction for heat flow from 5-m tank to the soil would increase coefficient a few percent.

Figure 5-5, derived in much the same manner as Fig. 5-2, shows the relative effect α_p of net advection on Class A pan evaporation [28]. Advection by means of water added to the pan is normally unimportant, but advection of sensible heat through the pan walls is sufficient to produce moderate variation in the pan coefficient under varying climatic regimes. From the Bowen ratio concept and the empirical relation of Fig. 5-4, a function can be derived for estimating heat transfer through the Class A pan from observations of air and water temperature, wind movement, and atmospheric pressure. This is the underlying relation of the chart in Fig. 5-6, which further assumes a 0.7 pan coefficient when air and water temperatures are equal. Lake evaporation computed from the chart should be adjusted for any appreciable net advection to the lake.

5-7 Summary and Appraisal of Techniques for Estimating Reservoir Evaporation

There are relatively few reservoirs for which reliable evaporation estimates can be derived from a water budget on a continuous basis, but values for selected periods may often serve to calibrate or check other techniques. When conditions are such that satisfactory results cannot be obtained by applying a water budget, evaporation from an existing reservoir can be determined by either the empirical aerodynamic or energy-budget approach. Both instrumentation and maintenance of continuing observations are expensive for these two approaches, and their widespread use may not be economically feasible for some time. However, the purposes to be served may justify their application for a short period to calibrate a less costly method.

The operation of a pan station (near the reservoir, but not close enough to be materially affected by it) is relatively inexpensive and should provide reasonably good estimates of annual reservoir evaporation (Fig. 5-6). Some reliability would be gained if net reservoir advection were estimated, but this item is rarely of much importance. If monthly or seasonal distribution of annual evaporation is considered necessary, net advection to the reservoir must be evaluated. Even though evaporation pans are normally inoperative during freezing weather, pan evaporation is small during such periods and can usually be estimated with sufficient accuracy. Reservoir evaporation can be relatively large during early winter, however, because of changes in energy storage.

For reservoir-design studies, Fig. 5-1 or 5-6 [also Eq. (5-18)] may be applied if representative data are available. All relevant data for the area should be analyzed using all techniques for which the data are suitable if the economic aspects of design so justify. There is seldom justifiable reason

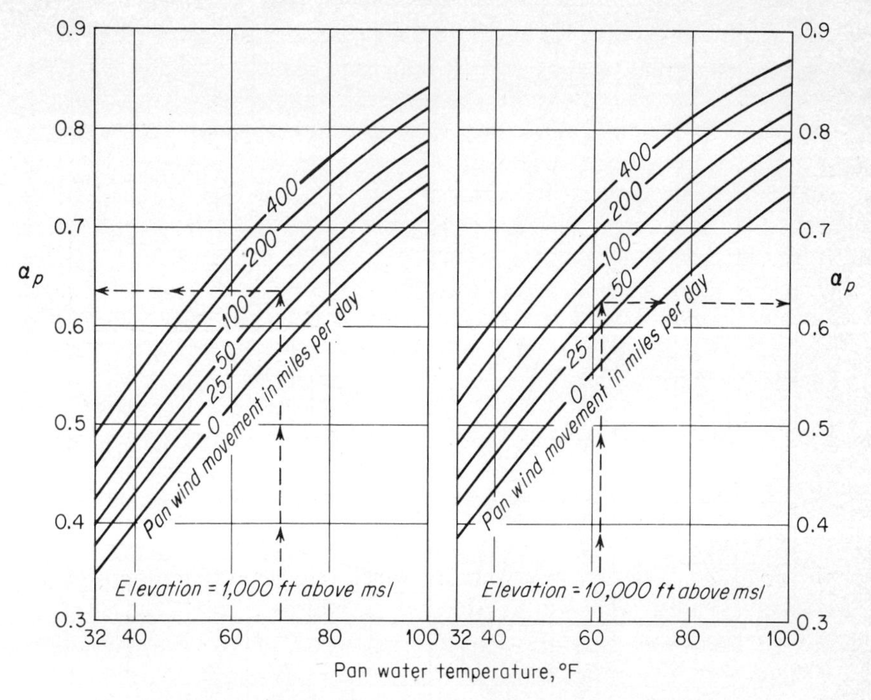

FIGURE 5-5
Portion of advected energy (into a Class A pan) utilized for evaporation. For metric units see Fig. C-3. (*U.S. National Weather Service.*)

for constructing a major reservoir prior to the collection of at least 1 or 2 y of pan and related meteorological data at the site for checking purposes. Figure 5-7, which presents generalized estimates of mean annual evaporation from shallow lakes for the conterminous United States [27], is considered sufficiently reliable for preliminary estimates and for design of projects in which evaporation is not important. Similar maps are available for a number of other countries [39–41], and Fig. 2-2 is a world map of heat used by evaporation.

In some design problems, it is necessary to estimate the monthly distribution of annual evaporation. While several of the techniques discussed will yield monthly values of free-water evaporation, adjustment is required for net advection (or heat-storage changes, at least) [42].

In reservoir design, the engineer is really concerned with the increased loss over the reservoir site resulting from the construction of the dam, i.e., reservoir evaporation less evapotranspiration under natural conditions. In humid areas construction of a dam causes only a nominal increase in water loss.

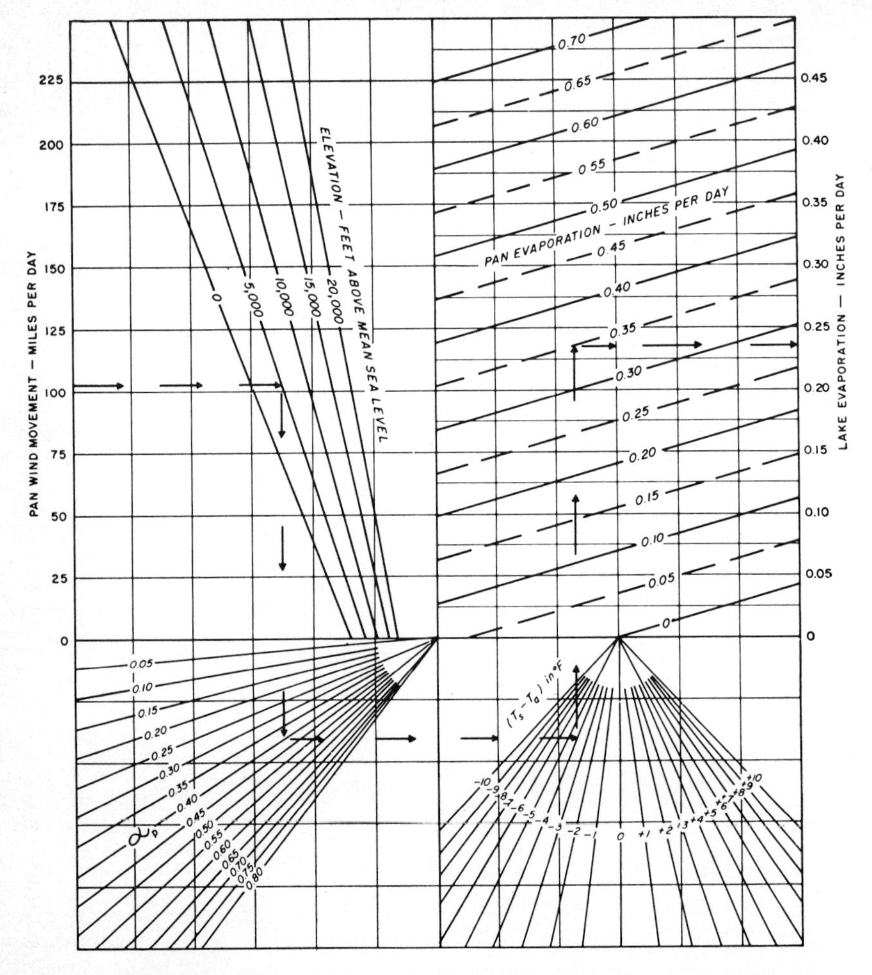

FIGURE 5-6
Shallow-lake evaporation as a function of Class A pan evaporation and heat transfer through the pan. For metric units see Fig. C-4. (*U.S. National Weather Service.*)

5-8 Increased Water Supplies through Reduced Evaporation

Any steps which can be taken to reduce reservoir evaporation per unit of storage provide a corresponding increase in usable water supply. Selecting the site and design yielding a minimum of reservoir area per unit of storage is advantageous. Designing the outlet works so that the warmer surface water can be released to meet demands will reduce evaporation from a reservoir.

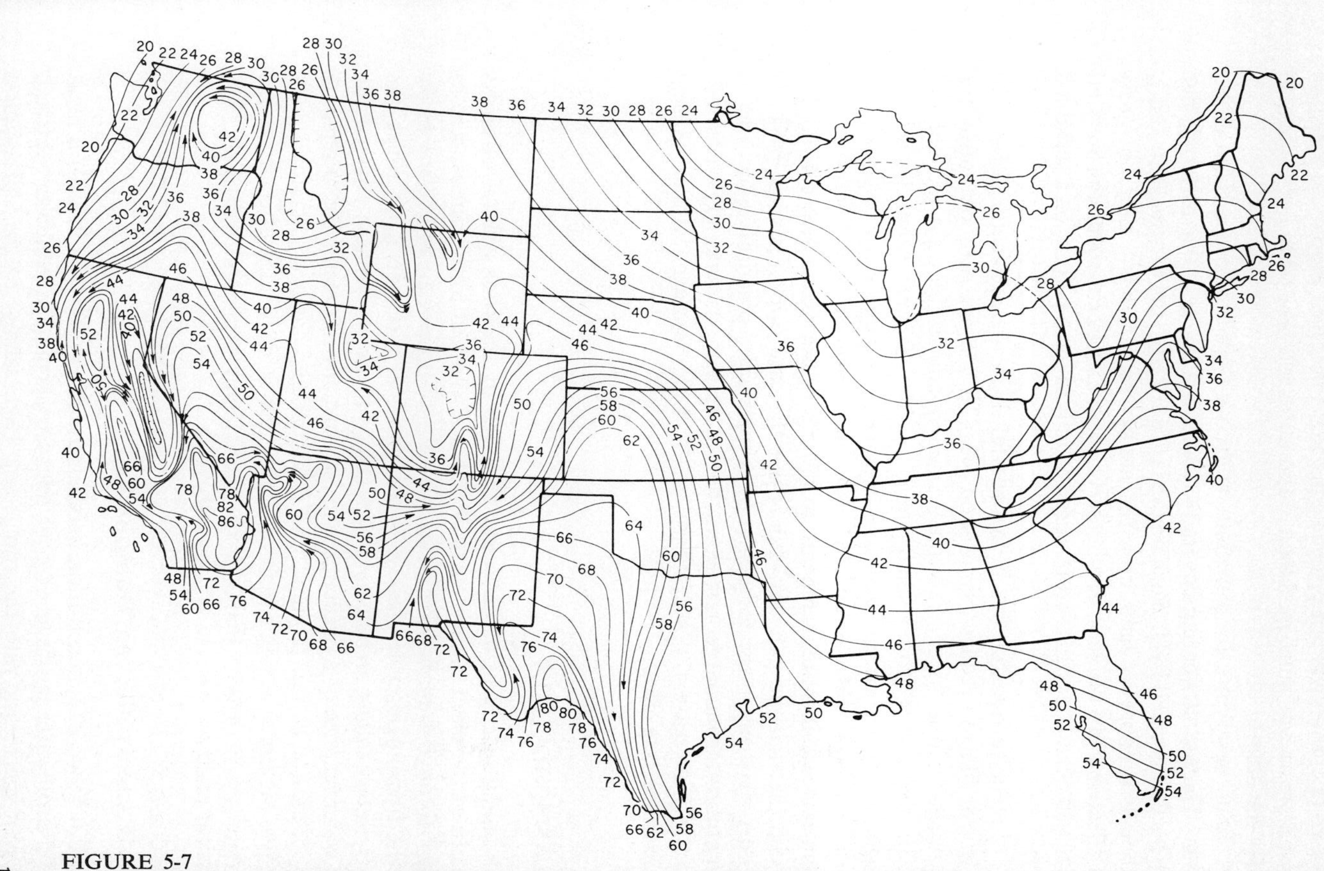

FIGURE 5-7
Average annual evaporation (inches) from shallow lakes. (*U.S. National Weather Service.*)

Such an operation will not provide equivalent increased supplies at a distant point downstream, since increased evaporation will take place in the channel below the dam.

Small reservoirs are sometimes entirely covered to reduce evaporation losses. The use of floating covers [43] and floating granular materials has also been advocated [44]. Such methods are effective but expensive to apply. Although windbreaks have been repeatedly recommended, they obviously are effective only on very small reservoirs: a 25 percent reduction in wind speed will normally reduce evaporation by only about 5 percent on a sustained basis.

Extensive research has been undertaken in the application of monomolecular films to reduce evaporation [45–47]. Despite early optimism, the approach is little used. Extremely small quantities of cetyl alcohol will reduce evaporation from a small pan as much as 40 percent, but it is seldom possible to maintain more than 10 to 20 percent coverage over a reservoir on a continuing basis. Moreover, any reduction in evaporation is accompanied by an increase in water temperature. Much of the excess heat can be dissipated through the pan, but this is not the case for a reservoir [48]. It appears that any hope for achieving appreciable, practical reductions in evaporation from large reservoirs lies in finding a material which will effectively increase the reflectivity of the water surface without causing undesirable side effects.

TRANSPIRATION

Only minute portions of the water absorbed by the root systems of plants remain in the plant tissues; virtually all is discharged to the atmosphere as vapor through *transpiration*. This process constitutes an important phase of the hydrologic cycle since it is the principal mechanism by which the precipitation falling on land areas is returned to the atmosphere. In studying the water balance of a drainage basin, it is usually impracticable to separate evaporation and transpiration, and the practicing engineer therefore treats the two factors as a single item. Nevertheless, a knowledge of each process is required to assure that the techniques employed are consistent with physical reality.

5-9 Factors Affecting Transpiration

The difference in concentration between the sap in the root cells of a plant and the soil water causes an osmotic pressure which moves soil water through the root membrane into the root cells. High salinity of the soil solution

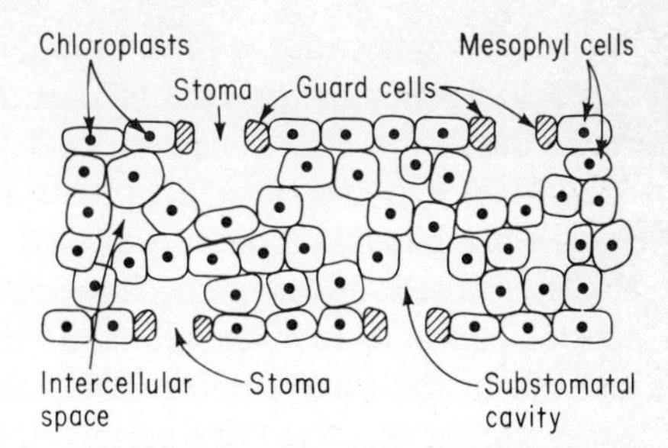

FIGURE 5-8
Internal structure of a leaf.

and/or high moisture tension in the soil may prevent or greatly reduce the osmotic transfer. Once inside the root, the water is transferred [49] through the plant to the *intercellular space* within the leaves (Fig. 5-8). Air enters the leaf through the *stomata*, openings in the leaf surface, and the *chloroplasts* within the leaf use carbon dioxide from the air and a small portion of the available water to manufacture carbohydrates for plant growth (*photosynthesis*). As air enters the leaf, water escapes through the open stomata; this is the process of *transpiration*. The ratio of the water transpired to that used in forming plant matter is very large—up to 800 or more.

The rate of transpiration is largely independent of plant type, provided there is adequate soil water and the surface is entirely covered by vegetation. Since photosynthesis is highly dependent on the radiation received, about 95 percent of daily transpiration occurs during daylight hours [50], compared to 75 to 90 percent for soil evaporation [51]. Plant growth normally ceases when temperatures drop to near 40°F (4°C), and transpiration becomes very small.

Transpiration is limited by the rate at which moisture becomes available to the plants. Although there is little doubt that the rate of soil evaporation, under fixed meteorological conditions, decreases quasi-exponentially with time, divergent views persist with respect to transpiration (see Sec. 5-15). It is believed that the controversy and apparent discrepancies can be attributed to the varied methods of deriving supporting data and the nondescript terminology used to describe the results. Some investigators believe that transpiration is independent of available moisture until it has receded to the *wilting point* (moisture content at which permanent wilting of plants occurs), while others assume that transpiration is roughly proportional to the moisture remaining in the soil and available to the plants. The *field capacity* is defined as the amount of water held in the soil after excess gravitational water has drained, and the range in soil moisture from field capacity to wilting point (*available water*) is a measure of the maximum quantity of water available for plant use without replenishment (Sec. 6-3).

Available water varies with soil type, ranging from about 0.5 in. per foot of depth (0.5 mm/cm) for sand to over 2 in. per foot of depth (2 mm/cm) for clay loams.

Plant type becomes an important factor in controlling transpiration when available soil moisture is limited. As the upper layers of the soil dry out, shallow-rooted species can no longer obtain water and wilt, while deep-rooted species continue to transpire until the soil moisture at greater depths is reduced to the wilting point. Thus deep-rooted vegetation transpires more water during sustained dry periods than shallow-rooted species. Transpiration per unit area also depends on the density of vegetative cover. With widely spaced plants (low cover density), not all solar radiation reaches the plants, and some of it is absorbed at the soil surface. The relative transpiration is not proportional to cover density, however, for two reasons: (1) an isolated plant receives radiation on the side facing the sun which would fall on an adjacent plant were there solid cover, and (2) a portion of the radiation reaching the ground is subsequently transmitted to the plants (*oasis effect*).

Plant type also influences transpiration during drought conditions, even with specified soil-moisture conditions. *Xerophytes*, desert species, which have fewer stomata per unit area and less surface area exposed to radiation, transpire relatively little water. *Phreatophytes*, on the other hand, have root systems reaching to the water table and transpire at rates largely independent of moisture content in the zone of aeration. All plants can control stomatal opening to some extent, and thus even *mesophytes*, plants of the temperate zones, have some ability to reduce transpiration during periods of drought. Note, however, that the ability to control is only to reduce transpiration. Even aquatic plants, *hydrophytes*, cannot pump water into the atmosphere at rates in excess of those controlled by available radiant and sensible energy. A pond covered with aquatic plants does not lose water at a rate appreciably different from a pond free of vegetation. Any difference in sensible-heat transfer brought about by increased surface roughness tends to be counterbalanced by the increase in albedo.

Rainfall intercepted by vegetation is subsequently evaporated and thereby utilizes some of the energy otherwise available for transpiration. Experiments with grass cover [52] indicate that the transpiration reduction may be equivalent to the interception, while other experiments with small potted pine trees [53] seem to show that the reduction is much less than interception losses.

5-10 Measurement of Transpiration

Since it is not possible to measure transpiration loss from an appreciable area under natural conditions, determinations are restricted to studies of small samples under laboratory conditions. One method involves placing one or more potted plants in a closed container and computing transpiration as the increase in moisture content of the confined space. Most measurements are made with a *phytometer*, a large vessel filled with soil in which one or more plants are rooted. The only escape of moisture is by transpiration (the soil surface is sealed to prevent evaporation), which can be determined by weighing the plant and container at desired intervals of time. By providing aeration and additional water, a phytometer study can be carried through the entire life cycle of a plant. Since it is virtually impossible to simulate natural conditions, the results of phytometer observations are mostly of academic interest to the hydrologist, constituting little more than an index to water use by a crop under field conditions.

Ceramic and Piche atmometers have been widely used in transpiration studies. Such instruments [34] automatically feed water from a reservoir to an exposed, wetted surface. The change in content of the reservoir serves as an index to transpiration. With proper care and exposure, atmometers are useful in experimental work for estimating time and spatial variations in potential transpiration.

EVAPOTRANSPIRATION

In studying the hydrologic balance for a catchment area, one is usually concerned only with the *total evaporation* (or *evapotranspiration*), the evaporation from all water, soil, snow, ice, vegetation, and other surfaces plus transpiration. *Consumptive use* is the total evaporation from an area plus the water used directly in building plant tissue. The distinction between the two terms is largely academic, falling well within the error of measurement, and they are now generally treated as synonymous [54].

On the assumption that any reduction in evapotranspiration brought about by a deficiency of soil moisture is independent of meteorological conditions, the concept of *potential evapotranspiration* introduced by Thornthwaite [55] is widely used. He defined the term as "the water loss which will occur if at no time there is a deficiency of water in the soil for the use of vegetation." It has since been found that evapotranspiration depends on the density of cover and its stage of development. To serve the purpose, potential evapotranspiration must either be independent of the nature and condition of the surface, except with respect to available moisture, or it must be defined in terms of a particular surface. Penman [56] suggested that the original definition be modified to include the stipulation that the surface be fully covered by green vegetation. This modified definition is generally satisfactory, but it becomes meaningless during winter in higher latitudes.

In the interest of clarity and reproducible results, there is good reason to consider potential evapotranspiration to be equivalent to the evaporation from a free-water surface of extended proportions but with negligible heat-storage capacity [57]. Potential evapotranspiration as defined by Thornthwaite approaches free-water evaporation provided there is complete vegetal cover, and the effects of meteorological factors on the two are sufficiently alike to be converted into actual evapotranspiration in the same manner.

There are numerous approaches to estimation of actual and potential evapotranspiration, none of which is generally applicable for all purposes. The type of data required depends on the intended use. In some hydrologic studies, mean basin evapotranspiration is required, while in other cases we are interested in water use of a particular crop cover or the change in water use resulting from changed vegetal cover.

5-11 Water-Budget Determination of Mean Basin Evapotranspiration

Assuming that storage and all items of inflow and outflow except evapotranspiration can be measured, the volume of water (usually expressed in units of depth) required to balance the continuity equation for a basin represents evapotranspiration. Among other things, the reliability of a water-budget computation hinges largely on the time increments considered. As a rule, normal annual evapotranspiration can be reliably computed as the difference between long-time averages of precipitation and streamflow, since the change in storage over a long period of years is inconsequential [58]. Any deficiencies in such computations are usually attributable either to inadequate precipitation or runoff data or to subterranean flow into or out of the basin. Estimates of annual evapotranspiration are subject to appreciable errors if changes in storage are neglected, except where moisture storage in a basin is nearly the same on a given date each year. Generally it is necessary to evaluate soil moisture, groundwater, and surface storage at the beginning of each year.

The water-budget method can also be applied to short time periods [59]. In Fig. 5-9 over 6 in. (150 mm) of rain fell in a 3-d period, resulting in the rise of June 21, and 4.60 in. (117 mm) more fell by June 29. The runoff (June 26–30) produced by the second storm was 2.37 in. (60 mm). If it is assumed that the soil was equally near saturation at the end of rain on June 21 and 29, the evapotranspiration during the period was 2.23 in. (4.60 − 2.37), or 0.28 in./d (7 mm/d). The error in computations for periods as short as a week can be appreciable, and it is advisable to use longer periods when feasible. If the computations are carried through July 16, the estimated

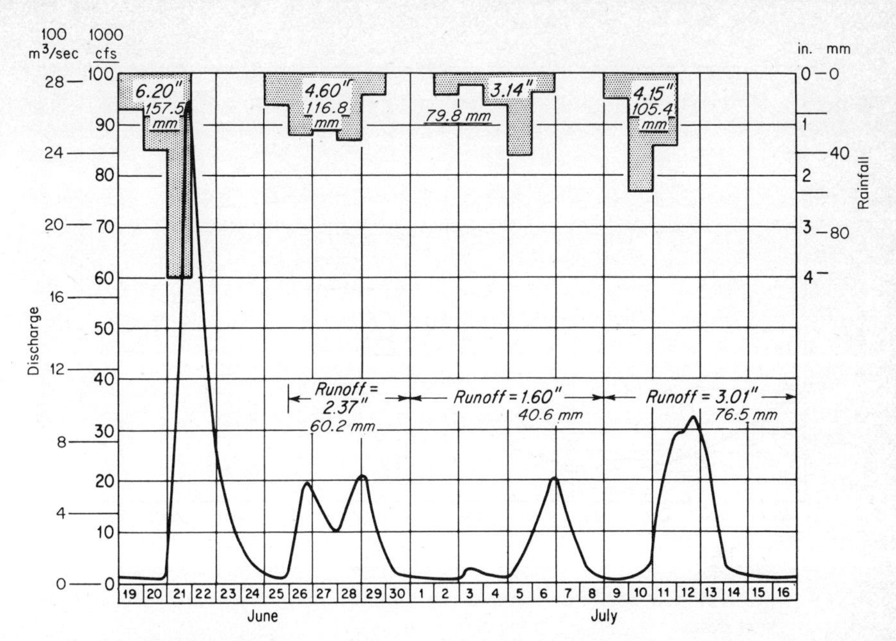

FIGURE 5-9
Derivation of short-period evapotranspiration estimates.

evapotranspiration averages about 0.23 in./d (6 mm/d), undoubtedly a more realistic value. The computations must be based on total runoff (Sec. 7-5). This procedure is best adapted to regions where depth to groundwater is relatively small and precipitation is evenly distributed throughout the year. Although evapotranspiration estimates derived in the manner described must be geared to the fortuitous occurrence of large storms or relatively wet periods, sufficient determinations can be made from many years of record to define the seasonal distribution (Fig. 5-10). If the resulting curve is to represent normal annual evapotranspiration, the computations must be carried on a continuous basis, since omitting dry periods will bias the results. If the curve is to represent potential evapotranspiration, computations should be made only for those periods during which potential conditions existed.

5-12 Field-Plot Determination of Evapotranspiration

Application of a water budget to field plots produces satisfactory results only under ideal conditions, which are rarely attained. Precise measurement of percolation is not possible, and consequent errors tend to be accumulative. If the groundwater table lies at great depth, accretions may be inconsequential

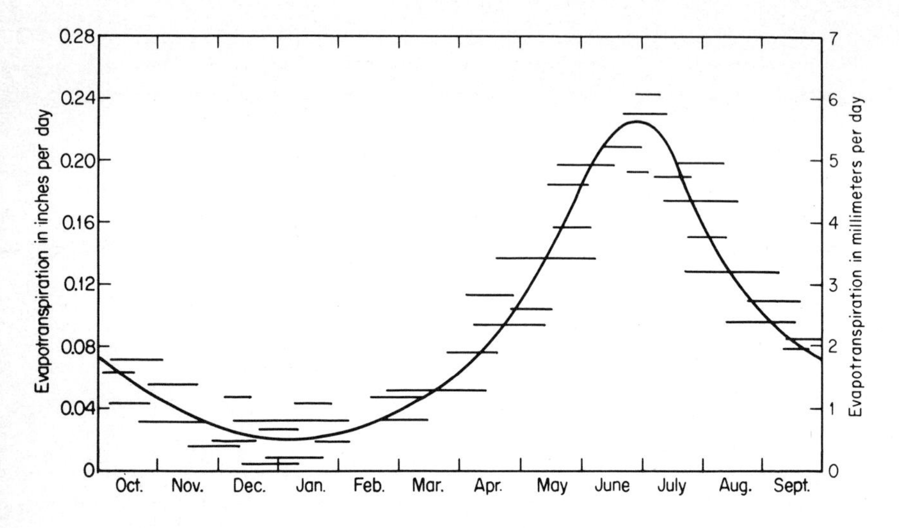

FIGURE 5-10
Mean evapotranspiration curve derived through analysis of rainfall and streamflow data.

but not necessarily so. If accretions are inconsequential, soil-moisture measurements become the principal source of error—random in nature but large enough to preclude computation of short-period evapotranspiration. Reasonable seasonal estimates may be feasible, however [60, 61].

The energy budget can be applied to the determination of evapotranspiration from a field plot much as for a lake (Sec. 5-3). Instead of being concerned with heat storage in a mass of water, however, we must compute that stored in the soil profile. Specific heat of soil varies from about 0.2 to 0.8 cal/cm^3 depending upon moisture content and soil type, and so one must know the specific heat as well as temperature throughout the profile. Alternatively, the heat flux can be measured at some convenient depth (say 10 cm or 4 in.) in lieu of determining the change in heat storage below that depth [62]. In applying Bowen's ratio [Eq. (5-4)] to a lake, temperature and vapor pressure of the surface are used. Measuring the vapor pressure of a vegetated surface presents quite another problem, and it becomes necessary to measure the temperature and vapor-pressure gradients between two levels above the surface. Advected energy, in terms of water-budget items, is usually small and customarily neglected.

The Thornthwaite-Holzman turbulent-transport equation has been used to estimate evapotranspiration [63, 64], although there is some question whether it has been adequately tested. Instrumental requirements are not easily satisfied under field conditions, since computed evapotranspiration is proportional to the differences in measured wind and vapor pressure at two levels near the surface under investigation.

The energy-budget and turbulent-transport techniques are also suited to the determination of potential evapotranspiration, the only added requirement being that the area under observation have sufficient water at all times. Since extreme care is required in the application of these techniques, their use has been largely limited to experimental purposes.

5-13 Lysimeter Determination of Evapotranspiration

Many observations of evapotranspiration are made in soil containers [65–70], variously known as *tanks*, *evapotranspirometers*, and *lysimeters*. The first two terms customarily refer to containers with sealed bottoms, while there has been an attempt to restrict the word lysimeter to containers with pervious bottoms or with a mechanism for maintaining negative pressure at the bottom. Evapotranspiration is computed by maintaining a water budget for the container.

Like evaporation pans, small evapotranspirometers provide only an index to potential evapotranspiration. Accordingly, instrumental and operational standardization is of extreme importance. In summarizing the results of worldwide observations, Mather [65] states:

> The evapotranspirometer, when properly operated, i.e., when watered sufficiently so that there is no moisture deficiency and no appreciable moisture surplus in the soil of the tank, and when exposed homogeneously within a protective buffer area of the proper size to eliminate the effect of moisture advection, is an instrument which should give reasonably reliable values of potential evapotranspiration. Great care must be taken in the operation of the instrument, and standardized soil, vegetation, cultivation, and watering practices must be maintained on the tanks in order to insure comparable results from one installation to another.

Reliable tank observations of actual evapotranspiration (when appreciably less than potential) are seldom attained, since it is virtually impossible to maintain comparable soil moisture and vegetal cover in and adjacent to the tank under such conditions. Experimental results indicate [71] that reliable measurements of evapotranspiration can be made with large lysimeters 5 m (15 ft) or more in diameter if provision is made for applying a suction force at the base which is comparable to that in the natural soil profile. It is necessary that root development not be inhibited by limited dimensions of the lysimeter, and cover characteristics (density, height, and vigor) must be the same over and adjacent to the lysimeter.

5-14 Estimating Potential Evapotranspiration from Meteorological Data

Several empirical techniques have been developed for estimating potential evapotranspiration from readily available climatological data and latitude (duration of possible sunshine). Thornthwaite [72] has derived a somewhat involved procedure using only temperature and duration of possible sunshine. Blaney's approach [54] involves the same two factors but was designed primarily to transpose observed consumptive-use data for irrigated areas to other localities on the basis of derived coefficients. Using average yearly data, Lowry and Johnson [73] found high correlations between consumptive use and accumulated degree-days during the growing season. Procedures which rely on temperature as the sole index to heat supply at a particular latitude and which neglect cloudiness, humidity, wind, and other factors are subject to rather large errors under adverse circumstances.

Potential evapotranspiration and evaporation from a thin free-water surface are affected by the same meteorological factors: radiation, humidity, wind, and temperature. Even though the differences in surface roughness, albedo, and possibly other factors are involved, free-water evaporation should be a better index to potential evapotranspiration than air temperature is. Thus it would seem that Eq. (5-16) or (5-18) constitutes the best approach to computing potential evapotranspiration, even though there remains a question concerning the need for a reduction factor.

Conflicting experimental results for reduction factors appear in the literature, and the definition of potential evapotranspiration becomes involved (see Sec. 5-13). Using observations from one of the largest, most reliable lysimeters in the United States, Pruitt and Lourence [74] found the average annual potential evapotranspiration from fescue grass to be 68 in. (173 cm). The evaporation from a 20-m^2 sunken evaporation pan for the same 3-y period was actually 2 percent less. These and other results [75, 76] lead to the conclusion that the annual reduction factor is much nearer unity than the 0.75 found by Penman [11], and pending further research an assumed value of unity is perhaps satisfactory when considering catchment areas of varied vegetal cover. Any of the techniques for estimating free-water evaporation may therefore be applied, including the pan-coefficient approach.

5-15 Estimating Actual Evapotranspiration from Potential

The effect of moisture deficiency on the relation between actual and potential evapotranspiration has long been a subject for debate [77–83]. Some investigators contend that evapotranspiration from a homogeneous plot continues at an undiminished rate until moisture content throughout the root zone is reduced to near the wilting point; others cite experimental results to show that the rate (relative to potential) is approximately proportional to the remaining available water; a third point of view is that the rate is a complex function of available water (only) but limited to the potential rate. Regardless of the functional relationship for a homogeneous plot, the rate of depletion from an initially saturated heterogeneous catchment does soon decrease with time (for constant potential evapotranspiration) because of variations in root-zone capacity and other relevant factors.

The assumption that the ratio of actual to potential evapotranspiration is proportional to the remaining available water would perhaps be satisfactory for basin accounting if each storm saturated the soil. Unfortunately, this simple function cannot adequately provide for the increased evapotranspiration immediately following a moderate storm on relatively dry soil. This difficulty can be overcome by arbitrary separation of moisture storage into two categories [84]. In this approach it is visualized that *upper-zone* moisture is always depleted at the potential rate and that any deficiency in this zone must be satisfied before rainfall begins to recharge the *lower zone*. Depletion from the lower zone occurs only when there is no remaining moisture in the upper zone, in which case the rate of evapotranspiration is assumed to be proportional to the available moisture in the lower zone. By applying this simple model to observations of precipitation and total runoff (Sec. 7-5) daily values of evapotranspiration can be derived by accounting procedures. It should be pointed out that the more complex models used for simulation of streamflow also provide derived values of evapotranspiration (Chap. 10).

5-16 Controlling Evapotranspiration

Following reported success in reducing evaporation from water surfaces by the monomolecular film technique (Sec. 5-8), experiments were undertaken to reduce transpiration from plants by mixing fatty alcohols into the soil [85]. Some positive results have been reported, while other experiments have indicated no significant effects [86] or even increased transpiration. Detailed analysis of many independent experiments leads to the conclusions [87] that concentrations of fatty alcohols sufficient to reduce transpiration also reduce plant growth and that such materials are not suitable as antitranspirants.

A major and continuing research effort has been under way since the early part of this century to determine and predict the hydrologic effects of land-use changes. There is little doubt that land-use changes can have an appreciable effect on annual evapotranspiration and on its seasonal distribution as well [88, 89]. Differences in albedo, aerodynamic roughness, and plant behavior have some effect, but primary factors are those pertaining to the availability of water supply and extent to which the area is covered by freely growing vegetation.

The availability of water supply is largely determined by the extent of the root zone and the climatic regime. If extended periods without rainfall during the growing season are characteristic of an area, then a deep-rooted forest cover will be transpiring freely much of the time when the supply to shallow-rooted plants has been exhausted. In areas where climatic conditions are such that shallow-rooted plants have adequate supply most of the time, evapotranspiration is less affected by the extent of the root zone. Changes in land use reflecting differences in duration of the growing season also have an effect.

Attempting to reduce evapotranspiration through changing land use should be undertaken only after thorough study of possible related side effects. Clear-cut logging will decrease evapotranspiration and increase streamflow but may result in unacceptable rates of erosion [90] and increased peak flows [91].

REFERENCES

1. B. A. Hutchinson, A Comparison of Evaporation from Snow and Soil Surfaces, *Bull. Int. Assoc. Sci. Hydrol.*, vol. 11, pp. 34–42, March 1966.
2. R. D. Tabler, Design of a Watershed Snow Fence System and First-Year Snow Accumulation, *Proc. West. Snow Conf., 39th Annu. Meet.*, pp. 50–55, 1971.
3. G. E. Harbeck, The Effect of Salinity on Evaporation, *U.S. Geol. Surv. Prof. Pap.* 272-A, 1955.
4. L. J. Turk, Evaporation of Brine: A Field Study on the Bonneville Salt Flats, Utah, *Water Resour. Res.*, vol. 6, pp. 1209–1215, 1970.
5. W. B. Langbein, C. H. Hains, and R. C. Culler, Hydrology of Stock-Water Reservoirs in Arizona, *U.S. Geol. Surv. Circ.* 110, 1951.
6. G. E. Harbeck and F. W. Kennon, The Water-Budget Control, in Water-Loss Investigations: Lake Hefner Studies, *Tech. Rep.*, *U.S. Geol. Surv. Prof. Pap.* 269, p. 34, 1954.
7. G. E. Harbeck and others, Utility of Selected Western Lakes and Reservoirs for Water-Loss Studies, *U.S. Geol. Surv. Circ.* 103, 1951.
8. E. R. Anderson, L. J. Anderson, and J. J. Marciano, A Review of Evaporation Theory and Development of Instrumentation, *U.S. Navy Electron. Lab. Rep.* 159, pp. 41–42, 1950.
9. I. S. Bowen, The Ratio of Heat Losses by Conduction and by Evaporation from Any Water Surface, *Phys. Rev.*, vol. 27, pp. 779–787, 1926.
10. See [8], pp. 43–48.
11. H. L. Penman, Natural Evaporation from Open Water, Bare Soil, and Grass, *Proc. R. Soc. (Lond.)*, ser. A, vol. 193, pp. 120–145, April 1948.
12. E. A. Anderson and D. R. Baker, Estimating Incident Terrestrial Radiation

under All Atmospheric Conditions, *Water Resour. Res.*, vol. 3, pp. 975–988, 1967.

13. G. E. Harbeck, Cummings Radiation Integrator, in Water-Loss Investigations: Lake Hefner Studies, Technical Report, *U.S. Geol. Surv. Prof. Pap.* 269, pp. 120–126, 1954.
14. E. L. Peck and R. Farnsworth, A Dual-Purpose Evaporimeter, U.S. National Weather Service (in process).
15. See [8], pp. 3–37.
16. G. E. Harbeck, Water-Loss Investigations: Lake Mead Studies, *U.S. Geol. Surv. Prof. Pap.* 298, pp. 29–37, 1958.
17. R. K. Linsley, M. A. Kohler, and J. L. H. Paulhus, " Applied Hydrology," pp. 165–169, McGraw-Hill, New York, 1949.
18. J. Dalton, Experimental Essays on the Constitution of Mixed Gases; on the Force of Steam or Vapor from Waters and Other Liquids, Both in a Torricellian Vacuum and in Air; on Evaporation; and on the Expansion of Gases by Heat, *Mem. Proc. Manch. Lit. Phil. Soc.*, vol. 5, pp. 535–602, 1802.
19. E. K. Webb, Evaporation from Lake Eucumbene, *Div. Meteorol. Phys. Tech. Pap.* 10, C.S.I.R.O., Australia, 1960.
20. See [8], p. 10.
21. H. Lettau and F. Dorffel, Der Wasserdampfübergang von einer nassen Platte an strömende Luft, *Ann. Hydrogr. Marit. Meteorol.*, vol. 64, pp. 342, 504, 1936.
22. G. E. Harbeck, A Practical Field Technique for Measuring Reservoir Evaporation Utilizing Mass-Transfer Theory, *U.S. Geol. Surv. Prof. Pap.* 272-E, pp. 101–105, 1962.
23. See [11], pp. 125–126.
24. J. Ferguson, The Rate of Natural Evaporation from Small Ponds, *Aust. J. Sci. Res.*, vol. 5. pp. 315–330, 1952.
25. R. O. Slatyer and I. C. McIlroy, "Practical Microclimatology," UNESCO, Paris, 1961.
26. M. A. Kohler and L. H. Parmele, Generalized Estimates of Free-Water Evaporation, *Water Resour. Res.*, vol. 3, pp. 997–1005, 1967.
27. M. A. Kohler, T. J. Nordenson, and D. R. Baker, Evaporation Maps for the United States, *U.S. Weather Bur. Tech. Pap.* 37 1959.
28. M. A. Kohler, T. J. Nordenson, and W. E. Fox, Evaporation from Pans and Lakes, *U.S. Weather Bur. Res. Pap.* 38, 1955.
29. See [26], p. 999.
30. See [26], p. 1001.
31. T. J. Nordenson and D. R. Baker, Comparative Evaluation of Evaporation Instruments, *J. Geophys. Res.*, vol. 67, pp. 671–679, 1962.
32. A. A. Young, Some Recent Evaporation Investigations, *Trans. Am. Geophys. Union*, vol. 28, pp 279–284, April 1947.
33. U.S. National Weather Service Substation Observations, *Obs. Handb.*, no. 2, pp. 38–55, 1970; rev. December 1972.

34. World Meteorological Organization, Measurement and Estimation of Evaporation and Evapotranspiration, *WMO* no. 201, *Tech. Note* 83, 1966.
35. Instructions for Hydrometeorological Stations and Posts, no. 7, pt. II, Hydrometeorological Publishing House, Leningrad, 1963.
36. See [17], pp. 168-169.
37. R. W. Hamon, L. L. Weiss, and W. T. Wilson, Insolation as an Empirical Function of Daily Sunshine Duration, *Mon. Weather Rev.*, vol. 82, pp. 141–146, June 1954.
38. S. Fritz and T. H. MacDonald, Average Solar Radiation in the United States, *Heat. Vent.*, vol. 46, pp. 61–64, July 1949.
39. H. L. Ferguson, A. D. J. O'Neill, and H. F. Cork, Mean Evaporation over Canada, *Water Resour. Res.*, vol. 6, pp. 1618–1633, 1970.
40. V. Stoenescu, La Répartition de l'évaporation à la surface de l'eau sur la territorie de la Roumanie (Distribution of Water Surface Evaporation in Romania), *Int. Assoc. Sci. Hydrol. Publ.* 78, pp. 326–337, 1968.
41. J. Němec, "Engineering Hydrology," p. 207, McGraw-Hill, London, 1972.
42. L. R. Beard and R. G. Willey, An Approach to Reservoir Temperature Analysis, *Water Resour. Res.*, vol. 6, pp. 1335–1345, 1970.
43. K. R. Cooley, Energy Relationships in the Design of Floating Covers for Evaporation Reduction, *Water Resour. Res.*, vol. 6, pp. 717–727, 1970.
44. L. E. Myers and G. W. Frasier, Evaporation Reduction with Floating Granular Materials, *J. Irrig. Drain Div. ASCE*, vol. 96, pp. 425–436, December 1970.
45. W. W. Mansfield, Influence of Monolayers on the Natural Rate of Evaporation of Water, *Nature*, vol. 175, p. 247, 1955.
46. N. N. Gunaji, Evaporation Investigations at Elephant Butte Reservoir in New Mexico, *Int. Assoc. Sci. Hydrol. Publ.* 78, pp. 308–325, 1968.
47. V. F. Pushkarev and G. P. Levchenko, Use of Monomolecular Films to Reduce Evaporation from the Surface of Bodies of Water, *Tr. GGI* 142, pp. 84–107, 1967 (*Sov. Hydrol.: Sel. Pap.*, no. 3, pp. 253–272, 1967).
48. J. F. Bartholic, J. R. Runkels, and E. B. Stenmark, Effects of a Monolayer on Reservoir Temperature and Evaporation, *Water Resour. Res.*, vol. 3, pp. 173–180, 1967.
49. D. W. Hendricks and V. E. Hansen, Mechanics of Transpiration, *J. Irrig. Drain. Div. ASCE*, vol. 88, pp. 67–82, June 1962.
50. C. H. Lee, Transpiration and Total Evaporation, chap. 8, p. 280, in O. E. Meinzer (ed.), " Hydrology," McGraw-Hill, New York, 1942.
51. H. Landsberg, "Physical Climatology," 2d ed., p. 190, Pennsylvania State University, University Park, Pa., 1958.
52. W. D. McMillan and R. H. Burgy, Interception Loss from Grass, *J. Geophys. Res.*, vol. 65, pp. 2389–2394, 1960.
53. D. B. Thorud, The Effect of Applied Interception on Transpiration Rates of Potted Ponderosa Pine, *Water Resour. Res.*, vol. 3, pp. 443–450, 1967.

54. H. F. Blaney, Consumptive Use of Water, *Trans. ASCE*, vol. 117, pp. 949–973, 1952.
55. C. W. Thornthwaite, Report of the Committee on Transpiration and Evaporation, 1943–1944, *Trans. Am. Geophys. Union*, vol. 25, pt. 5, p. 687, 1944.
56. H. L. Penman, Estimating Evaporation, *Trans. Am. Geophys. Union*, vol. 37, pp. 43–50, 1956.
57. M. A. Kohler and M. M. Richards, Multicapacity Basin Accounting for Predicting Runoff from Storm Precipitation, *J. Geophys. Res.*, vol. 67, pp. 5187–5197, 1962.
58. C. E. Knox and T. J. Nordenson, Average Annual Runoff and Precipitation in the New England–New York Area, *U.S. Geol. Surv. Hydrol. Invest. Atlas* HA-7, undated.
59. W. E. Fox, Computation of Potential and Actual Evapotranspiration, U.S. Weather Bureau (processed), 1956.
60. R. S. Johnson, Evapotranspiration from Bare, Herbaceous, and Aspen Plots: A Check on a Former Study, *Water Resour. Res.*, vol. 6, pp. 324–327, 1970.
61. H. E. Brown and J. A. Thompson, Summer Water Use by Aspen, Spruce, and Grassland in Western Colorado, *J. For.*, vol. 63, pp. 756–760, 1965.
62. V. E. Suomi and C. B. Tanner, Evapotranspiration Estimates from Heat-Budget Measurements over a Field Crop, *Trans. Am. Geophys. Union*, vol. 39, pp. 298–304, 1958.
63. C. W. Thornthwaite and B. Holzman, Measurement of Evaporation from Land Water Surfaces, *U.S. Dept. Agric. Tech. Bull.* 817, 1942.
64. N. E. Rider, Evaporation from an Oat Field, *Q. J. R. Meteorol. Soc.*, vol. 80, pp. 198–211, April 1954.
65. J. R Mather (ed.), The Measurement of Potential Evapotranspiration, *Johns Hopkins Univ. Lab. Climatol., Seabrook, N.J., Publ. Climatol.*, vol. 7, no. 1, 1954.
66. L. L. Harrold and F. R. Dreibelbis, Agricultural Hydrology as Evaluated by Monolith Lysimeters, *U.S. Dept. Agric. Tech. Bull.* 1050, 1951.
67. W. O. Pruitt and D. E. Angus, Large Weighing Lysimeter for Measuring Evapotranspiration, *Trans. Am. Soc. Agric. Eng.*, vol. 3, pp. 13–18, 1960.
68. O. V. Popov, Lysimeters and Hydraulic Soil Evaporimeters, *Int. Assoc. Sci. Hydrol. Publ.* 49, pp. 26–37, 1959.
69. I. C. McIlroy and D. E. Angus, The Aspendale Multiple Weighed Lysimeter Installation, *Div. Meteorol. Phys. Tech. Pap.* 14, C.S.I.R.O., Australia, 1963.
70. See [41], pp. 62–66.
71. W. O. Pruitt and F. J. Lourence, Tests of Aerodynamic, Energy Budget, and Other Evaporation Equations over a Grass Surface, in "Investigations of Energy, Momentum, and Mass Transfer near the Ground," U.S. Army Electronics Command Atmospheric Laboratory, Ft. Huachuca, Ariz., pp. 37–63, 1966.
72. C. W. Thornthwaite, An Approach toward a Rational Classification of Climate, *Geograph. Rev.*, vol. 38, pp. 55–94, 1948.

73. R. L. Lowry and A. F. Johnson, Consumptive Use of Water for Agriculture, *Trans. ASCE,* vol. 107, p. 1252, 1942.
74. W. O. Pruitt and F. J. Lourence, Correlation of Climatological Data with Water Requirements of Crops, *Univ. Calif. Water Sci. Eng. Pap.* 9001, Davis, Calif., June 1968.
75. C. H. M. Van Bavel, Potential Evaporation: The Combination Concept and Its Experimental Verification, *Water Resour. Res.*, vol. 2, pp. 455–467, 1966.
76. H. F. Blaney, Discussion of paper by H. L. Penman, Estimating Evaporation, *Trans. Am. Geophys. Union*, vol. 37, pp. 46–48, February 1956.
77. F. J. Veihmeyer and A. H. Hendrickson, Does Transpiration Decrease as Soil Moisture Decreases?, *Trans. Am. Geophys. Union*, vol. 36, pp. 425–448, 1955.
78. E. R. Lemon, Some Aspects of the Relationship of Soil, Plant, and Meteorological Factors to Evapotranspiration, *Soil Sci. Soc. Am. Proc.*, vol. 21, pp. 464–468, 1957.
79. W. O. Pruitt, Correlation of Climatological Data with Water Requirements of Crops, *Univ. Calif. (Davis) Dept. Irrig. 1959–1960 Ann. Rep.*, p. 91, September 1960.
80. W. R. Gardner and D. I. Hillel, The Relation of External Evaporative Conditions to the Drying of Soils, *J. Geophys. Res.*, vol. 67, pp. 4319–4325, October 1962.
81. J. R. Philip, Evaporation, and Moisture and Heat Fields in the Soil, *J. Meteorol.*, vol. 14, pp. 354–366, 1957.
82. W. E. Marlatt, A. V. Havens, N. A. Willits, and G. D. Brill, A Comparison of Computed and Measured Soil Moisture under Snap Beans, *J. Geophys. Res.*, vol. 66, pp. 535–541, February, 1961.
83. F. J. Molz, I. Remson, A. A. Fungaroli, and R. L. Drake, Soil Moisture Availability for Transpiration, *Water Resour. Res.*, vol. 4, pp. 1161–1170, December 1968.
84. M. A. Kohler, Meteorological Aspects of Evaporation Phenomena, *Gen. Assemb. Int. Assoc. Sci. Hydrol., Toronto*, vol. 3, pp. 421–436, 1957.
85. W. J. Roberts, Reduction of Transpiration, *J. Geophys. Res.*, vol. 66, pp. 3309–3312, 1961.
86. H. W. Anderson, A. J. West, R. R. Zeimer, and F. R. Adams, Evaporative Loss from Soil, Native Vegetation, and Snow as Affected by Hexadecanol, *Int. Assoc. Sci. Hydrol. Publ.* 62, pp. 7–12, 1963.
87. J. Gale, E. B. Roberts, and R. M. Hagan, High Alcohols as Antitranspirants, *Water Resour. Res.*, vol. 3, no. 2, pp. 437–441, 1967.
88. A. R. Hibbert, Forest Treatment Effects on Water Yields, in W. E. Sopper and H. W. Lull (eds.), "International Symposium on Forest Hydrology," pp. 527–544, Pergamon, New York, 1967.
89. H. C. Pereira, The Influence of Man on the Hydrological Cycle, in "World

Water Balance," *Proc. Reading Symp., July 1970*, pp. 553–569, IASH-Unesco-WMO, Gentbrugge, Paris, Geneva, 1972.

90. C. T. Dryness, Erosion Potential of Forest Watersheds, in W. E. Sopper and H. W. Hull (eds.), "International Symposium on Forest Hydrology," pp. 599–610, Pergamon, New York, 1967.
91. H. Nakano, Effects of Changes of Forest Conditions on Water Yield, Peak Flow, and Direct Runoff of Small Watersheds in Japan, in W. E. Sopper and H. W. Lull (eds.), "International Symposium on Forest Hydrology," pp. 551–564, Pergamon, New York, 1967.
92. D. W. Pritchard, unpublished notes.

BIBLIOGRAPHY

AUSTRALIAN WATER RESOURCES COUNCIL: Evaporation from Water Storages, *Dept. Natl. Dev., Canberra, Hydrol Ser.* 4, 1970.

BLANEY, H. F., and others: Consumptive Use of Water: A Symposium, *Trans. ASCE*, vol. 117, pp. 949–1023, 1952.

BUDAGOVSKII, A. I.: Isparenije Poche venos Vlagi (in Russian)—Evaporation of Soil Moisture, Acad. Sci. USSR, Moscow, 1964.

BUDYKO, M. I.: Isparenie v Estestveniich usloviach (in Russian)—Evaporation in Natural Conditions, Gidrometeoizdat, Leningrad, 1948.

HOUNAM, C. E.: Problems of Evaporation Assessment in the Water Balance, *Rep. WMO/IHD Proj.*, no. 13, World Meteorological Organization, Geneva, 1971.

HUMPHRYS, C. R.: Evapotranspiration Bibliography, *Mich. State Univ., Dept. Resour. Dev., Agric. Exp. Stn. Water Bull.* 7, 1961.

JONES, F. E.: Evaporation of Water: A Review of Pertinent Laboratory Research, *Natl. Bur. Stand. Rep.* 10 235, May 1970.

KONSTANTINOV, A. R.: "Evaporation in Nature," Israel Program for Scientific Translations, Jerusalem, 1966; available from U.S. Dept. of Commerce National Technical Information Center, Springfield, Va.

MAGIN, G. B., and RANDALL, L. E.: Review of Literature on Evaporation Suppression, *U.S. Geol. Surv. Prof. Pap.* 272-C, 1960.

PANARA, R.: Bibliography on Evaporation Measurement, *Meteorol. Abstr. Bibliogr.*, vol. 10, no. 8, pp. 1234–1262, 1959.

PECK, E. L., and HELY, A.: Precipitation, Runoff and Water Loss in the Lower Colorado River–Salton Sea Area, *U.S. Geol. Surv. Prof. Pap.* 486-B, 1964.

PEREIRA, H. C. (compiler): Hydrological Effects of Changes in Land Use in Some East African Catchment Areas, *East Afr. Agric. For. J.*, vol. 77, special issue, March 1962.

RIJTEMA, P. E.: "Analysis of Actual Evapotranspiration," Wageningen, Netherlands, 1965.

ROBINSON, T. W., and JOHNSON, A. I.: Selected Bibliography on Evaporation and Transpiration, *U.S. Geol. Surv. Water-Supply Pap.* 1539-R, 1961.

THORNTHWAITE, C. W., and MATHER, J. R.: The Water Balance, *Drexel Inst. Lab. Climatol. Centerton, N.J., Publ. Climatol*, vol. 8, no. 1, 1955.

THURONYI, G.: Bibliography on Evapotranspiration (–1955), *Meteorol. Abstr. Bibliogr.*, vol. 10, no. 9, pp. 1394–1426, 1959.

———: Bibliography on Evapotranspiration (1956–1959), *Meteorol. Abstr. Bibliogr.*, vol. 10, no. 10, pp. 1552–1595, 1959.

WORLD METEOROLOGICAL ORGANIZATION, "Guide to Hydrological Practices," 3d ed., WMO no. 168, Geneva, 1974.

UNITED STATES DATA SOURCES

Observations of pan evaporation and other data for the station are published in monthly and annual issues of *Climatological Data* (NOAA, Environmental Data Service). The data are also available on magnetic tape from the National Climatic Center, Federal Building, Asheville, N.C. 28801.

Summarized evapotranspiration data can be found in the following publications:

LOWRY, R. L., and JOHNSON, A. F.: Consumptive Use of Water for Agriculture, *Trans. ASCE*, vol. 107, pp. 1243–1302, 1942.

WILLIAMS, G. R., and others: Natural Water Loss in Selected Drainage Basins, *U.S. Geol. Surv. Water-Supply Pap.* 846, 1940.

YOUNG, A. A., and BLANEY, H. F.: Use of Water by Native Vegetation, *Calif. Div. Water Resour. Bull.* 50, 1942.

PROBLEMS

5-1 Entering the upper left-hand relation of Fig. 5-4 with evaporation, wind movement, and dewpoint in reverse of the order indicated provides an estimate of surface-water temperature in the pan (on the axis labeled "air temperature"). In this manner, compute water temperature for each of the 10 meteorological conditions enumerated in Table 5-1. Describe those conditions which result in a water temperature higher than that of the overlying air, and vice versa.

5-2 Using Fig. 5-1, compute lake evaporation (thus neglecting advection and changes in heat storage) for each set of data presented in Table 5-1. Also compute the pan coefficient for each case. Is the range of coefficient displayed here indicative of the range in annual coefficients to be expected over the United States? Why?

5-3 An indication of the gross variation of lake evaporation with elevation can be obtained from Fig. 5-1 for specified conditions if values of the respective parameters are known for two elevations. Considering the data in Table 5-1 to constitute a series of sea-level observations, compute the evaporation at 3000 ft, mean sea level, using the following elevation gradients (per 1000 ft): air temperature, -3 Fahrenheit degree; dewpoint, -1 Fahrenheit degree; wind, 10 percent; and radiation, 2 percent. If these results are compared with those of Prob. 5-2, is the elevation effect reasonably constant for these selected circumstances? Discuss.

5-4 Given the data tabulated below, compute monthly and annual lake evaporation from Fig. 5-1. Assuming that a proposed reservoir will experience this computed amount of evaporation per year (and noting that precipitation less runoff is natural evapotranspiration), what would be the net anticipated loss from the reservoir per acre of surface area? On the basis of the computed lake evaporation and that provided for a Class A pan, compute monthly, mean monthly, and annual pan coefficients. Discuss qualitatively the effects of heat storage in a deep reservoir on the derived monthly pan coefficients.

Period	Air temperature, °F	Dewpoint, °F	Wind, mi/d	Radiation, Ly/d	Pan evaporation, in.	Precipitation, in.	Runoff, in.
Oct.	57	46	52	290	3.2	2.7	0.8
Nov.	46	35	59	200	1.7	2.4	1.1
Dec.	37	28	68	150	1.0	3.1	1.3
Jan.	34	25	81	150	0.8	3.4	1.6
Feb.	36	24	90	230	1.0	2.9	1.7
Mar.	44	30	92	310	2.8	3.5	1.8
Apr.	54	39	83	400	4.8	3.6	1.7
May	65	51	66	480	6.0	3.7	1.4
June	72	60	50	510	6.4	3.9	1.2
July	77	65	44	500	7.1	4.0	0.8
Aug.	75	64	41	440	5.8	4.6	1.0
Sept.	69	58	44	370	4.2	3.6	0.7
Sum	...	...	...	...	44.8	41.4	15.1
Mean	56	44	64	336			

5-5 Using data published in *Water-Supply Papers* and *Climatological Data*, plot hydrographs of mean daily flow and bar charts of daily precipitation (several years) for a selected small basin. The basin and period should be selected to include reasonably saturated conditions on a number of occasions. Compute mean daily evapotranspiration for several periods delineated by times of assumed basin saturation (Fig. 5-9). Which of the periods analyzed do you believe to be indicative of potential conditions?

6

SUBSURFACE WATER

About one-fifth of the total water used in the United States (exclusive of hydroelectric power generation) is drawn from groundwater. Subsurface water is relatively free of pollution and is especially useful for domestic use in small towns and isolated farms. In arid regions groundwater is often the only reliable source of water for irrigation. Groundwater temperatures are relatively low, and large quantities are used for cooling in warm regions.

Aside from its direct use, groundwater is also an important phase of the hydrologic cycle. Most of the flow of perennial streams originates from subsurface water, while a large part of the flow of ephemeral streams may percolate beneath the surface. Thus, no treatment of surface water hydrology can ignore subsurface processes. Since the occurrence and movement of subsurface water are intimately related to geological structure, an understanding of geologic controls is a prerequisite to a comprehension of groundwater hydrology. This chapter stresses interrelations between surface and subsurface water and presumes only an elementary knowledge of geology.

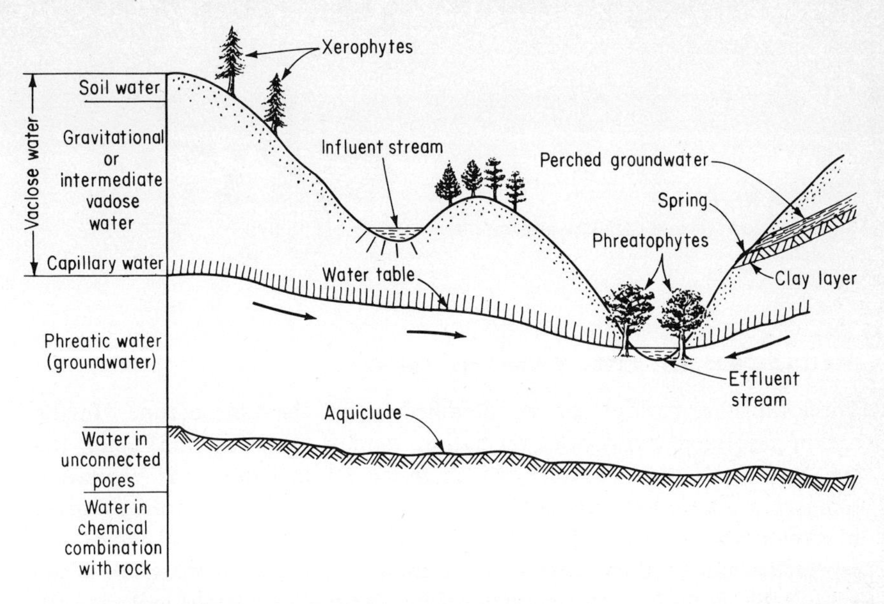

FIGURE 6-1
Schematic cross section showing the occurrence of groundwater.

6-1 Occurrence of Subsurface Water

Figure 6-1 is a schematic cross section of the upper portion of the earth's crust with an idealized column showing a suggested classification of subsurface water [1]. The two major subsurface zones are divided by an irregular surface called the *water table.* The water table is the locus of points (in unconfined material) where hydrostatic pressure equals atmospheric pressure. Above the water table, in the *vadose zone*, soil pores may contain either air or water; hence it is sometimes called *zone of aeration.* In the *phreatic zone*, below the water table, interstices are filled with water; sometimes this is called the *zone of saturation.* The phreatic zone may extend to considerable depth, but as depth increases, the weight of overburden tends to close pore spaces and relatively little water is found at depths greater than 10,000 ft (3 km).

Local saturated zones sometimes exist as *perched groundwater* above a small impervious body. Sometimes groundwater is overlain by an impervious stratum to form *confined*, or *artesian*, water. Confined groundwater is usually under pressure because of the weight of the overburden and the hydrostatic head. If a well penetrates the confining layer, water will rise to the *piezometric level*, the artesian equivalent of the water table. If the piezometric level is above ground level, the well discharges as a *flowing well.*

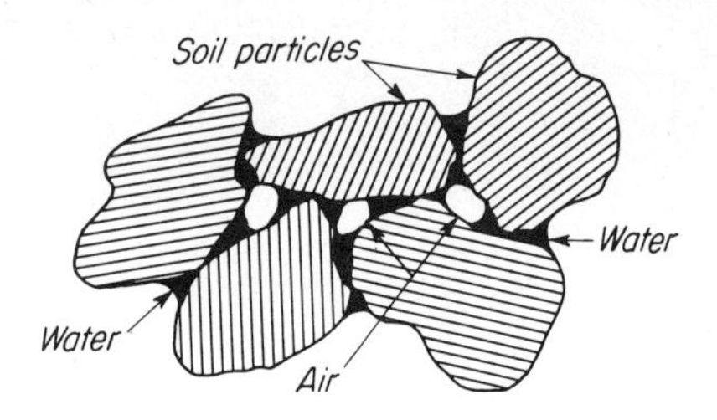

FIGURE 6-2
The occurrence of capillary moisture in soil.

MOISTURE IN THE VADOSE ZONE

Three moisture regions can be identified within the vadose zone. In the region penetrated by roots of vegetation, ranging to 30 ft (10 m) below the soil surface, is the *soil water*, which fluctuates in amount as vegetation removes moisture between rains. Above the water table, moisture is raised by capillarity into the *capillary fringe*, which may have a vertical extent of a few inches to several feet depending on the pore sizes of the material. If the water table is close to the ground surface, the capillary fringe and the soil-moisture region may overlap, but where the water table is deep, an *intermediate* region exists where moisture levels remain constant at the field capacity of the soil and rock of the region.

6-2 Soil-Water Relationships

Soil moisture may be present as *gravity water* in transit in the larger pore spaces, as *capillary water* in the smaller pores (Fig. 6-2), as *hygroscopic moisture* adhering in a thin film to soil grains, and as water vapor. Gravity water is in a transient state. After a rain, water may move downward in the larger pores, but this water must either be dispersed into capillary pores or pass through the vadose zone to the groundwater or to a stream channel. Hygroscopic water, on the other hand, is held by molecular attraction and is not normally removed from the soil under usual climatic conditions. The important variable element of soil moisture, therefore, is capillary water.

If a soil-filled tube is placed with its open lower end in a container of water, some water will move up into the soil. The rate of upward movement becomes progressively smaller with time, until it eventually approaches zero. Measurement of the moisture content of the soil at various levels will show that moisture in the soil column decreases with height above the water surface (Fig. 6-3). If a sample of soil is saturated with water and then subjected to successively greater negative pressures and the moisture content is noted after each change in pressure, a similar curve (Fig. 6-4) results.

Buckingham [2] first proposed characterizing soil-moisture phenomena

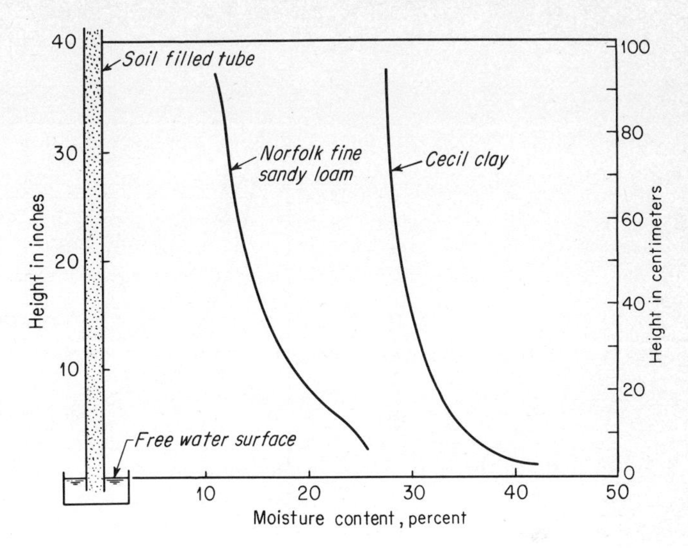

FIGURE 6-3
Moisture-content versus height curves for typical soils. (*From E. Buckingham, Studies on the Movement of Soil Moisture, U.S. Dept. Agric. Bur. Soils Bull.* 38, 1907.)

on the basis of energy relationships. He introduced the concept of capillary potential to describe the attraction of soil for water. With a free-water surface taken as reference, *capillary potential* is defined as the work required to move a unit mass of water from the reference plane to any point in the soil column. Thus, capillary potential is the potential energy per unit mass of water. By definition, capillary potential is negative since water will move upward by capillarity without external work. Capillary potential ψ is related to the acceleration of gravity g and height above datum y (negative) by the equation

$$\psi = gy \qquad (6\text{-}1)$$

Curves like Fig. 6-3 provide a basis for relating capillary potential and moisture content for a particular soil. Schofield [3] suggested the term pF to represent the common logarithm of the capillary potential in centimeters of water. This is analogous to the use of pH in chemistry.

6-3 Equilibrium Points

Visualizing several states of water in soil, early soil scientists tried to define limits of these states by *equilibrium points.* Figures 6-3 and 6-4 indicate that no clear-cut boundaries exist, but equilibrium points are convenient for

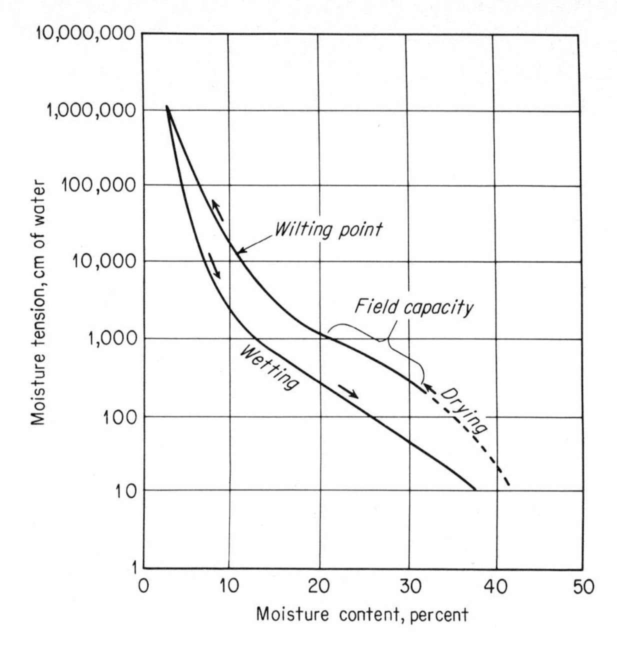

FIGURE 6-4
Moisture-tension curves for a typical soil. (*From R. K. Schofield, The pF of Water in Soil, Trans. 3d Int. Congr. Soil. Sci., vol.* 2, *pp.* 37–48, 1935.)

discussing soil moisture. The two of greatest interest are field capacity and wilting point. *Field capacity* is defined as the moisture content of soil after gravity drainage is complete. Colman [4] has shown that field capacity is essentially the water retained in soil at a tension of $\frac{1}{3}$ atm. Veihmeyer and Hendrickson [5] found that the *moisture equivalent*, water retained in a soil sample, $\frac{3}{8}$ in. (9.5 mm) deep after being centrifuged for 30 min at a speed equivalent to a force of 1000*g*, was also nearly the field capacity of fine-grained soils.

The *wilting point* represents the soil-moisture level when plants cannot extract water from soil. It is the moisture held at a tension equivalent to the osmotic pressure in the plant roots. For years the wilting point was determined by growing sunflower seedlings in a soil sample, but it is now commonly assumed to be equivalent to the moisture content at a tension of 15 atm. The difference between the moisture content at field capacity and at wilting point is called *available moisture*. It represents the useful storage capacity of the soil (Secs. 5-9 and 5-15) and the maximum water available to plants. Typical values of moisture content at field capacity and wilting point and available moisture are given in Table 6-1.

6-4 Measurement of Soil Moisture

The standard determination of soil moisture is the loss in weight when a soil sample is oven-dried. A *tensiometer* (Fig. 6-5) consists of a porous ceramic cup which is inserted in the soil, filled with water, and connected to a manometer. The tensiometer remains in equilibrium with the soil. If the soil moisture falls below saturation, water is drawn from the cup and a negative pressure indicated by the manometer. A tensiometer [6] can indicate soil-moisture tension from saturation to a tension of about 1 atm.

In the resistivity [7] method, a pair of electrodes embedded in a porous dielectric (plaster of paris, nylon, Fiberglas) is buried in the soil. The dielectric maintains a moisture equilibrium with the soil, and the resistance between the electrodes varies with the moisture content of the material. Resistance is measured with an ac bridge to avoid polarizing the element. The block must be in intimate contact with the soil. Calibration is best achieved by taking periodic soil samples from the area surrounding the installation and correlating moisture content of the samples with concurrent resistivity readings.

The neutron-scattering method of soil-moisture measurement [8, 9] uses a source of fast neutrons which is lowered into an aluminum access tube in the soil. Fast neutrons lose energy and are converted to slow neutrons through collision with atoms of low atomic weight. The hydrogen in water is normally the only atom of low atomic weight in soil. The slow neutrons are

Table 6-1 TYPICAL MOISTURE VALUES FOR VARIOUS SOIL TYPES

	Percent dry weight of soil			Specific weight	
Soil type	Field capacity	Wilting point	Available water	lb/ft^3	kg/m^3
Sand	5	2	3	95	1520
Sandy loam	12	5	7	90	1440
Loam	19	10	9	85	1360
Silt loam	22	13	9	80	1280
Clay loam	24	15	9	80	1280
Clay	36	20	16	75	1200
Peat	140	75	65	20	320

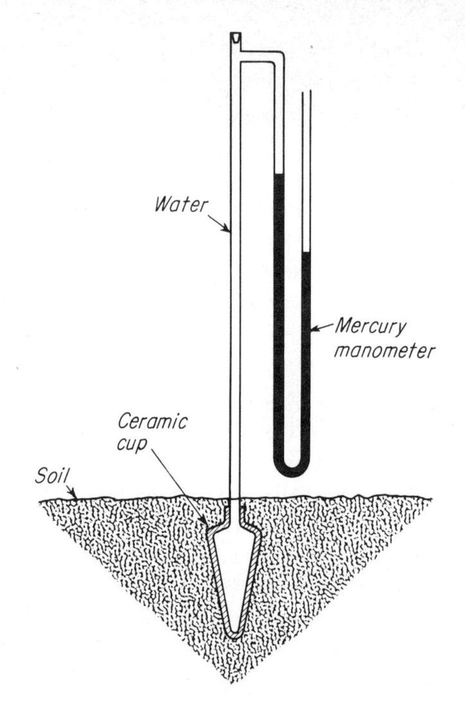

FIGURE 6-5
Schematic drawing of a simple tensiometer.

detected and counted with a counter, which is part of the device lowered into the access hole. A high count of slow neutrons indicates a high moisture content. Once calibrated, the device is easily and quickly used. However, it measures moisture content in a sphere of soil surrounding the source and is less accurate near the surface, where the sample volume is distorted. Organic matter in soil also causes errors in this method.

Aerial observation of natural gamma radiation from the soil may also prove useful in determining soil-moisture variations over large areas.

6-5 Movement of Soil Moisture

Infiltration is the movement of water through the soil surface into the soil as distinguished from *percolation*, the movement of water through the soil. When water is first applied to the soil surface, gravity water moves down through the larger soil openings while the smaller surface pores take in water by capillarity. The downward-moving gravity water is also taken in by capillary pores. As capillary pores at the surface are filled and intake capacity reduced, the infiltration rate decreases. In homogeneous soil, infiltration decreases gradually until the zone of aeration is saturated. Normally, the soil is stratified, and subsoil layers are often less permeable than the surface soil. In this case, the infiltration rate is eventually limited to the rate of percolation through the least pervious subsoil stratum.

Infiltration from rainfall is distinguished by very shallow depths of water on the soil surface but is extensive over large areas. Quantities of water infiltrated are usually very small (a few inches per day) and rarely sufficient to saturate a great depth of soil. When rain stops, gravity water remaining in the soil continues to move downward and at the same time is taken up in capillary-pore spaces. Usually the infiltrated water is distributed within the upper few feet of soil, with little or no contribution to groundwater unless the soil is highly permeable or the vadose zone very thin. Infiltration from rainfall is discussed in greater detail in Sec. 8-2.

For irrigation and artificial recharge of groundwater (Sec. 6-17), water is ponded to a considerable depth over limited areas for long periods of time. The aim of recharge operations is to saturate the soil down to the water table. Under these conditions the time variation of infiltration is complex, with temporary increases in rate superimposed on a gradually declining trend. Escape of soil air around the infiltration basin, bacterial action, changes in water temperature, changes in soil structure, and other factors influence these variations.

Movement of moisture in soil is governed by the moisture potential following the equation

$$q = -K_w \frac{\partial \Lambda}{\partial x} \tag{6-2}$$

where q is flow per unit time through unit area normal to the direction of flow, x is distance along the line of flow, K_w is conductivity, and Λ is potential. After gravity water has left the soil, the principal component of total potential is the capillary potential. Equation (6-2) states that flow is from a region of high potential to a region of lower potential. Quantitative determination of the conductivity is difficult, although it has been shown to increase with moisture content and decrease with pore size. Thus capillary movement decreases as soil dries and is least in fine-grained soil. Fortunately, a qualitative understanding of these phenomena is normally sufficient for engineering hydrology.

Vapor pressure in the soil is controlled by temperature. Vapor movement is from high temperature (high vapor pressure) to low temperature. Vapor transport is an important factor in moisture movement when the moisture content is lowered to the point where capillary moisture is discontinuous. Under this condition, however, moisture-content and temperature gradients are usually so small that the quantity of moisture moved is negligible. When the surface soil is frozen, the vapor-pressure gradient is upward and is accentuated by the lower vapor pressure of ice relative to water at the same temperature. Thus when frozen soil thaws, its moisture content may be greater than at the time of freezing. Conversely, during summer, vapor-pressure gradients would be downward were it not for evaporation and transpiration.

Electrical and chemical factors also contribute to total potential, but under natural conditions these factors are hydrologically unimportant.

MOISTURE IN THE PHREATIC ZONE

Within the phreatic zone all pore spaces are filled with water, and the different states of moisture, moisture tension, etc., are of little concern. Interest is centered on the amount of water present, the amount which can be removed, and the movement of this water.

6-6 Aquifers

A geologic formation which contains water and transmits it from one point to another in quantities sufficient to permit economic development is called an *aquifer*. In contrast, an *aquiclude* is a formation which contains water but cannot transmit it rapidly enough to furnish a significant supply to a well or spring. An *aquifuge* has no interconnected openings and cannot hold or transmit water. The ratio of the pore volume to the total volume of the formation is called *porosity*. The *original porosity* of a material is that which existed at the time the material was formed. *Secondary porosity* results from fractures and solution channels.

Table 6-2 APPROXIMATE AVERAGE POROSITY, SPECIFIC YIELD, AND PERMEABILITY OF VARIOUS MATERIALS

Material	Porosity, %	Specific yield, %	Permeability		Intrinsic permeability, D
			Meinzer units	$m^3/d\,m^2$	
Clay	45	3	0.01	0.0004	0.0005
Sand	35	25	1,000	41	50
Gravel	25	22	100,000	4,100	5,000
Gravel and sand	20	16	10,000	410	500
Sandstone	15	8	100	4.1	5
Dense limestone and shale	5	2	1	0.041	0.05
Quartzite, granite	1	0.5	0.01	0.0004	0.0005

Secondary porosity cannot be measured without an impossibly large sample. Original porosity is usually measured by oven-drying an undisturbed sample and weighing it. It is then saturated with some liquid and weighed again. Finally, the saturated sample is immersed in the same liquid, and the weight of displaced liquid is noted. The weight of liquid required to saturate the sample divided by the weight of liquid displaced is the porosity as a decimal. If the material is fine-grained, the liquid may have to be forced into the sample under pressure to assure complete saturation.

High porosity does not necessarily indicate a productive aquifer, since much of the water may be retained in small pore spaces under capillary tension as the material is dewatered. The *specific yield* of an aquifer is the ratio of the water which will drain freely from the material to the total volume of the formation and is always less than the porosity. The relation between specific yield and porosity depends on the size of particles in the formation. Specific yield of a fine-grained aquifer will be small, whereas coarse-grained material will yield a greater amount of its contained water. Table 6-2 lists approximate average values of porosity and specific yield for some typical materials. Large variations from these average values must be expected. Note that although clay has a high porosity, it has a very low specific yield. Sand and gravel, which make up most of the more productive aquifiers in the United States, yield about 80 percent of their total water content.

6-7 Movement of Groundwater

In 1856 Darcy confirmed the applicability of principles of fluid flow in capillary tubes, developed several years earlier by Hagen and Poiseuille, to the flow of water in permeable media. *Darcy's law* is

$$v = ks \tag{6-3}$$

where v is the velocity of flow, s is the slope of the hydraulic gradient, and k is a coefficient having the units of v (feet per day or meters per day). The discharge q is the product of area A and velocity. The effective area is the gross area times the porosity p of the media. Hence

$$q = kpAs = K_pAs \tag{6-4}$$

The coefficient K_p is called the *coefficient of permeability* or the *hydraulic conductivity*. It is dependent on properties of the fluid and the medium and can be expressed as

$$K_p = k\frac{w}{\mu} = Cd^2\frac{w}{\mu} \tag{6-5}$$

where k is the *intrinsic permeability* of the medium, w is the specific weight of the fluid, μ is its absolute viscosity, C is a factor involving the shape, packing, porosity, and other characteristics of the medium, and d is the average pore size of the medium.

In petroleum engineering the intrinsic permeability is expressed in *darcys*, which have the dimensions of area.† The hydraulic conductivity K_p has the dimensions of velocity and is stated in a variety of units by different disciplines and in different countries. In the United States for hydrologic purposes K_p is usually given in *Meinzer units*,‡ the flow in gallons per day through an area of one square foot under a gradient of one foot per foot at 60°F (Table 6-2).

It is convenient to use the *transmissibility* T to represent the flow in gallons per day through a section 1 ft wide and the thickness of the aquifer under a unit head (slope of 1 ft per ft):

$$T = K_p Y \tag{6-6}$$

where Y is the saturated thickness of the aquifer. With this coefficient Eq. (6-4) becomes

$$q = TBs \tag{6-7}$$

where B is the width of the aquifer.

6-8 Determination of Permeability

Laboratory measurements of permeability are made with *permeameters* (Fig. 6-6). A sample of the material is subjected to water under a known head, and the flow through the sample in a known time is measured. Such tests have limited practical value because of the difficulty of placing samples of unconsolidated materials in the permeameter in their natural state and the uncertainty whether a sample is truly representative of the aquifer. Flow in solution cavities or rock fractures and the effect of large boulders in gravel aquifers cannot be duplicated in a permeameter.

The earliest field techniques for determining permeability involved introducing salt into the aquifer at one well and timing its movement to a downstream well [10]. Fluorescein dye [11], detectable at a concentration of 0.03 ppm by the unaided eye and in concentrations as low as 0.0001 ppm under ultraviolet light, has also been used as a tracer. More recently, radioactive materials have been used [12]. Tracer techniques have encountered numerous difficulties [12, 13]. Chemical reactions between the tracer

† One darcy (D) = 1.062×10^{-11} ft^2 = 0.987×10^{-8} cm^2.
‡ One Meinzer unit = 0.0408 m^3/d m^2 with unit gradient.

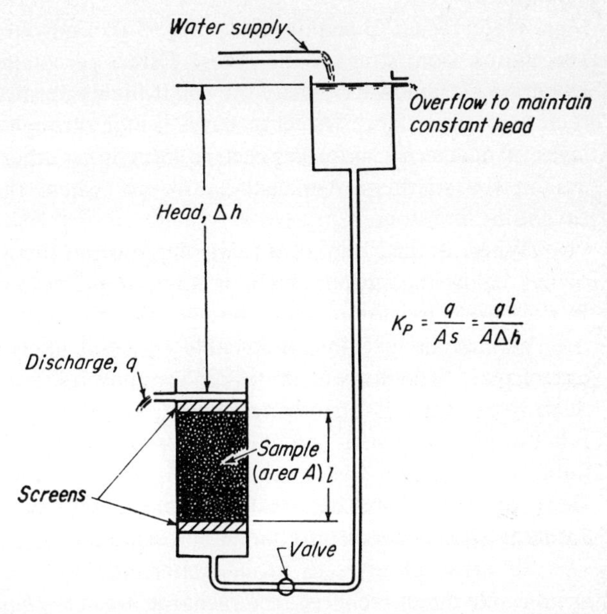

FIGURE 6-6
A simple upward-flow permeameter.

elements and the formation sometimes occur. Because of diffusion, tests must be conducted over short distances in order to have detectable concentrations at the downstream well, and even then it is difficult to determine a representative time of arrival. Tracers are most useful for determining path of flow, e.g., when it is necessary to locate a source of pollution.

Today permeability is most commonly determined by pumping tests. By using the principles of well hydraulics (Secs. 6-11 and 6-12) it is possible to estimate average permeability of an aquifer for a large distance around the test well.

6-9 Sources of Groundwater

Almost all groundwater is *meteoric water* derived from precipitation. *Connate water*, present in the rock at its formation and frequently highly saline, is found in some areas. *Juvenile water*, formed chemically within the earth and brought to the surface in intrusive rocks, occurs in small quantities. Connate and juvenile waters are sometimes important sources of undesirable minerals in groundwater. Groundwater in the San Joaquin Valley, California, contains boron brought to the surface from great depths.

Water from precipitation reaches groundwater by infiltration and percolation from streams and lakes. Direct percolation is most effective in recharging groundwater where the soil is highly permeable or the water table is close to the surface. Direct recharge is high through the permeable basaltic lavas of northern California, eastern Oregon, southern Idaho, and Hawaii, and in the southern Appalachian region, where thin soil cover overlies cavernous limestone.

Where annual rainfall is relatively low and the water table is hundreds of feet below the surface, little or no recharge from rain can be expected. In such areas irrigation water may provide for some recharge, but seepage from stream channels into permeable gravels is likely to be the main source of recharge. Streams contributing to groundwater are called *influent streams.* Such streams are frequently *ephemeral*; i.e., they go dry during protracted rainless periods when percolation depletes all flow. Streams are rarely influent throughout their entire length. Most of the percolation occurs in short reaches where bed materials are highly permeable. Considerable recharge often occurs from channels crossing coarse gravels in an alluvial fan.

In areas of artesian groundwater, the overlying aquiclude prevents appreciable direct recharge; the recharge area (*forebay*) may be far removed from the artesian area.

6-10 Discharge of Groundwater

Without interference by man, a groundwater basin fills and discharges excess water by several routes until a quasi equilibrium is reached. Streams intersecting the water table and receiving groundwater flow are called *effluent streams.* Perennial streams are generally effluent through at least a portion of their length.

Where an aquifer intersects the earth's surface, a spring or seep will form (Fig. 6-7). There may be a concentrated flow constituting the source of a small stream or merely seepage which evaporates from the ground surface. Most springs are small and of little hydrologic significance, although even a small spring may provide water for a single farmstead. First-magnitude springs [14] discharge 100 ft^3/s (2.8 m^3/s), and, according to Meinzer [15], there are 65 such springs in the United States: 38 in volcanic rocks of California, Oregon, and Idaho; 24 in limestone in the Ozarks, the Balcones Fault area of Texas, and Florida; and 3 sandstone springs in Montana. The Fontaine de Vaucluse in France has a discharge often exceeding 4000 ft^3/s (113 m^3/s). It is the world's largest spring and is in a limestone formation.

Where the water table is close to the surface, groundwater may be discharged by direct evaporation or by transpiration from the capillary fringe. Plants deriving their water from groundwater, called *phreatophytes*,

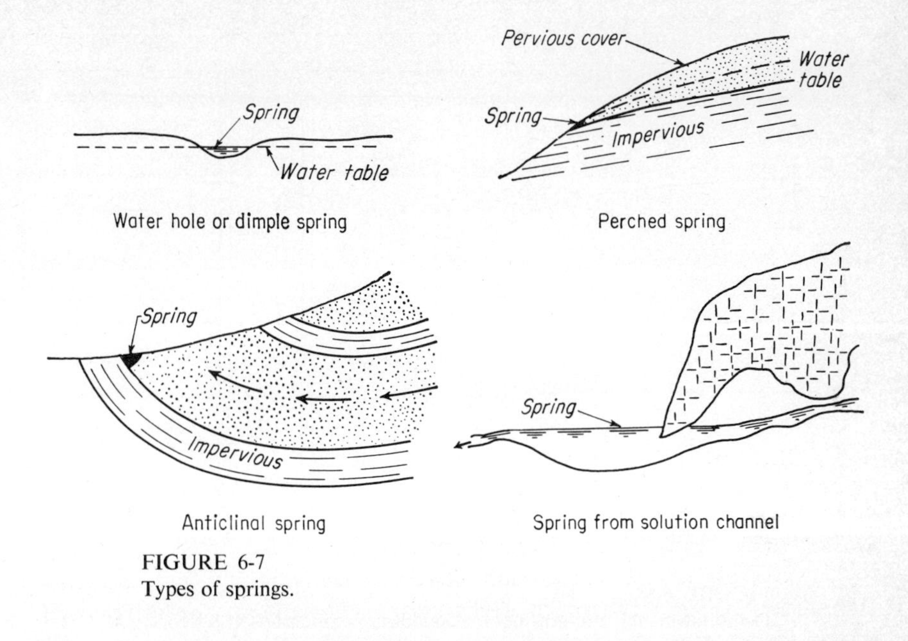

FIGURE 6-7
Types of springs.

often have root systems extending to depths of 40 ft (12 m) or more. This invisible evapotranspiration loss may be quite large. At a rate of 3 ft/y the loss would be 1920 acre-ft/mi^{2}y. (Similarly 1 m/y = 1 hm^3/km^{2}y.)

The various channels of groundwater discharge may be viewed as spillways of the groundwater reservoir. When groundwater is high, discharge through natural spillways tends to maintain a balance between inflow and outflow. During dry periods natural discharge is reduced as groundwater levels fall, and outflow may even cease. Artesian aquifers may not reflect this natural balance as rapidly as water-table aquifers, but sustained drought will decrease water levels in the recharge area and decrease discharge from the aquifer.

6-11 Equilibrium Hydraulics of Wells

Figure 6-8 shows a well in a homogeneous aquifer of infinite extent with an initially horizontal water table. For flow to occur to the well there must be a gradient toward the well. The resulting water-table form is called a *cone of depression*. If the decrease in water level at the well (*drawdown*) is small with respect to the total thickness of the aquifer, and if the well completely penetrates the aquifer, the streamlines of flow to the well may be assumed to be horizontal. In this case an approximate formula relating well discharge and aquifer characteristics can be derived.

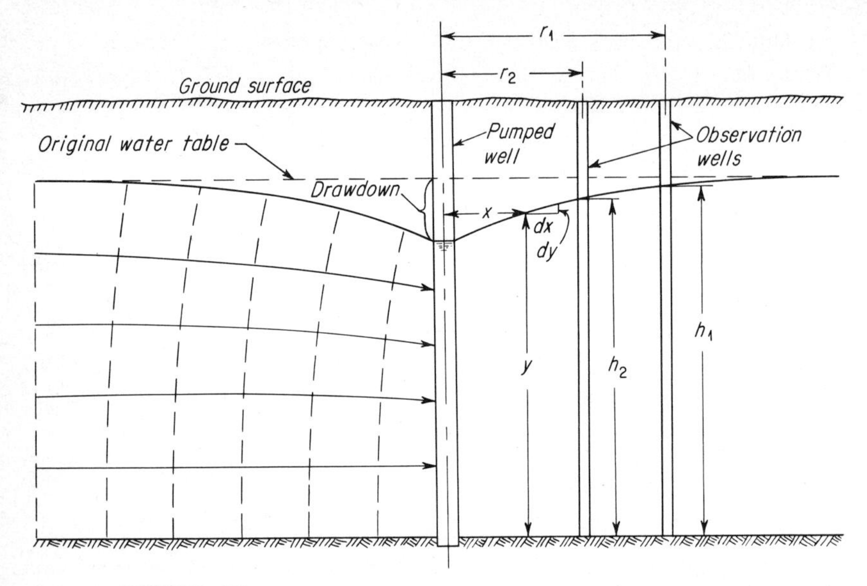

FIGURE 6-8
Definition sketch and flow net for equilibrium conditions at a well.

Flow toward the well through a cylindrical surface at radius x must equal the discharge of the well, and from Darcy's law [Eq. (6-4)]

$$q = 2\pi xyK_p \frac{dy}{dx} \tag{6-8}$$

where $2\pi xy$ is the area of the cylinder and dy/dx is the slope of the water table. Integrating with respect to x from r_1 to r_2 and y from h_1 to h_2 yields

$$q = \frac{\pi K_p (h_1^2 - h_2^2)}{\ln (r_1/r_2)} \tag{6-9}$$

where h is the height of the water table above the base of the aquifer at distance r from the pumped well and ln is the logarithm to the base e.

This equation was first proposed by Dupuit in 1863 and subsequently modified by Thiem [16] in 1906. A serious restriction results because the low velocities of groundwater flow mean that true equilibrium conditions occur only after a very long period of pumping at constant rate.

6-12 Nonequilibrium Hydraulics of Wells

During the initial period of pumping from a new well, much of the discharge is derived from storage in the portion of the aquifer unwatered as the cone of depression develops. Equilibrium analysis indicates a permeability which is too high, because only part of the discharge comes from flow through the aquifer to the well. This leads to an overestimate of the potential yield of the well.

In 1935 Theis [17] presented a formula based on the heat-flow analogy which accounts for the effect of time and the storage characteristics of the aquifer. His formula is

$$Z_r = \frac{q}{4\pi T} \int_u^{\infty} \frac{e^{-u}}{u}\, du \qquad (6\text{-}10)$$

where Z_r is the drawdown in an observation well at distance r from the pumped well, q is the flow in cubic feet per day, T is transmissibility in cubic feet per day per foot, and u is given by

$$u = \frac{r^2 S_c}{4Tt} \qquad (6\text{-}11)$$

In Eq. (6-11) t is the time in days since pumping began, and S_c is the storage constant of the aquifer, or the volume of water removed from a column of aquifer 1 ft square when the water table or piezometric surface is lowered 1 ft. For water-table aquifers it is essentially the specific yield. The integral in Eq. (6-10), commonly written $W(u)$ and called the *well function of u*, can be evaluated from the series

$$W(u) = -0.5772 - \ln u + u - \frac{u^2}{2 \cdot 2!} + \frac{u^3}{3 \cdot 3!} \cdots \qquad (6\text{-}12)$$

Values of $W(u)$ for various values of u are given in Table 6-3.

Equation (6-10) is usually solved graphically by first plotting a *type curve* of u versus $W(u)$ on logarithmic paper (Fig. 6-9). From Eq. (6-11),

$$\frac{r^2}{t} = \frac{4T}{S_c} u \qquad (6\text{-}13)$$

If q is constant, Eq. (6-10) indicates that Z_r equals a constant times $W(u)$. Thus a curve of r^2/t versus Z_r should be similar to the type curve of u versus $W(u)$. After the field observations are plotted, the two curves are superimposed with their axes parallel and adjusted until some portions of the two curves coincide. The coordinates of a common point taken from the region where the curves coincide are used to solve for T and S_c, using Eqs. (6-10) and (6-13). Values of Z_r and r^2/t may come from one well with various values of t, from several wells with different values of r, or a combination of both. Metric units may be used in these equations without change in the constants.

When u is small, the terms of Eq. (6-12) following ln u are small and may be neglected. Equation (6-11) indicates that u will be small when t is

Table 6-3 VALUES OF $W(u)$ FOR VARIOUS VALUES OF u

u	1.0	2.0	3.0	4.0	5.0	6.0	7.0	8.0	9.0
$\times 1$	0.219	0.049	0.013	0.0038	0.00114	0.00036	0.00012	0.000038	0.000012
$\times 10^{-1}$	1.82	1.22	0.91	0.70	0.56	0.45	0.37	0.31	0.26
$\times 10^{-2}$	4.04	3.35	2.96	2.68	2.48	2.30	2.15	2.03	1.94
$\times 10^{-3}$	6.33	5.64	5.23	4.95	4.73	4.54	4.39	4.26	4.14
$\times 10^{-4}$	8.63	7.94	7.53	7.25	7.02	6.84	6.69	6.55	6.44
$\times 10^{-5}$	10.95	10.24	9.84	9.55	9.33	9.14	8.99	8.86	8.74
$\times 10^{-6}$	13.24	12.55	12.14	11.85	11.63	11.45	11.29	11.16	11.04
$\times 10^{-7}$	15.54	14.85	14.44	14.15	13.93	13.75	13.60	13.46	13.34
$\times 10^{-8}$	17.84	17.15	16.74	16.46	16.23	16.05	15.90	15.76	15.65
$\times 10^{-9}$	20.15	19.45	19.05	18.76	18.54	18.35	18.20	18.07	17.95
$\times 10^{-10}$	22.45	21.76	21.35	21.06	20.84	20.66	20.50	20.37	20.25
$\times 10^{-11}$	24.75	24.06	23.65	23.36	23.14	22.95	22.81	22.67	22.55
$\times 10^{-12}$	27.05	26.36	25.95	25.66	25.44	25.26	25.11	24.97	24.86
$\times 10^{-13}$	29.36	28.66	28.26	27.97	27.75	27.56	27.41	27.28	27.16
$\times 10^{-14}$	31.66	30.97	30.56	30.27	30.05	29.87	29.71	29.58	29.46
$\times 10^{-15}$	33.96	33.27	32.86	32.58	32.35	32.17	32.02	31.88	31.76

SOURCE: Adapted from [18].

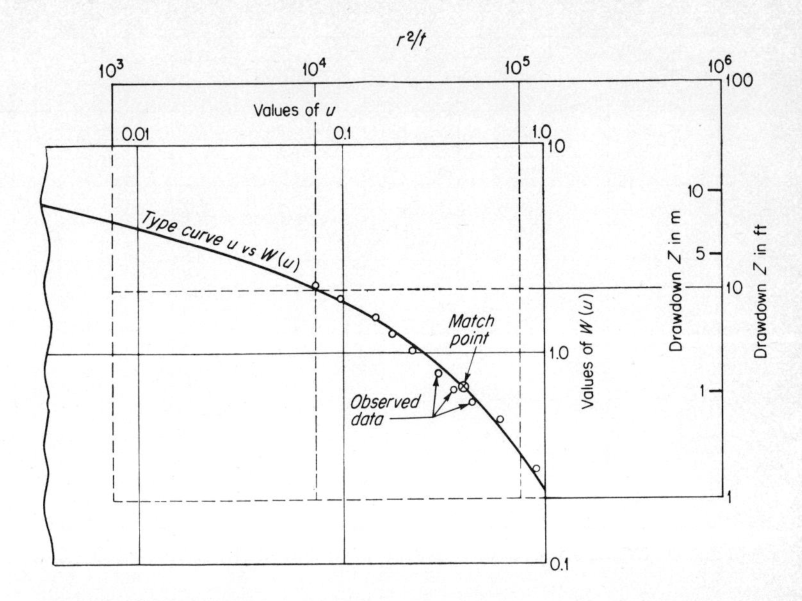

FIGURE 6-9
Using the Theis method to solve a well problem. Match point coordinates $u = 0.4$, $W(u) = 0.7$, $Z = 3.4$, $r^2/t = 5.3 \times 10^4$.

large, and in this case a modified solution of the Theis method is possible [19] by writing

$$T = \frac{2.3q}{4\pi \, \Delta Z} \log \frac{t_2}{t_1} \tag{6-14}$$

where ΔZ is the change in drawdown between times t_1 and t_2. The drawdown Z is plotted on an arithmetic scale against time t on a logarithmic scale (Fig. 6-10). If ΔZ is taken as the change in drawdown during one log cycle, $\log_{10} (t_2/t_1) = 1$, and T is easily determined from Eq. (6-14). When $Z = 0$,

$$S_c = \frac{2.25Tt_0}{r^2} \tag{6-15}$$

where t_0 is the intercept (in days) obtained if the straight-line portion of the curve is extended to $Z = 0$.

As in the Thiem equation, Theis assumes parallel streamlines, i.e., small drawdown and full penetration of the well. While Theis adjusts for the effect of storage in the aquifer, he does assume instantaneous unwatering of the aquifer material as the water table drops. These conditions are reasonably

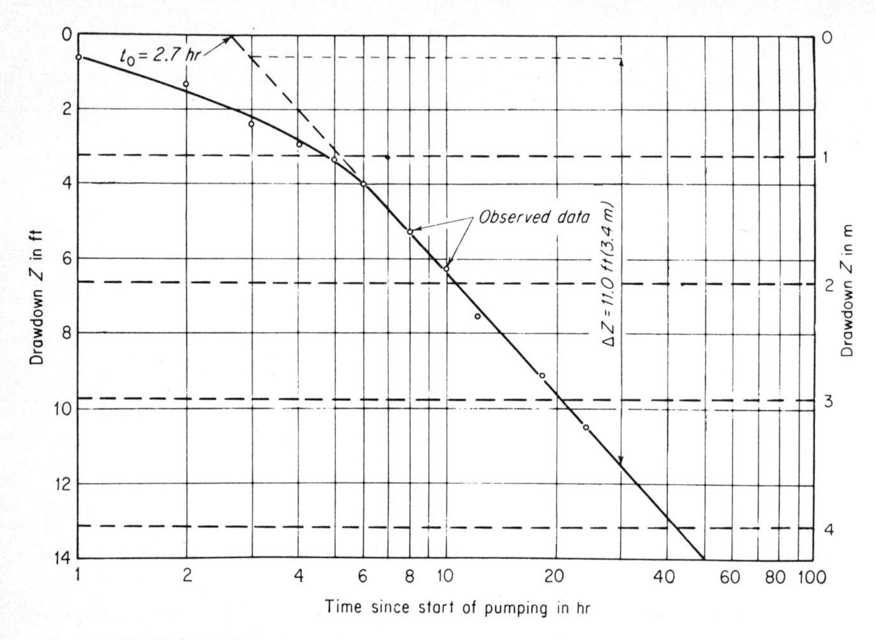

FIGURE 6-10
Use of the modified Theis method.

well satisfied in artesian aquifers. However, the procedure should be used with caution in thin or poorly permeable water-table aquifers.

6-13 Boundary Effects

The assumption of a symmetrical cone of depression implies a homogeneous aquifer of great extent. Such an ideal aquifer is rarely encountered, although in many cases the condition is approximated closely enough for reasonable accuracy. When several wells are close together, their cones of depression may overlap, or *interfere*, and the water table appears as in Fig. 6-11. Where the cones of depression overlap, the drawdown at a point is the sum of the drawdowns caused by the individual wells. The two-dimensional analysis is greatly simplified, but it should be evident that when wells are located too close together, the flow from the wells is impaired and the drawdowns increased.

Figure 6-12 shows an aquifer with a positive boundary in the form of an intersecting surface stream. The gradient from the stream to the well causes influent seepage from the stream. If the streamflow is more than the seepage,

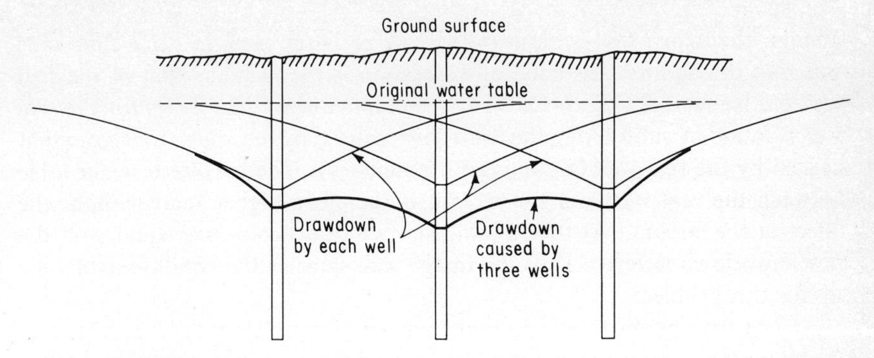

FIGURE 6-11
Effect of interference between wells.

so that flow continues in the stream, the cone of depression of the well must coincide with the water surface in the stream. A rigorous analysis would require that the channel be the full depth of the aquifer to avoid vertical flow components. However, if the well is not too close to the stream, no serious error is introduced if this condition is not satisfied. The method of images devised by Lord Kelvin for electrostatic theory is a convenient way to treat boundary problems. An image well is assumed to have all the properties of the real well but to be located on the opposite side of the stream and at the same distance from it as the real well. Since the stream adds water to the

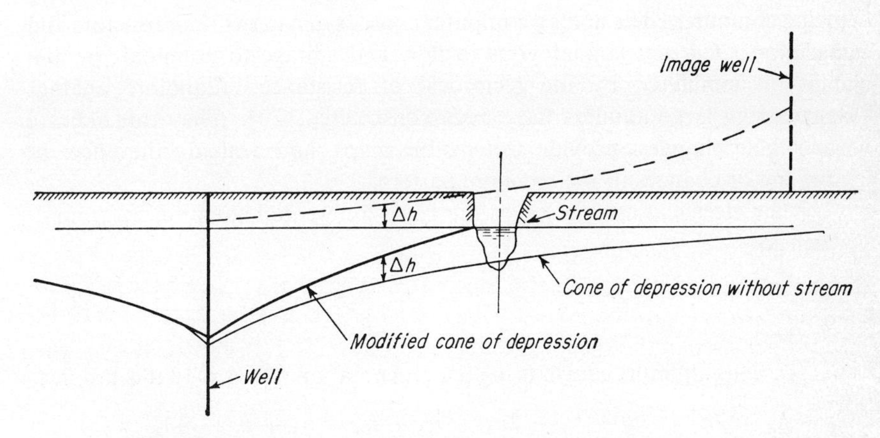

FIGURE 6-12
Image well simulating the effect of seepage from a stream on water levels adjacent to a pumped well.

aquifer, the image well is assumed to be a recharge well, i.e., one that adds water to the aquifer. Its cone of depression is the same as that of the real well but is inverted (Fig. 6-12). The resultant cone of depression for the real well is found by subtracting the drawdown caused by the image well from that caused by the real well (assuming no boundary). The corrected water table between the real well and the stream is therefore higher than without the effect of the stream. At the stream, the two drawdowns are equal, and the new drawdown is zero. Thus the image well satisfies the conditions first set up for this problem.

Negative boundaries, i.e., faults and similar structures across which no groundwater is transmitted, can be analyzed in a similar manner. More complicated boundary problems require judicious selection of multiple image wells. Most geologic boundaries are neither abrupt nor straight. Unless the aquifer is very small relative to the zone of influence of the well, the assumption of sharp discontinuities is not serious.

6-14 Aquifer Analysis

Techniques of the previous sections are suitable for analysis of single wells or a small well field, but study of a large aquifer generally requires more efficient computational systems. A Hele-Shaw apparatus consisting of closely spaced glass plates with a viscous fluid between them is often convenient for solving two-dimensional groundwater-flow problems [1]. The equations governing groundwater flow are the same as those for the viscous flow of the analog.

Three-dimensional problems are commonly treated with a digital or analog computer. The analog computer consists of a network of resistors and capacitors. Current is analogous to flow and voltage to potential; permeability is simulated by the reciprocal of resistance. Elaborate analogs representing large aquifers have been constructed [20]. They can indicate water-table changes, provide water-table maps, and evaluate the effect of pumping or changes in the recharge pattern.

The electrical analog solves the basic differential equation of groundwater flow

$$\frac{\partial^2 h}{\partial x^2} + \frac{\partial^2 h}{\partial y^2} = \frac{S_c}{T}\frac{\partial h}{\partial t} \tag{6-16}$$

In finite-difference form using the grid notation of Fig. 6-13 this becomes

$$\frac{h_2 + h_4 - 2h_1}{a^2} + \frac{h_3 + h_5 - 2h_1}{a^2} = \frac{h_2 + h_4 + h_3 + h_5 - 4h_1}{a^2}$$

$$= \frac{S_c}{T}\frac{h}{\partial T} \tag{6-17}$$

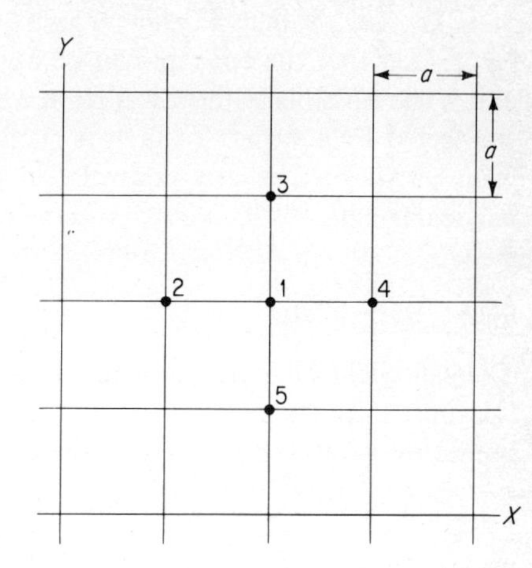

FIGURE 6-13
Grid notation for Eq. (6-17).

where a is the grid size. This can also be solved with a digital computer [21] and without constructing an elaborate analog. If repeated solutions are desired over a long period of time, the analog may be the most effective device. If a relatively limited analysis is expected to suffice, the digital computer will generally prove more efficient. Grid sizes for both analog and digital solutions may vary from 100 ft (30 m) to 10,000 ft (3000 m), depending on the nature of the problem. The solutions can be expected to be as precise as the aquifer description used. Aquifer properties are required at each grid point (thickness, permeability, and coefficient of storage). If they are not accurately defined, the solution will be in error.

POTENTIAL OF A GROUNDWATER RESERVOIR

A basic problem in engineering groundwater studies is the question of the permissible rate of withdrawal from a groundwater basin. This quantity, commonly called the *safe yield*, is defined by Meinzer [14] as "the rate at which water can be withdrawn for human use without depleting the supply to such an extent that withdrawal at this rate is no longer economically feasible." Many other definitions of safe yield have been suggested, and alternative terms such as sustained yield, feasible rate of withdrawal, and optimum yield have been proposed. The concept of safe yield has received considerable criticism. Kazmann [22] has suggested that it be abandoned because of its frequent interpretation as a permanent limitation on the permissible withdrawal. Safe yield must be recognized as a quantity determined for a specific set of controlling conditions and subject to change as a result of changing economic or physical conditions. It should also be recognized that the concept can be applied only to a complete groundwater unit. The possible withdrawal from a single well or group of wells in a field is affected by a variety of factors such as size, construction, and spacing of wells as well as by any controls on the flow of groundwater toward the particular field.

6-15 Safe Yield

The safe yield of a groundwater basin is governed by many factors, one of the most important being the quantity of water available. This hydrologic limitation is often expressed by the equation

$$G = P - Q_S - E_T + Q_g - \Delta S_g - \Delta S_s \qquad (6\text{-}18)$$

where G is safe yield, P is precipitation on the area tributary to the aquifer, Q_S is surface streamflow from the same area, E_T is evapotranspiration, Q_g is net groundwater inflow to the area, ΔS_g is change in groundwater storage, and ΔS_s is change in surface storage. If the equation is evaluated on a mean annual basis, ΔS_s will usually be zero.

All terms of Eq. (6-18) are subject to artificial change, and G can be computed only by assuming the conditions regarding each item. Artificial-recharge operations can reduce Q_S. Irrigation diversion from influent streams may increase evapotranspiration. Lowering the water table by pumping may increase groundwater inflow (or reduce groundwater outflow) and may make otherwise effluent streams into influent streams.

The permanent withdrawal of groundwater from storage is called *mining*, the term being used in the same sense as for mineral resources. If the storage in the aquifer is small, excessive mining may be disastrous to any economy dependent on the aquifer for water. On the other hand, many large groundwater basins contain vast reserves of water, and planned withdrawal of this water at a rate that can be sustained over a long period may be a wise use of this resource. The annual increment of mined water, ΔS_g of Eq. (6-18), increases the yield. Thus Eq. (6-18) cannot properly be considered an equilibrium equation or solved in terms of mean annual values. It can be solved correctly only on the basis of specified assumptions for a stated period of years.

The factors which control the assumptions on which Eq. (6-18) is solved are primarily economic. The feasibility of artificial recharge or surface diversion is usually determined by economics. If water levels in the aquifer are lowered, pumping costs are increased. Theoretically, there is a water-table elevation at which pumping costs equal the value of the water pumped and below which water levels should not be lowered. Practically, the increased cost is often passed on to the ultimate consumer, and the minimum level is never attained. Excessive lowering of the water table may result in contamination of the groundwater by inflow of undesirable waters. This hazard is especially prevalent near seacoasts, where seawater intrusion (Sec. 6-16) may occur. A similar problem may develop wherever an aquifer is adjacent to a source of saline groundwater.

Transmissibility of an aquifer may also place a limit on safe yield. Although Eq. (6-18) may indicate a potentially large draft, this can be realized only if the aquifer is capable of transmitting the water from the source area to the wells at a rate high enough to sustain the draft. This problem is especially likely to develop in long artesian aquifers.

6-16 Seawater Intrusion

Since seawater (specific gravity about 1.025) is heavier than fresh water, the groundwater under a uniformly permeable circular island would appear as shown in Fig. 6-14. The lens of fresh water floating on salt water is known as a *Ghyben-Herzberg lens*, after the codiscoverers [23] of the principle. About 40 ft of fresh water is required below sea level for each foot of fresh water above sea level to maintain hydrostatic equilibrium. True hydrostatic equilibrium does not exist with a sloping water table since flow must occur. Thus, there is likely to be a seepage face for freshwater flow to the ocean and a zone of mixing along the saltwater-freshwater interface. Areally variable recharge, pumping of wells, and tidal action also disturb the equilibrium. A hydrodynamic balance governs the form of the interface. If velocities are low, the 1/40 ratio may be a reasonable first approximation but more adequate methods of analysis are available [24, 25].

When a cone of depression is formed about a pumping well in fresh water, an inverted cone of salt water will rise into the fresh water (Fig. 6-14*b*). A saltwater rise of approximately 40 ft per foot (40 m per meter) of freshwater drawdown may occur, depending on the local situation. Horizontal skimming wells are commonly used to avoid this effect.

Salt water sometimes enters an aquifer through damaged well casings. If a well passes through an aquifer containing undesirable water or through salt water into an underlying aquifer, salt water can enter through a leaky casing and drop down to the fresh water. Jacob [26] pointed out that several aquifers may exist at a coastline (Fig. 6-15), and if saltwater intrusion is permitted in the upper aquifer, the deeper aquifer may be contaminated by leakage.

6-17 Artificial Recharge

The yield of an aquifer may be increased artificially by introducing water into it. In most cases this is equivalent to reducing the surface runoff from the area [Q_s, Eq. (6-18)]. The methods employed for artificial recharge are

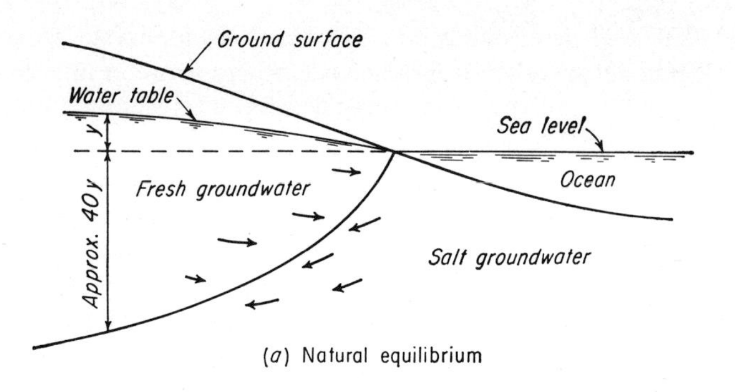

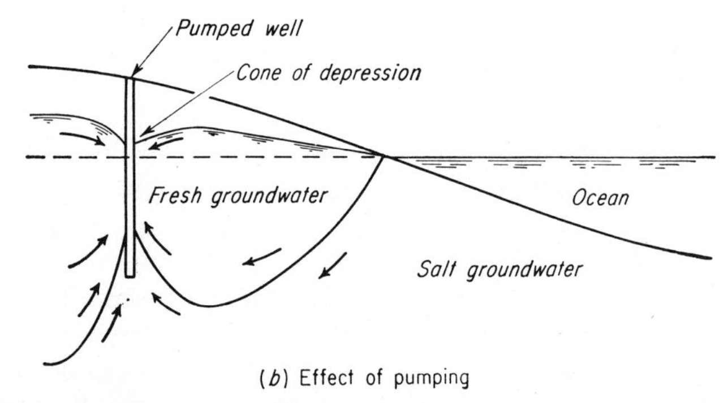

FIGURE 6-14
Saltwater-freshwater relations adjacent to a coastline.

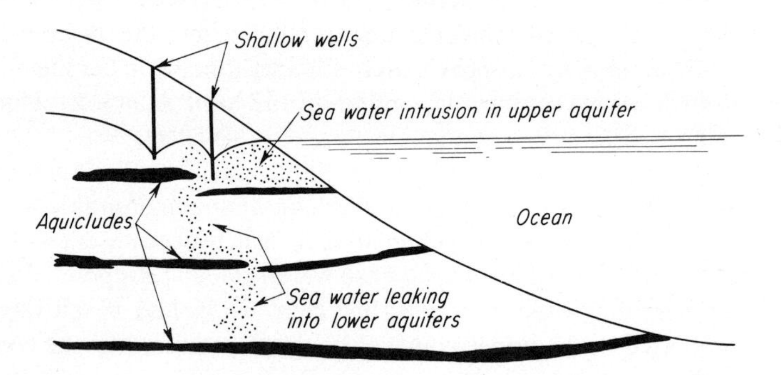

FIGURE 6-15
Contamination of deep aquifers from above.

controlled by the geologic situation of an area and by economic considerations. Some possible methods include:

1 Storing floodwaters in reservoirs constructed over permeable areas
2 Storing floodwaters in reservoirs for later release into the stream channel at rates approximating the percolation capacity of the channel
3 Diverting streamflow to spreading areas located in a highly permeable formation
4 Excavating recharge basins to reach permeable formations
5 Pumping water through recharge wells into the aquifer
6 Overirrigating in areas of high permeability
7 Construction of wells adjacent to a stream to induce percolation from streamflow

Where conditions are favorable, use of an aquifer as a reservoir may eliminate evaporation losses, protect against pollution, provide a low-cost distribution system and result in a cost saving compared with a surface reservoir. Artificial recharge with wells has also been employed to create a groundwater mound along a coast as a barrier to saltwater intrusion and as a means of disposing of waste materials. The latter process must be used with great caution to avoid polluting useful aquifers. Todd [25] discusses methods and results of recharge operations in some detail.

6-18 Artesian Aquifers

The evaluation of the potential of an artesian aquifer involves some special factors. The confining strata are commonly assumed to be watertight. If the permeability of the aquiclude is 1 Meinzer unit and the hydraulic gradient is unity, the daily seepage would amount to about 28 million gallons (86 acre-ft)/mi^2, or 40,900 m^3/km^2. A quantity of this magnitude would be quite significant in the groundwater exchange of an aquifer. Hantush [27] has demonstrated a procedure which accounts for such leakage in the analysis of pumping tests on artesian aquifers.

Artesian aquifers demonstrate considerable compressibility. This is evident from cases where fluctuations in tide level, barometric pressure, or even the superimposed load of trains are reflected in fluctuations of water level in wells penetrating the aquifer. If the pressure in an artesian aquifer is relieved locally by removal of water, compression of the aquifer may result, with subsidence of the ground above it. Such subsidence has been observed [28] in areas subject to heavy withdrawal of groundwater, with ground-surface elevations declining more than 10 ft (3.0 m). Aside from the disrupting effects of the surface subsidence, the phenomenon suggests that pumping tests on such aquifers may be misleading because of the flow derived from storage as a result of the compression. Although the small fluctuations appear to exhibit elastic behavior, there is no evidence that the ground levels in regions of pronounced subsidence will recover if the aquifers are repressurized.

6-19 Time Effects in Groundwater

Flow rates in the groundwater are normally extremely slow, and considerable time may be involved in groundwater phenomena. A critical lowering of the water table adjacent to a coast may not bring immediate saltwater intrusion because of the time required for the salt water to move inland. Werner [29] suggests that several hundred years might be required for a sudden increase in water level in the recharge area of an extensive artesian aquifer to be transmitted through the aquifer. Jacob [30] found that water levels on Long Island were related to an effective precipitation which was the sum of the rainfalls for the previous 25 y, each weighted by a factor which decreased with time. McDonald and Langbein [31] found long-term fluctuations in streamflow in the Columbia basin which they believe are related to groundwater fluctuations. Thus in interpreting groundwater data it is important to give full weight to the influence of time. Observed variations in groundwater levels must be correctly related to causal factors if serious misconceptions are not to result.

REFERENCES

1. Stanley N. Davis and R. J. DeWiest, "Hydrogeology," p. 39, Wiley, New York, 1966.
2. E. Buckingham, Studies on the Movement of Soil Moisture, *U.S. Dept. Agric. Bur. Soils Bull.* 38, 1907.
3. R. K. Schofield, The pF of the Water in Soil, *Trans. 3d Int. Congr. Soil Sci.*, vol. 2, pp. 37–48, 1935.
4. E. A. Colman, A Laboratory Procedure for Determining the Field Capacity of Soils, *Soil Sci.*, vol. 63, p. 277, 1947.
5. F. J. Veihmeyer and A. H. Hendrickson, The Moisture-Equivalent as a Measure of the Field Capacity of Soils, *Soil Sci.*, vol. 32, pp. 181–193, 1931.
6. K. K. Watson, Some Operating Characteristics of a Rapid Response Tensiometer System, *Water Resour. Res.*, vol. 1, pp. 577–586, 1965.
7. W. M. Broadfoot and others, Some Field, Laboratory and Office Procedures for Soil-Moisture Measurement, *South. Forest Exp. Stn. Occ. Pap.* 135, 1954.
8. D. J. Belcher, T. R. Cuykendall, and H. S. Sack, The Measurement of Soil Moisture and Density by Neutron and Gamma-Ray Scattering, *U.S. Civ. Aeronaut. Admin. Tech. Develop. Eval. Cent. Rep.* 127, 1950.

9. W. Gardner and D. Kirkham, Determination of Soil Moisture by Neutron Scattering, *Soil Sci.*, vol. 73, pp. 391–401, 1952.
10. C. S. Slichter, Field Measurements of Rate of Movement of Underground Water, *U.S. Geol. Surv. Water-Supply Pap.* 140, 1905.
11. R. B. Dole, Use of Fluorescein in Study of Underground Waters, *U.S. Geol. Surv. Water-Supply Pap.* 160, pp. 73–85, 1906.
12. E. Halevy and A. Nir, The Determination of Aquifer Parameters with the Aid of Radioactive Tracers, *J. Geophys. Res.*, vol. 67, pp. 2403–2409, 1962; see also "Radioisotopes in Hydrology," International Atomic Energy Agency, Vienna, 1963.
13. W. J. Kaufman and G. T. Orlob, An Evaluation of Ground-Water Tracers, *Trans. Am. Geophys. Union*, vol. 37, pp. 297–306, June, 1956.
14. O. E. Meinzer, Outline of Groundwater Hydrology, *U.S. Geol. Survey Water-Supply Pap.* 494, 1923.
15. O. E. Meinzer, Large Springs in the United States, *U.S. Geol. Survey Water-Supply Pap.* 557, 1927.
16. G. Thiem, "Hydrologische Methoden," Gebhardt, Leipzig, 1906.
17. C. V. Theis, The Relation between the Lowering of the Piezometric Surface and the Rate and Duration of Discharge of a Well Using Ground-Water Storage, *Trans. Am. Geophys. Union*, vol. 16, pp. 519–524, 1935.
18. L. K. Wenzel, Methods for Determining the Permeability of Water-bearing Materials, *U.S. Geol. Surv. Water-Supply Pap.* 887, 1942.
19. C. E. Jacob, Drawdown Test to Determine the Effective Radius of Artesian Well, *Trans. ASCE*, vol. 112, pp. 1047–1070, 1947.
20. W. C. Walton, "Groundwater Resource Evaluation," pp. 518–598, McGraw-Hill, New York, 1970.
21. T. A. Prickett and C. G. Lonnquist, Selected Digital Computer Techniques for Groundwater Resource Evaluation, *Ill. State Water Surv. Urbana, Ill., Bull.* 55, 1971.
22. R. G. Kazmann, "Safe Yield" in Ground-Water Development, Reality or Illusion?, *J. Irrig. Drain. Div. ASCE*, vol. 82, November 1956; see also discussion by McGuinness, Ferris, and Kramsky, in ibid., vol. 82, May 1957.
23. W. Badon Ghyben, Nota in Verband met de Voorgenomen Putboring naby Amsterdam, *Tijdschr. K. Inst. Ing.* (*The Hague*), 1888–1889, p. 21.
24. H. H. Cooper, A Hypothesis Concerning the Dynamic Balance of Fresh Water and Salt Water in a Coastal Aquifer, *J. Geophys. Res.*, vol. 64, pp. 461–468, April 1959.
25. D. K. Todd, "Ground-Water Hydrology," Wiley, New York, 1959.
26. C. E. Jacob, Full Utilization of Groundwater Reservoirs, *Trans. Am. Geophys. Union*, vol. 38, p. 417, June 1957.
27. M. S. Hantush, Analysis of Data from Pumping Tests in Leaky Aquifers, *Trans. Am. Geophys. Union*, vol. 37, pp. 702–714, December 1956.
28. J. F. Poland and G. H. Davis, Subsidence of the Land Surface in the Tulare-

Wasco (Delano) and Los Banos–Kettleman City Area, San Joaquin Valley, California, *Trans. Am. Geophys. Union*, vol. 37, pp. 287–296, June 1956.
29. P. W. Werner, Notes on Flow-Time Effects in the Great Artesian Aquifers of the Earth, *Trans. Am. Geophys. Union*, vol. 27, pp. 687–708, October 1946.
30. C. E. Jacob, Correlation of Ground-Water Levels and Precipitation on Long Island, New York, *Trans. Am. Geophys. Union*, vol. 24, pt. 2, pp. 564–580, 1943, vol. 25, pt. 6, pp. 928–939, 1944.
31. C. C. McDonald and W. B. Langbein, Trends in Runoff in the Pacific Northwest, *Trans. Am. Geophys. Union*, vol. 29, pp. 387–397, June 1948.

BIBLIOGRAPHY

AMERICAN SOCIETY OF CIVIL ENGINEERS: "Manual on Ground Water Basin Management," *ASCE Man.* 40, New York, 1961.

BEAR, J.: "Dynamics of Fluids in Porous Media," American Elsevier, New York, 1972.

BUTLER, S. S.: "Engineering Hydrology," Prentice-Hall, Englewood Cliffs, N.J., 1957.

DOMENICO, A.: "Concepts and Models in Groundwater Hydrology," McGraw-Hill, New York, 1972.

HANTUSH, M. S.: Hydraulics of Wells, *Advan. Hydrosci.*, vol. 1, 1964.

HUBBERT, M. K.: The Theory of Ground-Water Motion, *J. Geol.*, vol. 48, pp. 785–944, 1940.

MEINZER, O. E.: Ground Water, chap. 10 in O. E. Meinzer (ed.), "Hydrology," vol. 9, Physics of the Earth Series, McGraw-Hill, New York, 1942; reprinted Dover, New York, 1949.

POLUBARINOVA-KOCHINA, P. YA.: "Theory of Groundwater Movements," trans. from Russian, Princeton University Press, Princeton, 1972.

THOMAS, H. E.: "The Conservation of Ground Water," McGraw-Hill, New York, 1951.

PROBLEMS

6-1 An undisturbed rock sample has an oven-dry weight of 652.47 g. After saturation with kerosene its weight is 731.51 g. It is then immersed in kerosene and displaces 300.66 g. What is the porosity of the sample?

6-2 At station *A* the water-table elevation is 642 ft above sea level, and at *B* the elevation is 629 ft. The stations are 1100 ft apart. The aquifer has a permeability of 300 Meinzer units and a porosity of 14 percent. What is the actual velocity of flow in the aquifer?

6-3 If the root zone in clay-loam soil is 3 ft thick, what quantity of available moisture (in inches depth) should it hold? Use Table 6-1.

6-4 A soil sample has a coefficient of permeability of 250 Meinzer units. What would its permeability be at 50°F? What is its instrinsic permeability?

6-5 A 12-in.-diameter well penetrates 80 ft below the static water table. After 24 h of pumping at 1100 gal/min the water level in a test well at 320 ft is lowered 1.77 ft, and in a well 110 ft away the drawdown is 3.65 ft. What is the transmissibility of the aquifer? Use Eq. (6-9).

6-6 The time-drawdown data for an observation well 296 ft from a pumped well (500 gal/min) are tabulated below. Find the transmissibility and storage constant of the aquifer. Use the Theis method.

Time, h	Drawdown, ft	Time, h	Drawdown, ft
1.9	0.28	9.8	1.09
2.1	0.30	12.2	1.25
2.4	0.37	14.7	1.40
2.9	0.42	16.3	1.50
3.7	0.50	18.4	1.60
4.9	0.61	21.0	1.70
7.3	0.82	24.4	1.80

6-7 Tabulated below are the time-drawdown data for an observation well 150 ft from a well pumped at 350 gal/min. Find the transmissibility and storage constant by the modified Theis method.

Time, h	1.8	2.7	5.4	9.0	18.0	54.0
Drawdown, ft	1.8	2.4	3.6	4.3	5.8	8.1

6-8 A well 250 ft deep is planned in an aquifer having a transmissibility of 10,000 gal/d per foot width and a storage coefficient of 0.010. The well is expected to yield 500 gal/min and will be 12 in. in diameter. If the static water level is 50 ft below the ground surface, estimate the pumping lift at the end of 1 and 3 y of operation.

6-9 After pumping a new 12-in. well for 24 hr at 150 gal/min, the drawdowns in a number of nearby observation wells as are as given below. Find the storage coefficient and transmissibility of the aquifer.

Well no.	Distance, ft	Drawdown, ft	Well no.	Distance, ft	Drawdown, ft
1	100	10.5	5	283	2.4
2	141	7.5	6	347	1.4
3	190	6.2	7	490	0.6
4	200	4.0			

6-10 An 18-in. well is in an aquifer with a transmissibility of 8000 gal/d per foot width and a storage coefficient of 0.07. What pumping rate can be adopted so that the maximum drawdown after 2 y will not exceed 20 ft?

6-11 A 24-in. well is in an aquifer with a transmissibility of 10,000 gal/d per foot width and a storage coefficient of 0.05. Draw the profile of the cone of depression after 1 y of pumping at 500 gal/m. If a fault is located 1000 ft from this well, what would be the profile of the cone of depression?

6-12 A well is 30 cm in diameter and penetrates 50 m below the static water table. After 36 h of pumping at 4 m^3/min (use Eq. [6-9]) the water level in a test well 200 m distant is lowered by 1.2 m and in a well 40 m away the drawdown is 2.7 m. What is the transmissibility of the aquifer?

6-13 Using the data for Prob. 6-6 but assuming the drawdowns to be given in meters, find the transmissibility and storage constant of the aquifer by the Theis method. The observation well is 100 m from the pumped well and the pumping rate is 2000 l/min.

6-14 Using the data from Problem 6-7, determine the transmissibility and storage constant of the aquifer if the indicated drawdowns are in meters. The observation well is at a distance of 50 m, and the pumping rate is 2800 l/min. Use the modified Theis method.

6-15 The well of Prob. 6-11 is 800 ft from a stream which flows all year. How much is the drawdown midway between the well and the stream decreased because of this seepage?

6-16 Using data from the *Water-Supply Papers* or other source, find out what you can about the trend of groundwater levels in your area. What explanation can you see for the observed trends? What is the source of the groundwater? Is an overdraft of the available supply indicated? Are there any possible ways of improving the yield?

6-17 For a basin selected by your instructor, make an estimate of the safe yield, assuming no change in present conditions.

7

STREAMFLOW HYDROGRAPHS

Engineering hydrology is concerned primarily with three characteristics of streamflow: monthly and annual volumes available for storage and use, low-flow rates which restrict in-stream uses of water, and floods. Detailed analysis of flood hydrographs is usually important in flood damage mitigation, flood forecasting, or establishing design flows for the many structures which must convey floodwaters.

CHARACTERISTICS OF THE HYDROGRAPH

The water which constitutes streamflow may reach the stream channel by any of several paths from the point where it first reaches the earth as precipitation. Some water flows over the soil surface as *surface runoff* and reaches the stream soon after its occurrence as rainfall. Other water infiltrates through the soil surface and flows beneath the surface to the stream. This water moves more slowly than the surface runoff and contributes to the sustained flow of the stream during periods of dry weather. In hydrologic studies involving rate of flow in streams it is necessary to distinguish between these components of total flow. The first step in such studies is to divide observed hydrographs of streamflow into components before analyzing the relation between rainfall and runoff (Chap. 8), determining the characteristic shape of hydrographs for a basin, or studying drought conditions (Sec. 11-17).

7-1 Components of Runoff

The route followed by a water particle from the time it reaches the ground until it enters a stream channel is devious. It is convenient to visualize three main routes of travel: overland flow, interflow, and groundwater flow.

Overland flow, or *surface runoff*, is that water which travels over the ground surface to a channel. The word channel as used here refers to any depression which may carry a small rivulet of water in turbulent flow during a rain and for a short while after. Such channels are numerous, and the distance water must travel as overland flow is relatively short, rarely more than a few hundred feet. Therefore overland flow soon reaches a channel, and if it occurs in sufficient quantity, is an important element in the formation of flood peaks. The amount of surface runoff may be quite small, however, since surface flow over a permeable soil surface can occur only when the rainfall rate exceeds the infiltration capacity (Chap. 8). In many small and moderate storms, surface runoff may occur only from impermeable surfaces within the basin or from precipitation which falls directly on the water surface of the basin. Except in urban areas, the total of impermeable area and water surface is usually a small part of the basin area. Hence, surface runoff is an important factor in streamflow only as the result of heavy or high-intensity rains.

Some of the water which infiltrates the soil surface may move laterally through the upper soil layers until it enters a stream channel. This water, called *interflow* or *subsurface flow*, moves more slowly than the surface runoff and reaches the streams later. The proportion of total runoff which occurs as interflow depends on the geology of the basin. A thin soil cover overlying rock, hardpan, or plowbed a short distance below the soil surface favors substantial quantities of interflow, whereas uniformly permeable soil encourages downward percolation to groundwater. Although traveling more slowly than overland flow, interflow may be much larger in quantity, especially in storms of moderate intensity, and hence may be the principal factor in the smaller rises of streamflow.

Some precipitation may percolate downward until it reaches the water table (Chap. 6). This groundwater accretion may eventually discharge into the streams as *groundwater flow* (also called *base flow* and *dry-weather flow*) if the water table intersects the stream channels of the basin. The groundwater contribution to streamflow cannot fluctuate rapidly because of its very low flow velocity. In some regions more than 2 y is required [1] for the effect of a given accretion to groundwater to be discharged into the streams.

Basins having permeable surface soils and large, effluent groundwater bodies show sustained high flow throughout the year, with a relatively small ratio between flood flow and mean flow. Basins with surface soils of low permeability or influent groundwater bodies have higher ratios of peak to average flows and very low or zero flows between floods. Hydrographs for each type of basin are shown in Fig. 7-1. Hat Creek drains volcanic terrain with a large groundwater contribution, while the Santa Ynez River is influent throughout most of its length.

The distinctions drawn between the three components of flow are arbitrary. Water may start out as surface runoff, infiltrate from the sheet of overland flow, and complete its trip to the stream as interflow. On the other hand, interflow may surface where a relatively impervious stratum intersects a hillside and finish its journey to the stream as overland flow. The description of interflow is similar in many ways to that of perched groundwater (Sec. 6-1). Certainly what is described as interflow varies from groundwater only in speed of travel. In limestone terrain, groundwater frequently moves at relatively high velocities as turbulent flow through solution channels and fractures in the limestone. Streams in limestone country often exhibit a high ratio of flood-peak flows to average flow, a condition characteristic of streams having small groundwater contributions. In such terrain, groundwater flow actually has some of the characteristics ascribed to interflow. For convenience it has been customary to consider the total flow to be divided into only two parts: *storm*, or *direct*, *runoff* and *base flow*. The distinction is actually on the basis of time of arrival in the stream rather than on the path followed. Direct runoff is presumed to consist of overland flow and a substantial portion of the interflow, whereas base flow is considered to be largely groundwater. Computer simulation techniques (Chap. 10) commonly use all components.

7-2 Streamflow Recessions

A typical hydrograph resulting from an isolated period of rainfall (Fig. 7-2) consists of a *rising limb*, *crest segment*, and *falling limb*, or *recession*. The shape of the rising limb is influenced mainly by the character of the storm which caused the rise. The point of inflection on the falling side of the hydrograph is commonly assumed to mark the time at which surface inflow to the channel system ceases. Thereafter, the recession curve represents withdrawal of water from storage within the basin. The shape of the recession is largely independent of the characteristics of the storm causing the rise.

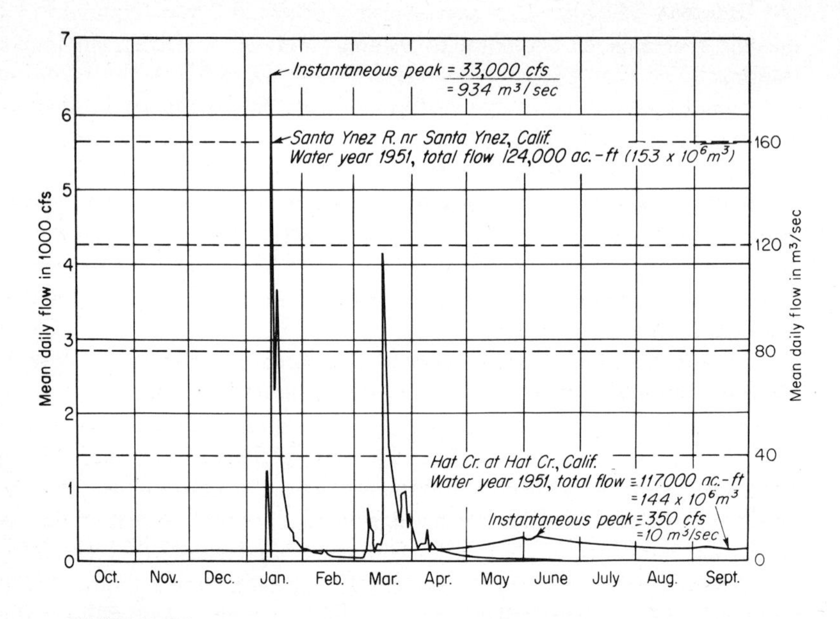

FIGURE 7-1
Comparison of hydrographs from two streams of differing geologic characteristics.

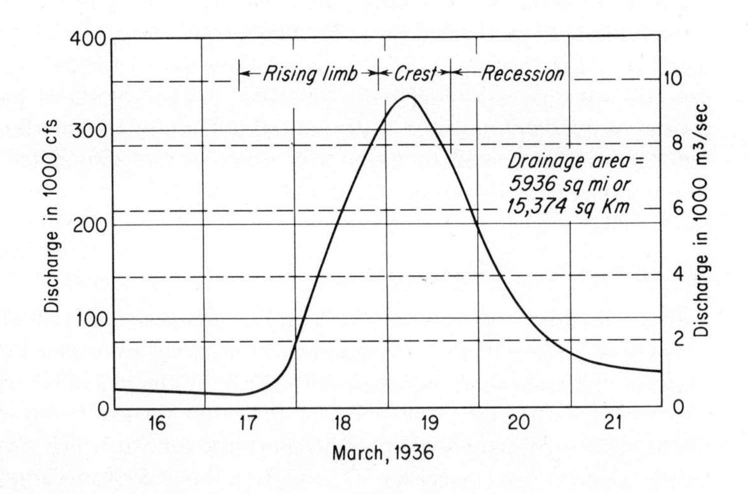

FIGURE 7-2
A typical hydrograph (Potomac River at Shepherdstown, West Virginia).

On large basins subject to runoff-producing rainfall over only a part of the basin, the recession may vary from storm to storm, depending on the particular area of runoff generation. If rainfall occurs while the recession from a previous storm is in progress, the recession will naturally be distorted. However, the recession curve for a basin is a useful tool in hydrology.

The *recession curve* (sometimes called a *depletion curve* because it represents depletion from storage) is described by a characteristic depletion equation

$$q_1 = q_0 K_r \qquad (7\text{-}1)$$

where q_0 is the flow at any time, q_1 is the flow one time unit later, and K_r is a recession constant which is less than unity. Equation (7-1) can be written in the more general form as

$$q_t = q_0 K_r^t \qquad (7\text{-}2)$$

where q_t is the flow t time units after q_0. The time unit is frequently taken as 24 h although on small basins a shorter unit may be necessary. The numerical value of K_r depends on the time unit selected. Integrating Eq. (7-2) and remembering that the volume of water discharged during time dt is $q\,dt$ and is equal to the decrease in storage $-dS$ during the same interval, we see that the storage S_t remaining in the basin at time t is

$$S_t = -\frac{q_t}{\ln K_r} \qquad (7\text{-}3)$$

Equation (7-2) will plot as a straight line on semilogarithmic paper with q on the logarithmic scale. If the recession of a stream rise is plotted on semilogarithmic paper (Fig. 7-3), the result is usually not a straight line but a curve with gradually decreasing slope, i.e., increasing values of K_r. The reason for this is that the water is coming from three different types of storage—stream channels, surface soil, and the groundwater—each having different lag characteristics. Barnes [2] suggests that the recession can be approximated by three straight lines on a semilogarithmic plot. The transition from one line to the next is often so gradual that it is difficult to select the points of change in slope. Considering the heterogeneity of the typical watershed, this is not surprising. Several aquifers may be contributing groundwater while influent seepage is occurring at other points in the stream. In most cases runoff occurs in varying amounts over the watershed.

The slope of the last portion of the recession should represent the characteristic K_r for groundwater since, presumably, both interflow and surface runoff have ceased. By projecting this slope backward in time (Fig. 7-3) and replotting the difference between the projected line and the total hydrograph, a recession which for a time consists largely of interflow

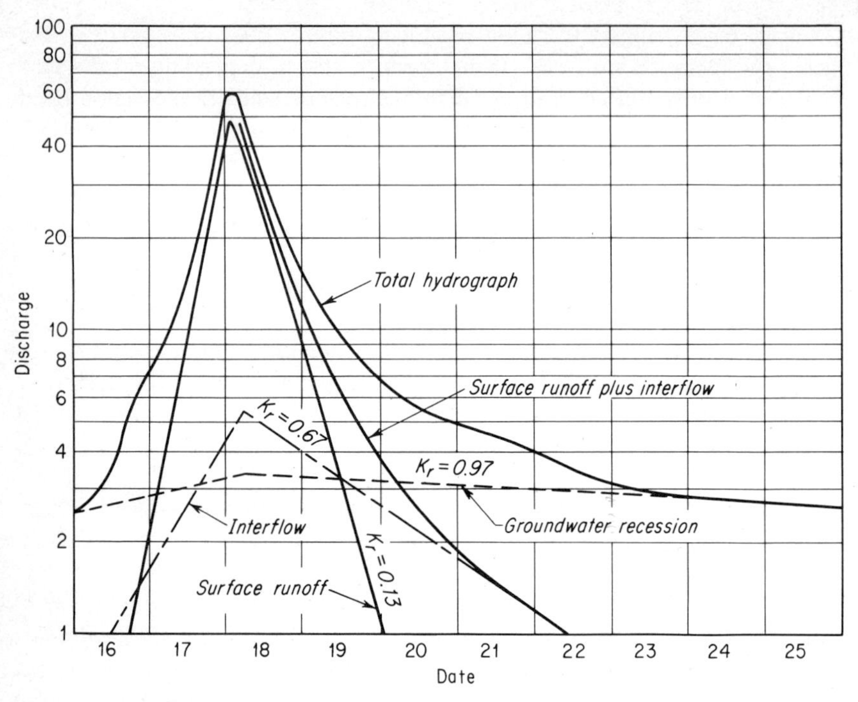

FIGURE 7-3
Semilogarithmic plotting of a hydrograph, showing method of recession analysis.

is obtained. With the slope applicable to interflow thus determined, the process can be repeated to establish the recession characteristics of surface runoff.

The technique described above represents a degree of refinement rarely used for engineering problems. A recession curve may be developed [3] by plotting values of q_0 against q_t some fixed time t later (Fig. 7-4). If Eq. (7-1) were strictly correct, the plotted data would indicate a straight line. Normally, however, a curve indicating a gradual change in the value of K_r results. This curve becomes asymptotic to a 45° line as q approaches zero.

The method illustrated in Fig. 7-4 may be used to construct recessions for base flow or direct runoff. For a base-flow recession, data should be selected from periods several days after the peak of a flood so that it is reasonably certain that no direct runoff is included. After the base-flow recession has been established, it can be projected back under the hydrograph immediately following a flood peak and the difference between projected base

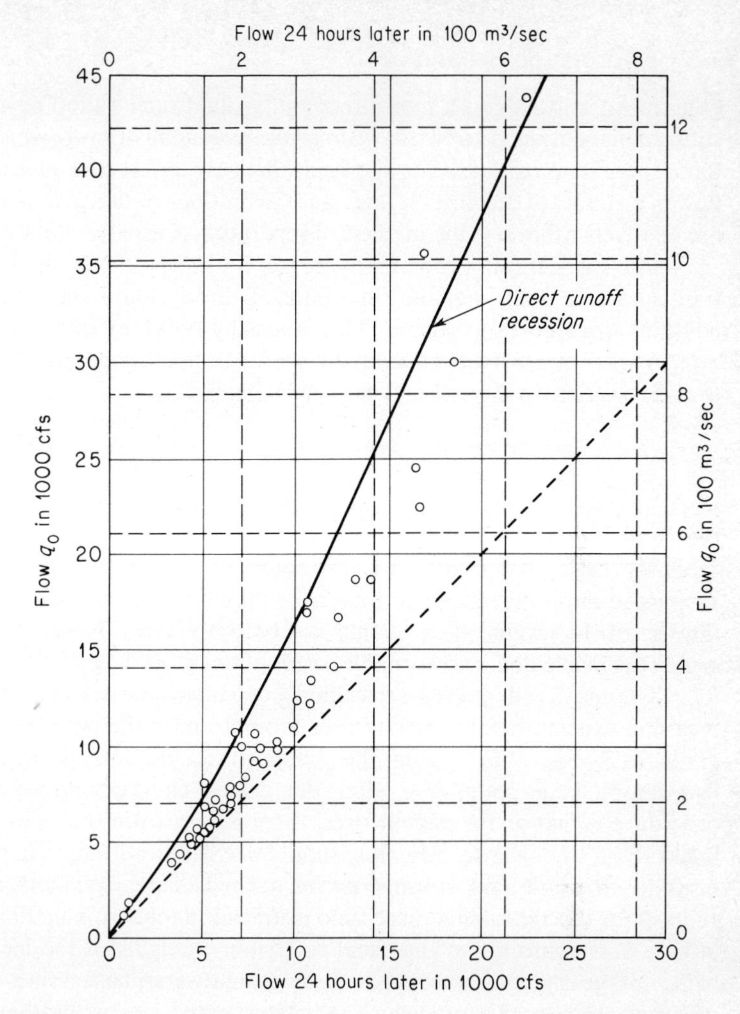

FIGURE 7-4
Recession curve in the form q_0 versus q_1 for the American River at Fair Oaks, California.

flow and the total hydrograph used to develop a direct-runoff recession curve. It is customary to draw the base-flow curve to envelop the plotted data on the right, because such a curve represents the slowest recession (high K_r) and points deviating to the left may include direct runoff. By a similar argument, data for the direct-runoff recession are enveloped on the left.

7-3 Hydrograph Separation

Division of a hydrograph into direct and groundwater runoff as a basis for subsequent analysis is known as *hydrograph separation* or *hydrograph analysis.* Since there is no real basis for distinguishing between direct and groundwater flow in a stream at any instant, and since definitions of these two components are relatively arbitrary, the method of separation is usually equally arbitrary.

For application of the unit hydrograph concept, the method of separation should be such that the time base of direct runoff remains relatively constant from storm to storm. This is usually provided by terminating the direct runoff at a fixed time after the peak of the hydrograph. As a rule of thumb, the time in days N may be approximated by

$$N = A^{0.2} \qquad (7\text{-}4)$$

where A is the drainage area in square miles. With area in square kilometers the computed values of N should be multiplied by 0.8. However, N is probably better determined by inspection of a number of hydrographs, keeping in mind that the total time base should not be excessively long and the rise of the groundwater should not be too great. Figure 7-5 illustrates some reasonable and unreasonable assumptions regarding N.

The most widely used separation procedure consists of extending the recession existing before the storm to a point under the peak of the hydrograph (AB, Fig. 7-6). From this point a straight line is drawn to the hydrograph at a point N d after the peak. The reasoning behind this procedure is that as the stream rises, there is flow from the stream into the banks (Fig. 7-7). Hence, base flow should decrease until stages in the stream begin to drop and bank storage returns to the channel. While there is some support for this reasoning, there is no justification for assuming that decrease in base flow conforms to the usual recession. Actually, if the increment of bank storage is greater than inflow from groundwater, base flow is effectively negative. Hence, this procedure is arbitrary and no better than line AC (Fig. 7-6), which is simply a straight line from the point of rise to the hydrograph N d after the peak. The difference in volume of base flow by these two methods is quite small and probably unimportant as long as one method is used consistently.

A third method of separation is illustrated by line ADE (Fig. 7-6). This line is constructed by projecting the groundwater recession after the storm back under the hydrograph to a point under the inflection point of the falling limb. An arbitrary rising limb is sketched from the point of rise of the hydrograph to connect with the projected base-flow recession. This type of separation may have some advantages where groundwater is relatively plentiful and reaches the stream fairly rapidly, as in limestone terrain.

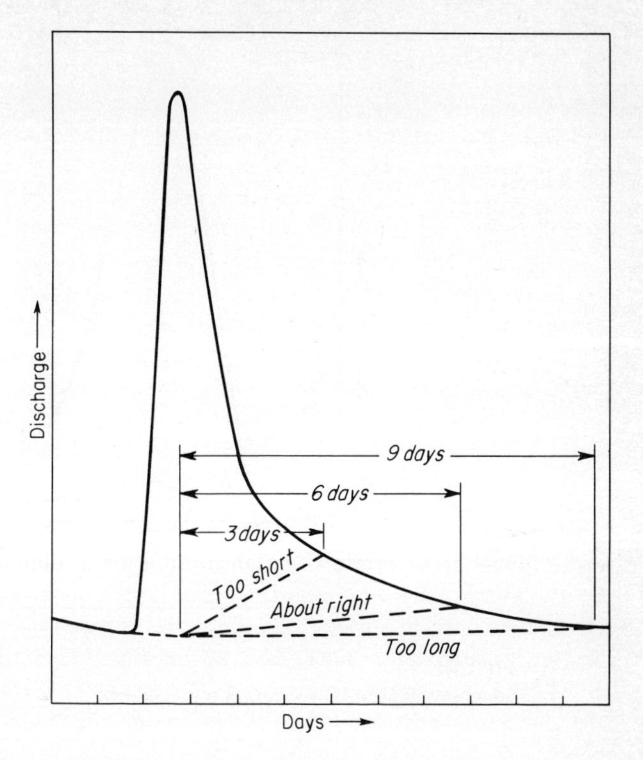

FIGURE 7-5
Selection of the time base for the surface-runoff hydrograph.

7-4 Analysis of Complex Hydrographs

The discussion of hydrograph separation in Sec. 7-3 assumed an isolated streamflow event without subsequent rainfall until after direct runoff had left the basin. This type of event is easier to analyze than the complex hydrographs resulting from two or more closely spaced bursts of rainfall (Fig. 7-8). Often, however, analysis of the more complex cases cannot be avoided. In these cases it is necessary to separate the runoff caused by individual bursts of rainfall in addition to separating direct runoff from base flow.

If a simple base-flow separation such as line *ABC* or *AC* of Fig. 7-6 is to be used, division between bursts of rain is usually accomplished by projecting the small segment of recession between peaks, using a total-flow recession curve for the basin (line *AB*, Fig. 7-8). Base-flow separation is then completed by drawing *CDB* and *EF*. Direct runoff for the two periods of rain is given

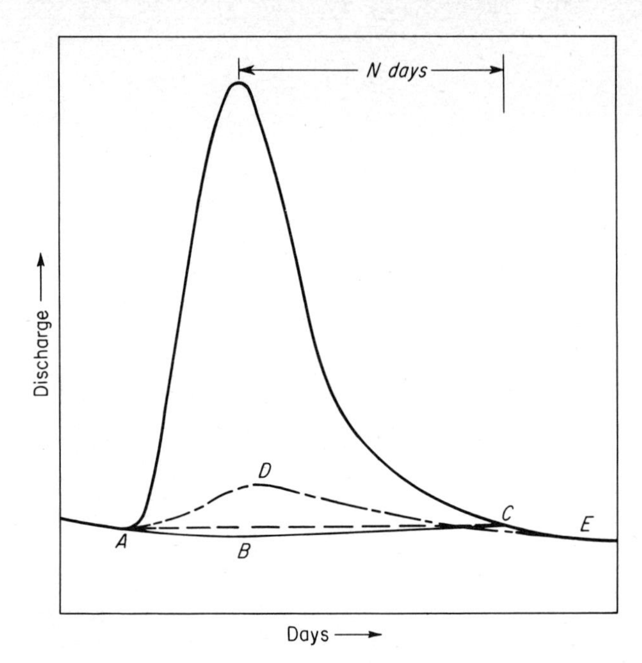

FIGURE 7-6
Some simple base-flow separation procedures.

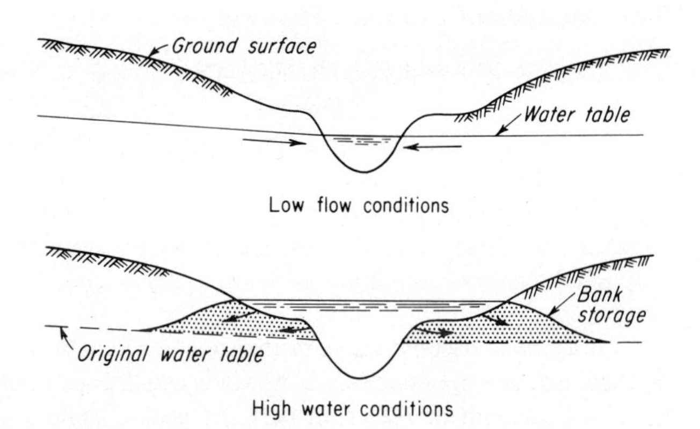

FIGURE 7-7
Bank-storage variations during a flood.

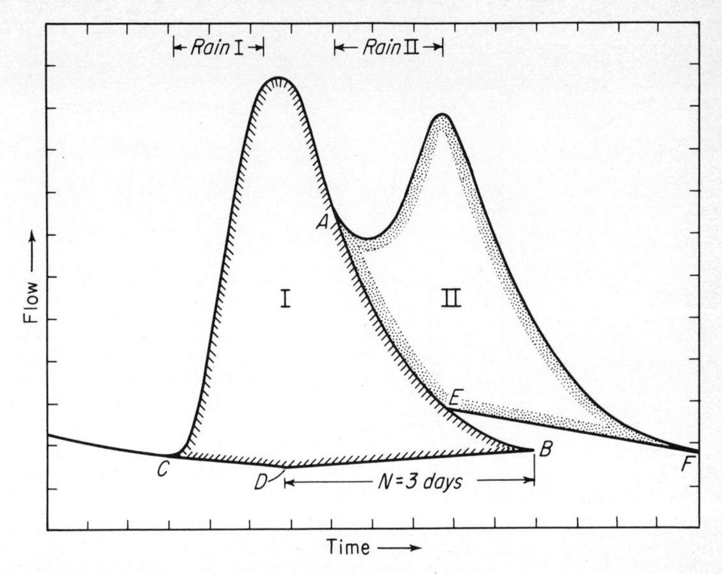

FIGURE 7-8
Separation of complex hydrograph using recession curve.

by shaded areas I and II. A separation of this type is impracticable unless there are two clearly defined peaks with a short segment of recession following the first. If such a separation is in error, its consequences are usually only compensating errors in estimated runoff volume for the two events. Such errors rarely cause serious trouble in subsequent analysis.

7-5 Determination of Total Runoff

In some types of analysis (see Sec. 5-11) there is need to determine the total streamflow (direct runoff plus groundwater) resulting from a particular storm or group of storms. This can be done by computing the total volume of flow occurring during a period beginning and ending with the same discharge and encompassing the rise under consideration, provided that groundwater-recession conditions prevail at both times. In Fig. 7-9, the area under the hydrograph between times A and A' or B and B' represents the total runoff from the storm causing the first rise. Similarly, that between times C and C' constitutes the total runoff produced by the last three storms. If groundwater-recession conditions do not prevail at the beginning and end of the rise under study, the recession must be extended as described earlier. Proof of the foregoing is found in the fact that, from continuity, total runoff

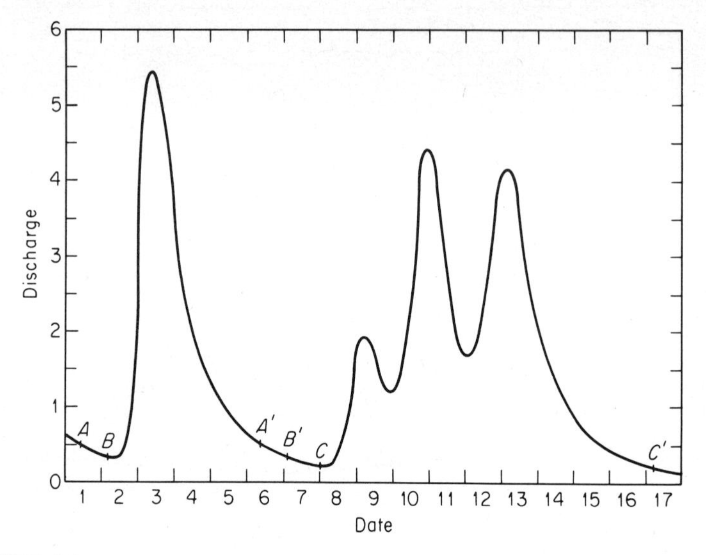

FIGURE 7-9
Determination of total runoff volume.

for the period AA' must equal the observed streamflow plus any change in storage between A and A'. But from Eq. (7-3) storage is a function of flow, and since the flows at A and A' are equal, the change in storage must be zero. Hence the observed streamflow must equal total runoff.

HYDROGRAPH SYNTHESIS

The earliest method of estimating peak flows was by using empirical formulas considered unacceptable for engineering applications today. In 1932, Sherman [4] introduced the unit hydrograph as a tool for estimating hydrograph shape. The unit hydrograph has been the mainstay of the flood hydrologist, but flood-routing methods (Chap. 9) offer greater flexibility and accuracy in many applications. Methods of estimating runoff volume are discussed in Chap. 8.

7-6 The Elemental Hydrograph

If a small, impervious area is subjected to rainfall at a constant rate, the resulting runoff hydrograph will appear much as in Fig. 7-10. Since sheet flow over the surface cannot occur without a finite depth of water on the surface, some of the rainfall goes into temporary storage, or *surface detention.*

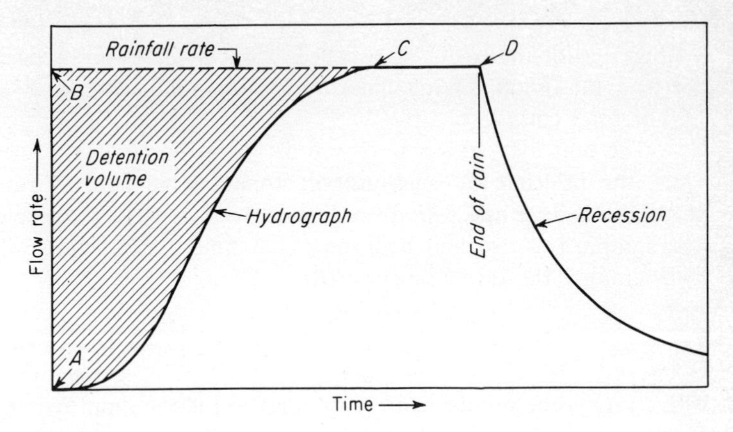

FIGURE 7-10
The elemental hydrograph.

At any instant the quantity of water in such storage is equal to the difference between total inflow to the area (rain) and total outflow from the area. When equilibrium has been reached, rate of outflow equals rate of inflow (point *C*) and the volume in detention is *ABC*. The water is in constant motion, and any given element of rainfall may pass through the system in a fairly short time, but the volume difference between inflow and outflow remains constant.

When rainfall ends (point *D*), there is no further inflow to sustain detention, and rate of outflow and detention volume decrease. The outflow follows a recession with flow decreasing at a decreasing rate; that is, d^2q/dt^2 is negative.

Theoretically, an infinite time is required both for the rising portion of the hydrograph to reach equilibrium and for the recession to return to zero. Practically, both the rising and falling curves approach their limits rapidly. The results of experimental study of the hydrographs from small areas are discussed in Sec. 7-14.

7-7 The Unit-Hydrograph Concept

The hydrograph of outflow from a small basin is the sum of the elemental hydrographs from all the subareas of the basin modified by the effect of transit time through the basin and storage in the stream channels. Since the physical characteristics of the basin—shape, size, slope, etc.—are constant, one might expect considerable similarity in the shape of hydrographs from storms of similar rainfall characteristics. This is the essence of the unit hydrograph as proposed by Sherman [4]. The unit hydrograph is a typical hydrograph for the basin. It is called a unit hydrograph because, for convenience, the runoff volume under the hydrograph is commonly adjusted to 1.00 in. (or 1 cm).

The unit hydrograph is viewed as a unit impulse in a linear system. Thus, the principle of superposition applies, and 2 in. of runoff would produce a hydrograph with all ordinates twice as large as those of the unit hydrograph, i.e., two unit hydrographs summed. Mathematically the unit hydrograph is the kernel function $U(t - T)$ in

$$q(t) = \int i(t)U(t - T)\, dt \qquad (7\text{-}5)$$

where $q(t)$ is the output hydrograph and $i(t)$ is the input hyetogram. The convolution of the unit hydrograph and rainfall excess yields the direct runoff hydrograph of a storm.

It would be wrong to imply that one typical hydrograph would suffice for any basin. Although the physical characteristics of the basin remain relatively constant, the variable characteristics of storms cause variations in the shape of the resulting hydrographs. The storm characteristics are rainfall duration, time-intensity pattern, areal distribution of rainfall, and amount of runoff. Their effects are discussed below.

Duration of rain The unit hydrograph may be employed in two ways. A unit hydrograph may be developed for a short duration (say 1 h) and all storms treated by dividing the rainfall excess into similar intervals. The alternate approach is to derive a series of unit hydrographs covering the range of durations experienced on the watershed. Because of a lack of data on hourly rainfall, the second approach was widely used when the unit hydrograph was first introduced. Theoretically an infinite number of unit hydrographs would be required to cover the range of durations. Actually the effect of small differences in duration is slight, and a tolerance of ± 25 percent in duration is usually acceptable. Thus only a few unit hydrographs are actually required. Where a computer solution is used, a short-duration unit hydrograph is preferable.

Time-intensity pattern If one attempted to derive a separate unit hydrograph for each possible time-intensity pattern, an infinite number of unit hydrographs would be required. Practically, unit hydrographs can be based only on an assumption of uniform intensity of runoff. However, large variations in rain intensity (and hence runoff rate) during a storm are reflected in the shape of the resulting hydrograph. The time scale of intensity variations that are critical depends mainly on basin size. Rainfall bursts lasting only a few minutes may cause clearly defined peaks in the hydrograph from a basin of a few hectares, while on basins of several hundred square kilometers only changes in storm intensity lasting for hours will cause distinguishable effects on the hydrograph. If the unit hydrographs for a basin are applicable to storms of shorter duration than the critical time for the basin, hydrographs of longer storms can be synthesized quite easily (Sec. 7-10). A basic duration of about one-fourth of the basin lag (Sec. 7-11) is generally satisfactory.

Areal distribution of runoff The areal pattern of runoff can cause variations in hydrograph shape. If the area of high runoff is near the basin outlet, a rapid rise, sharp peak, and rapid recession usually result. Higher runoff in the upstream portion of the basin produces a slow rise and recession and a lower, broader peak. Unit hydrographs have been developed for specific runoff patterns, e.g., heavy upstream, uniform, or heavy downstream. This is not wholly satisfactory because of the subjectivity of classification. A better solution is to apply the unit-hydrograph method only to basins small enough to ensure that the usual areal variations will not be great enough to cause major changes in hydrograph shape. The limiting basin size is fixed by the accuracy desired and regional climatic characteristics. Generally, however, unit hydrographs should not be used for basins much over 2000 mi^2 (5000 km^2) unless reduced accuracy is acceptable. What has been said does not apply to rainfall variations caused by topographic controls. Such rainfall patterns are relatively fixed characteristics of the basin. It is departures from the normal pattern that cause trouble.

Amount of runoff Inherent in the assumption of a linear unit hydrograph is the assumption that ordinates of flow are proportional to volume of runoff for all storms of a given duration and that the time bases of all such hydrographs are equal. This assumption is obviously not completely valid, since from the character of recession curves, duration of recession must be a function of peak flow. Moreover, unit hydrographs for storms of the same duration but different magnitudes do not always agree. Peaks of unit hydrographs derived from very small events are commonly lower than those derived from larger storms. This may be because the smaller events contain less surface runoff and relatively more interflow and groundwater than the larger events or because channel flow time is longer at low flows.

In the range of floods frequently experienced on a watershed, it is relatively simple to verify the adequacy of the assumption of linearity by comparing hydrographs from storms of various magnitudes. If important nonlinearity exists, derived unit hydrographs should be used only for reconstructing events of similar magnitude. A series of unit hydrographs covering appropriate ranges of magnitude would be required for each duration. Much more critical is the extreme event, exceeding any which has occurred on the watershed. There is no way of obtaining empirical evidence about the changes in unit hydrograph peak. Many hydrologists increase the peaks of unit hydrographs derived from ordinary floods by 5 to 20 percent before using them for estimation of extreme floods. This increase is based on the belief that channel-flow time shortens as flood magnitude increases. However, if extreme floods overflow onto floodplains, the opposite effect might result. Caution should be exercised in using the unit hydrograph to extrapolate extreme events.

In the light of the foregoing discussion, the *unit hydrograph* can be defined as *the hydrograph of one inch (or one centimeter) of direct runoff from a storm of specified duration.* For a storm of the same duration but with a different amount of runoff, the hydrograph of direct runoff is assumed to have the same time base as the unit hydrograph and ordinates of flow approximately proportional to the runoff volume. The duration assigned to a unit hydrograph should be the duration of rainfall producing significant runoff, determined by inspection of hourly rainfall data.

7-8 Derivation of Unit Hydrographs

The unit hydrograph is best derived from the hydrograph of a storm of reasonably uniform intensity, duration of desired length, and a runoff volume near or greater than 1.0 in. (or 1 cm). The first step (Fig. 7-11) is to separate the base flow from direct runoff. The volume of direct runoff is then determined, and the ordinates of the direct-runoff hydrograph are divided by observed runoff depth. The adjusted ordinates form a unit hydrograph.

A unit hydrograph derived from a single storm may be in error, and it is desirable to average unit hydrographs from several storms of the same duration. This should not be an arithmetic average of concurrent coordinates, since if peaks do not occur at the same time, the average peak will be lower than the individual peaks. The proper procedure is to compute average peak flow and time to peak. The average unit hydrograph is then sketched to conform to the shape of the other graphs, passing through the computed average peak, and having a unit volume of 1 in. or 1 cm (Fig. 7-12).

7-9 Derivation of Unit Hydrograph from Complex Storms

The simple approach outlined in Sec. 7-8 often proves inadequate because no ideal storm is available in the record. It is then necessary to develop the unit hydrograph from a complex storm. If individual bursts of rain in the storm result in well-defined peaks, it is possible to separate the hydrographs of several bursts (Sec. 7-4) and use these hydrographs as independent storms. If the resulting unit hydrographs are averaged, errors in the separation are minimized.

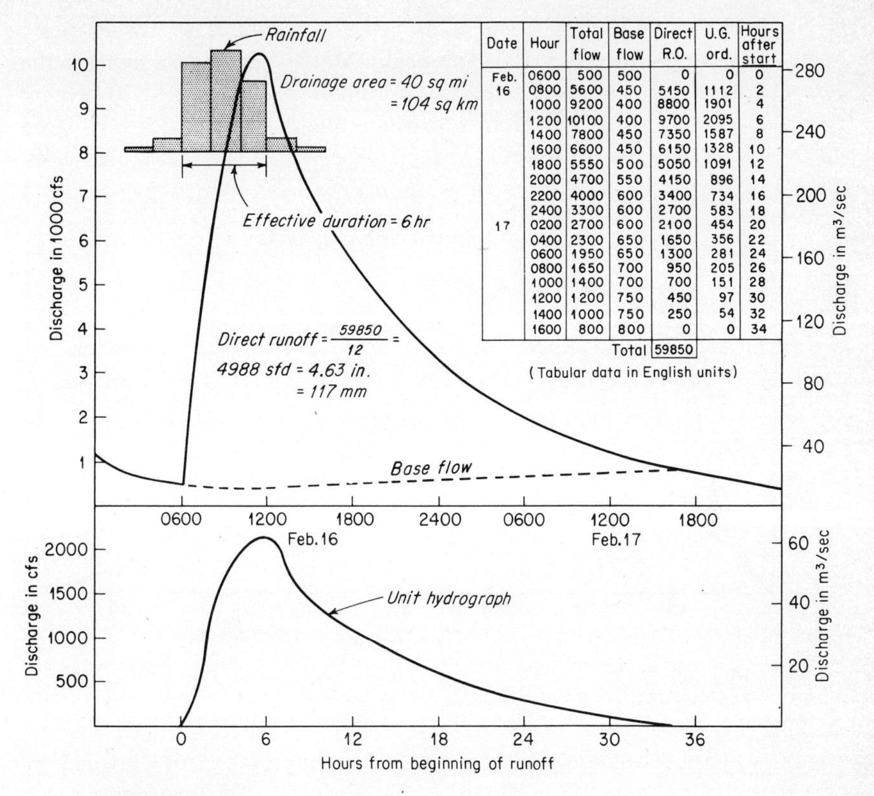

Date	Hour	Total flow	Base flow	Direct R.O.	U.G. ord.	Hours after start
Feb. 16	0600	500	500	0	0	0
	0800	5600	450	5150	1112	2
	1000	9200	400	8800	1901	4
	1200	10100	400	9700	2095	6
	1400	7800	450	7350	1587	8
	1600	6600	450	6150	1328	10
	1800	5550	500	5050	1091	12
	2000	4700	550	4150	896	14
	2200	4000	600	3400	734	16
	2400	3300	600	2700	583	18
17	0200	2700	600	2100	454	20
	0400	2300	650	1650	356	22
	0600	1950	650	1300	281	24
	0800	1650	700	950	205	26
	1000	1400	700	700	151	28
	1200	1200	750	450	97	30
	1400	1000	750	250	54	32
	1600	800	800	0	0	34
			Total	59850		

(Tabular data in English units)

FIGURE 7-11
Development of a unit hydrograph.

If the hydrograph does not lend itself to separation, analysis begins with estimates of direct-runoff volumes $Q_1, Q_2, \ldots, Q_n$ in successive periods during the storm. The equation for any ordinate of the total hydrograph q_n in terms of runoff Q and unit-hydrograph ordinate U is

$$q_n = Q_n U_1 + Q_{n-1} U_2 + Q_{n-2} U_3 + \cdots + Q_1 U_n \tag{7-6}$$

The first ordinate $q_1 = Q_1 U_1$, and since Q_1 is known (or estimated), U_1 can be computed (Fig. 7-13). The second ordinate is then

$$q_2 = Q_1 U_2 + Q_2 U_1 \tag{7-7}$$

The only unknown in this equation is U_2, which can be computed. All the ordinates can be determined in a similar way.

Although the procedure just outlined may seem simple, there are numerous difficulties. Each computation depends on all preceding computa-

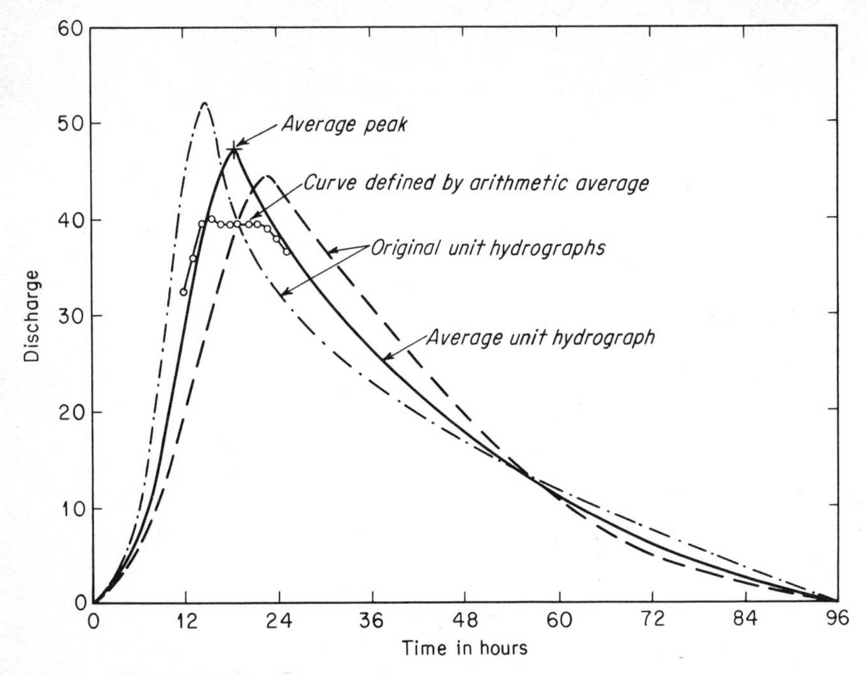

FIGURE 7-12
Construction of an average unit hydrograph.

tions for values of U. Errors in estimating runoff increments, errors in observed streamflow, and variations in intensity and areal distribution of rainfall during several storm periods can lead to cumulative errors which may become large in the latter portion of the unit hydrograph. Large negative ordinates sometimes develop.

A unit hydrograph can also be developed by successive approximations. A unit hydrograph is assumed and used to reconstruct the storm hydrograph (Fig. 7-19). If the reconstructed hydrograph does not agree with the observed hydrograph, the assumed unit hydrograph is modified and the process repeated until a unit hydrograph which seems to give the best fit is determined.

A more elegant but laborious method is the use of least squares [5], a statistical technique to find the constants a and b_i in an equation of the form

$$q = a + b_1Q_1 + b_2Q_2 + b_3Q_3 + \cdots + b_iQ_i \qquad (7\text{-}8)$$

Equations (7-6) and (7-8) are similar. Theoretically, for the unit hydrograph, a should be zero and the b's equivalent to the unit-hydrograph ordinates U_i. The least-squares technique would use data from a number of flood events

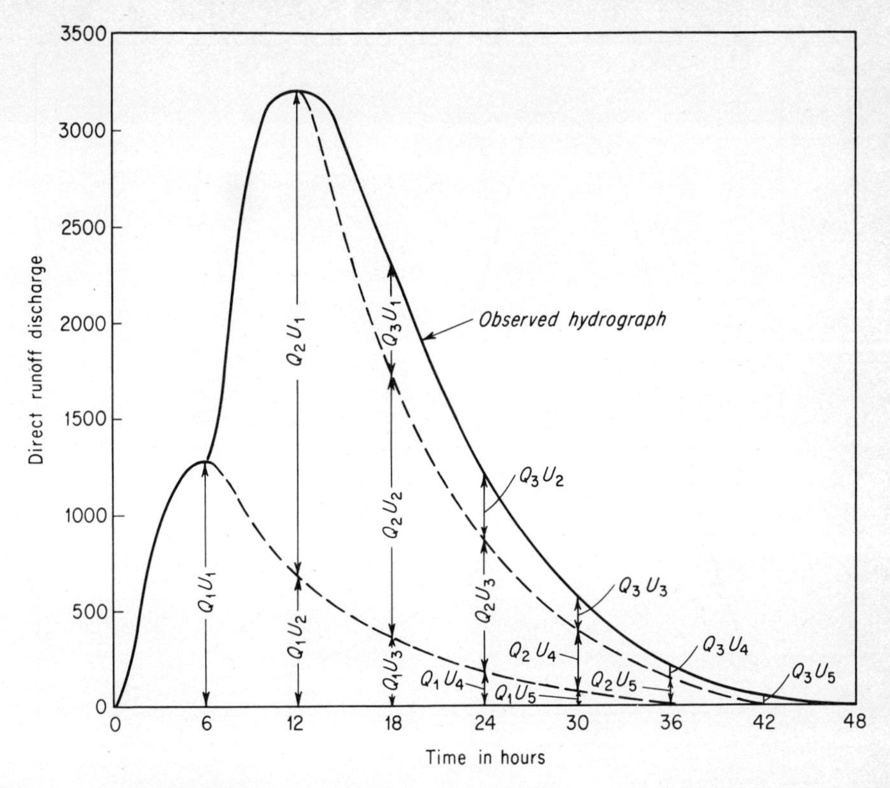

FIGURE 7-13
Unit hydrographs from a complex storm.

for which values of q and Q are established to develop a set of average values of U_i.

7-10 Unit Hydrographs for Various Durations

If a unit hydrograph for duration t h is added to itself lagged by t h (Fig. 7-14), the resulting hydrograph represents the hydrograph for 2 in. of runoff in $2t$ h. If the ordinates of this graph are divided by 2, the result is a unit hydrograph for duration $2t$ h. The final graph represents the flow from 1 in. of runoff generated at uniform intensity of $1/2t$ in./h in $2t$ h. This simple example illustrates the ease with which a unit hydrograph for a short duration can be converted into a unit hydrograph for any multiple of the original duration.

Construction of a short-duration unit hydrograph from a longer duration can be accomplished by the methods described in Sec. 7-9. A more

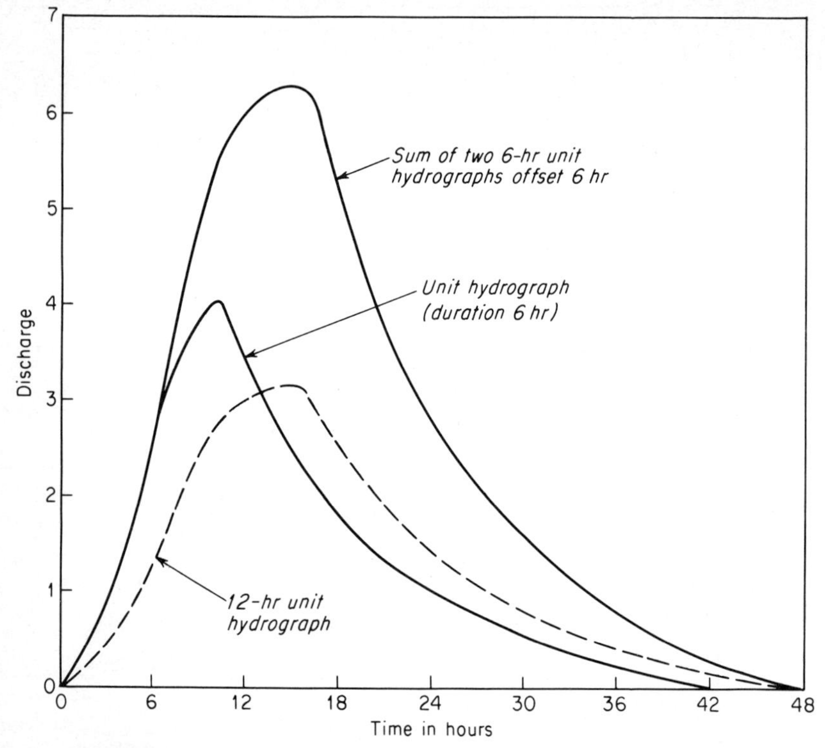

FIGURE 7-14
Construction of a unit hydrograph for duration $2t$.

convenient technique for conversion to either a shorter or longer duration is the S curve, or summation-curve, method. The *S curve* is the hydrograph that would result from an infinite series of runoff increments of 1 in. (or 1 cm) in t h. Thus, each S curve applies to a specific duration within which each inch of runoff is generated. The S curve is constructed by adding together a series of unit hydrographs, each lagged t h with respect to the preceding one (Fig. 7-15). If the time base of the unit hydrograph is T h, then a continuous rainfall producing 1 in. of runoff every t h would develop a constant outflow at the end of T h. Thus only T/t unit hydrographs need be combined to produce an S curve which should reach equilibrium at flow q_e:

$$q_e = \frac{24 \times 26.9A}{t} = \frac{645.6A}{t} \tag{7-9}$$

where A is drainage area in square miles, t is duration in hours, 24 is the

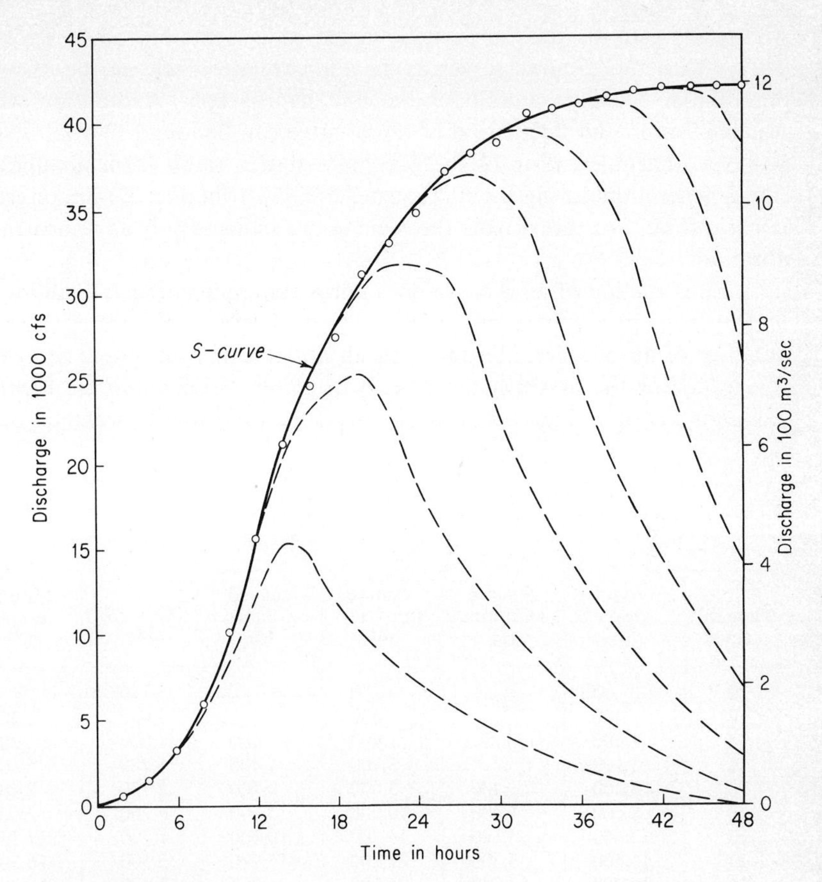

FIGURE 7-15
Graphical illustration of the *S* curve. See Table 7-1.

number of hours in a day, and 26.9 is the number of second-foot-days in 1 in. of runoff from 1 mi^2. In metric units 1 cm/h will produce an equilibrium flow of 2.78 $m^3/s\text{-}km^2$.

Commonly, the *S* curve tends to fluctuate about the equilibrium flow. This means that the initial unit hydrograph does not actually represent runoff at a uniform rate over time *t*. If a uniform rate of runoff is applied to a basin, equilibrium flow at the rate given by Eq. (7-9) must eventually develop. If the actual effective duration of runoff associated with the original unit hydrograph is not *t* h, the summation process results in a runoff diagram with either periodic gaps or periodic increases to a rate of 2 in. in t h (Fig. 7-16). Thus the S curve serves as an approximate check on the assumed duration of effective rainfall for the unit hydrograph. A duration which results in minimum fluctuation of the S curve can be found by trial. Note, however, that fluctuation of the S curve can also result from nonuniform runoff generation during the t h, unusual areal distribution of rain, or errors in basic data. For this reason, the S curve can indicate only an approximate duration.

Construction of an S curve does not require tabulating and adding T/t unit hydrographs with successive lags of t h. Table 7-1 illustrates the computation of an S curve, starting with an initial unit hydrograph for which $t = 6$ h. For the first 6 h the unit hydrograph and S curve are identical (columns 2 and 4). The S-curve additions (column 3) are the ordinates of the

Table 7-1 APPLICATION OF *S*-CURVE METHOD

Time, h (1)	6-h unit graph (2)	*S*-curve additions (3)	*S* curve (2) + (3) (4)	Lagged *S* curve (5)	(4) − (5) (6)	2-h unit graph (7)
0	0	...	0	...	0	0
2	400	...	400	0	400	1,200
4	1,400	...	1,400	400	1,000	3,000
6	3,100	0	3,100	1,400	1,700	5,100
8	5,400	400	5,800	3,100	2,700	8,100
10	8,600	1,400	10,000	5,800	4,200	12,600
12	12,600	3,100	15,700	10,000	5,700	17,100
14	15,400	5,800	21,200	15,700	5,500	16,500
16	14,600	10,000	24,600	21,200	3,400	10,200
18	11,800	15,700	27,500	24,600	2,900	8,700
20	9,900	21,200	31,100	27,500	2,400†	7,200
22	8,400	24,600	33,000	31,100	2,000†	6,000
24	7,200	27,500	34,700	33,000	1,800†	5,400
26	6,000	31,100	37,100	34,700	1,600†	4,800
28	5,100	33,000	38,100	37,100	1,400†	4,200
30	4,200	34,700	38,900	38,100	1,200†	3,600
32	3,400	37,100	40,500	38,900	1,000†	3,000
34	2,700	38,100	40,800	40,500	800†	2,400
36	2,100	38,900	41,000	40,800	600†	1,800
38	1,600	40,500	41,500†	41,000	400†	1,200
40	1,100	40,800	41,900	41,500	200†	600
42	700	41,000	42,000†	41,900	100	300
44	400	41,500†	42,000†	42,000	0	0
46	200	41,900	42,100	42,000	0†	
48	0	42,000	42,100†	42,100	0	

† Adjusted value.

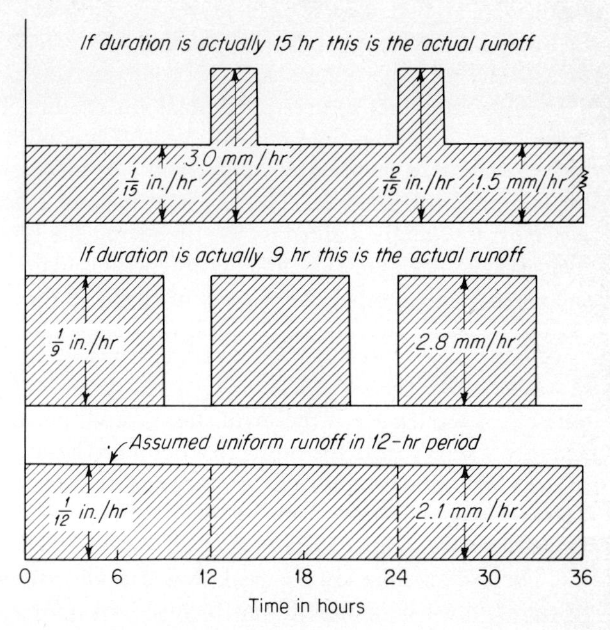

FIGURE 7-16
Influence of nonuniform rate of runoff on the *S* curve.

S curve set ahead 6 h. Since an *S*-curve ordinate is the sum of all concurrent unit-hydrograph ordinates, combining the *S*-curve additions with the initial unit hydrograph is the same as adding all previous unit hydrographs.

The difference between two *S* curves with initial points displaced by t' h gives a hydrograph for the new duration t' h. Since the *S* curve represents runoff production at a rate of 1 in. (or 1 cm) in t h, the runoff volume represented by this new hydrograph will be t'/t in. (or cm). Thus the ordinates of the unit hydrograph for t' h are computed by multiplying the *S*-curve differences by the ratio t/t'.

7-11 Synthetic Unit Hydrographs

Only a relatively small number of streams are gaged. Unit hydrographs can be derived as described in previous sections only if records are available. Hence some means of deriving unit hydrographs for ungaged basins is necessary. This requires a relation between the physical geometry of the basin and resulting hydrographs. Three approaches have been used: formulas relating hydrographs to basin characteristics, transposition of unit hydrographs from one basin to another (Sec. 7-12), and storage routing (Sec. 9-11).

Most attempts to derive formulas for the unit hydrograph have been aimed at determining time of peak, peak flow, and time base. These items plus the fact that the volume must equal 1.00 in. (or 1 cm) permit sketching the complete hydrograph. The key item in most studies has been the *basin lag*, most frequently defined as the time from the centroid of rainfall to the hydrograph peak.† The first synthetic-unit-hydrograph procedure was presented by Snyder [6]. In a study of basins in the Appalachian Mountain region, he found that the basin lag t_p (in hours) can be expressed by

$$t_p = C_t(LL_c)^{0.3} \qquad (7\text{-}10)$$

where L is the length of the main stream from outlet to divide in miles and L_c is the distance from the outlet to a point on the stream nearest the centroid of the basin. The product LL_c is a measure of the size and shape of the basin. The coefficient C_t varied from 1.8 to 2.2, with some indication of lower values for basins with steeper slopes.

Before an equation for peak flow can be written, a standard duration of rain t_r must be adopted, and Snyder took $t_r = t_p/5.5$. For rains of this duration he found that the unit-hydrograph peak q_p is given by

$$q_p = \frac{640C_pA}{t_p} \qquad (7\text{-}11)$$

where A is the drainage area in square miles, C_p is a coefficient ranging from 0.56 to 0.69, and 640 is a conversion factor to give q_p in cubic feet per second. For area in square kilometers and flow in cubic meters per second the constant becomes 7.0.

Snyder adopted the time base T (in days) of the hydrograph as

$$T = 3 + 3\frac{t_p}{24} \qquad (7\text{-}12)$$

The constants of Eq. (7-12) are fixed by the procedure used to separate base flow from direct runoff. Equations (7-10) to (7-12) define the three factors necessary to construct the unit hydrograph for duration t_r. For any other duration t_R the lag is

$$t_{pR} = t_p + \frac{t_R - t_r}{4} \qquad (7\text{-}13)$$

and this modified lag t_{pR} is used in Eqs. (7-11) and (7-12).

† Lag is also defined as the time difference between the centroid of rainfall and the centroid of runoff. This definition is more rigorous, but the one in the text is simpler to apply.

Snyder's synthetic-unit-hydrograph formulas have been tried elsewhere with varying success. The coefficients C_t and C_p are found to vary considerably. Other investigators have proposed formulas for synthetic unit hydrographs. Many have found the product LL_c significant with an exponent in the vicinity of 0.3. However, the best way any of the methods can be used is to derive coefficients for gaged streams in the vicinity of the problem basin and apply these to the ungaged stream. Perhaps because of approximations inherent in the unit-hydrograph concept, synthetic methods of constructing unit hydrographs seem to be of limited value.

7-12 Transposing Unit Hydrographs

From Eq. (7-10) a general expression for basin lag might be expected to take the form

$$t_p = C_t \left(\frac{LL_c}{\sqrt{s}} \right)^n \qquad (7\text{-}14)$$

If known values of lag are plotted against $LL_c/\sqrt{s}$ on logarithmic paper (Fig. 7-17), the resulting plot should define a straight line, provided that the values are taken from basins of similar hydrologic characteristics, that is, C_t = const. A relation such as Fig. 7-17 offers a means of estimating basin lag. The peak flow and shape of the unit hydrograph can be estimated by using a plot relating q_p to t_p or by using dimensionless hydrographs [7–9] such as Fig. 7-18. The dimensionless form eliminates the effect of basin size and much of the effect of basin shape. The similarity of the several graphs of Fig. 7-18 reflects a considerable similarity in flood hydrographs from various regions. Unfortunately, relationships between hydrograph shape and basin characteristics are so complex that no completely successful relationships have been derived.

7-13 Application of Unit Hydrographs

Application of a unit hydrograph has been demonstrated in Sec. 7-10 on the basis of uniform production of runoff. Figure 7-19 illustrates the use of a 3-h unit hydrograph to synthesize the storm hydrograph from a series of rainfall periods with varying intensity. The increments of runoff for successive 3-h periods are computed using runoff relations (Chap. 8). The hydrograph of direct runoff resulting from each 3-h increment is given by multiplying the unit hydrograph by the period runoff. The total hydrograph is the sum of all the incremental hydrographs and estimated base flow.

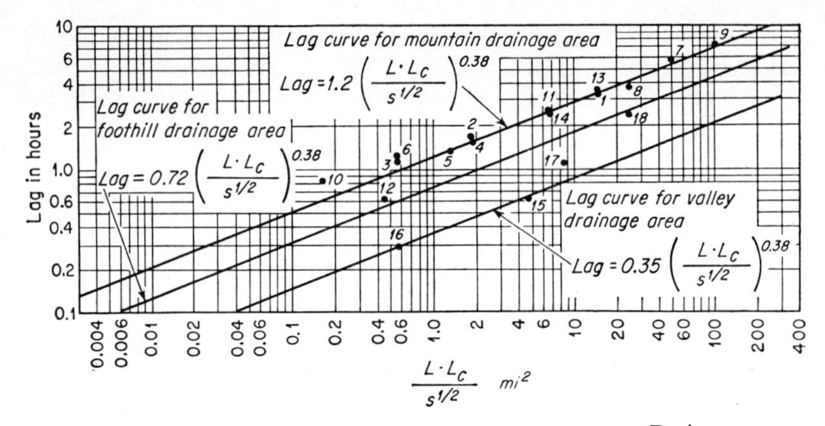

	Drainage area mi²	km²
1 San Gabriel River at San Gabriel dam	162.0	419.6
2 West Fork San Gabriel River at Cogswell dam	40.4	104.6
3 Santa Anita Creek at Santa Anita dam	10.8	28.0
4 San Dimas Creek at San Dimas dam	16.2	42.0
5 Eaton Wash at Eaton Wash dam	9.5	24.6
6 San Antonio Creek near Claremont	16.9	43.8
7 Santa Clara River near Saugus	355.0	919.4
8 Temecula Creek at Pauba Canyon	168.0	435.1
9 Santa Margarita River near Fallbrook	645.0	1670.6
10 Live Oak Creek at Live Oak dam	2.3	6.0
11 Tujunga Creek at Big Tujunga dam no. 1	81.4	210.8
12 East Fullerton Creek at Fullerton dam	3.1	8.0
13 Los Angeles River at Sepulveda dam	152.0	393.7
14 Pacoima Wash at Pacoima dam	27.8	72.0
15 Alhambra Wash above Short Street	14.0	36.3
16 Broadway drain above Raymond dike	2.5	6.5
17 Ballona Creek at Sawtelle Blvd.	88.6	229.5
18 San Jose Creek at Workman Mill Road bridge	81.3	210.6

FIGURE 7-17
Relationship between basin lag (beginning of rain to $\int q\,dt = Q/2$) and basin characteristics. (*U.S. Corps of Engineers.*)

7-14 Hydrograph of Overland Flow

Although the depth of flow in the overland sheet is quite small, the quantity of water temporarily detained in this sheet (*surface detention*) is relatively great. It is generally assumed that overland flow is laminar (Fig. 7-20). Hence,

$$\rho g(D - y)s = \mu \frac{dv}{dy} \qquad (7\text{-}15)$$

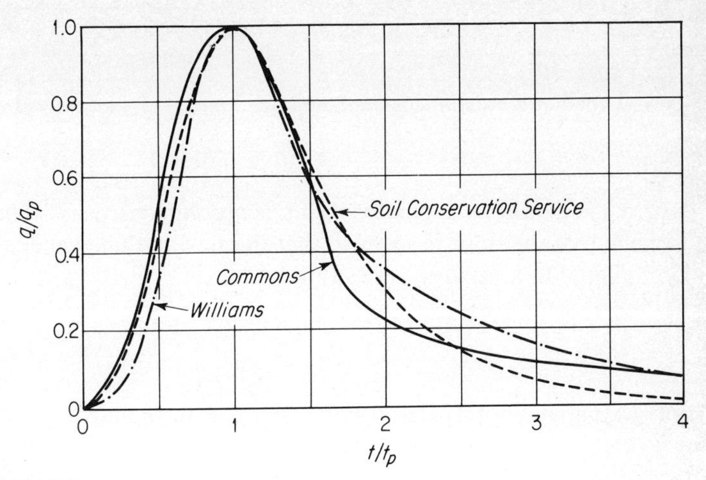

FIGURE 7-18
Some dimensionless unit hydrographs.

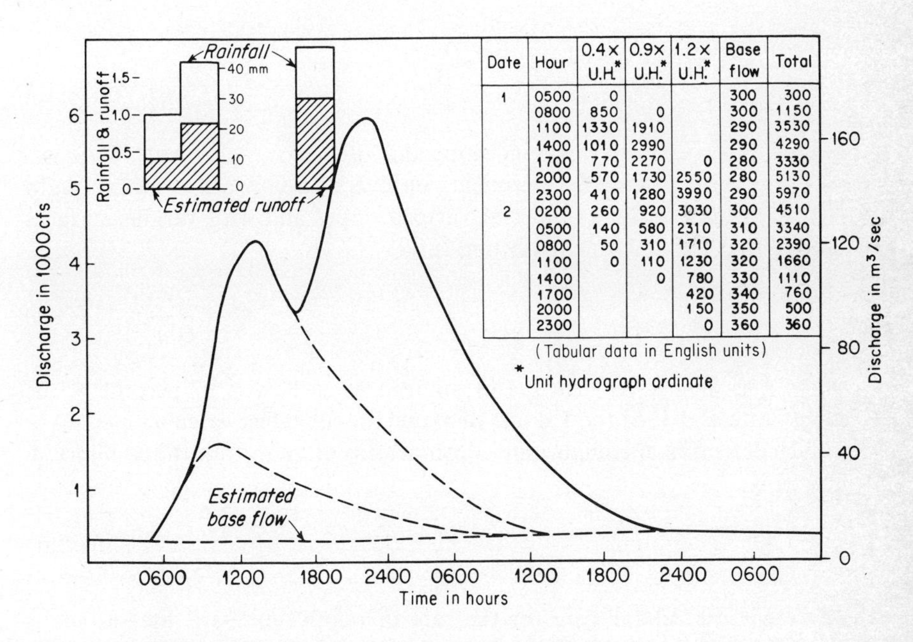

Date	Hour	0.4× U.H.*	0.9× U.H.*	1.2× U.H.*	Base flow	Total
1	0500	0			300	300
	0800	850	0		300	1150
	1100	1330	1910		290	3530
	1400	1010	2990		290	4290
	1700	770	2270	0	280	3330
	2000	570	1730	2550	280	5130
	2300	410	1280	3990	290	5970
2	0200	260	920	3030	300	4510
	0500	140	580	2310	310	3340
	0800	50	310	1710	320	2390
	1100	0	110	1230	320	1660
	1400		0	780	330	1110
	1700			420	340	760
	2000			150	350	500
	2300			0	360	360

FIGURE 7-19
Use of a unit hydrograph to synthesize a streamflow hydrograph.

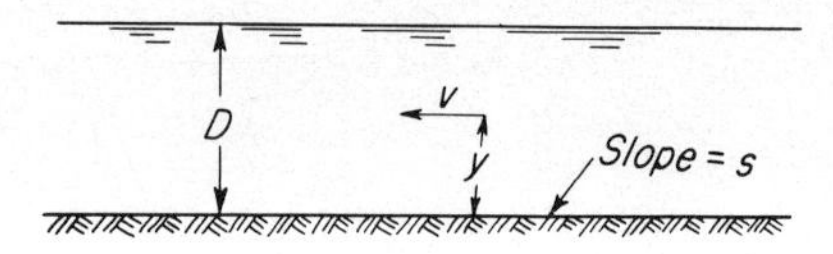

FIGURE 7-20
Definition sketch for laminar sheet flow.

where ρ is density, g is gravity, and μ is absolute viscosity. The assumption is made that the slope is so small that the sine and tangent are equal. Since μ/ρ is equal to kinematic viscosity ν,

$$dv = \frac{gs}{\nu}(D - y)\,dy \tag{7-16}$$

Integrating and noting that $v = 0$ when $y = 0$, we have

$$v = \frac{gs}{\nu}\left(yD - \frac{y^2}{2}\right) \tag{7-17}$$

Integrating from $y = 0$ to $y = D$ and dividing by D gives the mean velocity

$$v_m = \frac{gsD^2}{3\nu} \tag{7-18}$$

and the discharge per unit width $v_m D$ or

$$q = bD^3 \tag{7-19}$$

where b is a coefficient involving slope and viscosity.

The most extensive experiments on overland flow are those of Izzard [10]. His tests on long flumes at various slopes and with various surfaces showed that the time to equilibrium is

$$t_e = \frac{2V_e}{60q_e} \tag{7-20}$$

where t_e is defined as the time in minutes when flow is 97 percent of the supply rate and V_e is the volume of water (in cubic feet or cubic meters) in surface detention at equilibrium. From a strip of unit width the equilibrium flow q_e is

$$q_e = \frac{iL}{43{,}200} \tag{7-21}$$

where i is the rainfall rate (or the rate of rainfall excess if the surface is pervious) and L is the distance of overland flow. The constant 43,200 gives q_e in cubic feet per second when i is in inches per hour and L is in feet.

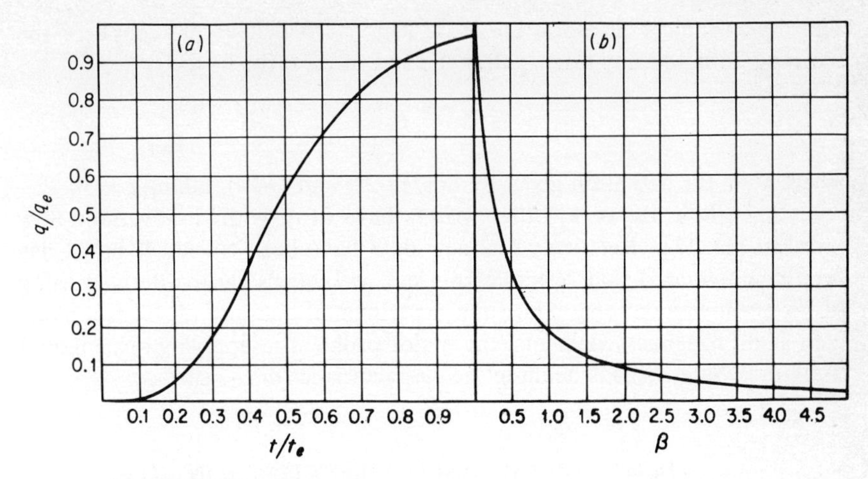

FIGURE 7-21
Dimensionless hydrograph of overland flow. (*From* [10].)

With i in millimeters per hour and L in meters the denominator becomes 3.6×10^6 to give q_e in cubic meters per second. When average depth on the strip V_e/L is substituted for outflow depth, Eq. (7-19) becomes

$$\frac{V_e}{L} = kq_e^{1/3} \tag{7-22}$$

Substituting Eq. (7-21) for q_e gives

$$V_e = \frac{kL^{4/3}i^{1/3}}{35.1} \tag{7-23}$$

where V_e is the volume of detention (in cubic feet) on the strip at equilibrium. In metric units the denominator becomes 50.2. Experimentally k was found to be given by

$$k = \frac{0.0007i + c}{s^{1/3}} \tag{7-24}$$

where s is the surface slope and the retardance coefficient c is given in Table 7-2. In metric units the multiplier for i is 2.8×10^{-5}.

Izzard found that the form of the overland-flow hydrograph can be presented as a dimensionless graph (Fig. 7-21). With t_e and q_e known, the q/q_e curve permits plotting of the rising limb of the overland-flow hydrograph.

The dimensionless recession curve of Fig. 7-21*b* defines the shape of the receding limb. At any time t_a after the end of rain, the factor β is

$$\beta = \frac{60 q_e t_a}{V_0} \qquad (7\text{-}25)$$

where V_0 is the detention given by Eqs. (7-22) and (7-24), taking $i = 0$.

Since Izzard's work, other writers have dealt with the overland-flow problem [11–13]. Relatively few new data have been collected, but it has been demonstrated that results equivalent to Izzard's can be derived using turbulent-flow equations. Since the experimental data are all from a plane with small roughness elements, the major problem in applying any method to natural watersheds is defining the characteristics of the surface.

Table 7-2 RETARDANCE COEFFICIENT c IN EQ. (7-24) [9]

Very smooth asphalt pavement	0.007
Tar and sand pavement	0.0075
Crushed-slate roofing paper	0.0082
Concrete	0.012
Tar and gravel pavement	0.017
Closely clipped sod	0.046
Dense bluegrass turf	0.060

REFERENCES

1. C. C. McDonald and W. B. Langbein, Trends in Runoff in the Pacific Northwest, *Trans. Am. Geophys. Union*, vol. 29, pp. 387–397, June 1948.
2. B. S. Barnes, Discussion of Analysis of Runoff Characteristics, *Trans. ASCE*, vol. 105, p. 106, 1940.
3. W. B. Langbein, Some Channel Storage and Unit Hydrograph Studies, *Trans. Am. Geophys. Union*, vol. 21, pt. 2, pp. 620–627, 1940.
4. L. K. Sherman, Streamflow from Rainfall by the Unit-Graph Method, *Eng. News-Rec.*, vol. 108, pp. 501–505, 1932.
5. W. M. Snyder, Hydrograph Analysis by Method of Least Squares, *Proc. ASCE*, vol. 81, sep. 793, September 1955.
6. F. F. Snyder, Synthetic Unit Hydrographs, *Trans. Am. Geophys. Union*, vol. 19, pt. 1, pp. 447–454, 1938.
7. G. G. Commons, Flood Hydrographs, *Civil Eng.*, vol. 12, p. 571, 1942.
8. H. M. Williams, Discussion of "Military Airfields," *Trans. ASCE*, vol. 110, p. 820, 1945.
9. V. Mockus, "Use of Storm and Watershed Characteristics in Synthetic Hydrograph Analysis and Application," U.S. Soil Conservation Service, 1957.

10. C. F. Izzard, Hydraulics of Runoff from Developed Surfaces, *Proc. High. Res. Board*, vol. 26, pp. 129–150, 1946.
11. F. M. Henderson and R. A. Wooding, Overland Flow and Groundwater Flow from a Steady Rainfall of Finite Duration, *J. Geophys. Res.*, vol. 69, pp. 1531–1540, April 1964.
12. J. R. Morgali and R. K. Linsley, Computer Analysis of Overland Flow, *J. Hydraul. Div. ASCE*, vol. 91, pp. 81–100, May 1965.
13. D. C. Woo and E. F. Brater, Spatially Varied Flow from Controlled Rainfall, *Proc. ASCE*, vol. 88, pp. 31–56, November 1962.

BIBLIOGRAPHY

CHOW, V. T.: Runoff, sec. 14 in V. T. Chow (ed.), "Handbook of Applied Hydrology," McGraw-Hill, New York, 1964.

EAGLESON, P. S.: "Dynamic Hydrology," McGraw-Hill, New York, 1970.

LINSLEY, R. K., KOHLER, M. A., and PAULHUS, J. L. H.: "Applied Hydrology," McGraw-Hill, New York, 1949.

SHERMAN, L. K.: The Unit Hydrograph, chap. 11E, in O. E. Meinzer (ed.), "Hydrology," McGraw-Hill, New York, 1942; reprinted Dover, New York, 1949.

PROBLEMS

7-1 Using the sample page from a *Water-Supply Paper* which appears as Fig. 4-16, construct direct- and base-flow recession curves of the type illustrated by Fig. 7-4. Are the values of K_r constant for these curves? What are the average values of K_r for direct and base flow? Using an average value of K_r, find the volume of groundwater storage when the flow is 500 ft^3/s.

7-2 Tabulated below are ordinates at 24-h intervals for a hydrograph. Assuming K_r for base flow = 0.9, separate the base flow from the direct runoff by each of the three methods illustrated in Fig. 7-6. Compute the volume of direct runoff in each case.

Time, d	Flow, ft^3/s	Time, d	Flow, ft^3/s
1	2,340	8	3,230
2	34,300	9	2,760
3	25,000	10	2,390
4	14,000	11	2,060
5	8,960	12	1,770
6	5,740	13	1,520
7	4,300	14	1,320

7-3 Plot the data of Prob. 7-2 on semilogarithmic paper and determine recession constants for surface runoff, interflow, and groundwater flow. What volume of each of the three components is present?

7-4 Tabulated below are the flows on a stream draining 2047 km². Using base-flow separation *ABC* of Fig. 7-6, determine the equivalent depth of direct runoff.

Time, d	Flow, m^3/s	Time, d	Flow, m^3/s
1	122	8	285
2	1137	9	223
3	950	10	185
4	627	11	161
5	531	12	149
6	429	13	140
7	347	14	132

7-5 Neglecting storage and assuming a linear rise and recession of the elemental hydrograph, sketch the outflow hydrograph from a basin in the shape of a 60° sector of a circle with outflow at the apex. Assume travel time proportional to distance and rainfall duration equal to time of concentration.

7-6 Repeat Prob. 7-5 for a semicircle with outflow at the midpoint of the boundary diameter. What change in shape would result if the rainfall (runoff) were to occur at unit rate for half the duration and 3 times that rate for the remainder?

7-7 How would the hydrograph of Prob. 7-6 be affected if runoff were to occur only from the outer half of the area?

7-8 Given below are the observed flows from a storm of 3 h duration on a stream with a drainage area of 122 mi². Derive the unit hydrograph. Assume constant base flow = 600 ft^3/s.

Hour	Day 1	Day 2	Day 3	Hour	Day 1	Day 2	Day 3
3 A.M.	600	4600	1700	3 P.M.	8000	2700	900
6 A.M.	550	4000	1500	6 P.M.	7000	2400	800
9 A.M.	6000	3500	1300	9 P.M.	6100	2100	700
Noon	9500	3100	1100	Midnight	5300	1900	600

7-9 Given below are three unit hydrographs derived from separate storms on a small basin. All are believed to have resulted from 4-h rains. Find the average unit hydrograph. Drainage area = 3.74 mi².

Hours	Storm 1	Storm 2	Storm 3	Hours	Storm 1	Storm 2	Storm 3
0	0	0	0	8	130	170	215
1	110	25	16	9	95	130	165
2	365	125	58	10	65	90	122
3	500	358	173	11	40	60	90
4	390	465	337	12	22	35	60
5	310	405	440	13	10	20	35
6	235	305	400	14	5	8	16
7	175	220	285	15	0	0	0

7-10 The hydrograph tabulated below resulted from three successive 6-h periods of rainfall having runoffs estimated as 0.6, 1.2, and 0.9 in., respectively. Using the method illustrated by Eq. (7-6), find the 6-h unit hydrograph for this basin. Drainage area = 50.4 mi^2. Base flow has been subtracted.

Time, h	Flow, ft^3/s	Time, h	Flow, ft^3/s
0	0	21	3140
3	750	24	1950
6	2800	27	930
9	2830	30	310
12	6620	33	90
15	4320	36	0
18	6450		

7-11 As an illustration of the effect of minor errors on the computation of Prob. 7-10, repeat it, using the same figures except for the ordinate at 6 h, which is changed to 2600 ft^3/s.

7-12 Using storm 2 of Prob. 7-9, construct the *S* curve and find the 2- and 6-h unit hydrographs. Smooth the *S* curve as required.

7-13 Take the flows given in Prob. 7-8 as being in cubic meters per second and the drainage area as 2231 km^2. Derive the unit hydrograph.

7-14 Given below is the 4-h unit hydrograph for a basin of 84 mi^2. Construct the *S* curve and find the 2- and 6-h unit hydrographs.

Time, h	Flow, ft^3/s	Time, h	Flow, ft^3/s
0	0	11	2700
1	400	12	2200
2	2500	13	1800
3	4400	14	1400
4	6000	15	1100
5	7000	16	800
6	6100	17	600
7	5200	18	400
8	4500	19	200
9	3800	20	100
10	3200	21	0

7-15 Using the unit hydrograph of storm 3, Prob. 7-9, find the peak flow resulting from four successive 4-h periods of rainfall producing 0.35, 0.87, 1.39, and 0.77 in. of runoff, respectively. Ignore base flow.

7-16 For a drainage basin selected by your instructor, derive a synthetic unit hydrograph using Snyder's method with $C_t = 2.0$ and $C_p = 0.62$.

7-17 A basin of 139 mi^2 has $L = 16$ mi, $L_c = 6$ mi. Using Snyder's method, coefficients as in Prob. 7-16, and $t_R = 3$ h, find the unit hydrograph.

7-18 Using actual streamflow data for a basin assigned by your instructor, find the unit hydrograph for a storm. What values of C_t and C_p for Snyder's method are indicated by the data?

7-19 A parking lot 150 ft long in the direction of the slope and 80 ft wide has a tar and gravel pavement on a slope of 0.0025. Assuming a uniform rainfall intensity of 2.75 in./h for 30 min, construct the outflow hydrograph, using Izzard's method.

7-20 A city lot 200 ft deep and 100 ft wide has a slope of 0.005 toward the street. The street is 60 ft wide and has a 6-in. crown. Assuming a rainfall intensity of 1.8 in./h, $c = 0.040$ for the lot and 0.007 for the street, and a rainfall duration of 60 min, find the peak flow into the gutter. What will the peak flow be if the rainfall duration is 10 min?

7-21 A paved area 50 m long by 80 m wide has a concrete surface on a slope of 0.005. What will the outflow hydrograph be for a uniform rainfall intensity of 60 mm/h for 30 min?

8

RELATIONS BETWEEN PRECIPITATION AND RUNOFF

The flow of streams is controlled primarily by variations in precipitation. Relationships between precipitation and runoff and techniques for distributing the runoff through time (Chap. 7) are the basis for efficient forecasting for the operation of hydraulic projects, for the extension of flow records on gaged streams, and for estimating the flow of ungaged streams.

THE PHENOMENA OF RUNOFF

Section 7-1 discussed the movement of runoff to the streams. The discussion which follows will be limited to processes which retain water in the watershed until it is removed by evapotranspiration.

8-1 Surface Retention

Much of the rain falling during the first part of a storm is stored on the vegetal cover as *interception* and in surface puddles as *depression storage*. As rain continues, the soil surface becomes covered with a film of water, known as *surface detention*, and flow begins downslope toward an established surface channel. That part of storm precipitation which does not appear either as infiltration or as surface runoff during or immediately following the storm is *surface retention.* In other words, surface retention includes interception, depression storage, and evaporation during the storm but does not include water temporarily stored en route to the streams.

Although the effect of vegetal cover is unimportant in major floods, interception by some types of cover may be a considerable portion of the annual rainfall. Interception-storage capacity is usually satisfied early in a storm so that a large percentage of the rain in the numerous small storms is intercepted. After the vegetation is saturated, interception would cease were it not for the fact that appreciable water may evaporate from the enormous wetted surface of the foliage. Once interception storage is filled, the amount of water reaching the soil surface is equal to the rainfall less evaporation from the vegetation. Interception-storage capacity is reduced as wind speed increases, but the rate of evaporation is increased. Apparently, high wind speeds tend to augment total interception during a long storm and to decrease it for a short storm.

Extensive experimental interception data have been accumulated by numerous investigators [1–5], but evaluation and application of the data to specific problems are made difficult because of varied experimental techniques employed. Data for forest cover, relatively more plentiful than those for crops and other low-level vegetation, are usually obtained by placing several rain gages on the ground under the canopy and comparing the average of their catch with that in the open. Sometimes the gages (*interceptometers*) are placed at random in an attempt to measure average interception for an area. In other cases, the interceptometers are placed at carefully selected points under the tree crown to measure interception on its projected area. In either case, application of the data requires detailed knowledge of the cover density over the area of interest. Most data for forests have been collected by placing interceptometers free of underbrush, grass, etc., and little is known of total interception as required in most hydrologic problems.

Trimble and Weitzman [6] found that mixed hardwood about 50 y old and typical of considerable area in the southern Appalachian Mountains intercepts about 20 percent of the rainfall both in summer and winter. No measurements were made of flow down the tree trunks, but it is likely that net interception is nearer 18 percent for storms with rainfall in the order of 0.5 in. (13 mm). Qualitatively, it can be said that annual interception by a well-developed forest canopy is about 10 to 20 percent of the rainfall and that the storage capacity of the canopy ranges from 0.03 to 0.06 in. (0.8 to 1.5 mm).

Horton [7] derived a series of empirical formulas for estimating interception (per storm) by various types of vegetal cover. Applying these formulas to 1-in. (25-mm) storms and assuming normal cover density give the values of interception in Table 8-1. Assuming sufficient rainfall to satisfy the interception-storage capacity, an equation for total-storm interception V_i can be written as

$$V_i = S_i + Et_R \qquad (8\text{-}1)$$

where S_i is storage capacity per unit of projected area, E is the evaporation rate, and t_R is the duration of rainfall. Assuming that the interception given by Eq. (8-1) is approached exponentially as the rainfall increases from zero to some high value, then

$$V_i = (S_i + Et_R)(1 - e^{-kP}) \qquad (8\text{-}2)$$

where e is the napierian base, P is the amount of rain, and k is a constant equal to $1/(S_i + Et_R)$.

Rainwater retained in puddles, ditches, and other depressions in the soil surface is termed *depression storage.* These depressions vary widely in area and depth; their size depends to a considerable degree on the definition of a depression. As soon as rainfall intensity exceeds the infiltration capacity (Sec. 8-2), the rainfall excess begins to fill surface depressions. Each depression has its own capacity, and when it is filled, further inflow is balanced by outflow plus infiltration and evaporation. Depressions of various sizes are both superimposed and interconnected. Almost immediately after the beginning of rainfall excess, the smallest depressions become filled and overland flow begins. Most of this water in turn fills larger depressions, but some of it follows an unobstructed path to the stream channel. This chain of events continues, with successively larger portions of overland flow contributing to the streams. Water held in depressions at the end of rain is either evaporated or absorbed by the soil through infiltration.

Table 8-1 VALUES OF INTERCEPTION AS COMPUTED WITH HORTON'S EQUATIONS FOR 1-in. (25-mm) STORMS

	Height		Interception	
Crop	ft	m	in.	mm
Corn	6	1.8	0.03	0.8
Cotton	4	1.2	0.33	8.4
Tobacco	4	1.2	0.07	1.8
Small grains	3	0.9	0.16	4.1
Meadow grass	1	0.3	0.08	2.0
Alfalfa	1	0.3	0.11	2.8

Meaningful observations of the magnitude of depression storage cannot be obtained easily, and values are highly dependent on definition of terms. Individual depressions of appreciable area relative to the drainage basin under consideration are usually called *blind drainage* and excluded from hydrologic analysis. The remaining depression storage is usually lumped with interception and treated as *initial loss* with respect to storm runoff. Nevertheless, depression storage may be of considerable magnitude and may play an important role in the hydrologic cycle. Stock ponds, terraces, and contour farming all tend to moderate the flood hydrograph by increasing depression storage, while land leveling and drainage reduce depression storage. Because most basins have some very large depressions, it is likely that the depression-storage capacity of the basin is never completely filled.

An expression for the volume of water in depression storage V_s in terms of the accumulated precipitation excess P_e can be written

$$V_s = S_d(1 - e^{-kP_e}) \qquad (8\text{-}3)$$

where S_d is the depression-storage capacity of the basin and k has the value $1/S_d$. Experience suggests that values of S_d for most basins will lie between 0.5 and 2.0 in. (10 and 50 mm). The equation neglects evaporation from the depression storage during a storm, a factor which is probably negligible.

8-2 Infiltration

Infiltration is the passage of water through the soil surface into the soil. Although a distinction is made between infiltration and *percolation*, the movement of water within the soil, the two phenomena are closely related since infiltration cannot continue unimpeded unless percolation removes infiltered water from the surface soil. The soil is permeated by noncapillary channels through which gravity water flows downward toward the groundwater, following the path of least resistance. Capillary forces continuously divert gravity water into capillary-pore spaces, so that the quantity of gravity water passing successively lower horizons is steadily diminished. This leads to increasing resistance to gravity flow in the surface layer and a decreasing rate of infiltration as a storm progresses. The rate of infiltration in the early phases of a storm is less if the capillary pores are filled from a previous storm.

The maximum rate at which water can enter the soil at a particular point under a given set of conditions is called the *infiltration capacity*. The actual infiltration rate f_i equals the infiltration capacity f_p only when the *supply rate* i_s (rainfall intensity less rate of retention) equals or exceeds f_p. Theoretical concepts presume that actual infiltration rates equal the supply rate when $i_s \leq f_p$ and are otherwise at the capacity rate (Fig. 8-1). The value of f_p is at a maximum f_0 at the beginning of a storm and approaches a low,

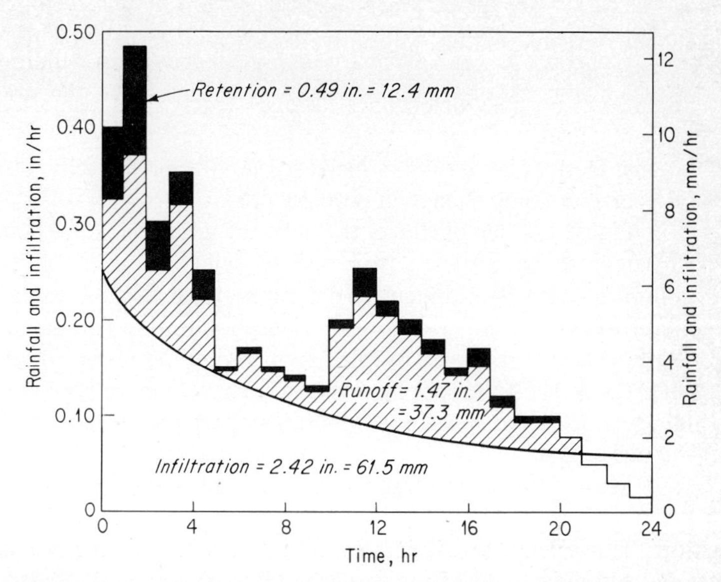

FIGURE 8-1
Simple separation of infiltration and surface runoff, using hourly rainfall data, estimated retention, and an infiltration-capacity curve.

constant rate f_c as the soil profile becomes saturated. The limiting value is controlled by subsoil permeability. Horton [8] found that infiltration-capacity curves approximate the form

$$f_p = f_c + (f_0 - f_c)e^{-kt} \tag{8-4}$$

where e is the napierian base, k is an empirical constant, and t is time from beginning of rainfall. The equation is applicable only when $i_s \geq f_p$ throughout the storm. Philip [9] suggested the equation

$$f_p = \frac{bt^{-1/2}}{2} + a \tag{8-5}$$

Integrating Eq. (8-5) with respect to time gives the cumulative infiltration F at time t as

$$F = bt^{1/2} + at \tag{8-6}$$

In Eqs. (8-5) and (8-6) a and b are empirical constants. Numerous other equations for infiltration have been suggested, but all reflect the capacity rate as an exponential function of time.

Infiltration capacity depends on many factors such as soil type, moisture content, organic matter, vegetative cover, and season. Of the soil characteristics affecting infiltration, noncapillary porosity is perhaps the most important. Porosity determines storage capacity and also affects resistance to flow. Thus, infiltration tends to increase with porosity. An increase in organic matter also results in increased infiltration capacity, largely because of a corresponding change in porosity [10].

Figure 8-2 demonstrates the effect of initial moisture content and the variations to be expected from soil to soil. The effect of vegetation on infiltration capacity is difficult to determine, for it also influences interception. Nevertheless, vegetal cover does increase infiltration compared with barren soil because (1) it retards surface flow, giving the water additional time to enter the soil; (2) the root systems make the soil more pervious; and (3) the foliage shields the soil from raindrop impact and reduces rain packing of the surface soil.

Most data on infiltration rates are derived from infiltrometer tests. An *infiltrometer* is a tube or other boundary designed to isolate a section of soil. The effective area varies from less than 1 ft^2 to several hundred square feet. Although many of the earlier tests were made by flooding the infiltrometer, this method is no longer recommended and has been supplanted by sprinkling techniques.† Since it is impossible to measure directly the quantity of water penetrating the soil surface, infiltration is computed by assuming it to equal the difference between water applied and measured surface runoff. In addition to the difficulties inherent in simulating raindrop size and velocity of fall with sprinklers, experiments using artificial rainfall have other features tending to cause higher infiltration rates in tests than under natural conditions.

Attempts have been made to derive infiltration data by analyzing rainfall and runoff data from small drainage basins with homogeneous soils [11], and Horton [12], Sherman [13], and others have presented techniques for deriving average "equivalent" infiltration-capacity data from records of heterogeneous basins. On natural basins allowance must be made for subsurface flow and surface retention.

8-3 The Runoff Cycle

The *runoff cycle* is the descriptive term applied to that portion of the hydrologic cycle between incident precipitation over land areas and subsequent discharge of this water through stream channels or evapotranspiration. Hoyt [14] has presented a comprehensive description of the hydrologic phenomena occurring at selected times during the runoff cycle by considering an idealized cross section of a basin.

† Tests to determine infiltration rates during flood or furrow irrigation are customarily made by flooding.

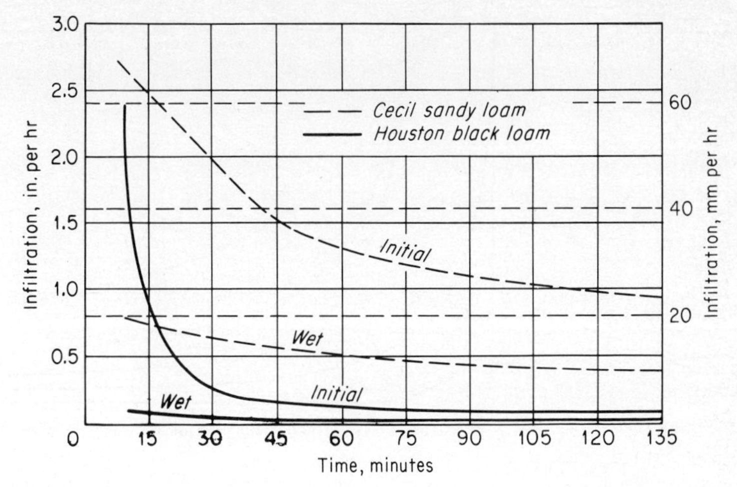

FIGURE 8-2
Comparative infiltration rates during initial and wet runs. (*From* [10].)

Figure 8-3 shows schematically the time variations of the hydrologic factors during an extensive storm on a relatively dry basin. The dotted area of the figure represents the portion of total precipitation which eventually becomes streamflow measured at the basin outlet. Channel precipitation is the only increment of streamflow during the initial period of rainfall. As streams rise, surface area and consequently the volume rate of channel precipitation increase.

The rate of interception is high at the beginning of rain, especially during summer and with dense vegetal cover. However, the available storage capacity is depleted rather quickly, so that the interception rate decreases to that required to replace water evaporated from the vegetation.

The rate at which depression storage is filled also decreases rapidly from a high initial value as the smaller depressions become filled and approaches zero at a relatively high value of total-storm rainfall. Depression storage is water retained in depressions until returned to the atmosphere through evaporation.

Except in very intense storms, the greater portion of the soil-moisture deficiency is satisfied before appreciable surface runoff takes place. However, some of the rain occurring late in the storm undoubtedly becomes soil moisture, since the downward movement of this water is relatively slow.

Water infiltrating the soil surface and not retained as soil moisture either moves to the stream as interflow or penetrates to the water table and eventually reaches the stream as groundwater. The rate of surface runoff starts at zero, increases slowly at first and then more rapidly, eventually

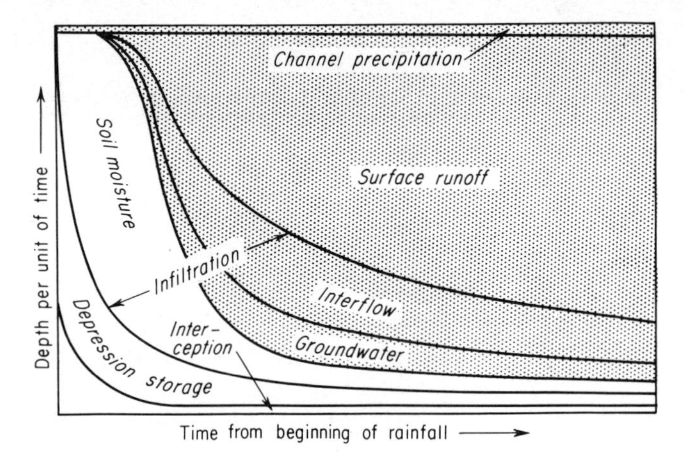

FIGURE 8-3
Schematic diagram of the disposition of storm rainfall.

approaching a relatively constant percentage of the rainfall rate. Both the percentage and the rate of runoff depend upon rainfall intensity.

Figure 8-3 illustrates only one of an infinite number of possible cases. A change in rainfall intensity would change the relative magnitude of all the factors. Further complications are introduced by varying rainfall intensity during the storm or by occurrence of snow or frozen ground. To appreciate further the complexity of the process in a natural basin, remember that all the factors of Fig. 8-3 vary from point to point within the basin during a storm. Nevertheless, the foregoing description should aid in understanding the relative time variations of hydrologic phenomena which are important in considering the runoff relations discussed later in the chapter.

Statistical evidence, field observation, and logic [15, 16] suggest that runoff is rarely generated uniformly over a watershed. Variations in rainfall amount and intensity, soil characteristics, vegetal cover, antecedent moisture, and topography all act to create a complex pattern of behavior in which runoff from most storms derives from a relatively small portion of the watershed closest to the stream channels. Exact definition of these *source areas* would require vastly more detail in the data employed for rainfall-runoff estimation than is ever likely to be reasonably available, thus frustrating a wholly theoretical approach to the problem. Relative statistical constancy of the source areas appears to exist, and thus the reliability of empirically derived rainfall-runoff relations is far better than might be expected from a physical analysis of the problem.

ESTIMATING THE VOLUME OF STORM RUNOFF

It will be evident from Fig. 8-3 that the rainfall-runoff process is relatively complex. Despite this, the practice of estimating runoff as a fixed percentage of rainfall is the most commonly used method in design of urban storm-drainage facilities, highway culverts, and many small water-control structures. The method can be correct only when dealing with a surface which is completely impervious so that the applicable runoff coefficient is 1.00.

Computer simulation techniques (Chap. 10) offer the most reliable method of computing runoff from rainfall because they permit a relatively detailed analysis using short time intervals. The type of computation used in computer simulation would be virtually impossible to carry through by hand because of the detailed computations required. The constraints of hand calculation led to methods using longer time intervals and a correspondingly less rigorous model. The following sections discuss some of the more successful approaches for manual solution.

8-4 Initial Moisture Conditions

The quantity of runoff from a storm depends on (1) the moisture conditions of the watershed at the onset of the storm and (2) the storm characteristics—rainfall amount, intensity, and duration. The storm characteristics are defined by the data from the precipitation-gage network, but no single observation serves to define the antecedent moisture conditions in the watershed. Much of the investigation of rainfall-runoff relations was directed at finding a simple index of basin moisture conditions.

In humid areas, where streams flow continuously, groundwater discharge at the beginning of the storm has been found to be a good index to initial moisture conditions. In a study of the Valley River, North Carolina, Linsley and Ackermann [17] found that field-moisture deficiency at any time was approximately equal to 90 percent of the total Class A pan evaporation since the ground was last saturated less any additions made to field moisture by intervening rains. Basin-accounting techniques (Sec. 5-15) applied on a daily basis provide a reasonably accurate estimate of moisture deficiency which can be used as an index to runoff [18]. This approach is laborious but should yield excellent results.

The most common index is based on antecedent precipitation. The rate at which moisture is depleted from a particular basin under specified meteorological conditions is roughly proportional to the am
(Sec. 5-15). In other words, the soil moisture should decrease
with time during periods of no precipitation [19],

$$I_t =$$

where I_0 is the initial value of *antecedent-precipitation index*, I_t is the reduced value t days later, and k is a recession factor ranging normally between 0.85 and 0.98. Letting t equal unity gives

$$I_1 = kI_0 \qquad (8\text{-}8)$$

Thus, the index for any day is equal to that of the previous day multiplied by the factor k. If rain occurs on any day, the amount of rain is added to the index (Fig. 8-4). Since storm runoff does not add to the residual moisture of the basin, an index of precipitation minus runoff, i.e., *basin recharge*, should be more satisfactory than the precipitation index alone. Commonly, however, the minor improvements gained do not justify the added computation.

Equation (8-7) assumes that the daily depletion of soil moisture (primarily evapotranspiration) is

$$I_0 - I_1 = I_0(1 - k) \qquad (8\text{-}9)$$

Since actual evapotranspiration is a function of the potential value and the available moisture (I_0), k should be a function of potential evapotranspiration. The variation in potential evapotranspiration is largely seasonal, and Eq. (8-7) has been found to be reasonably satisfactory when used jointly with calendar date (Sec. 8-6). There is an added advantage in using the date as a parameter because it also reflects variations in surface conditions related to farming practices, vegetation, etc.

The value of the index on any day theoretically depends on precipitation over an infinite antecedent period, but if a reasonable initial value is assumed, the computed index will closely approach the true value within a few weeks. The index value applicable to a particular storm is taken as that at the beginning of the first day of rain. Thus a value of 1.8 in. would be used for the storm of the ninth and tenth in Fig. 8-4.

8-5 Storm Analysis

In any statistical correlation, it is extremely important that the basic data be as consistent and reliable as possible. Methods of storm analysis should be rigorous and objective. Only that storm rainfall which produced the runoff being considered should be included. Small showers occurring after the hydrograph had started to recede should not be included if they had little effect upon the amount of runoff. Similarly, showers occurring before the main storm should be excluded from the storm rainfall and included in the antecedent-precipitation index. Long, complex storms should be separated into as many short storm periods as possible by hydrograph analysis.

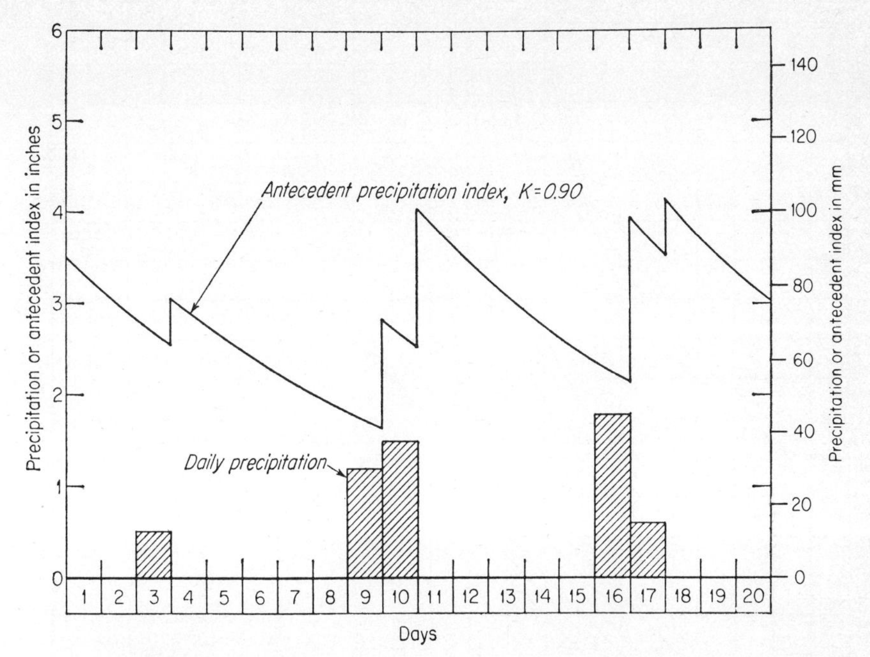

FIGURE 8-4
Variation of antecedent index with daily precipitation.

Runoff also depends upon rainfall intensity, but for basins of 100 mi^2 (250 km^2) or more, an average intensity as reflected by amount and duration is usually adequate. In this case duration can be estimated with sufficient accuracy from 6-h rainfall data. An objective rule is preferable, such as "the sum in hours of those 6-h periods with more than 0.2 in. of rain plus one-half the intervening periods with less than 0.2 in." Although experimental infiltration data indicate rates commonly in excess of 0.1 in./h (2.5 mm/h), relations such as Fig. 8-5 consistently show the effect of duration on storm runoff to be of the order of 0.01 in./h (0.25 mm/h). The difference is largely caused by intercorrelations and the inclusion of interflow with surface runoff.

8-6 Multivariate Relations for Total Storm Runoff

If storm characteristics and basin conditions are to be represented adequately in a runoff relation, a number of independent variables must be included. The relationship is not an additive one, and the usual multivariate linear correlation is not satisfactory. The coaxial graphical method (Sec. A-3)

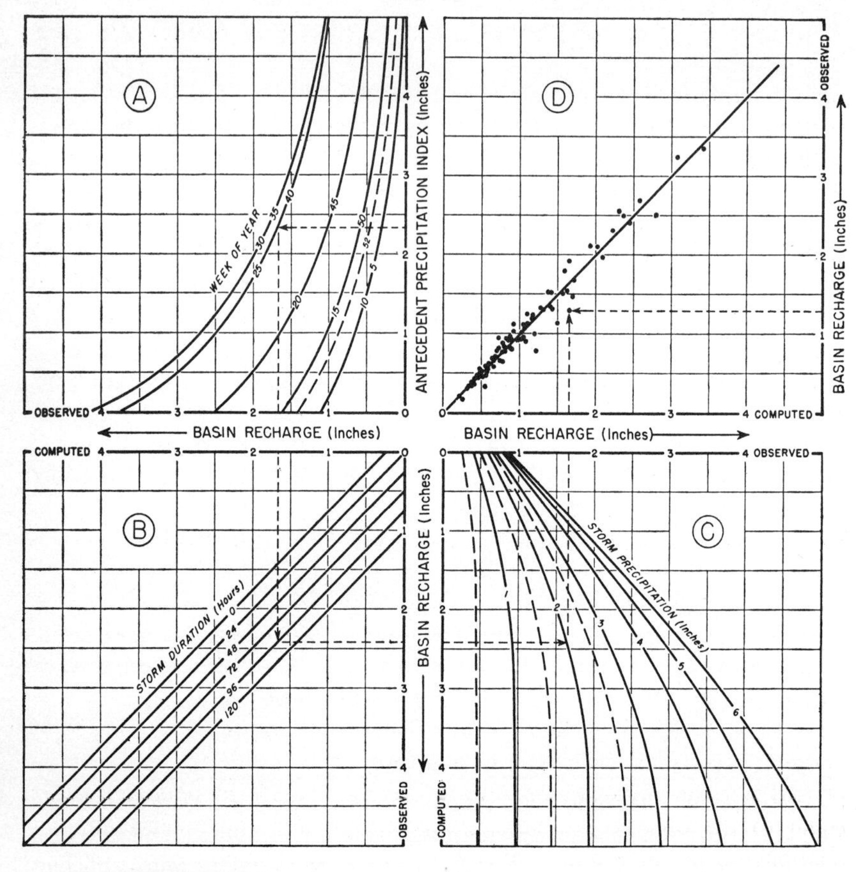

FIGURE 8-5
Basin-recharge relation for Monocacy River at Jug Bridge, Maryland. (*U.S. National Weather Service.*)

was first shown to be particularly useful for this work [19]. Betson et al. [20] subsequently demonstrated an analytical correlation technique which yields equivalent results.

To illustrate the coaxial method assume that a relation for estimating runoff is desired, using antecedent precipitation, date (or week number), and rainfall amount and duration as parameters. Values of these parameters are compiled for 50 or more storms. With the exception of rainfall amount, the parameters should be more closely related to the fraction of rainfall which does not run off than to the runoff volume. It is therefore convenient to calculate an auxiliary parameter equal to the storm rainfall minus the storm runoff. This parameter is called the *recharge* in the subsequent discussion.

A three-variable relation is developed first (Fig. 8-5, chart *A*) by (1) plotting antecedent precipitation versus recharge, (2) labeling the points with week number, and (3) fitting a smooth family of curves representing the various weeks. Chart *B* is placed with its horizontal scale matching that of chart *A* to facilitate plotting. Points are plotted in chart *B* with observed recharge as ordinate and recharge computed from chart *A* as abscissa, and these points are labeled with duration. A family of smooth curves is drawn to represent the effect of duration on recharge. Charts *A* and *B* together are a graphical relation for estimating recharge from antecedent index, week, and duration. Storm precipitation is then introduced (chart *C*) by (1) plotting recharge computed from charts *A* and *B* against observed recharge, (2) labeling the points with rainfall amount, and (3) fitting a family of curves. Charts *A*, *B*, and *C* constitute the *first approximation* to the desired relation. Chart *D* indicates the overall accuracy of the derived charts.

Since the parameters are intercorrelated and the first charts are developed independently of factors subsequently introduced, revision of the charts may improve the overall relation. The process is one of successive approximations. To check the week curves, the other curve families are assumed to be correct and the adjusted abscissa for a point in chart *A* is determined by entering charts *B* and *C* in reverse order with observed recharge, rainfall amount, and duration. The ordinate for the adjusted point is the observed antecedent-precipitation index. In other words, the week curves should be revised to fit this adjusted point if the relation is to yield a computed recharge equal to the observed recharge. The second (and subsequent) approximations for duration and rainfall are made in the same manner. In each case the points are plotted by entering the chart sequence from both ends with observed values to determine the adjusted coordinates.

Once a satisfactory relation for estimating the recharge is completed, it is a simple matter to revise chart *C* so that the final answer is in terms of runoff since rainfall minus recharge should equal runoff.

Although the method presented in the previous paragraphs is general and can be used as described, certain modifications simplify the procedure and require fewer approximations. Since storm rainfall is extremely important, the first plotting of chart *A* may show little correlation and the construction of the curves will be difficult. However, an important advantage in having the rainfall parameter in the last chart is that the possibility of computing runoff in excess of rainfall or of computing negative values of runoff is eliminated. Moreover, the arrangement of Fig. 8-5 results in the determination of a unified index of moisture conditions in the first chart, which is a decided advantage in river forecasting. If the plotting of chart *A* is limited to storms having rainfall within a specified class interval (2 to 4 in., for example), the construction of the curves is simplified, provided that there are sufficient data. Only limited data are required since the general curvature and convergence are always as shown in the example. The relations are quite similar throughout a geographic region, and charts *A* and *B* for one basin may be used as a first approximation for another basin in the area.

The analytical technique [20] uses the equations

$$R_I = c + (a + dS_I)e^{-bI} \tag{8-10}$$

$$Q = (P^n + R_I^n)^{1/n} - R_I \tag{8-11}$$

where R_I is a runoff index approximating the first quadrant of a coaxial plot, S_I is a fixed function of week number ranging between $+1$ and -1, I is the antecedent-precipitation index, e is the base of napierian logarithms, P is storm rainfall, Q is direct runoff, and a, b, c, d, and n are statistically derived coefficients. With only five constants the functions are quickly derived on a computer but are more constrained in form than the graphical solution. In comparative tests on watersheds in the Tennessee River basin, the analytic method gave results generally slightly better than the graphical method. McCallister [21] describes another method of using the computer to develop a runoff relation based on the parameters used in the coaxial method.

Since it is impossible to segregate the water passing a gaging station according to the portion of the basin in which it fell, statistical runoff relations must be based on basin averages of the parameters. Unfortunately, a relation based on storms of uniform areal distribution will yield runoff values which are too low when applied to storms with extremely uneven distributions. This can be demonstrated by computing the runoff for 4, 6, and 8 in. of rainfall, assuming all other factors to remain fixed. While 6 is the average of 4 and 8, the average runoff from the 4- and 8-in. rainfalls is not equal to that from a 6-in. rain. An uneven distribution of antecedent precipitation produces similar results. Runoff relations based on uniform areal conditions can be used to compute the runoff in the vicinity of each rainfall station, and the average of these runoff values will, in general, more nearly approach the observed runoff from the basin when either the storm or the antecedent precipitation is quite variable.

8-7 Relations for Incremental Storm Runoff

In order to determine increments of runoff throughout a storm for application of a unit hydrograph, Fig. 8-5 may be used with accumulated rainfall up to the end of each period and the successive values of runoff subtracted to obtain increments. When applied to small watersheds, however, there is a marked tendency to underestimate the peak flows, because the relation does not properly account for the time distribution of rainfall. Some of the errors for larger basins are certainly caused by the same factor.

The chief difficulty in developing a relation for hourly increments of runoff lies in the difficulty of determining short-period increments of runoff by hydrograph analysis. A process of successive iterations is necessary [22]. To illustrate, assume that a unit hydrograph and a relation such as Fig. 8-5 are applicable to a small basin and that an incremental relation such as Fig. 8-6 is desired. The analysis would proceed as follows:

1 For each of a number of short storms compute the accumulated storm runoff, hour by hour, with Fig. 8-5. Obtain hourly increments by subtraction of successive accumulated values and adjust these increments to yield the correct total-storm runoff.
2 Compute storm hydrographs by applying the increments just derived to the 1-h unit hydrograph (Sec. 7-13). If hydrograph reproduction is not satisfactory, adjust incremental runoff values as required.
3 Derive the antecedent-precipitation index for each hour by adding the storm rainfall up to the designated hour to the initial value of the index for the storm.
4 Entering Fig. 8-6† with the antecedent index and week number, plot the corresponding hourly runoff and label the point with hourly rainfall. Points should be plotted so that those for each storm are readily discernible (color or symbol). Since hourly increments are less reliable than total-storm runoff, the analyst may make compensating changes in the points for a given storm to obtain better agreement among storms.
5 Construct hourly rainfall curves to fit the plotted data.
6 Using the first-approximation curves, compute hourly runoff values by entering with antecedent index, week, and rainfall, and adjust each set of values to the correct storm total.
7 To check the week curves, enter the relation with hourly runoff (step 6) and rainfall, plot at the corresponding antecedent index, and label each point with the week number. Revise the week curves as indicated by the plotted data.
8 If the week curves require revision (step 7), the first-approximation rainfall curves should be checked. Enter the relation with antecedent index and week, plot at the corresponding runoff (step 6), label points with rainfall, and revise the curves as required.

† The curve families of this relation have been superimposed for illustrative purposes. In derivation they should be separated to avoid confusion.

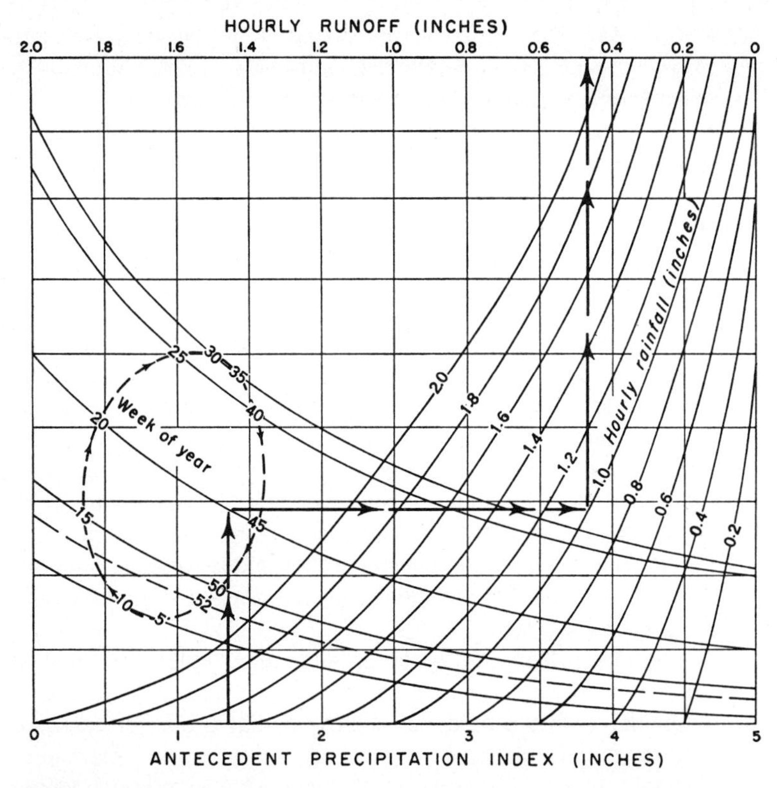

FIGURE 8-6
Hourly-runoff relation for Little Falls Branch at Bethesda, Maryland. (*U.S. National Weather Service.*)

8-8 Runoff Estimates Using Infiltration

The infiltration approach assumes that the surface runoff from a given storm is equal to that portion of the rainfall which is not disposed of through (1) interception and depression storage, (2) evaporation during the storm, and (3) infiltration. If items 1 and 2 are invariable or insignificant or can be assigned reasonable values, one need be concerned only with rainfall, infiltration, and runoff. In the simplest case, where the supply rate i_s is at or in excess of the infiltration capacity, surface runoff is equivalent to the storm rainfall less surface retention and the area under the capacity curve.

The procedure appears to be quite simple and to offer a solution to the estimation of short period increments of runoff. Experience has shown otherwise. If the rainfall intensity is always above the infiltration-capacity curve (Fig. 8-1) the problem is merely one of defining the infiltration curve which is a function of the antecedent moisture conditions. If rainfall intensities fluctuate above and below the infiltration curve, the matter is confused, since the curve inherently assumes that the infiltration capacity decreases because a fixed amount of water was added to the soil moisture during an interval. If $i_s < f_p$, the increment of soil moisture is less than assumed and the drop in the infiltration curve correspondingly less.

The time-intensity pattern of rainfall is rarely uniform across a watershed in any storm, and it is likely that the applicable infiltration-capacity curve varies from point to point within a watershed depending on soils, vegetation, and antecedent moisture. Finally, the infiltration approach leads to an estimate of surface runoff. Interflow and groundwater accretion must be determined in some other way. For these and other reasons the infiltration approach never proved satisfactory as a tool for hydrograph prediction. The detailed computations required could not be handled manually.

8-9 Infiltration Indexes

Difficulties with the theoretical approach to infiltration led to the use of infiltration indexes [23]. The simplest of these is the Φ *index*, defined as that rate of rainfall above which the rainfall volume equals the runoff volume (Fig. 8-7). The *W index* is the average infiltration rate during the time rainfall intensity exceeds the capacity rate; i.e.,

$$W = \frac{F}{t} = \frac{1}{t}(P - Q - S) \qquad \text{(8-12)}$$

where F is total infiltration, t is time during which rainfall intensity exceeds infiltration capacity, P is total precipitation corresponding to t, Q is surface runoff, and S is the effective surface retention. The W index is essentially equal to the Φ index minus the average rate of retention by interception and depression storage. While the segregation of infiltration and retention would seem to be a refinement, the task of estimating the retention rate is such that combining it with infiltration is probably equally satisfactory.

With very wet conditions, when the infiltration capacity is at a minimum and the retention rate is very low, the values of W and Φ are almost equal. Under these conditions, the W index becomes the W_{min} index by definition. This index is used principally in studies of maximum flood potential.

The derivation of infiltration indexes is simple, and the approach is cloaked in an aura of logic through the name *infiltration index*. Actually the indexes are no more than average loss rates, and their magnitude is highly dependent on antecedent conditions, so that they are in no way

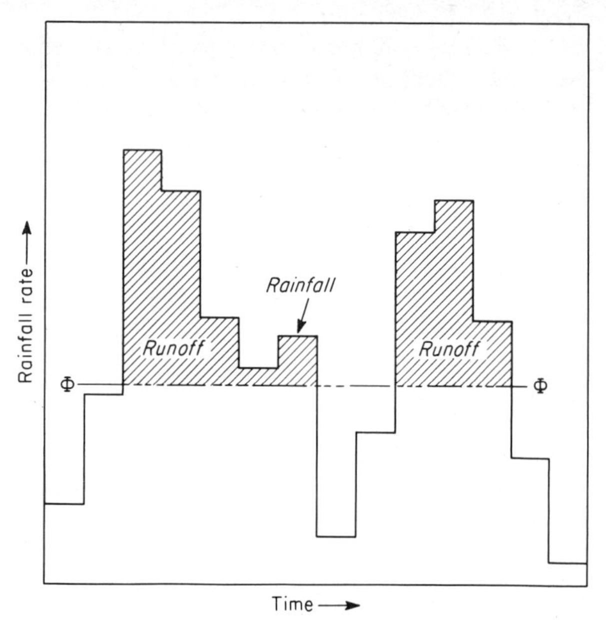

FIGURE 8-7
Schematic diagram showing the meaning of the Φ index.

superior to the multivariate relations. The Φ index has been used in some unit-hydrograph studies to define the time pattern of excess rainfall. In such cases the actual runoff volume is known, and there is no problem in calculating Φ. Since it is known that the actual infiltration is not constant, however, the runoff pattern derived with the Φ index cannot be correct.

ESTIMATING SNOWMELT RUNOFF

8-10 Physics of Snowmelt

The hydrologist is concerned with estimating the rates of snowmelt from his knowledge of the heat supplied to the snow. Then he must determine how the resulting meltwater, perhaps in combination with concurrent rainfall, will affect streamflow. From a physical point of view, the snowmelt and evaporation processes are quite similar. Both are thermodynamic processes and can be treated by the energy-budget approach. For each inch of water melted from snow at 32°F (0°C) heat must be supplied in the amount of 750 Btu/ft^2 or 203 cal/cm^2 (80 cal/cm^2 per centimeter of melt). Snowmelt computations are simplified by the fact that the surface is always at the freezing point during melt, and at the same time they are made more complex by variations in *albedo*, the percentage of solar and diffuse radiation reflected by the surface (Sec. 2-2). The energy for snowmelt is derived from (1) net radiation, (2) conduction and convective transfer of sensible heat from the overlying air, (3) condensation of water vapor from the overlying air, (4) conduction from the underlying soil, and (5) heat supplied by incident rainfall.

Components of radiation exchange are discussed in Sec. 5-3 in connection with evaporation. Although only 5 to 10 percent of incident shortwave radiation is reflected by a free-water surface, 80 to 90 percent is reflected by a clean, dry snow surface. As snow ages, its albedo drops to 50 percent or less because of changes in crystalline structure, density, and amount of dirt on the surface. Snow radiates essentially as a blackbody, and the outgoing longwave radiation at 32°F or 0°C during a 24-h period is equivalent to about 3.3 in. (84 mm) of negative melt. The net loss by longwave radiation is materially less than this amount since the atmosphere reradiates a portion back to the earth. Net longwave radiation depends primarily on air and snow temperatures, atmospheric vapor pressure, and extent and type of cloud cover. With clear skies net loss by longwave radiation is equal to about 0.8 in. (20 mm) of melt per day when dewpoint and air and snow temperatures are near freezing. Numerous empirical formulas have been derived to estimate this factor [24, 25] and net all-wave radiation, but they are not particularly satisfactory. Radiometer observations of net radiation could provide the needed data if an adequate network of stations were established.

Heat exchange between a snowpack and the atmosphere is also effected by conduction, convection, condensation, and evaporation. Although it is readily shown that conduction in still air is very small, convective exchange can be an important factor [26]. If the dewpoint of the air is above freezing, condensation on the snow occurs, with consequent release of latent heat. If the dewpoint is less than the snow-surface temperature, the vapor pressure of the snow is greater than that of the air and evaporation occurs. The rate of transfer of sensible heat (convection) is proportional to the temperature difference between the air T_a and the snow T_s, while vapor transport is proportional to the vapor-pressure gradient $e_a - e_s$. Both processes are proportional to wind velocity v. Since the latent heat of vaporization is about 7.5 times the latent heat of fusion, condensation of 1 in. (25 mm) of water vapor on the snow surface produces 8.5 in. (216 mm) of water, including condensate. The two processes can be described by similar equations

$$M_{\text{conv}} = k_c(T_a - 32)v \qquad (8\text{-}13)$$

$$M_{\text{cond}} = k_v(e_a - 6.11)v \qquad (8\text{-}14)$$

where M is melt, k is an exchange coefficient, and 6.11 is the vapor pressure (in millibars) at 32°F (0°C). If T_a of Eq. (8-13) is in degrees Celsius, the -32 is omitted.

Temperatures beneath the soil surface do not vary as much or as rapidly as surface temperatures, and there is usually an increase of temperature with depth during the winter and early spring. This results in a transfer of heat from the soil to the snowpack above. While the heat transfer is small on a daily basis, it may be equivalent to several inches of melt during an entire season—sufficient to keep the soil saturated and to permit a prompt response of streamflow when melting from other causes takes place.

Raindrop temperatures correspond closely to the surface wet-bulb temperature. As the drops enter a snowpack, their temperature is reduced to 32°F (0°C) and an equivalent amount of heat is imparted to the snow. The melt from rain is given by

$$M_{\text{rain}} = \frac{P(T_w - 32)}{144} \tag{8-15}$$

where P is the rainfall in inches, T_w is the wet-bulb temperature in degrees Fahrenheit, and 144 is the latent heat of fusion in Btu per pound. The heat available in 1 in. of rain at 50°F will melt about $\frac{1}{8}$ in. of water from the snow. There has been a tendency to overemphasize the importance of rainfall melt because warm rains are accompanied by high humidity and temperature and often by moderate to strong winds. In metric units Eq. (8-15) becomes

$$M_{\text{rain}} = \frac{PT_w}{80} \tag{8-15a}$$

where M and P are in millimeters, T_w is in degrees Celsius, and 80 is the latent heat of fusion in calories per gram.

From a theoretical point of view, it would appear that snowmelt computations could be made by applying the energy-budget approach, and this is basically true for snowmelt at a point. On a natural basin, however, the hydrologist is confronted with numerous complications, among the more obvious being the effects of variations in elevation, slope, aspect, and forest cover on radiation exchange, temperature, wind, and other key factors in snowmelt. The heat required to melt snow depends on its thermal quality (Sec. 3-16), which may vary within a basin. Finally, since the hydrologist is concerned with the runoff from snowmelt, water retained in the snowpack is not effective. Limited data indicate that snow can retain from 2 to 5 percent liquid water by weight. These complexities and the scarcity of pertinent data have made the theoretical approach generally inapplicable, but computer simulation (Chap. 10) is usually based on the energy budget if adequate data are available.

8-11 Estimating Snowmelt Rates and Consequent Runoff

Air temperature is the single most reliable index to snowmelt. It so completely reflects radiation, wind, and humidity that residual errors are usually not materially correlated with these factors. Since snowmelt does not occur with temperatures appreciably below freezing, the temperature data are commonly converted to degree-days or degree-hours above some base, usually 32°F (0°C). A day with a mean temperature of 52°F represents 20 degree-days above 32°F. If the minimum temperature for the day is above freezing, there are 20 × 24 or 480 degree-hours. On the other hand, there are actually more than 480 degree-hours in a day with a mean temperature of 52°F if the temperature drops below 32°F for a portion of the day.

A variety of different relationships have been suggested for forecasting snowmelt. Most commonly, however, a *degree-day factor* or the ratio of snowmelt to concurrent degree-days is utilized. If the actual rate of snowmelt were known, the degree-day factor might well be substantially constant. Actually, the rate of runoff must be used in lieu of rate of melt, and a plot of accumulated snowmelt runoff versus accumulated degree-days tends to an ogive shape. Early in a period of melt some heat may be required to bring the snow temperature up to the freezing point. In addition, some of the initial meltwater will be stored in the snow or soil so that the degree-day factor will appear to be low. As time passes, the degree-day factor will increase as an increasing fraction of each day's meltwater reaches the stream gage. Near the end of the melting period the snow cover may become thin and patchy, and the apparent melt per degree-day will again decrease unless the runoff is calculated for the actual area of snow surface.

In many basins of high relief, winter conditions are typified by semi-permanent snow cover in the headwaters while the lower portions of the basins have only intermittent snow cover. The lower limit of snow cover (*snow line*) is a dynamic feature, moving up and down the slope. In practice, the snow line is often considered as a contour of elevation, a reasonably realistic assumption following a storm. As melting proceeds, however, snow cover recedes more rapidly on southerly and barren slopes, and the snow line can be defined only as an average elevation of the lower limit of snow. Since surface air temperature is an inverse function of elevation, the rate of snowmelt decreases with elevation. If the freezing isotherm is below the snow line, there is no melting within the basin. Thus, temperature at an index station must be considered in conjunction with the extent of snow cover in estimating snowmelt. Mean daily surface air temperature in mountain regions drops about 3 to 5 Fahrenheit degrees per 1000-ft increase in elevation (0.5 to 1 Celsius degrees per 100 m) (Sec. 2-14). Given the temperature at an index station, an area-elevation curve for the basin, and the average snow-line elevation, one can compute the area subject to melting on the basis of an assumed rate of change of temperature with elevation. The sum $\sum DA$, where D is the degree-days for each small increment of elevation above the snow line and A is the percent of basin area in each elevation zone, gives a weighted value of degree-days over the melting zone [27].

Values of the degree-day factor generally range from near zero to 0.2 in./Fahrenheit degree-day (9 mm/Celsius degree-day), values between 0.05 and 0.1 in./F degree-day (2 and 5 mm/C degree-day) being most common.

SEASONAL AND ANNUAL RUNOFF RELATIONS

The hydrologist is frequently required to estimate monthly, seasonal, or annual volumes of runoff. Such estimates may be needed for operational purposes or to provide a data base for evaluating reservoir storage requirements. The most reliable basis for such estimates is the calculation of short-term runoff (daily or storm period) with relations such as those discussed in Sec. 8-6. However, if it is necessary to estimate flows for many years, such computations would be extremely tedious. Simulation methods (Chap. 10) offer the most promising combination of reliability and ease of calculation. In relatively humid regions, where precipitation is reasonably uniformly distributed through the year, and in cases where the runoff is largely from the melting of a seasonal snowpack, variations in antecedent conditions and distinctions between direct and groundwater runoff may be less important and simpler relationships may prove adequate.

Inasmuch as meteorological forecasts of precipitation are not yet feasible on a long-term basis, forecasts of seasonal or annual runoff are possible only where precipitation accumulates as snow during the winter and melts during the subsequent spring and early summer.

8-12 Precipitation-Runoff Relations

A simple plotting of annual precipitation versus annual runoff will often display good correlation (Fig. 8-8), particularly in areas where the major portion of the precipitation falls in the winter months. Numerous refinements can be introduced if a moderate increase in reliability justifies the additional effort required in analysis [28]. First, there is the question of consistent data throughout the period of record under study. Observed runoff should be adjusted for storage changes in reservoirs and, in some cases, for increased evaporation losses from the reservoirs. Adjustment may also be needed for

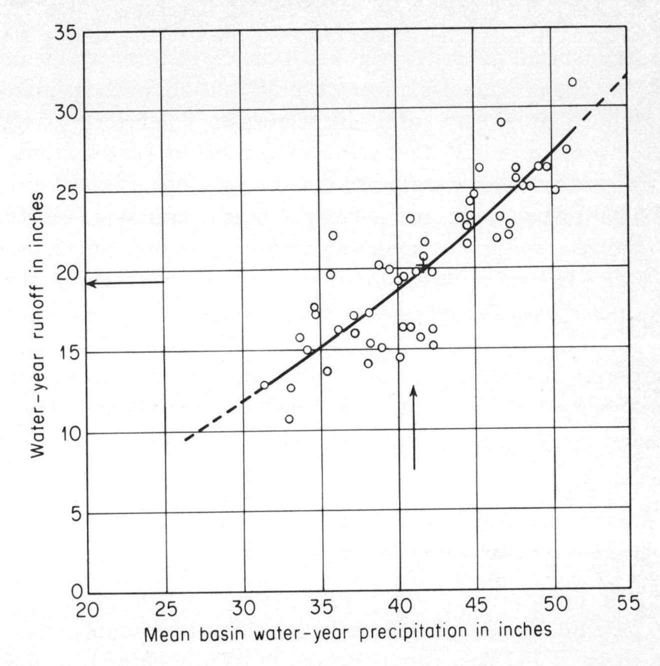

FIGURE 8-8
Relation between annual precipitation and runoff for the Merrimack River above Lawrence, Massachusetts (4460 mi^2 or 11,550 km^2). Correlation coefficient = 0.87 and standard error of estimate = 2.3 in. (58 mm).

consumptive loss from irrigation or other water uses. Finally, changes in observational techniques, discharge section, gage location, and other factors have contributed to discharge records which in some cases show definite time trends unexplained by precipitation. Testing precipitation and runoff data by double-mass analysis (Chaps. 3 and 4) may disclose inconsistencies which can be eliminated through appropriate adjustments.

Figure 8-8 is based on averages of precipitation observed at 30 stations in and near the basin. If records are available for only a few stations, application of station weights may improve results. Summer precipitation and a portion of that occurring in the fall and spring are in the form of rain and fall on bare ground, and much of this precipitation is lost to soil moisture and evapotranspiration. Winter snows accumulate as a snowpack, and evapotranspiration losses are relatively small. The effectiveness of precipitation in producing streamflow thus depends on when it occurs, and the application of monthly weights may improve the reliability of the relation.

Station and monthly weights can be derived through least-squares analysis. The high intercorrelation of precipitation observed at stations within a limited area may result in regression coefficients (weights) which overemphasize the relative value of the better stations, and if so, the results should be tempered toward equal weighting. Derived monthly weights may tend to be erratic, particularly if based on a short record, but they can be adjusted to provide a smooth transition throughout the year.

There is often a substantial lag between precipitation and the subsequent discharge of that portion which recharges the groundwater. Where groundwater is an appreciable part of the total flow, introduction of precipitation during the previous year as an additional parameter may improve the reliability of the relation. Streamflow during the previous year and flow during one or more winter months [29] have also been used as indexes of groundwater carry-over.

8-13 Use of Snow Surveys

The application of snow-survey data to the preparation of water-supply forecasts is appealing because of the rather simple relation envisioned. If the seasonal flow results primarily from melting of a mountain snowpack, measurements of the water in the snowpack before melt begins should indicate the volume of runoff to be expected. For many years snow surveys were made near the end of the snow-accumulation season, but now they are usually made monthly or more often during the accumulation season. Daily observations telemetered from snow pillows are also used. Because of drifting, variations in winter melt from point to point, etc., snow cover is usually more variable than precipitation within a given area. It is not economically feasible to take a sufficient number of samples to determine directly the volume of water stored in the pack, and the surveys (Sec. 3-16) constitute nothing more than an index. Accordingly, samples must be taken at the same points and in a consistent manner each year. Care must be taken to avoid changes in exposure brought about by forest fires, growth of timber and underbrush, etc.

Although there is good correlation between snow-survey data and seasonal runoff, it is now recognized that reliable water-supply forecasts cannot be made from snow surveys alone [30, 31]. Runoff subsequent to the surveys is also dependent upon (1) groundwater storage, (2) antecedent soil-moisture deficiency, and (3) precipitation during the runoff period. It has been found [32, 33] that snow-survey data can best be treated as an independent measure of winter precipitation in a multiple correlation (Fig. 8-9).

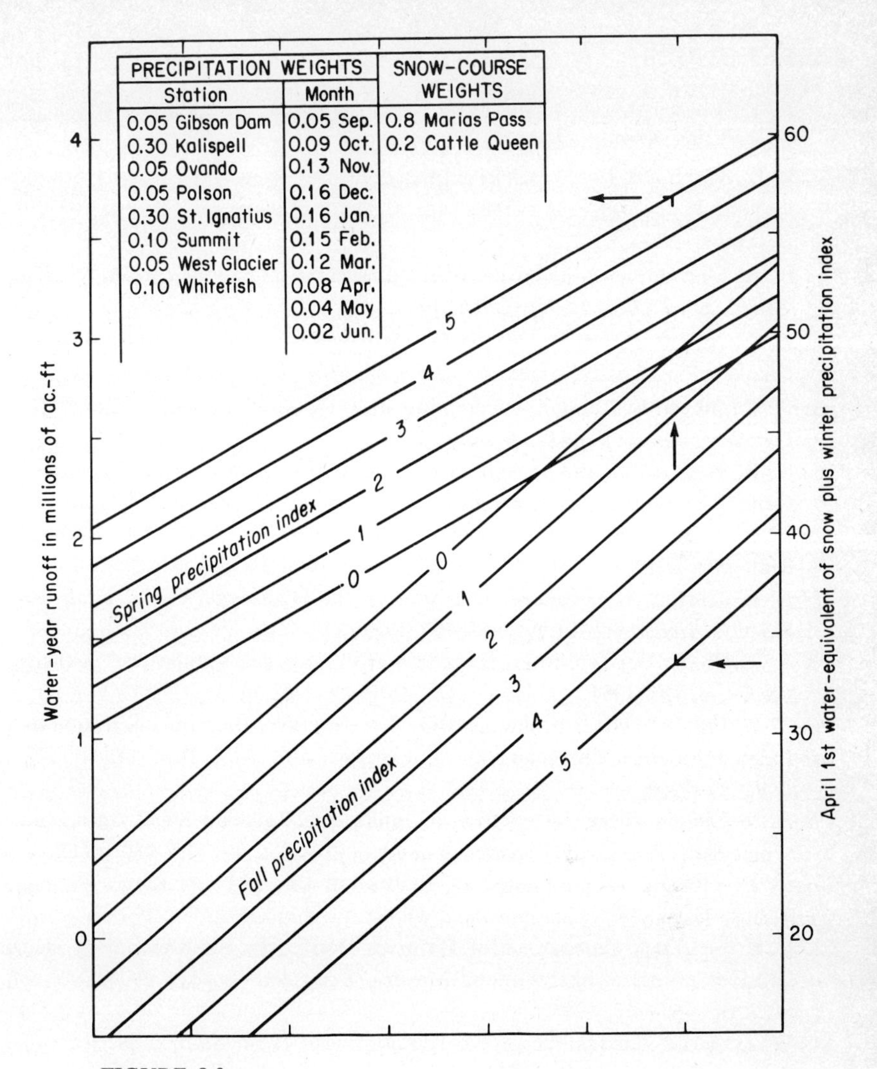

FIGURE 8-9
Annual runoff relation for South Fork, Flathead River, above Hungry Horse Reservoir, Montana. (*U.S. National Weather Service.*)

REFERENCES

1. R. K. Linsley, M. A. Kohler, and J. L. H. Paulhus, "Applied Hydrology," pp. 263–268, McGraw-Hill, New York, 1949.
2. P. B. Rowe and T. M. Hendrix, Interception of Rain and Snow by Second-Growth Ponderosa Pine, *Trans. Am. Geophys. Union*, vol. 32, pp. 903–908, December 1951.
3. R. A. Merriam, A Note on the Interception Loss Equation, *J. Geophys. Res.*, vol. 65, pp. 3850–3851, November 1960.
4. E. S. Corbett and R. P. Crouse, Rainfall Interception by Annual Grass and Chaparral . . . Losses Compared, *U.S. For. Serv. Res. Pap.* PSW-48, 1968.
5. R. S. Johnston, Rainfall Interception in a Dense Utah Aspen Clone, *U.S. For. Serv. Res. Note* INT-143, July 1971.
6. G. R. Trimble, Jr., and S. Weitzman, Effect of a Hardwood Forest Canopy on Rainfall Intensities, *Trans. Am. Geophys. Union*, vol. 35, pp. 226–234, April 1954.
7. R. E. Horton, Rainfall Interception, *Mon. Weather Rev.*, vol. 47, pp. 603–623, September 1919.
8. R. E. Horton, The Role of Infiltration in the Hydrologic Cycle, *Trans. Am. Geophys. Union*, vol. 14, pp. 446–460, 1933.
9. J. R. Philip, An Infiltration Equation with Physical Significance, *Soil Sci.*, vol. 77, p. 153, 1954.
10. G. R. Free, G. M. Browning, and G. W. Musgrave, Relative Infiltration and Related Physical Characteristics of Certain Soils, *U.S. Dept. Agric. Tech. Bull.* 729, 1940.
11. A. W. Zingg, The Determination of Infiltration Rates on Small Agricultura Watersheds, *Trans. Am. Geophys. Union*, vol. 24, pt. 2, pp. 476–480, 1943.
12. R. E. Horton, Determination of Infiltration Capacity for Large Drainage Basins, *Trans. Am. Geophys. Union*, vol. 18, pt. 2, pp. 371–385, 1937.
13. L. K. Sherman, Comparison of F-Curves Derived by the Methods of Sharp and Holtan and of Sherman and Mayer, *Trans. Am. Geophys. Union*, vol. 24, pt. 2, pp. 465–467, 1943.
14. W. G. Hoyt, An Outline of the Runoff Cycle, *Penn. State Coll. Sch. Eng. Tech. Bull.* 27, pp. 57–67, 1942.
15. R. D. Black, Partial Area Contributions to Storm Runoff in a Small New England Watershed, *Water Resour. Res.*, vol. 6, pp. 1296–1311, October 1970.
16. R. P. Betson and J. P. Marius, Source Areas of Storm Runoff, *Water Resour. Res.*, vol. 5, pp. 574–582, June 1969.
17. R. K. Linsley and W. C. Ackermann, Method of Predicting the Runoff from Rainfall, *Trans. ASCE*, vol. 107, pp. 825–846, 1942.
18. M. A. Kohler, Meteorological Aspects of Evaporation Phenomena, *Gen. Assemb. Int. Assoc. Sci. Hydrol., Toronto*, vol. 3, pp. 421–436, 1957.
19. M. A. Kohler and R. K. Linsley, Predicting the Runoff from Storm Rainfall, *U.S. Weather Bur. Res. Pap.* 34, 1951.

20. R. P. Betson, R. L. Tucker, and F. M. Haller, Using Analytical Methods to Develop a Surface-Runoff Model, *Water Resour. Res.*, vol. 5, pp. 103–111, February, 1969.
21. J. P. McCallister, Role of Digital Computers in Hydrologic Forecasting and Analysis, *Int. Assoc. Sci. Hydrol. Publ.* 63, pp. 68–77, 1963.
22. J. F. Miller and J. L. H. Paulhus, Rainfall-Runoff Relation for Small Basins, *Trans. Am. Geophys. Union*, vol. 38, pp. 216–218, April 1957.
23. H. L. Cook, The Infiltration Approach to the Calculation of Surface Runoff, *Trans. Am. Geophys. Union*, vol. 27, pp. 726–747, October 1946.
24. E. R. Anderson, Energy-Budget Studies, Water-Loss Investigations, vol. 1, Lake Hefner Studies, *U.S. Geol. Surv. Prof. Pap.* 269 (reprint of *U.S. Geol. Surv. Circ.* 229), pp. 71–119, 1954.
25. "Snow Hydrology, Summary Report of the Snow Investigations," pp. 145–166, North Pacific Division, U.S. Corps of Engineers, Portland, Oreg., June 1956.
26. P. Light, Analysis of High Rates of Snow Melting, *Trans. Am. Geophys. Union*, vol. 22, pt. 1, pp. 195–205, 1941.
27. R. K. Linsley, A Simple Procedure for the Day-to-Day Forecasting of Runoff from Snowmelt, *Trans. Am. Geophys. Union*, vol. 24, pt. 3, pp. 62–67, 1943.
28. M. A. Kohler and R. K. Linsley, Recent Developments in Water Supply Forecasting, *Trans. Am. Geophys. Union*, vol. 30, pp. 427–436, June 1949.
29. E. L. Peck, Low Winter Streamflow as an Index to the Short- and Long-Term Carryover from Previous Years, *Proc. West. Snow. Conf.*, pp. 41–48, 1954.
30. U.S. Soil Conservation Service, Snow Survey and Water Supply Forecasting, "SCS National Engineering Handbook," sec. 22, April 1972.
31. "Review of Procedures for Forecasting Inflow to Hungry Horse Reservoir, Montana," Water Management Subcommittee, Columbia Basin Interagency Committee, June 1953.
32. J. Hannaford, Multiple-graphical Correlation for Water Supply Forecasting, *Proc. West. Snow Conf.*, pp. 26–32, 1956.
33. M. A. Kohler, Water-Supply Forecasting Developments, *Proc. West. Snow Conf.*, pp. 62–68, 1957.

BIBLIOGRAPHY

GARTSKA, W. U.: Snow and Snow Survey, sec. 10 in V. T. Chow (ed.), "Handbook of Applied Hydrology," McGraw-Hill, New York, 1964.

JENS, S. T., and MCPHERSON, M. B.: Hydrology of Urban Areas, sec. 20 in V. T. Chow (ed.), "Handbook of Applied Hydrology," McGraw-Hill, New York, 1964.

NORDENSON, T. J., and RICHARDS, M. M.: River Forecasting, sec. 25-IV in V. T. Chow (ed.), "Handbook of Applied Hydrology," McGraw-Hill, New York, 1964.

OGROSKY, H. O., and MOCKUS, V.: Hydrology of Agricultural Lands, sec. 21 in V. T. Chow (ed.), "Handbook of Applied Hydrology," McGraw-Hill, New York, 1964.

STRAUSS. F. A.: Forecasting Water Supply Through Snow Surveys, *J. Am. Water Works Assoc.*, vol. 46, pp. 853–863, 1954.

WORLD METEOROLOGICAL ORGANIZATION, "Guide to Hydrological Practices," 3d ed., Geneva, 1974.

PROBLEMS

8-1 Derive a mean rainfall-versus-runoff curve from the data shown in the table on page 285, for the Ramapo River at Pompton Lakes, New Jersey (drainage area = 160 mi^2). Compute the average error of the relation and the bias for the tabulated summer (May to October) and winter (November to April) storms.

8-2 Derive an average loss rate (recharge divided by duration) for each of the storms in the table on page 285. Using the average of these loss rates, compute the runoff for each storm and the average error of estimate and bias of the estimates.

8-3 Compare the runoff for several assumed storms as computed from Figs. 8-5 and 8-6. Under what circumstances does Fig. 8-6 yield appreciably more runoff?

8-4 Compute the Φ and W indexes for the storm depicted in Fig. 8-1.

8-5 Find the contribution to recharge and runoff (combined) for a day on which net radiation = 150 cal/cm^2; convective transfer to the snowpack = 75 cal/cm^2; condensation = 0.02 in. depth; rainfall (at 45°F) = 1.00 in.; and water equivalent of the snowpack = 2.00 in.

8-6 Given maximum and minimum daily temperatures of 55 and 35°F, respectively, and assuming temperature to follow a sine curve, compute the degree-days and degree-hours for a 24-h period using a base of 32°F. Repeat the computations with temperatures of 45 and 25°F. Comment on the results.

8-7 Assuming the temperature-index station to be at an elevation of 1000 ft, the variation of temperature with elevation to be 4 Fahrenheit degrees per 1000 ft, and the following area-elevation characteristics,

Elevation, ft	Percent of area	Elevation, ft	Percent of area
1000	0	5000	80
2000	5	6000	90
3000	30	7000	96
4000	60	8000	100

DATA FOR PROBLEMS 8-1, 8-2, and 8-10.

Storm no.	Date	Antecedent-precipitation index	Week of year	Storm duration, h	Storm precipitation, in.	Storm runoff, in.	Basin recharge, in.
1	3/15/40	0.90	11	12	2.03	1.23	0.80
2	4/22/40	2.20	16	18	1.95	1.05	0.90
3	5/17/40	0.58	20	9	1.50	0.46	1.04
4	9/ 2/40	2.70	35	12	1.81	0.33	1.48
5	10/ 3/40	1.00	40	12	1.07	0.10	0.97
6	12/17/40	0.75	51	12	1.13	0.33	0.80
7	2/ 8/41	0.50	6	15	1.95	0.99	0.96
8	4/ 7/41	0.29	14	20	1.30	0.53	0.77
9	12/25/41	1.20	52	15	1.60	0.53	1.07
10	2/ 1/42	0.65	5	9	1.06	0.30	0.76
11	2/18/42	1.03	7	12	0.79	0.22	0.57
12	3/ 4/42	0.45	9	9	1.83	0.89	0.94
13	8/10/42	1.26	32	15	2.87	0.49	2.38
14	8/18/42	3.29	33	6	2.81	1.06	1.75
15	9/28/42	1.00	39	18	3.58	1.01	2.57
16	2/ 7/43	1.06	6	18	0.67	0.17	0.50
17	6/ 2/43	1.07	22	9	1.47	0.30	1.17
18	11/10/43	2.05	45	21	2.33	0.92	1.41
19	3/ 8/44	0.45	10	12	0.95	0.50	0.45
20	4/25/44	1.57	17	21	1.94	1.10	0.84
21	1/ 2/45	1.25	1	6	1.30	0.74	0.56
22	7/23/45	5.45	30	12	3.55	2.39	1.16
23	8/ 7/45	2.56	32	9	1.61	0.42	1.19
24	11/30/45	1.42	48	24	2.25	0.92	1.33
25	5/28/46	1.44	22	15	2.90	1.44	1.46
26	9/25/46	1.03	39	12	2.27	0.43	1.84
27	6/ 9/47	1.27	23	10	1.54	0.31	1.23
28	5/14/48	1.11	20	27	2.27	0.71	1.56
29	5/30/48	1.56	22	6	1.18	0.34	0.84
30	7/14/48	1.00	28	9	1.83	0.22	1.61
31	12/31/48	0.70	52	36	4.67	2.77	1.90
32	11/26/50	1.12	48	15	3.30	1.12	2.18
33	3/31/51	1.00	13	36	5.25	3.51	1.74
34	10/ 8/51	0.40	41	18	1.85	0.11	1.74
35	3/12/52	0.81	11	12	2.83	2.09	0.74
36	4/ 6/52	0.89	14	18	3.10	1.55	1.55
37	6/ 2/52	2.17	22	15	4.10	2.07	2.03
38	8/17/52	2.66	33	15	3.08	0.83	2.25
39	9/ 2/52	1.20	35	6	3.94	1.11	2.83
40	12/ 4/52	2.11	49	12	1.50	0.71	0.79
41	3/12/53	1.01	11	36	3.50	2.15	1.35
42	3/23/53	2.33	12	12	2.00	1.08	0.92

compute weighted degree days for index-station temperatures of 60 and 80°F and snow-line elevations of 3000 and 5000 ft.

8-8 For a basin selected by the instructor, compile the necessary data and derive a precipitation-runoff relation of the type shown in Fig. 8-8.

8-9 Assuming $S_i = 0.1$ in. and $E = 0.002$ in./h, what would be the accumulated interception loss at the end of 4 and 20 h for the storm depicted in Fig. 8-1? Supposing $S_d = 0.35$ in.; what would be the depression storage at the same times?

8-10 Using the data for Prob. 8-1, develop a coaxial relation similar to Fig. 8-5. Use the two left-hand quadrants of Fig. 8-5 as a first approximation. What is the average error of this relation?

9

STREAMFLOW ROUTING

As discharge in a channel increases, stage also increases and with it the volume of water in temporary storage in the channel. During the falling portion of a flood an equal volume of water must be released from storage. As a result, a flood wave moving down a channel appears to have its time base lengthened and (if volume remains constant) its crest lowered. The flood wave is said to be *attenuated. Streamflow routing* is the technique used in hydrology to compute the effect of channel storage on the shape and movement of a flood wave.

Given the flow at an upstream point, routing can be used to compute the flow at a downstream point. The principles of routing apply also to computation of the effect of a reservoir on the shape of a flood wave. Hydraulic storage occurs not only in channels and reservoirs but also as water flowing over the ground surface. Hence, storage is effective at the very inception of the flood wave, and routing techniques may be used to compute the hydrograph which will result from a specified pattern of rainfall excess.

9-1 Wave Movement

One of the simplest wave forms is the *monoclinal rising wave* in a uniform channel. Such a wave (Fig. 9-1) consists of an initial steady flow, a period of uniformly increasing flow, and a continuing steady flow at the higher rate. Superimposing on this wave system a velocity v equal and opposite to the wave celerity u causes the wave configuration to become stationary, and a steady flow q' takes place from right to left, with the velocities shown. This flow, known as the *overrun*, is

$$q' = (u - v_1)A_1 = (u - v_2)A_2 \tag{9-1}$$

where A is the cross-sectional area of the channel. Solving Eq. (9-1) for the wave celerity gives

$$u = \frac{A_1 v_1 - A_2 v_2}{A_1 - A_2} = \frac{q_1 - q_2}{A_1 - A_2} \tag{9-2}$$

The celerity of a monoclinal wave is thus a function of the area-discharge relation for the stream (Fig. 9-2). Since velocity usually increases with stage, area-discharge curves are usually concave upward. The slopes of the secants OA and OB represent the water velocities at sections 1 and 2, respectively ($v_1 = q_1/A_1 = \tan\theta_1$), while the slope of the secant AB represents the wave celerity [Eq. (9-2)]. It may be concluded that (1) wave celerity is greater than the water velocity in most channels; (2) for a given peak flow, the wave having the highest initial flow will travel fastest; and (3) for a wave of very small height

$$u = \frac{dq}{dA} = \frac{1}{B}\frac{dq}{dy} \tag{9-3}$$

where B is the channel width. Equation (9-3) is known as *Seddon's law* after the man who first demonstrated its validity on the Mississippi River [1]. Theoretical aspects of the law were independently derived by Kleitz (1858) and others [2], but Seddon was unaware of these works.

From the Chézy formula for flow in a wide, open channel (assuming depth equal to hydraulic radius),

$$v = Cy^{1/2}s^{1/2} \tag{9-4}$$

and

$$q = Av = vBy = CBy^{3/2}s^{1/2} \tag{9-5}$$

where s is water-surface slope. Differentiating gives

$$\frac{dq}{dy} = \tfrac{3}{2}CBy^{1/2}s^{1/2} = \tfrac{3}{2}Bv \tag{9-6}$$

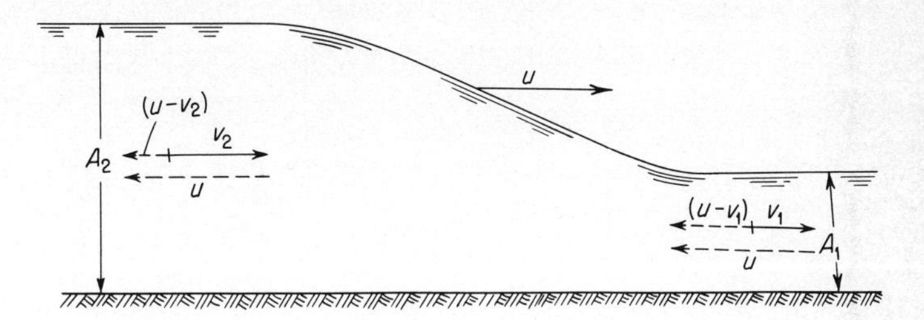

FIGURE 9-1
Definition sketch for analysis of a monoclinal rising wave.

Substituting Eq. (9-6) in Eq. (9-3), we have

$$u = \tfrac{3}{2}v \qquad (9\text{-}7)$$

The derived ratio between water velocity and wave celerity depends on channel shape and the flow formula used. Values shown in Table 9-1 may be used as rough guides for estimates of wave celerity.

A second type of wave is shown in Fig. 9-3. This is an *abrupt wave*,

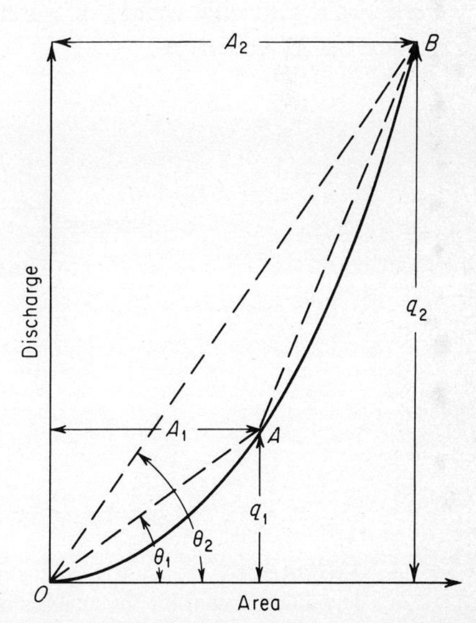

FIGURE 9-2
Typical area-discharge relation for a stream, and its influence on wave celerity.

Table 9-1 THEORETICAL RATIO BETWEEN WAVE CELERITY AND WATER VELOCITY FOR TYPICAL SECTIONS

Shape	Manning	Chézy
Triangle	1.33	1.25
Wide rectangle	1.67	1.50
Wide parabola	1.44	1.33

and the figure illustrates conditions 1 s after the instantaneous opening of the gate. The volume of water entering the channel in this time is $q_2 = A_2 v_2$ (area *acfd*). The increased volume *abhg* is

$$q_2 - q_1 = u(A_2 - A_1) \tag{9-8}$$

Substituting $Av = q$ gives

$$v_2 = (A_1 v_1 + A_2 u - A_1 u) \frac{1}{A_2} \tag{9-9}$$

The volume *dfjg* has been accelerated from v_1 to v_2 by the force F:

$$F = \frac{W}{g}(v_2 - v_1) = \frac{(u - v_2)(v_2 - v_1) A_2 w}{g} \tag{9-10}$$

where w is the specific weight of water. Since F also equals the difference in pressure on A_1 and A_2,

$$F = wA_2\bar{y}_2 - wA_1\bar{y}_1 \tag{9-11}$$

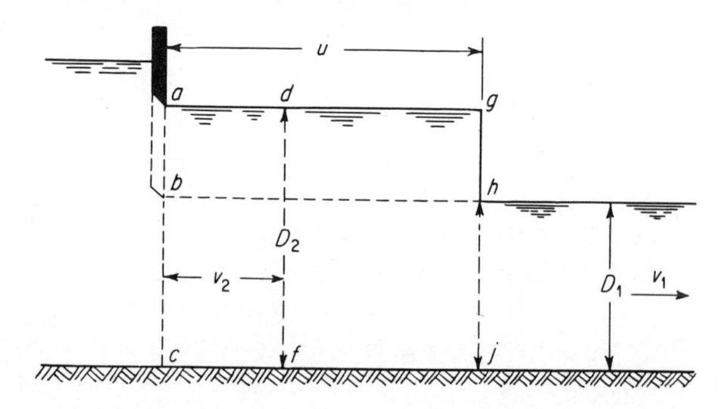

FIGURE 9-3
Definition sketch for the analysis of an abrupt translatory wave.

where $\bar{y}$ is the depth to center of gravity of the section. Equating Eqs. (9-10) and (9-11), inserting v_2 from Eq. (9-9), and solving for u give

$$u = v_1 \pm \sqrt{g \frac{A_2\bar{y}_2 - A_1\bar{y}_1}{A_1(1 - A_1/A_2)}} \tag{9-12}$$

In unit width of rectangular channel we may substitute $D = A$ and $D/2 = \bar{y}$. Hence

$$u = v_1 \pm \sqrt{\frac{gD_2}{2D_1}(D_2 + D_1)} \tag{9-13}$$

For very small wave height, $D_1 \approx D_2$ and

$$u = v_1 \pm \sqrt{gD} \tag{9-14}$$

Equation (9-12) is a general equation applying to any channel. Equation (9-13) applies only to rectangular channels, and Eq. (9-14) only to waves of very small height in rectangular channels. Abrupt translatory waves occur as tidal bores in many estuaries, as surges in power canals and tailraces, as seiches in lakes, and occasionally as flood waves caused by intense, small-area storms.

9-2 Dynamic and Kinematic Waves

Examination of Eqs. (9-3) and (9-14) shows that the celerity of the two types of waves considered are seemingly unrelated: in one case, the wave can propagate in either direction, while in the other it can move only in the downstream direction. To clarify apparent inconsistencies it is necessary to consider the more basic equations governing wave movement. Assuming constant bed slope s_b and no lateral inflow, it can be shown [3] that

$$q = AC\left[R\left(s_b - \frac{\partial y}{\partial x} - \frac{v}{g}\frac{\partial v}{\partial x} - \frac{1}{g}\frac{\partial v}{\partial t}\right)\right]^{1/2} \tag{9-15}$$

and

$$\frac{\partial q}{\partial x} + B\frac{\partial y}{\partial t} = 0 \tag{9-16}$$

where C is the Chézy coefficient and R the hydraulic radius. Equation (9-16) is a form of the continuity equation.

The derivation of Eq. (9-3) assumes that the wave does not subside or disperse but will change shape due to the dependence of u on q. The equation also implies that q is a function of y alone [4] and that u follows directly from the continuity equation. Lighthill and Whitham [2] have accordingly termed such waves *kinematic*, whereas those waves depending also upon inertial influences are termed *dynamic*. Kinematic wave motion requires that the three slope terms other than s_b in Eq. (9-15) be negligible, i.e., that the energy grade line parallels the channel bottom. This condition is satisfied in many natural channels with bed slopes of 10 ft/mi (2 m/km) or more. In general the term $\partial y/\partial x$ will be small, and the other terms are usually negligible. With a celerity of 10 ft/s (3.0 m/s) and a rate of rise of 5 ft/h (1.5 m/h), $\partial y/\partial x = 1/7200$. Thus, only when the channel slope is very flat or extraordinary rates of change of flow exist (as for a dam break) are the kinematic assumptions violated.

9-3 Waves in Natural Channels

Controlled experiments [5, 6] in flumes of regular cross section have confirmed the equations developed in Sec. 9-1. Reasonable checks have also been obtained in natural streams where the effect of local inflow is negligible, as in Seddon's work on the lower Mississippi [1] and Wilkinson's [7] study of waves downstream from TVA dams. Equation (9-14) gives very good estimates of the celerity of impulse waves in still water.

Simple mathematical treatment of flood waves is necessarily limited to uniform channels with fairly regular cross section. The hydrologist must deal with nonuniform channels of complex section with nonuniform slope and varying roughness. The formulas of Sec. 9-1 apply to waves generated at a point on a channel, but most flood waves are generated by nonuniform lateral inflow along all the channels of the stream system. Thus natural flood waves are considerably more complex than the simplified cases which yield to mathematical analysis and experimental verification.

Theoretical treatment is useful in studies of surges in canals, impulse waves in still water (including seiches and tides), and waves released from dams. Until recently, wave movement in natural channels has been treated (in design and prediction) using *hydrologic-routing* procedures [8]. Such procedures solve the equation of continuity, or *storage equation*, for an extended reach of the river, usually bounded by judiciously selected gaged points. Improvements and widespread availability of electronic computers has led to renewed interest in *hydraulic-routing* methods, which deal directly with the hydraulic characteristics of the channel and may also take dynamic effects into account [9–15]. Full treatment of such techniques is beyond the scope of this text, but they are discussed in Sec. 9-10.

Natural flood waves are generally intermediate between pure translation and pondage, which occurs in a broad reservoir or lake. Figure 9-4 shows an example of a flood wave moving with nearly pure translation, i.e., little change in shape. Figure 9-5 illustrates the great modifications which can occur when a flood wave moves through a reservoir in which discharge

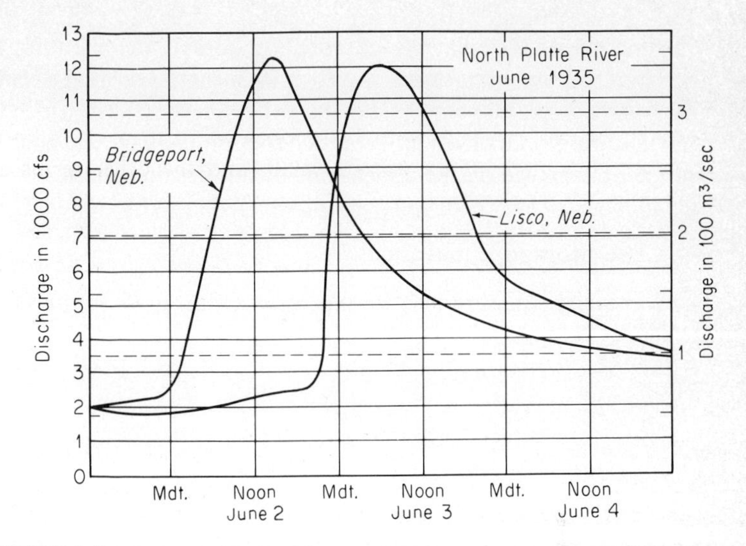

FIGURE 9-4
Example of translatory wave movement, North Platte River between Bridgeport and Lisco, Nebraska.

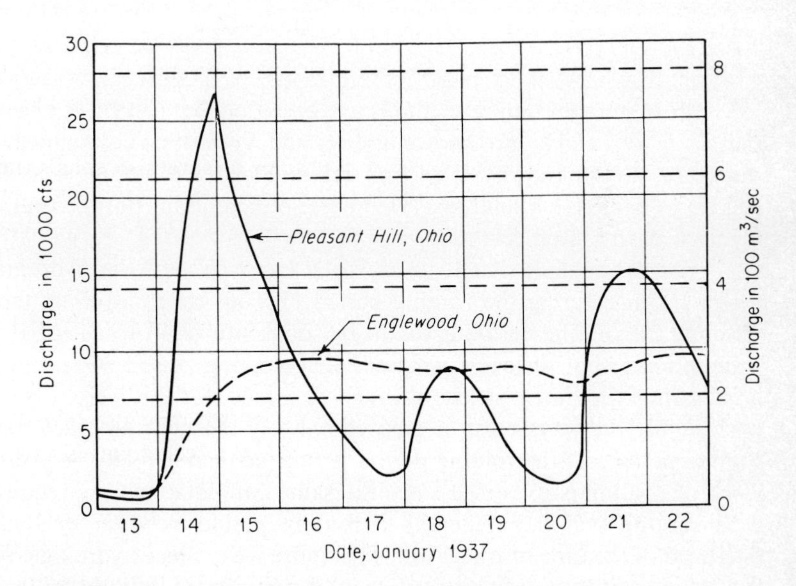

FIGURE 9-5
Reduction of discharge by storage in the Englewood retarding basin, Stillwater River, Ohio.

is a function of the quantity of water in storage. Momentum forces predominate in pure translatory waves, and such waves have relatively short time bases compared with the dimensions of the system in which they move. Most natural flood waves move under friction control and have time bases considerably exceeding the dimensions of the stream system.

9-4 The Storage Equation

The continuity equation may be expressed as in Eq. (9-16) or as

$$I - O = \frac{dS}{dt} \tag{9-17}$$

or

$$\bar{I} - \bar{O} = \frac{\Delta S}{t} \tag{9-18}$$

where I is inflow rate, O is outflow rate, and S is storage, all for a specific reach of a stream. To provide a form more convenient for hydrologic routing, it is commonly assumed that the average of the flows at the beginning and end of a short time period t (*routing period*) equals the average flow during the period. Using subscripts 1 and 2 to represent the beginning and end of the period, respectively, we have

$$\frac{I_1 + I_2}{2} t - \frac{O_1 + O_2}{2} t = S_2 - S_1 \tag{9-19}$$

Most storage-routing methods are based on Eq. (9-19). It is assumed that I_1, I_2, O_1, and S_1 are known and O_2 and S_2 must be determined. Since there are two unknowns, a second relation between storage and flow is needed to complete a solution. The major difficulties in storage routing are involved in this latter relation.

The assumption that $(I_1 + I_2)/2 = \bar{I}$ implies that the hydrograph is a straight line during the routing period t. Thus the controlling factor in selecting the routing period is that it be sufficiently short to ensure that this assumption is not seriously violated. The routing period should never be greater than the time of travel through the reach, for if it were, it would be possible for the wave crest to pass completely through the reach during a routing period. If the routing period is shorter than is really necessary, the work of routing is increased since the same computations are required for each routing period. Generally a routing period between one-half and one-third of the time of travel will work quite well. Since hydrologic routing is a solution of the continuity equation, the computed outflow volume for a flood must equal the inflow volume adjusted for any residual change in storage. If the volumes do not check, a serious error in the procedure exists.

Minor computational errors usually are compensated quite rapidly. If outflow in an interval is overestimated, the storage at the end of the interval will be too low and the outflow in the next interval will be somewhat too low. Such errors rarely lead to instability in the solution.

9-5 Determination of Storage

Before a relation between storage and flow can be established, it is necessary to determine the volume of water in the stream at various times. The obvious method for finding storage is to compute volumes in the channel from cross sections by using the prismoidal formula. The water surface is usually assumed to be level between cross sections. Total storage in the reach for any given flow conditions is the sum of the storage increments between successive cross sections. For the summation, the elevation in any subreach is the elevation indicated by a backwater curve for the midpoint of the subreach (Fig. 9-6). This method requires extensive surveys to provide adequate cross sections and computation of water-surface profiles for many unsteady, nonuniform flow situations in order to represent the range of conditions expected. The method is difficult and relatively costly and is used where no alternative is possible. It would be used, for example, to compute storage in a reach in which channel alteration or levee construction is planned, since conditions after construction would be quite unlike those existing before construction.

Storage-elevation curves for reservoirs are usually computed by planimetering the area enclosed within successive contours on a topographic map. The measured area multiplied by the contour interval gives the increment of volume from the midpoint of one contour interval to the midpoint of the next higher interval. A level water surface is usually assumed, a condition which is satisfied in most reservoirs. In reservoirs with relatively small cross-sectional area, the water surface may be far from level when large flows occur (Fig. 9-7). Under such conditions a computation similar to that described above for natural channels must be used.

The common method of determining storage in a reach of natural channel is to use Eq. (9-18) or (9-19) with observed flows. Figure 9-8 shows the inflow and outflow hydrographs for a reach of river. When inflow exceeds outflow, ΔS is positive, and when outflow exceeds inflow, ΔS is negative. Since routing involves only ΔS, absolute storage volumes are not necessary and the point of zero storage can be taken arbitrarily. Thus, the storage at any time is the sum of the positive and negative storage increments since the selected zero point. The computation is illustrated in the figure.

One of the most annoying problems in flood routing is the treatment of the *local inflow* which enters the reach between the inflow and outflow stations.

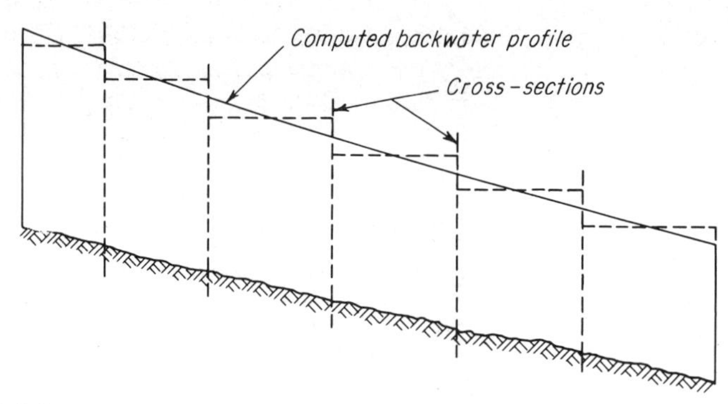

FIGURE 9-6
Computation of reach storage from channel cross sections.

If the local inflow enters mainly near the upstream end of the reach, it is usually added to the main-stream inflow to obtain total inflow. At a major junction, inflow stations (for which the flows are added) should be upstream from backwater effects. If local inflow occurs primarily near the downstream end of the reach, it may be *subtracted* from the outflow before storage is

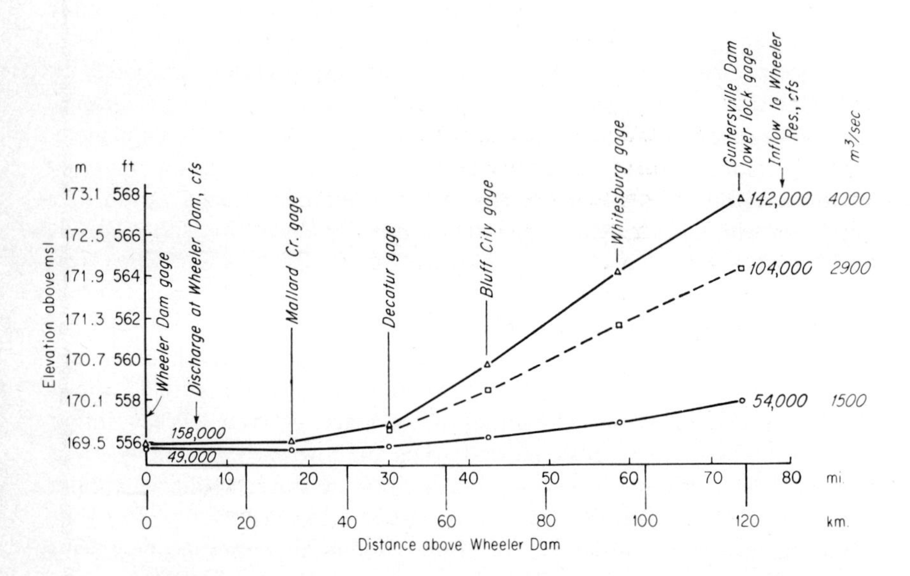

FIGURE 9-7
Profiles of the water surface in Wheeler Reservoir on the Tennessee River. (*Data from TVA*.)

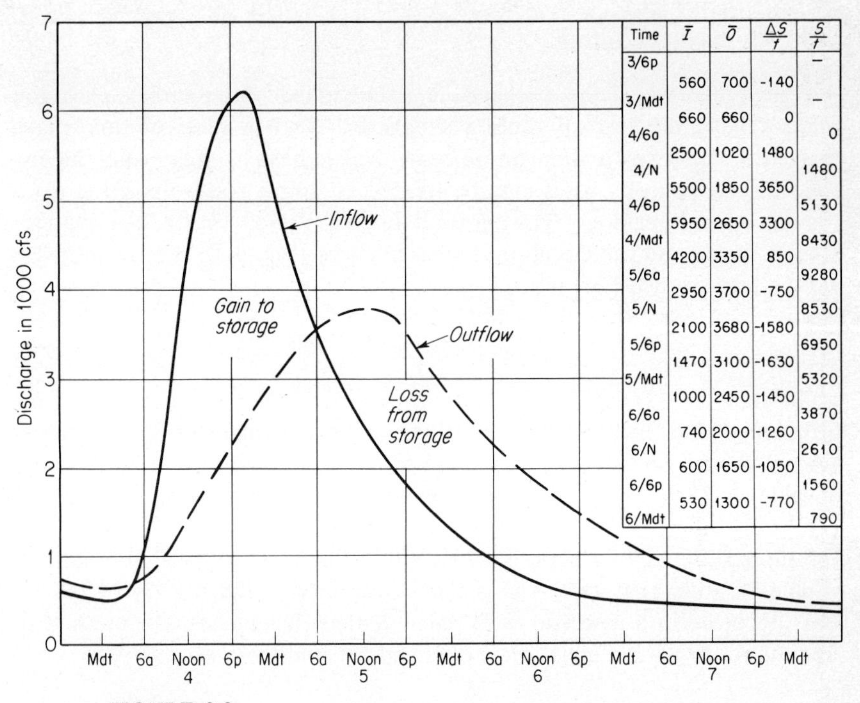

Time	$\bar{I}$	$\bar{O}$	$\frac{\Delta S}{t}$	$\frac{S}{t}$
3/6p				–
	560	700	-140	
3/Mdt				–
	660	660	0	
4/6a				0
	2500	1020	1480	
4/N				1480
	5500	1850	3650	
4/6p				5130
	5950	2650	3300	
4/Mdt				8430
	4200	3350	850	
5/6a				9280
	2950	3700	-750	
5/N				8530
	2100	3680	-1580	
5/6p				6950
	1470	3100	-1630	
5/Mdt				5320
	1000	2450	-1450	
6/6a				3870
	740	2000	-1260	
6/N				2610
	600	1650	-1050	
6/6p				1560
	530	1300	-770	
6/Mdt				790

FIGURE 9-8
Calculation of channel storage from inflow and outflow hydrographs.

computed. In this case, the main-stream flow is routed through the reach and the local inflow added after routing is complete. Between these two extremes lie many possibilities of combining various percentages of the local with the main-stream inflow before routing and adding the remainder to the outflow after routing. If the local inflow is relatively small compared with the main-stream inflow, any reasonable treatment consistently applied should give satisfactory results. If local inflow is large, consideration should be given to reducing the length of the reach.

The total volume of ungaged local inflow can be found by subtracting measured outflow from measured inflow for a period beginning and ending at about the same low flow, that is, $\Delta S = 0$. The time distribution of the ungaged local inflow is usually assumed to agree with the observed flows in a small tributary similar in size and character to the typical streams of the ungaged area. This procedure throws all errors of flow measurement into the ungaged local inflow, and the resulting flows may not appear altogether reasonable. If influent seepage is large, the computed ungaged inflow may be negative.

9-6 Reservoir Routing

A reservoir in which the discharge is a function of water-surface elevation offers the simplest of all routing situations. Such a reservoir may have ungated sluiceways and/or an uncontrolled spillway. Reservoirs having sluiceway or spillway gates may be treated as simple reservoirs if the gates remain at fixed openings. The known data on the reservoir are the elevation-storage curve and the elevation-discharge curve (Fig. 9-9). Equation (9-19) can be transformed [16] into

$$I_1 + I_2 + \left(\frac{2S_1}{t} - O_1\right) = \frac{2S_2}{t} + O_2 \qquad (9\text{-}20)$$

Solution of Eq. (9-20) requires a routing curve showing $2S/t + O$ versus O (Fig. 9-9). All terms on the left-hand side of the equation are known, and a value of $2S_2/t + O_2$ can be computed. The corresponding value of O_2 can be determined from the routing curve. The computation is then repeated for succeeding routing periods. Table 9-2 illustrates a typical solution. It should be noted that $2S/t - O$ is easily computed as $(2S/t + O) - 2O$.

Routing in a reservoir with gated outlets depends on the method of operation. A general equation is obtained by modifying Eq. (9-19) to

$$\frac{I_1 + I_2}{2}t - \frac{O_1 + O_2}{2}t - \bar{O}_R t = S_2 - S_1 \qquad (9\text{-}21)$$

where O is uncontrolled outflow and O_R is regulated outflow. If O is zero, Eq. (9-21) becomes

$$\bar{I}t - \bar{O}_R t + S_1 = S_2 \qquad (9\text{-}22)$$

which can be readily solved for S_2 and reservoir elevation. If O is not zero, the routing equation becomes

$$I_1 + I_2 - 2\bar{O}_R + \left(\frac{2S_1}{t} - O_1\right) = \left(\frac{2S_2}{t} + O_2\right) \qquad (9\text{-}23)$$

The solution of Eq. (9-23) is identical with that for Eq. (9-20), except for the inclusion of O_R.

If the gates are set at fixed openings so that the discharge is a function of head, the solution requires a family of $2S/t + O$ curves for various gate openings. The routing method is again the same as that for Eq. (9-20) except that the curve appropriate to the existing gate opening is used each time.

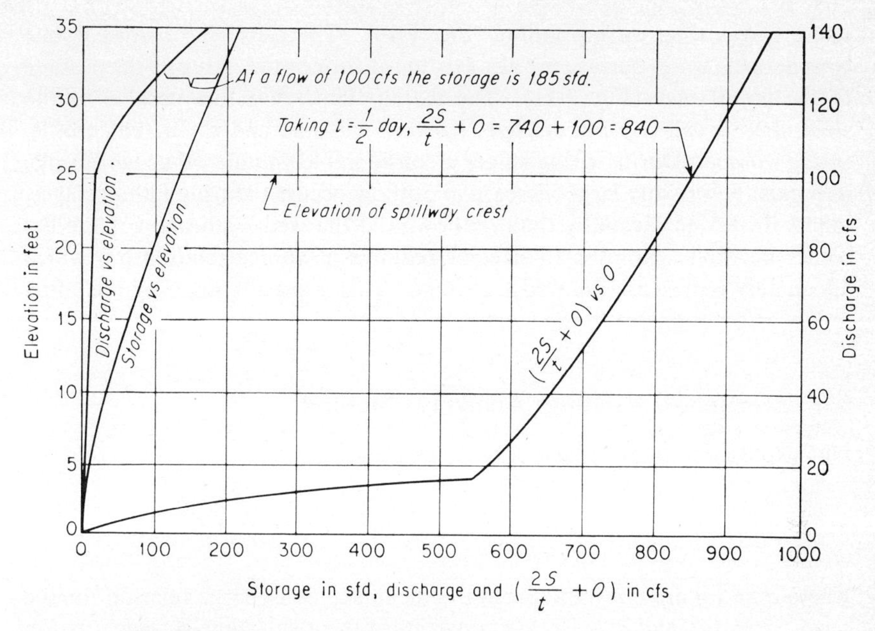

FIGURE 9-9
Routing curves for a typical reservoir.

9-7 Routing in River Channels

Routing in natural river channels is complicated by the fact that storage is not a function of outflow alone. This is illustrated when the storage computed in Fig. 9-8 is plotted against simultaneous outflow. The resulting curve is usually a wide loop indicating greater storage for a given outflow during

Table 9-2 ROUTING WITH THE $2S/t + O$ CURVE OF FIG. 9-9
Data available at start of routing shown in boldface

Date	Hour	I, ft³/s	$\frac{2S_1}{t} - O$, ft³/s	$\frac{2S_2}{t} + O$, ft³/s	O, ft³/s
1	Noon	**20**	470	500	**15**
	Midnight	**50**	508	540	16
2	Noon	**100**	578	658	40
	Midnight	**120**	632	798	83
3	Noon	**80**	642	832	95
	Midnight	**40**	620	762	71

rising stages than during falling (Fig. 9-10). The cause is obvious if one considers the water-surface profiles existing at various times during the passage of the flood wave (Fig. 9-11). The storage beneath a line parallel to the streambed is called *prism storage*; between this line and the actual profile, *wedge storage*. During rising stages a considerable volume of wedge storage may exist before any large increase in outflow occurs. During falling stages, inflow drops more rapidly than outflow, and the wedge-storage volume becomes negative. Routing in streams requires a storage relationship which adequately represents the wedge storage. This is usually done by including inflow as a parameter in the storage equation.

9-8 Streamflow Routing: Analytical Method

One expression for storage in a reach of a stream is

$$S = \frac{b}{a}\left[xI^{m/n} + (1 - x)O^{m/n}\right] \qquad (9\text{-}24)$$

where a and n are constants from the mean stage-discharge relation for the reach, $q = ag^n$, and b and m are constants in the mean stage-storage relation for the reach, $S = bg^m$. In a uniform rectangular channel, storage would vary with the first power of stage ($m = 1$) and discharge would vary as the $\frac{5}{3}$ power (Manning formula). In a natural channel with overbank floodplains the exponent n may approach or become less than unity. The constant x expresses the relative importance of inflow and outflow in determining storage. For a simple reservoir, $x = 0$ (inflow has no effect), while if inflow and outflow were equally effective, x would be 0.5. For most streams, x is between 0 and 0.3 with a mean value near 0.2.

The Muskingum method [17] assumes that $m/n = 1$ and lets $b/a = K$. Equation (9-24) then becomes

$$S = K[xI + (1 - x)O] \qquad (9\text{-}25)$$

The constant K, known as the *storage constant*, is the ratio of storage to discharge and has the dimension of time. It is approximately equal to the travel time through the reach and in the absence of better data is sometimes estimated in this way. If flow data on previous floods are available, K and x are determined by plotting S versus $xI + (1 - x)O$ for various values of x (Fig. 9-10). The best value of x is that which causes the data to plot most nearly as a single-valued curve. The Muskingum method assumes that this curve is a straight line with slope K. The units of K depend on the units of flow and storage. If storage is in second-foot-days and flow in cubic feet per second, K is in days.

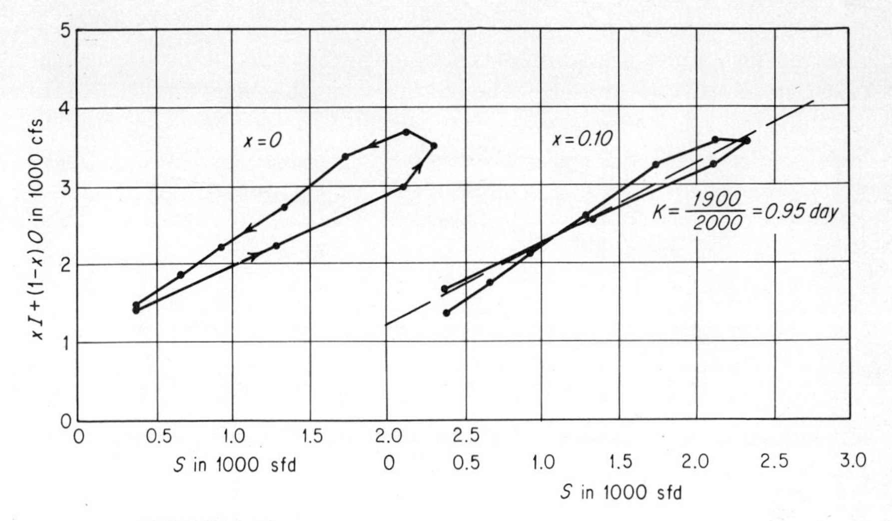

FIGURE 9-10
Determination of the Muskingum storage constants [Eq. (9-25)].

If Eq. (9-25) is substituted for S in Eq. (9-19) and like terms are collected, the resulting equation reduces to

$$O_2 = c_0 I_2 + c_1 I_1 + c_2 O_1 \tag{9-26}$$

where

$$c_0 = -\frac{Kx - 0.5t}{K - Kx + 0.5t} \tag{9-26a}$$

$$c_1 = \frac{Kx + 0.5t}{K - Kx + 0.5t} \tag{9-26b}$$

$$c_2 = \frac{K - Kx - 0.5t}{K - Kx + 0.5t} \tag{9-26c}$$

Combining Eqs. (9-26*a*) to (9-26*c*) gives

$$c_0 + c_1 + c_2 = 1 \tag{9-26d}$$

In these equations t is the routing period in the same time units as K. With K, x, and t established, values c_0, c_1, and c_2 can be computed. The routing operation is simply a solution of Eq. (9-26) with the O_2 of one routing period becoming the O_1 of the succeeding period. Table 9-3 illustrates a typical computation.

Since most routing procedures involve computation of cumulative storage, the outflow at anytime can be determined only by routing from the

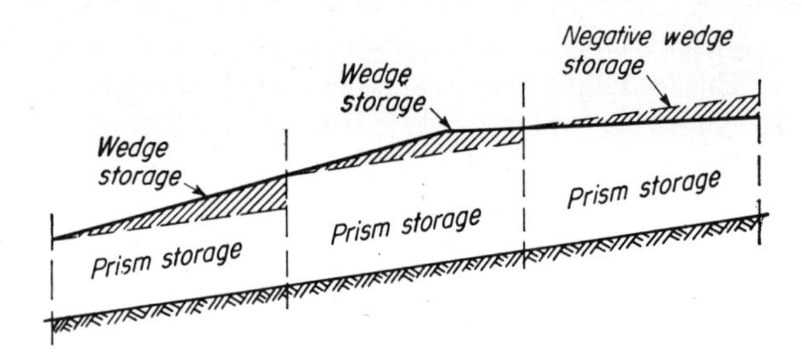

FIGURE 9-11
Some possible water-surface profiles during the passage of a flood wave.

last known value of outflow. An expression for O_4 can be written from Eq. (9-26) as

$$O_4 = c_0 I_4 + c_1 I_3 + c_2 c_0 I_3 + c_1 c_2 I_2 + c_2^2 c_0 I_2 + c_2^2 c_1 I_1 + c_2^3 O_1 \qquad (9\text{-}27)$$

Since c_2 is less than unity, c_2^3 will usually be negligible, and combining coefficients gives

$$O_4 = aI_4 + bI_3 + cI_2 + dI_1 \qquad (9\text{-}28)$$

Equation (9-28) provides a means of computing outflow at any time if the preceding inflows are known.

Since x is dependent upon the relative importance of wedge storage, it also depends on the length of the reach. The distance between gaging stations is usually such that x is significantly greater than zero. The time delay from channel storage can be simulated, however, by successive routings

Table 9-3 APPLICATION OF THE MUSKINGUM METHOD
Based on $K = 11$ h, $t = 6$ h, $x = 0.13$; hence, $c_0 = 0.124$, $c_1 = 0.353$, and $c_2 = 0.523$; values known at the beginning of routing are shown in boldface

Date	Hour	I, ft³/s	$c_0 I_2$	$c_1 I_1$	$c_2 O_1$	O, ft³/s
1	6 A.M.	**10**	...	...	...	**10**
	Noon	**30**	3.7	3.5	5.2	12.4
	6 P.M.	**68**	8.4	10.6	6.5	25.5
	Midnight	**50**	6.2	24.0	13.3	43.5
2	6 A.M.	**40**	5.0	17.7	22.7	45.4
	Noon	**31**	3.8	14.1	23.7	41.6
	6 P.M.	**23**	2.9	10.9	21.8	35.6

[18] through a number of increments of reservoir-type storage ($x = O$). This technique may be visualized as dividing the reach into a number of unit lengths for which wedge storage is small relative to the prismatic storage. The optimum number of unit reaches and corresponding value of K are usually determined by trial and error, but approximate formulas have been derived [19].

The Muskingum method assumes that K is constant at all flows. While this assumption is generally adequate, in some cases the storage-flow relation is nonlinear and an alternative method must be found. One approach is to assume that K is a function of inflow [18, 20]. It is apparent that Eq. (9-20) can be modified for channel routing by deriving a curve relating $2S/t + aI + O$ and $aI + O$.

9-9 Streamflow Routing: Graphical Methods

A variety of graphical methods for solving the routing equation have been suggested. The Muskingum storage equation, when $x = O$, can be expressed as

$$\frac{dS}{dt} = K\frac{dO}{dt} \tag{9-29}$$

Combining Eqs. (9-17) and (9-29) gives

$$\frac{I - O}{K} = \frac{dO}{dt} \tag{9-30}$$

Equation (9-30) is the basis for a very simple graphical routing method [21]. Given an inflow hydrograph and the initial segment of the outflow hydrograph, the latter can be extended by placing a straightedge as indicated in Fig. 9-12. No routing period is involved, and it will be noted that K need not be constant but can be expressed as a function of O. The procedure is therefore suitable for routing through an ungated reservoir for which a K curve (dS/dO versus O) can be constructed.

It is not necessary to determine K by the procedure described in Sec. 9-8. Instead, K may be found by reversing the routing procedure described above. A straight line conforming to the slope of the outflow hydrograph at any time t is projected to a discharge value equal to the inflow at that time. The time difference between the inflow and this projection is K. When the value of K for a number of selected points on the rising and falling limbs of a series of historical events has been determined, the relation between K and O can readily be derived.

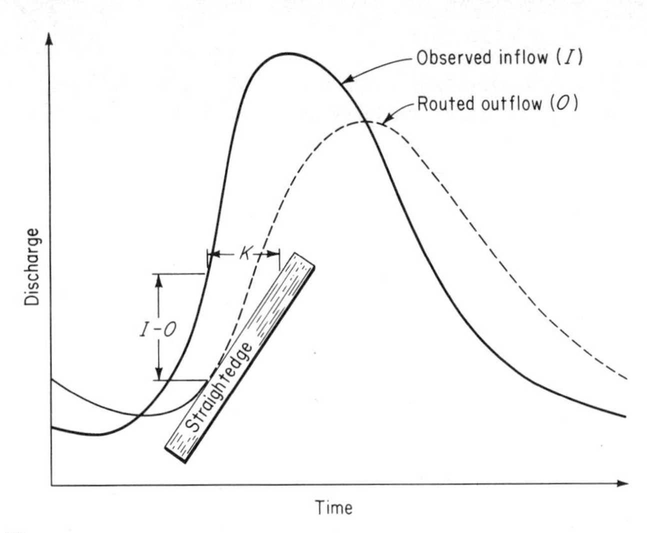

FIGURE 9-12
A graphical routing method.

The graphical procedure described above assumes pure reservoir action ($x = 0$), and the peak of the outflow hydrograph must fall on the receding limb of the inflow hydrograph. A graphical construction introducing the equivalent of the Muskingum x could be derived, but a simpler solution is available. The factor x may be viewed as a measure of the translatory component of the wave motion. Figure 9-13 shows that, with a constant K, the translation of the hydrograph increases as x increases. Thus, the effect of increasing x can be introduced by lagging the inflow hydrograph and decreasing K. If the lag T_L is constant, it is immaterial whether the inflow is lagged and then routed or the routed flow is lagged. A completely flexible procedure would utilize both variable K and T_L as functions of flow [22]. Since, with no translation, the outflow peak would fall on the inflow recession, a measure of T_L is the time difference between the outflow peak and the occurrence of an equal flow during the recession of inflow (Fig. 9-14). Values of T_L can be determined from the hydrographs of several historic floods and plotted as a T_L-versus-I curve (Fig. 9-15). Using the historic data, inflow is lagged according to the T_L-versus-I curve, and a K-versus-O curve is constructed from the lagged inflow and observed outflow as described earlier. Routing is accomplished by first lagging the inflow hydrograph (Fig. 9-15) and then routing with a straightedge as illustrated in Fig. 9-12. Templates can be devised to achieve both operations simultaneously [22], and the process is readily computerized [23].

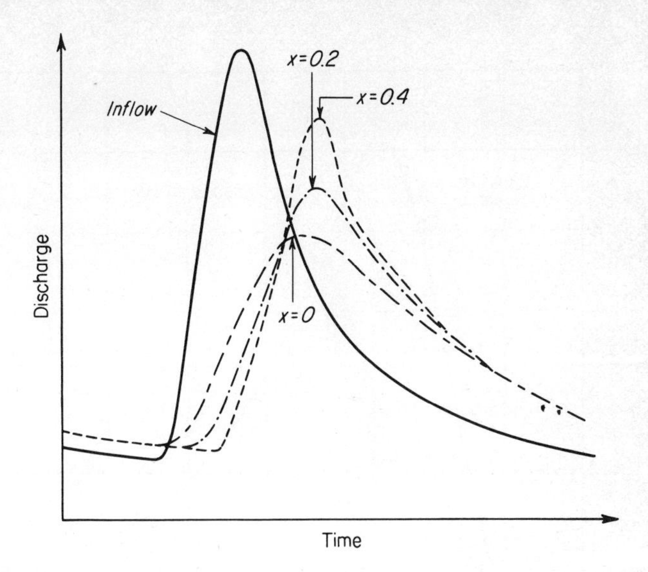

FIGURE 9-13
Effect of changes in the Muskingum x on the outflow hydrograph.

9-10 Dynamic and Kinematic Routing

The hydrologic routing techniques discussed in previous sections rely on empirical methods of fitting historical streamflow records and are therefore unsuited for the derivation of flow and stage at intermediate, ungaged points. Difficulties are also experienced in treating flows exceeding the maximum

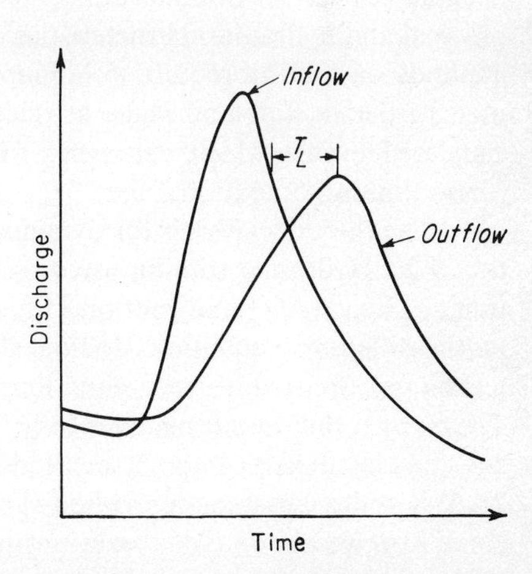

FIGURE 9-14
Determination of the lag T_L.

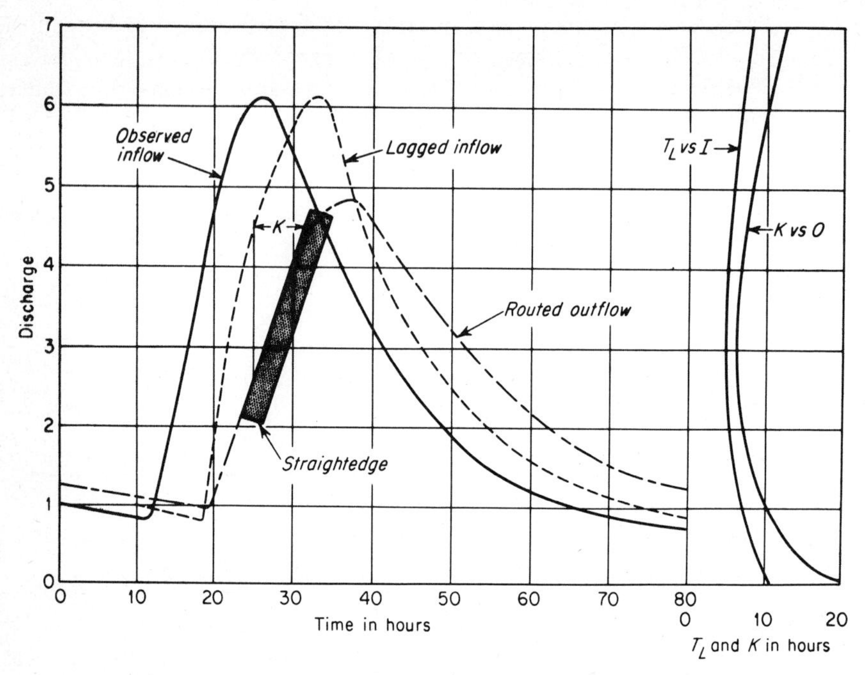

FIGURE 9-15
Graphical flood routing with a variable lag and storage factor.

flood of record or cases of extremely rapid change in flow, e.g., failure of a dam. If either of these deficiencies is considered unacceptable it becomes necessary to use hydraulic-routing techniques which take into account the physical and hydraulic characteristics of the channel system. Since hydraulic methods of routing require voluminous computations, they are sometimes used to derive flow and stage at relevant points for a series of actual and simulated events which can serve for development of simpler techniques for continuing operational use.

The theoretical basis for dynamic and kinematic routing is discussed in Sec. 9-2. Hydraulic routing involves the solution of the equations of continuity [Eq. (9-16)] and motion [Eq. (9-15)]. Analytical solutions to these partial differential equations do not exist, but they can be solved by numerical techniques (finite-difference equations) with the aid of a digital computer. The many techniques for numerically integrating the unsteady-flow equations may be classified as implicit methods [9–12, 24–27], explicit methods [25, 28–31], and characteristic methods [11, 13, 29, 32–34]. Certain initial-flow conditions and boundary conditions must be known or assumed, regardless of the method used.

In the characteristic method, the basic equations are transformed into four total differential (characteristic) equations. This results in a set of characteristic directions (or curves) in the *xt* plane along which a differential relationship exists between depth and velocity. The four equations can be numerically integrated to determine the four unknowns—depth, velocity, distance, and time—at each point in the curvilinear net. It is generally much easier to program a solution using a rectangular net, i.e., solving for depth and velocity at specified mesh points with fixed intervals of x. This method seems definitely preferable when the problem involves very large transients, as in the case of dam breaks.

In the explicit method, the basic equations are directly expressed in finite-difference form. The *xt* plane is partitioned by a rectangular net. While this approach seemingly constitutes the simplest method, both convergence and stability place stringent limitations on the time increment and the selection of a finite-difference formula is critical for meaningful results.

The implicit method also uses the finite-difference form of the basic equations and a rectangular net in the *xt* plane. However, the finite-difference formulas used do not permit solving for depth and velocity at each point in the net, as in the explicit method. In the implicit method, the depth and velocity for all points for a particular t value are determined by simultaneous solution of $2n$ equations, where n is the number of x points. The implicit method is unconditionally stable, but convergence requires that the time step be limited to a degree which is largely dependent on the time rate of change of flow. When the rate of change of flow is low, the implicit method should permit use of longer reaches and hence require less computational time than the other methods.

When kinematic assumptions are valid, and this is frequently the case as noted in Sec. 9-2, important programming simplifications are possible with consequent savings in computer capacity and time.

There are many ways in which the kinematic-wave concept might be applied to routing. In all cases the cross sections at the ends of the reach and the slope and roughness of the channel must be known, although the detail in which these are described may vary. For example, cross sections may be specified in detail with floodplains defined or simply as equivalent trapezoids. For any routing interval I_1, I_2, O_1 and the corresponding stages and mean velocities are assumed to be known. Lateral inflow must also be given or included in the inflows. A simple solution is to assume that $O_2 = O_1$ and calculate ΔS from the continuity equation and the resulting stage at the outflow section at the end of the interval. O_2 can then be calculated from Manning's equation. If the computed O_2 does not equal the assumed value, an iterative process must be used to determine its final value. Iterations may be reduced by weighting the initial assumption for O_2 in the direction of the change in inflow. It should be realized, however, that the underlying assumptions are the same for kinematic and hydrologic routing, and the increased computational requirements may not be justified unless the problem requires (1) interpolation between gaged points, (2) extrapolation above observed flows, or (3) predicting the effects of channel changes.

A procedure which might be called a *modified kinematic routing* is in use in connection with a hydrologic-simulation model. This procedure assumes that each reach is prismatic with a cross section as given for the lower end. This greatly reduces the computations for stage. It further assumes that the velocity remains unchanged during the interval so that the change in outflow is determined by the change in cross-sectional area. This usually yields a very close estimate of O_2 on the first trial, and no iterations are made. If the routing period is sufficiently short, the error compensation noted in Sec. 9-4 acts to avoid serious error or instability. Computation time may be further shortened by bypassing most of the calculations when the change in inflow is very small, as during sustained low-flow periods. Finally the programming can be made to adjust the computed flow to the water-surface slope defined by the stage at the outflow section and a stage either upstream or downstream. This step departs from the kinematic assumption by using the dy/dx term, but it does permit the program to adjust for large rates of change of inflow or for backwater caused by tides, a downstream reservoir, or tributary inflow downstream. This makes the procedure considerably more useful than a simple kinematic or hydrologic routing and yet avoids the computational detail of the dynamic routing methods.

9-11 Deriving Basin Outflow by Routing

The shape of the hydrograph from a basin depends on the travel time through the basin and on the shape and storage characteristics of the basin. When excess rainfall (runoff) is considered to be inflow and the hydrograph to be outflow, the problem is analogous to storage routing. The similarity [35, 36] of Eqs. (7-6) and (9-28) shows that the unit hydrograph itself is basically a set of average routing coefficients.

The nature of the problem suggests the use of lag-and-route methods (Sec. 9-9). Inflow can be lagged by dividing the basin into zones by isochrones of travel time from the outlet. The area between isochrones is then measured, and a time-area diagram (Fig. 9-16) is plotted. This diagram may be viewed as inflow to a hypothetical reservoir with storage characteristics equivalent to those of the basin and located at the basin outlet. Thus routing the time-area diagram by the Muskingum method (Sec. 9-8) with $x = 0$ (or some other suitable method) yields the outflow hydrograph after adjustment for units. Because of the method of constructing the time-area diagram, such a hydrograph would be the result of an instantaneous rainfall (duration $= 0$ h),

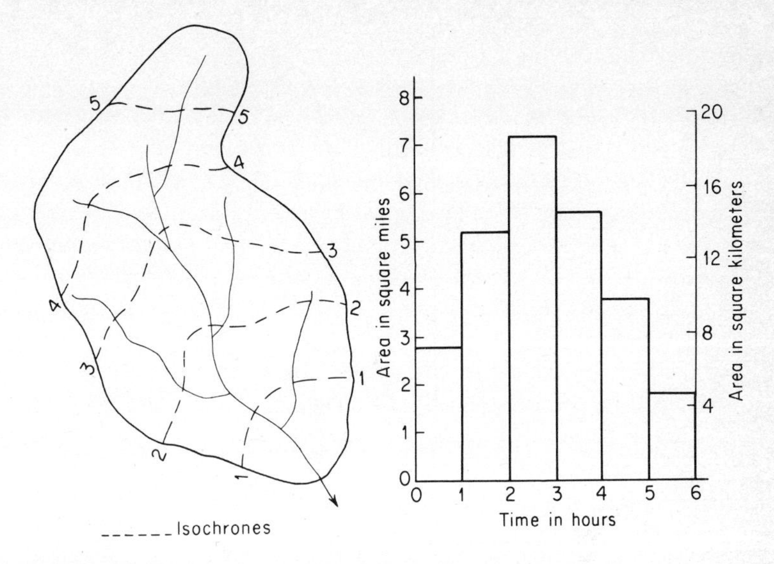

FIGURE 9-16
Derivation of the time-area diagram for a basin.

and it is called an *instantaneous unit hydrograph* [37]. It can be converted into a unit hydrograph for any duration t by averaging ordinates t units of time apart and plotting the average at the end of the period (Fig. 9-17).

The technique outlined above need not be limited to deriving unit hydrographs. For a storm of duration equal to the interval between isochrones, the average runoff can be estimated for each time zone and expressed in cubic feet per second.† The resulting time-runoff diagram is then routed through storage to give the actual outflow hydrograph. If rain lasts for several time periods, the time-runoff diagrams are lagged and superimposed (Fig. 9-18) and the summation is routed. The method accounts for time-intensity variations and areal distribution of rainfall, two factors which the unit hydrograph cannot readily consider. For this reason the routing approach can be applied to much larger basins than the simple unit-hydrograph approach can. This approach is commonly used in hydrologic simulation (Sec. 10-2).

The routing of excess rainfall over the basin, often referred to as the *isochrone* method, is not as readily applied as one might think. There is no simple, rigorous means of deriving the time-area diagram: it is usually

† q in cubic feet per second $= (24 \times 26.9AQ)/t = 646AQ/t$, where A is in square miles, Q is in inches, and t is in hours ($q = 2.78AQ/t$ m^3/s with A in square kilometers and Q in centimeters).

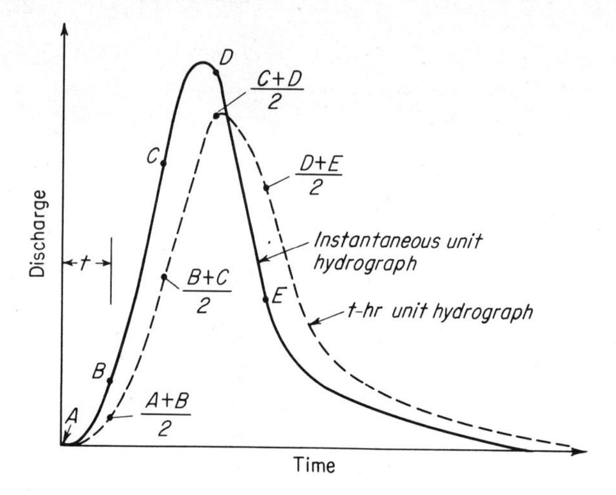

FIGURE 9-17
Converting an instantaneous unit hydrograph to one of finite duration.

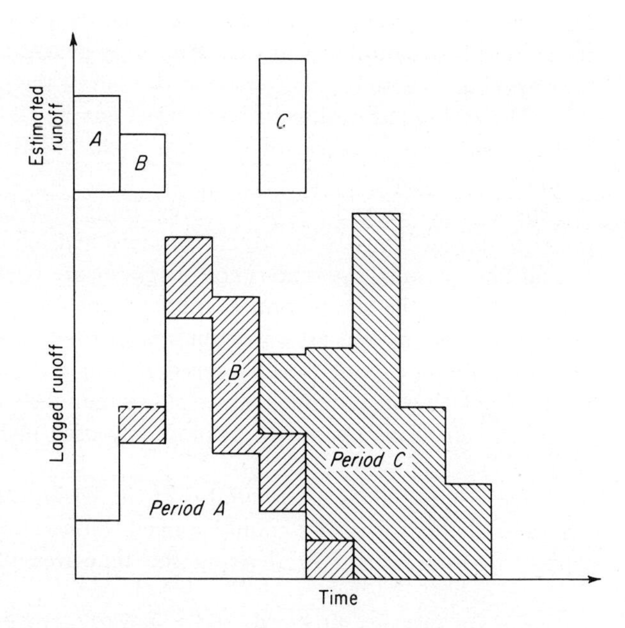

FIGURE 9-18
Time-runoff diagram for a long storm.

assumed that travel time is proportional to channel distance from each point to the outflow station, possibly taking variations in slope into account and noting that inflow must equal outflow at the time of the outflow peak. The time-area curve so derived is only a first approximation, which may require adjustment to yield an optimum fit when coupled with the best value of K. In other words, a trial-and-error approach is used in which various values are tried until a combination that gives a good fit to historical floods is found. Although laborious, this procedure is satisfactory for gaged streams but obviously unsuited to ungaged basins.

An estimate of K may be obtained from data on the recession of flow for a basin. From Eqs. (7-3) and (9-25), assuming $x = 0$,

$$K = \frac{1}{-\ln K_r} \tag{9-31}$$

where K_r is the recession constant for the stream.

Clark [37] suggested that K (in hours) be estimated from

$$K = \frac{cL}{\sqrt{s}} \tag{9-32}$$

where L is the length of the main stream in miles, s is the mean channel slope, and c varies from about 0.8 to 2.2 (0.5 to 1.4 with L in kilometers). Linsley, in a discussion of Clark's paper, suggested the formula

$$K = \frac{bL\sqrt{A}}{\sqrt{s}} \tag{9-33}$$

where A is the drainage area in square miles and b varies from about 0.04 to 0.08 (0.01 to 0.03 with L in kilometers and A in square kilometers) for the basins tested.

9-12 Gage Relations

A discussion of wave travel and routing would be incomplete without brief mention of a simple empirical solution which is often quite successful. *Gage relations* are graphs correlating an observed stage or discharge at one or more upstream stations with the resulting stage or discharge at a downstream station. Gage relations are most effective at crest (Fig. 9-19). If such a relation is to be reliable, the quantity of local inflow between the stations in each flood must bear a fixed relation to the reach inflow at the upstream station. Since such a proportional relation is unlikely, gage relations are most effective when the local inflow is relatively small compared with the main-stream inflow. It is also necessary that the peak of the local

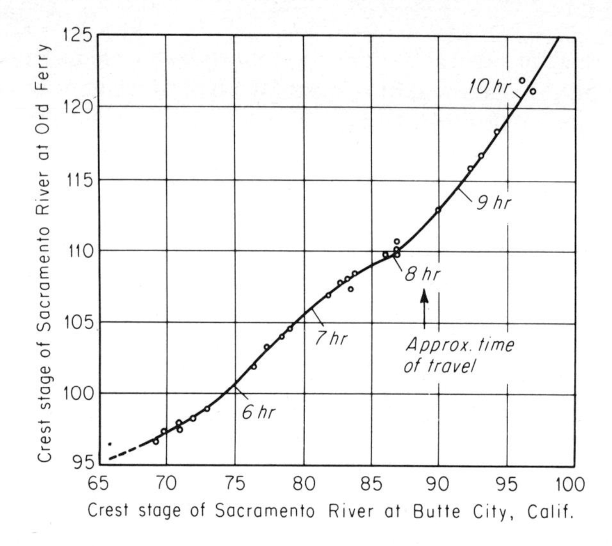

FIGURE 9-19
A simple crest-stage relation. (*U.S. National Weather Service.*)

inflow bear a fixed time relation to the peak of the main-stream inflow. If a slight difference in time of occurrence can cause a considerable difference in the resulting outflow (Fig. 9-20), gage relations will not be successful. Thus gage relations are most useful on large streams where local inflow is

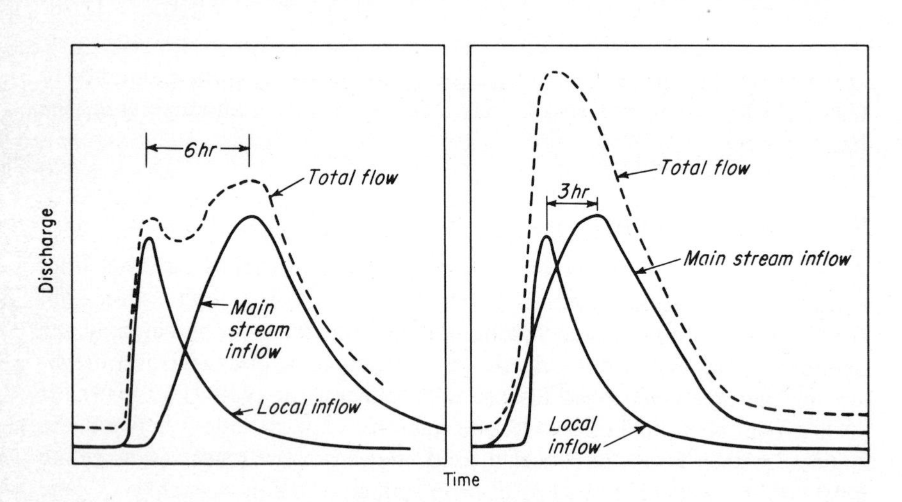

FIGURE 9-20
Effect of timing of flood peaks on reach outflow.

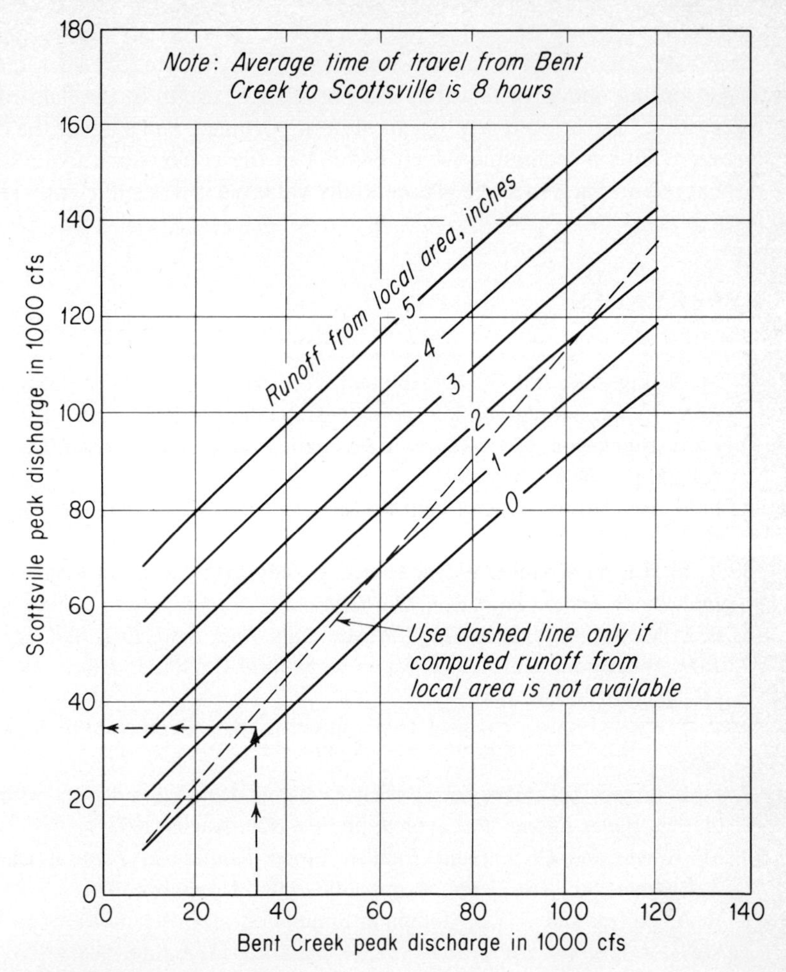

FIGURE 9-21
Gage relation for the James River from Bent Creek to Scottville, Virginia, with parameter for local inflow. (*U.S. National Weather Service.*)

small with respect to main-stream flow and rates of change of flow are relatively low.

More complex gage relations can be constructed to account for variable local inflow (Fig. 9-21). It is also possible to derive charts for routing in terms of stage [38]. Stage relations and stage routing are useful when dealing with streams for which discharge data are not available. It should be emphasized that any change in the channel, either natural or artificial, may result in changes in the stage-discharge and stage-storage relationships for the reach. An analysis in terms of stage may be invalidated by such changes. Stage routing and gage relations are particularly useful in the field of flood forecasting. Here speed is of paramount importance, and stage is the desired answer. Thus a technique which eliminates the conversions from stage to discharge and back to stage is potentially valuable if it is sufficiently reliable.

REFERENCES

1. J. Seddon, River Hydraulics, *Trans. ASCE,* vol. 43, pp. 217–229, 1900.
2. M. J. Lighthill and G. B. Whitham, I. Flood Movement in Long Rivers, *Proc. R. Sci.*, ser. A, vol. 229, pp. 281–316, 1955.
3. F. M. Henderson, Flood Waves in Prismatic Channels, *J. Hydraul. Div. ASCE,* vol. 89, pp. 39–67, July 1963.
4. F. M. Henderson, "Open Channel Flow," pp. 366–367, Macmillan, New York, 1966.
5. R. E. Horton, Channel Waves Subject Chiefly to Momentum Control, *Perm. Int. Assoc. Navig. Congr. Bull.* 27, 1939.
6. E. E. Moots, A Study in Flood Waves, *Univ. Iowa Stud. Eng. Bull.* 14, 1938.
7. J. H. Wilkinson, Translatory Waves in Natural Channels, *Trans. ASCE,* vol. 110, pp. 1203–1236, 1945.
8. V. T. Chow, "Open Channel Hydraulics," p. 586, McGraw-Hill, New York, 1959.
9. D. L. Fread, Technique for Implicit Dynamic Routing in Rivers with Tributaries, *Water Resour. Res.*, vol. 9, pp. 918–926, August 1973.
10. M. Amein and C. S. Fang, Implicit Flood Routing in Natural Channels, *J. Hydraul. Div. ASCE,* vol. 96, pp. 2481–2500, December 1970.
11. R. A. Baltzer and C. Lai, Computer Simulation of Unsteady Flows in Water-Ways, *J. Hydraul. Div. ASCE,* vol. 94, pp. 1083–1117, July 1968.
12. M. B. Abbott and F. Ionescu, On the Numerical Computation of Nearly Horizontal Flows, *J. Hydraul. Res.*, vol. 5, pp. 97–117, 1967.
13. D. L. Brakensiek, Kinematic Flood Routing, *Trans. ASAE,* vol. 10, no. 3, pp. 340–343, 1967.
14. M. Amein, Streamflow Routing on Computer by Characteristics, *Water Resour. Res.*, vol. 2, pp. 123–130, 1966.
15. See [8], pp. 587–607.
16. R. D. Goodrich, Rapid Calculation of Reservoir Discharge, *Civ. Eng.*, vol. 1, pp. 417–418, 1931.
17. G. T. McCarthy, The Unit Hydrograph and Flood Routing, presented at *Conf. North Atl. Div., U.S. Corps Eng., June 1938*; see also "Engineering Construction: Flood Control," pp. 147–156, The Engineer School, Ft. Belvoir, Va., 1940.

18. D. M. Rockwood, Application of Streamflow Synthesis and Reservoir Regulation—"SSARR"—Program to the Lower Mekong River, *Int. Assoc. Sci. Hydrol. Publ.* 80, pp. 329–334, 1968.
19. World Meteorological Organization, "Guide to Hydrometeorological Practices," 2d ed., pp. B.37–B.38, WMO no. 168, *Tech. Pap.* 82, Geneva, 1970.
20. R. Prasad, A Nonlinear Hydrologic System Response Model, *J. Hydraul. Div. ASCE*, vol. 93, pp. 201–222, July 1967.
21. W. T. Wilson, A Graphical Flood-Routing Method, *Trans. Am. Geophys. Union*, vol. 21, pt. 3, pp. 893–898, 1941.
22. M. A. Kohler, Mechanical Analogs Aid Graphical Flood Routing, *J. Hydraul. Div. ASCE*, vol. 84, pp. 1585.1–1585.14, April 1958.
23. Staff Hydrologic Research Laboratory, National Weather Service River Forecast System Forecast Procedures, *NOAA Tech. Mem.* NWS HYDRO-14, December 1972.
24. D. L. Fread, Effects of Time Step Size in Implicit Dynamic Routing, *Water Resour. Bull. AWRA*, vol. 9, no. 2, pp. 338–351, 1973.
25. J. J. Dronkers, Tidal Computations for Rivers, Coastal Areas, and Seas, *J. Hydraul. Div. ASCE*, vol. 95, pp. 29–77, January 1969.
26. T. Strelkoff, The One-dimensional Equations of Open-Channel Flow, *J. Hydraul. Div. ASCE*, vol. 96, pp. 223–252, January 1970.
27. J. W. Kamphuis, Mathematical Tidal Study of St. Lawrence River, *J. Hydraul. Div. ASCE*, vol. 96, pp. 643–664, March 1970.
28. E. Isaacson, J. J. Stoker, and A. Troesch, Numerical Solution of Flood Prediction and River Regulation Problems, *New York Univ. Inst. Math. Mech. Rep.* IMM-205 and IMM-235, 1956.
29. J. A. Liggett and D. A. Woolhiser, Difference Solution of the Shallow-Water Equations, *J. Eng. Mech. Div. ASCE*, vol. 95, pp. 39–71, 1967.
30. J. M. Garrison, J. P. Granju, and J. T. Price, Unsteady Flow Simulation in Rivers and Reservoirs, *J. Hydraul. Div. ASCE*, vol. 95, pp. 1559–1576, September 1969.
31. G. Terzidis and T. Strelkoff, Computation of Open Channel Surges and Shocks, *J. Hydraul. Div. ASCE*, vol. 96, pp. 2581–2610, December 1970.
32. V. L. Streeter and E. B. Wylie, "Hydraulic Transients," McGraw-Hill, New York, 1967.
33. V. Yevjevich and A. H. Barnes, Flood Routing through Storm Drains, pt. IV, Numerical Computer Methods of Solution, *Colo. State Univ., Fort Collins, Hydrol. Pap.* 43, 1970.
34. D. L. Fread and T. E. Harbaugh, Transient Hydraulic Simulation of Breached Dams, *J. Hydraul. Div. ASCE*, vol. 99, pp. 139–154, June 1973.
35. E. G. Popov, "Hydrological Forecasts," pp. 126–128, Gidrometeoizdat, Moscow, 1957.
36. See [19], pp. A.37–A.38 and B.34–B.36.

37. C. O. Clark, Storage and the Unit Hydrograph, *Trans. ASCE*, vol. 110, pp. 1419–1488, 1945.
38. M. A. Kohler, A Forecasting Technique for Routing and Combining Flow in Terms of Stage, *Trans. Am. Geophys. Union*, vol. 25, pt. 6, pp. 1030–1035, 1944.

BIBLIOGRAPHY

CHOW, V. T.: "Open Channel Hydraulics," pp. 586–621, McGraw-Hill, New York, 1959.

EAGLESON, P. S.: "Dynamic Hydrology," pp. 325–368, McGraw-Hill, New York, 1970.

GILCREST, B. R.: Flood Routing, chap. 10, in H. Rouse (ed.), "Engineering Hydraulics," pp. 635–709, Wiley, New York, 1950.

KEULEGAN, G. H.: Wave Motion, chap. 11, in H. Rouse (ed.), "Engineering Hydraulics," pp. 711–768, Wiley, New York, 1950.

KLEMEŠ, V.: Applications of Hydrology to Water Resources Management, *Opera. Hydrol. Rep.*, no. 4, World Meteorological Organization, Geneva, 1973.

LAWLER, E. A.: Flood Routing, sec. 25-II, in V. T. Chow (ed.), "Handbook of Applied Hydrology," pp. 25.34–25.59, McGraw-Hill, New York, 1964.

LINSLEY, R. K., KOHLER, M. A., and PAULHUS, J. L. H.: "Applied Hydrology," pp. 465–541, McGraw-Hill, New York, 1949.

NĚMEC, J.: "Engineering Hydrology," pp. 274–281, McGraw-Hill, London, 1972.

THOMAS, H. A.: The Hydraulics of Flood Movement in Rivers, *Carnegie Inst. Tech. Eng. Bull.*, 1935.

PROBLEMS

9-1 If the channel width for the stream whose stage-discharge relation is shown in Fig. 4-11 is 30 ft at a stage of 4 ft, what would be the wave celerity for a translatory monoclinal wave of small height?

9-2 Find the ratio between wave celerity and water velocity for a semicircular channel when $y = r$; when $y = 0.2r$. Use the Chézy formula.

9-3 A uniform rectangular channel 10 ft wide ($n = 0.015$) on a slope of 0.0004 is flowing at a depth of 5 ft. A sudden gate opening increases the depth to 6 ft. What is the celerity of the resulting abrupt wave?

9-4 Given the hydrographs tabulated below, find the storage in the reach and plot a curve showing storage at any instant as a function of simultaneous outflow. Ignore local inflow.

Date	Hour†	Inflow, ft³/s	Outflow, ft³/s	Date	Hour†	Inflow, ft³/s	Outflow, ft³/s
1	M	40	40	7	N	245	394
2	N	35	39		M	192	307
	M	37	37	8	N	144	235
3	N	125	52		M	118	180
	M	340	130	9	N	95	142
4	N	575	287		M	80	114
	M	722	472	10	N	67	93
5	N	740	624		M	56	77
	M	673	676	11	N	50	64
6	N	456	638		M	42	55
	M	320	574				

† M = midnight; N = noon.

9-5 A small reservoir has an area of 300 acres at spillway level, and the banks are essentially vertical for several feet above spillway level. The spillway is 15 ft long and has a coefficient of 3.75. Taking the inflow hydrograph of Prob. 9-4 as the inflow to the reservoir, compute the maximum pool level and maximum discharge to be expected if the reservoir is initially at the spillway level at midnight on the first.

9-6 Tabulated below are the elevation-storage and elevation-discharge data for a small reservoir. Taking the inflow hydrograph of Prob. 9-4 as the reservoir inflow, and assuming the pool elevation to be 875 at midnight on the first, find the maximum pool elevation and peak outflow rate.

Elevation	Storage, acre-ft	Discharge, ft³/s	Elevation	Storage, acre-ft	Discharge, ft³/s
862	0	0	882	1220	100
865	40	0	884	1630	230
870	200	0	886	2270	394
875	500	0	888	3150	600
880	1000	0			

9-7 Find the Muskingum K and x for the flood of Prob. 9-4.

9-8 Taking the outflow hydrograph of Prob. 9-4 as the inflow to a reach with $K = 27$ h and $x = 0.2$, find the peak outflow, using the Muskingum method of routing.

9-9 Write the routing equation for the case when storage is a function of $aI + bO$.

9-10 Using the graphical method of Sec. 9-9, find the K curve for the flood of Prob. 9-4.

9-11 Using the outflow hydrograph of Prob. 9-4 as the inflow to a reach for which the lag $T_L = 6$ h and $K = 18$ h, find the peak outflow and time of peak by the graphical method of Sec. 9-9.

9-12 Assume that the stages of Fig. 4-11 are in meters and the discharges in cubic meters per second. If the channel width at a stage of 4 m were 60 m, what would be the celerity of a monoclinal rising wave of small height?

9-13 A small reservoir has an area of 4000 hectares at spillway level and the banks are essentially vertical for several meters above spillway level. The spillway is 50 m long and has a coefficient of 3.75. Taking the outflow hydrograph of Prob. 9-4 as the inflow to the reservoir in cubic meters per second, compute the maximum discharge and pool level to be expected if the reservoir is at spillway level at midnight on the first.

9-14 Take the data of Prob. 9-6 as elevation in meters, storage in 10^5 m^3 and flows in cubic meters per second, and determine the maximum pool level and outflow rate.

10

COMPUTER SIMULATION OF STREAMFLOW

The preceding chapters discussed methodology for estimating various types of hydrologic data in the absence of observations. While the value of observed facts is undeniable, only time can provide the lengths of observed record required for adequate probability analysis of hydrologic variables. Considering the rate at which man-made changes are altering the hydrologic environment, time may not yield data which are homogeneous and suitable for frequency analysis. Because of the need for estimates defining the probabilities which will be applicable in the future, engineering hydrology has stressed techniques for predicting hydrologic variables by a variety of indirect methods.

The techniques described in previous chapters have all been applied in the solution of engineering problems. They have yielded results considered adequate or, perhaps more correctly, the best available. These methods have a common characteristic: they were derived before digital computers became available. Because computers can do vastly more detailed calculations than are possible manually, they offer a basis for improvement in hydrologic analysis. However, research in computer applications in hydrology dates from the mid 1950s, and the ultimate potential of computers in hydrology is far from being realized.

10-1 Philosophy of Simulation

Most of the conventional hydrologic methods discussed earlier can be used in a computer solution [1]. With hourly rainfall increments and hourly streamflow ordinates as input, it is possible to derive unit hydrographs and loss rates (Φ index). Conversely, it is possible to convolve unit hydrograph ordinates and increments of excess rainfall to produce a streamflow hydrograph. Routing of streamflow in a channel [2] or of runoff from a watershed is easily programmed for computer solution using the Muskingum method. If a statistical solution is preferred, computers will readily determine the regression equation, given the necessary input data.

In each case, the computer solution is probably faster and less subject to errors than the corresponding manual solution. With observed data increasingly available in punched-card or magnetic-tape form, data preparation may also be faster and less likely to involve mistakes. Despite these advantages, the basic assumptions and approximations included in the methodology to make it suitable for manual solution are still incorporated. If the potential value of the computer is to be realized, a new approach should be sought.

This new procedure should calculate on shorter time intervals and in greater areal detail within the basin. The mathematical functions employed should be designed to simulate the natural hydrologic processes as closely as present knowledge and computer constraints permit. The less important processes such as interception and overland flow should be included, and all components of runoff (surface, interflow, groundwater, and channel precipitation) should be calculated. Infiltration provides the most rational base for the computation. Finally, the procedure should be programmed to permit continuous reproduction of flow so that, if necessary, long "records" of flow (or other hydrologic variables) can be developed from observed rainfall data.

The most promising approach to computer application for the purposes described above is *simulation*, the representation of the hydrologic system on the computer so as to imitate the behavior of the natural system. Clearly many levels of approximation to the natural processes are possible. For example, one might employ a simple function to relate runoff volume to rainfall or, alternatively, a series of functions which attempt to reproduce all the steps in the runoff process—interception, infiltration, overland flow, etc. Provided adequate data are at hand, the more realistic programs will be the most accurate, and these programs are what are commonly referred to as simulation models. Given enough free parameters, almost any program can be fitted to watershed data to reproduce a historic flow sample with good accuracy, but extrapolation beyond the range of the historic data sample or transfer to another watershed may be quite unsatisfactory unless the program realistically describes the hydrologic processes.

Traditional approaches to hydrology would suggest that a special simulation model should be developed for each watershed. This, however, does not recognize that the hydrologic cycle is fundamentally the same on all watersheds. The relative magnitudes of the various processes change with climate, geology, soil, and topography, but there is no obvious reason why a general model cannot apply to all watersheds, and experience has confirmed this. Such a general model must contain parameters which can be adjusted to the special conditions of each basin and must even permit some of the processes to be shut off when appropriate. Use of a general program permits accumulation of information on parameters from many watersheds which will eventually permit relating parameters to measurable physical characteristics of the watersheds.

Simulation requires that relatively large quantities of data be read into computer storage. Thus a simulation model for practical application must include programming for data management so that the data can be stored and recovered rapidly as required by the simulation operation. Since the model will be used and reused many times, programming efficiency is also important. Computer running times tend to be long, and complicated or inefficient algorithms can be relatively costly. Thus, while many hydrologic processes can be solved by a finite-difference solution of differential equations, a simpler more empirical function may serve almost as well and at lower cost. The structure of the model should also be consistent with the available data. While it might seem desirable to subdivide the watershed into many subareas so that the variable characteristics of soils, topography, and vegetation may be correctly represented, little real advantage may result if the precipitation on all subareas must be estimated from a single rain gage.

Simulation is not the only application of computers in hydrology. Chapter 12 discusses the generation of stochastic data by computers, and specific applications to other problems are discussed elsewhere in this text.

10-2 Structure of a Flow-Simulation Program

A number of different simulation programs have been devised (see Bibliography). One of the earliest was the Stanford Watershed Model [3, 4], the general logic of which will be described. A flow diagram of the Stanford Watershed Model is shown in Fig. 10-1. The principal input data are hourly precipitation and daily potential evapotranspiration. The model is based on water-balance accounting. Separate accounts are kept for segments of the watershed representing differing soils, vegetation, land use, and precipitation. Each segment is described by a set of parameters representing specific

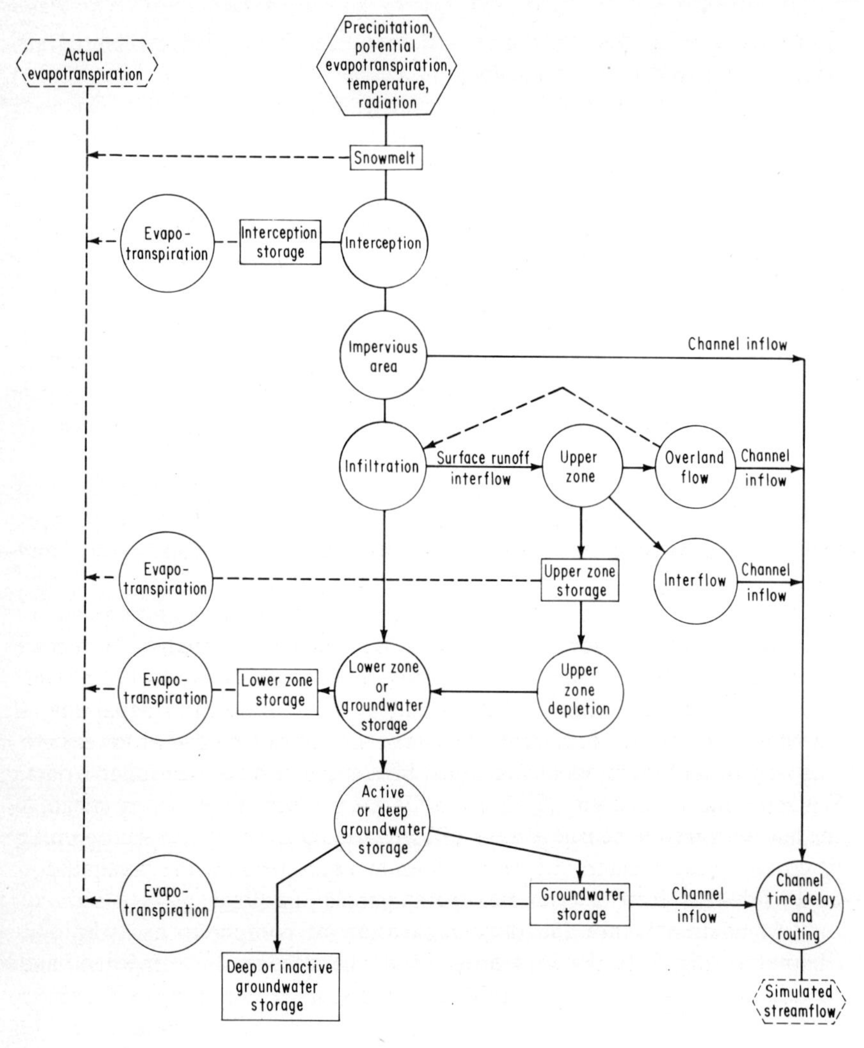

FIGURE 10-1
Flow diagram of Stanford Watershed Model IV.

physical features of the segment. Although increasing the number of segments permits improved representation of differing characteristics within the watershed, each additional segment utilizes additional computer time, and a compromise between detailed description of the watershed and cost is inevitable. Moreover, with mutiple segments, choice of parameter values

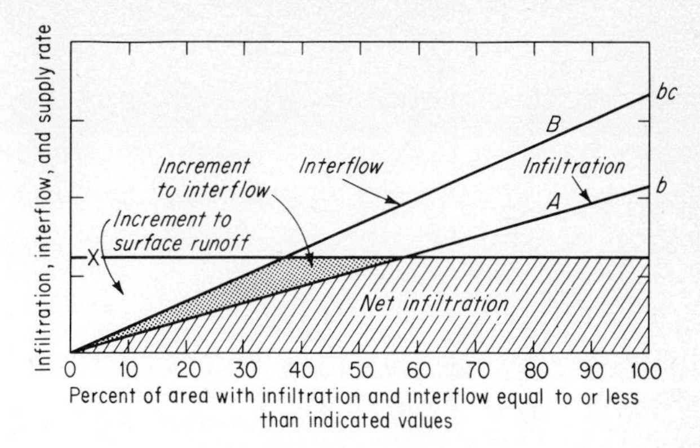

FIGURE 10-2
Infiltration-interflow function used in Stanford Watershed Model.

becomes more difficult. For simplicity, the following text discusses a single segment basin.

If a realistic water-balance accounting is to result, the point observations of rainfall must represent the areal average over the segment. Hence, the first step in the simulation is to multiply station rainfall by a constant equal to the ratio of basin normal annual rainfall to station normal annual rainfall. *Interception* loss is simulated by fixing an interception-storage capacity (usually between 0 and 0.2 in. or 0 to 5 mm). All rainfall is assumed to enter interception storage until it is filled. Interception storage is depleted by evaporation at the potential rate until reduced to zero. Evaporation during a storm increases interception loss (Sec. 8-1). Impervious area in the watershed is simulated by diverting a constant percentage of each hourly increment of rain to the stream as *impervious-area runoff*. The percentage is equal to the percentage of impervious area directly connected to the streams in the watershed. The channel surfaces of the stream system, lakes, and swamps are also treated as impervious area. For most rural basins impervious area is small (1 or 2 percent), but for urban areas this factor is usually more than 20 percent.

It is assumed that infiltration capacity at any time varies over the segment. For lack of better information this variation is assumed to be linear (Fig. 10-2). The position of this line is varied by varying the value of b, as a function of the ratio of moisture in the lower-zone storage S_L to the nominal capacity of this zone L (Fig. 10-3). The net infiltration to the lower zone is the shaded trapezoid of Fig. 10-2 and is a function of the current soil-moisture ratio (S_L/L) and the supply rate X (precipitation minus interception).

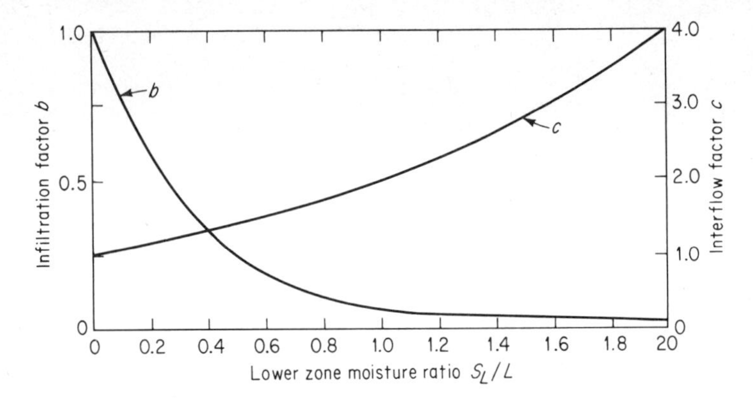

FIGURE 10-3
Variation of the infiltration factor b and interflow factor c as a function of S_L/L when INFILTRATION and INTERFLOW = 1.0.

Interflow is calculated by a similar process. Line B of Fig. 10-2 divides the rainfall excess triangle into two portions, surface runoff and interflow. The position of line B is fixed by multiplying b by a factor c, which is greater than 1 and is also a function of S_L/L. The fraction of interflow increases as soil moisture increases (Fig. 10-3). Interflow runoff is placed in interflow storage, from which a fixed fraction of the storage is released to the stream in each time interval. This produces an exponential decline in the rate of interflow in the stream. The fraction is set by an interflow recession constant K_I determined from observed flow data (Chap. 7).

Surface runoff enters the upper-zone function, from which a portion directed to *upper-zone storage* simulates depression storage and upper-soil storage. The fraction ΔS_U entering upper-zone storage is a function of the ratio S_U/U (Fig. 10-4). The balance of the surface runoff enters the overland-flow process. An empirical function derived from experimental data relates outflow from overland flow to the volume of water in overland-flow detention. A simple continuity computation calculates the detention volume from

$$D_2 = D_1 + \Delta D - q_o \tag{10-1}$$

where q_o is outflow at the beginning of the interval, ΔD is the increment to overland-flow detention, and D_1 and D_2 are the detention volumes at the beginning and end of the interval. There is an obvious approximation in using q_o for $\bar{q}$, but if the interval is short, the error is small and an iterative solution is not warranted. Any water left in detention at the end of the interval is returned to the rainfall supply to infiltration, simulating delayed infiltration from overland flow. For small watersheds, the time delay in overland flow

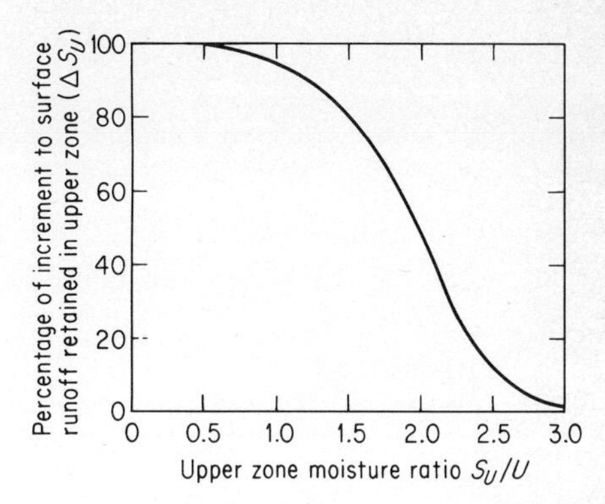

FIGURE 10-4
Fraction of rainfall excess retained in upper-zone storage.

may be very important, but for all watersheds the delayed infiltration is an important factor in the runoff process.

A portion of the upper-zone storage may pass on to the lower zone whenever S_U/U is greater than S_L/L (upper-zone depletion). This water and the net infiltration are divided between *lower-zone soil-moisture storage* and *groundwater storage* on the basis of a function of S_L/L (Fig. 10-5). More water goes to soil moisture when $S_L/L < 1$ and more to groundwater when the ratio exceeds unity. A fixed percentage of the water entering groundwater may be diverted to deep groundwater which does not contribute to streamflow. The remainder goes to groundwater storage from which water enters the stream in accord with an exponential recession (Chap. 7).

Water in upper-zone and interception storage is assumed to evaporate at the potential rate. Water in lower-zone storage is depleted by evapotranspiration at a rate which is a function of the ratio S_L/L. When the ratio is high, evapotranspiration is near the potential rate and decreases as the ratio decreases. If shallow groundwater exists, evaporation from it is permitted.

The runoff quantities calculated in the water-balance computation must be routed through the channels of the basin to calculate the outflow hydrograph. This can be accomplished by an application of lag and route techniques. For this purpose a time-area diagram converted to flow units and Muskingum type routing are convenient (Chap. 9). If several segments are used, the time-area diagram can reflect the differing runoff in each segment. Kinematic-wave routing is also a useful method. Kinematic routing requires that channel cross sections, slopes, and roughness be defined. Compensating for this, however, the kinematic routing permits calculation of flow and stages at the end of each reach. Further, the routing is defined by the physical characteristics of the channel with natural nonlinearities and discontinuities represented far better than is possible with an empirical routing procedure.

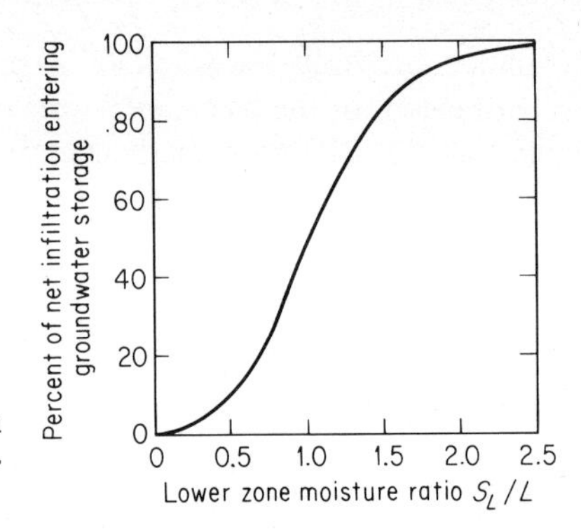

FIGURE 10-5
Division of infiltrated water between lower-zone moisture storage and groundwater.

10-3 Parameters

Table 10-1 lists the 15 parameters that are set to adapt the general simulation program to a specific watershed. A, LL, and SS can be determined from maps (Chap. 14). NN must be estimated from knowledge of the watershed, but its exact magnitude does not seem to be very critical. K3 can be estimated from aerial photography or by field inspection. A mean-annual-precipitation map for the watershed is required to estimate K1. The two groundwater parameters K24L and K24EL are often zero. When appropriate, they must be estimated from a priori knowledge of the groundwater situation in the watershed. If no information exists regarding possible deep groundwater storage, the value of K24L can sometimes be determined in the calibration process (Sec. 10-6). The recession parameters KK24 and KI can be estimated

Table 10-1 PARAMETERS OF THE STANFORD WATERSHED MODEL FOR RAINFALL RUNOFF

K1	Ratio of normal basin rainfall to normal station rainfall
EPXM	Interception storage
A	Percent of impervious area
INFILTRATION	Average infiltration rate
INTERFLOW	Interflow function
U	Upper-zone nominal capacity
LL	Average length of overland flow
SS	Average slope of overland flow
NN	Manning's *n* for overland flow
L	Lower-zone storage index
K24L	Portion of groundwater assigned to deep storage
K3	Evaporation-loss index (percent of area with deep-rooted vegetation)
K24EL	Percent of area with shallow groundwater subject to direct evapotranspiration
KK24	Groundwater recession rate
KI	Interflow recession constant

from a short streamflow record by the methods discussed in Chap. 7. If no record exists for the basin, estimates of these values for similar basins nearby may be used. EPXM is relatively unimportant and is usually assigned a value between 0.1 and 0.2 in. (2 and 5 mm), depending on vegetation density.

The remaining four parameters are determined by *calibration* or comparison with observed hydrographs. These four parameters are the storages L and U and the INTERFLOW and INFILTRATION parameters. The storages are defined as index values or nominal capacities rather than absolute limits to avoid the discontinuity which would occur if fixed storages were filled and no capacity remained. Actual values of water in storage can exceed the index capacities, but as the excess amount increases, the functions (Fig. 10-4 and 10-5) limit the input to the storages so that a limiting capacity is approached asymptotically. L is an index to the transient storage in the soil: essentially permanent soil moisture is of no concern in the simulation process. Thus L is related to the annual rainfall. Experience has shown that a first estimate of L is given by

$$L = \begin{cases} 4 + 0.25P & \text{seasonal precipitation} \\ 4 + 0.125P & \text{precipitation distributed} \\ & \text{throughout the year} \end{cases} \qquad (10\text{-}2)$$

where P is the mean annual precipitation in inches. U is commonly about one-tenth of L. The storages primarily affect the volume of runoff and have little influence on hydrograph shape. In metric units the constant 4 in Eq. (10-2) should be changed to 100.

The parameters INFILTRATION and INTERFLOW adjust the functions of Fig. 10-3 upward or downward to reflect the differences in infiltration and interflow characteristics between basins. An initial assumption of unity for both parameters is suitable. An overestimate of INFILTRATION will result in underestimating peak flows and will cause excessive groundwater flows. If INTERFLOW is too high, peaks will tend to be low and recessions immediately after the peaks will be too slow.

Routing parameters will depend on the method used (Chap. 9). If a Muskingum type routing is used, the watershed is divided into time zones representing fixed intervals of travel time (often 1 h) and the percentage of area in each zone is input. In addition an estimate of the Muskingum K must be provided. If kinematic routing is used, cross-section data, channel slopes, and estimates of Manning's n for each reach must be provided.

10-4 Simulation of Snowmelt

Simulation of snowmelt runoff involves all of the processes described in Sec. 10-2. In addition, simulation of the heat exchanges involved in snowmelt and simulation of the snowfall and snow-accumulation process must be added [5]. Because of significant temperature controls on snow accumulation and melt, the watershed must be divided into elevation zones if the range in elevation of the basin is greater than can be tolerated in one zone. Each additional segment increases run time on the computer. However, since one may expect temperature differences between 7 and 11 Fahrenheit (4 and 6 Celsius) degrees over an elevation difference of 2000 ft (600 m), this is probably the maximum difference in elevation that should be used and 1000 ft (300 m) may be better.

Temperature in each elevation zone must be estimated from the nearest temperature station by use of a lapse rate. The elevation difference (ELDIF) between the station and the median elevation of each zone which it represents is an input parameter (Table 10-2). The lapse rate to be used may be fixed at an average value of about 4 Fahrenheit degrees per 1000 ft (2.2 Celsius degrees per 300 m) or may be varied from 3.5 to 5.5 Fahrenheit (2 to 3 Celsius) degrees depending on whether precipitation is occurring or not. If there are two temperature stations, the lapse rate can be calculated at each observation as the temperature difference between the stations divided by the elevation difference.

Since records of snowfall-water equivalent are generally unavailable, the simulation program must distinguish between snowfall and rain and maintain a running account of the depth, density, and water equivalent of snow on the ground. The decision between rain and snow is based on current temperature in the segment with rain assumed to occur at temperatures above 32°F (0°C) and snow at temperatures of 32°F or below. If snow is assumed, an initial density DNS must be assigned, by specifying a parameter

Table 10-2 PARAMETERS OF THE SNOWMELT ROUTINE OF THE STANFORD WATERSHED MODEL

RADCON, CONDS-CONV	Adjust theoretical melt equations to field conditions; usual values near 1.0
SCF	Correction to compensate for deficiency of gage catch during snowfall
ELDIF	Elevation difference between temperature station and midpoint of a segment
INDS	Index density of new snow at 0°F (−18°C)
F	Percent of segment with forest cover
DGM	Daily ground melt, in./d
WC	Maximum water content of snow in percent
MPACK	Water equivalent of pack when segment is completely snow-covered
EVAPSNOW	Adjust theoretical evaporation equation to field conditions
MELEV	Mean elevation of segment

INDS as the average density of new snow at 0°F and a variation with Fahrenheit temperature T according to the equation

$$DNS = INDS + \left(\frac{T}{100}\right)^2 \qquad (10\text{-}3)$$

If there is new snow, the water equivalent of snow on the ground is the sum of the water equivalent at the end of the previous interval and the precipitation amount for the current interval. A snow-correction factor SCF, which is >1, allows for adjusting reported precipitation to account for the underregistration resulting from the effect of wind on the gage catch of snowfall (Sec. 3-5). Whenever the programming determines that precipitation in an interval is snow, the reported catch is multiplied by SCF.

Snow depth at any time is determined from the water equivalent of the snowpack and the density. If the density SDEN is less than 0.55, depth is decreased each hour by multiplying the current depth D_s by $1 - 0.00002D_s$ (0.55-SDEN). The new density is then calculated from the water equivalent and depth. This process simulates the consolidation of the snowpack under its own weight.

Liquid-water content of the snowpack is controlled by a parameter WC, which is the maximum permissible liquid-water content expressed as a fraction of the water equivalent of the pack (usually between 0.01 and 0.05). Liquid water (precipitation or melt) can be accumulated in the pack as water content up to the limit given by WC times water equivalent of the pack.

It is assumed that a segment is completely snow-covered if the current water-equivalent PACK exceeds a variable IPACK, which is defined as the largest value of PACK previously attained during the season. If PACK is less than IPACK, the fraction of snow cover in the segment is calculated as PACK/IPACK. IPACK is not permitted to exceed a maximum MPACK, an input parameter, which is the value of snow-water equivalent above which the segment is assumed to be completely snow-covered, regardless of any previous values of snowpack which have occurred.

Simulation of the melt process is very dependent on the data available. Given complete data—insolation, temperature, dewpoint, and wind—an energy-budget approach (Sec. 8-10) is used. An adjustment parameter is programmed for each function to make adjustments for data representivity. If data for the complete equations are missing, substitutes such as the use of a fixed value for wind, estimating dewpoint from minimum temperature, or estimating radiation from clear-sky values adjusted for cloudiness may be used. As a last resort, a simple degree-day approach (Sec. 8-11) can be used if only temperature data are available.

Heat loss from the snow may occur when insolation is small and back radiation large or air temperatures very low. This is simulated by a negative heat storage. The calculated heat loss in any interval is multiplied by 1.0 − NEGMELT/NEGMM and added to the current negative heat storage NEGMELT. NEGMM is the maximum possible negative heat storage calculated as a function of water equivalent and air temperature by assuming that temperature in the snow varies linearly from the ambient air temperature at the surface to 32°F (0°C) at the soil surface. No melt can occur as long as NEGMELT is greater than zero. Any heat available to the snowpack first offsets NEGMELT before melting can begin.

When melt starts, meltwater adds to the water content of the pack until the maximum (WC × water equivalent) is reached. Melt increments then enter the runoff phase of the simulation in the same manner as increments of rainfall (Fig. 10-1).

10-5 Application of Flow Simulation in Hydrology

Flow simulation produces output containing information similar to that obtained from a gaging station: mean daily flows, monthly and annual flow volumes, flood peaks, and details of the hydrographs. Thus, in one sense, simulation can be looked upon as a method of estimating streamflow data which can be subjected to probability analysis to determine design parameters. The value of simulation lies in its ability to produce such data for ungaged points or to extrapolate short-streamflow records. Possibly more important is the ability to project flows for assumed *future* conditions in the watershed. Changes in land use can be anticipated by appropriate parameter changes, and if kinematic routing is used, changes in the channel system can be simulated by specifying appropriate channel characteristics. Since any water planning is for the future, this capability is especially important.

Urban drainage To utilize simulation for design of urban drainage, the first step would be to lay out an approximate drainage system. The simulation model is calibrated against the nearest gaged stream. The impervious-area parameter is then adjusted to a value appropriate for the urban conditions existing or expected, and the parameters defining overland flow may be adjusted as necessary. Groundwater flow and interflow may also be eliminated from the output if it is felt they will not contribute to the storm flow. Using the nearest available rainfall record, 20 or more years can be simulated with flows determined at key points within the system. These flows can be analyzed to determine flow frequency, and design flows of the desired probability are selected. The tentative system is then revised to conform to these criteria. If snowfall is a common occurrence, it may be included in the simulation. Detention basins can be simulated in the system.

Floodplain mapping The procedure for floodplain mapping also begins with calibration to available records on the stream. Parameters should then be adjusted for anticipated future conditions (urbanization, deforestation, channel changes, etc.). A simulation of flows for the longest possible period (constrained by available precipitation data) is then made with as many flow points within the floodplain area as possible. If kinematic routing is used, stage data can be output. Frequency analysis of the output defines the probability of flow or stage at all flow points. With these data, the necessary floodplain map can be constructed. Stages along minor tributaries entering the floodplain can be determined at the same time as stages along the main stream. The incorrect assumption that stages at all points in the system during a storm have the same frequency is not necessary.

Reservoir design Reservoirs for flood-damage mitigation must have adequate storage capacity, discharge facilities, and operating procedures to operate effectively to reduce flood peaks downstream. Effectiveness is measured by the reduction in the damage-probability curve. The streamflow record should be extended as far as possible by simulation. Not only is the inflow hydrograph to the reservoir required but also all tributary flows downstream of the reservoir. Reservoir inflow must be routed through the reservoir and combined with downstream inflows to determine the effect of the reservoir at all critical points downstream. The effect of backwater on tributary stages must be calculated to determine the extent of flooding. The study would be laborious by manual methods but can be executed easily by computer simulation.

After a design has been selected by testing against historic floods, a simulated operating test would indicate the real effectiveness of the system. Such a test could be carried out by generating hourly rainfall inputs stochastically and converting the precipitation to discharge by computer simulation. The person playing the role of operator would not know what might be expected by way of future rainfall and therefore would be forced to make operating decisions under uncertainty, as in real life. One or more days of advance rainfall might be generated, modified by a random factor to simulate forecast errors and given to the operator as a quantitative rainfall forecast to further simulate reality. A test on historic data can never be completely valid since the operator can recognize the development of the historic sequence and anticipate the outcome.

Water-supply reservoirs are usually insensitive to short-term flow fluctuations, and use of stochastic methods (Chap. 12) to generate monthly flow sequences is usually adequate for design. However, relatively small reservoirs may require more detailed study, and simulation may be used to generate the necessary inflow data.

Other applications Simulation is obviously useful in any situation calling for calculation of streamflow data from precipitation (or snowmelt). Its greatest utility may be in evaluating watershed change. The consequences of augmenting precipitation can be simulated by comparing flows calculated with historic precipitation and those calculated with the historic values augmented by some appropriate procedure [6]. Similarly, the effect of land-use changes can be simulated throughout a long period of years by appropriate changes in parameters. As discussed earlier, urbanization modifies the impervious area and the conveyance system. Land terracing modifies overland-flow slope and distance. Vegetation changes are primarily effective through changes in the evapotranspiration although vegetation may also affect infiltration, roughness of overland flow, and interception. Simulation of 20 to 30 y with appropriate parameter changes will give a far better picture of the true long-term effects of land management than calculations for a few selected events or even a few years of experimental data. Experimental data may be quite useful in defining appropriate parameter changes. In this sense, simulation may prove to be a very effective device for transposing information gained on experimental watersheds to larger basins.

Forecasting Streamflow forecasting almost always involves an element of urgency. Computer simulation offers a high-speed and effective tool for forecasting. Simulation can be especially useful for systems involving one or more reservoirs or other units for which operating decisions must be made. Simulation can permit testing of several operating decisions to determine which one gives best results, whereas manual methods may not allow so many tests within the time available.

Since a simulation model for forecasting will be used to produce flows for a few hours to several days in advance, and since the forecast is periodically updated with new data, the model must be somewhat different in its operating characteristics from one used for generating long records for planning purposes [7]. The basic logic of the simulation programs, however, can be identical.

10-6 Calibration and Optimization

As indicated in Sec. 10-3, some parameters must be determined by trial in which observed and calculated flows are compared until a good fit is reached. Calibration is most easily done by first establishing a good fit of monthly and annual flow volumes. When a good volume fit has been obtained, the shapes of the computed and observed hydrographs are compared in detail. INFILTRATION and INTERFLOW are the main parameters affecting hydrograph shape if kinematic routing is used. If an empirical routing procedure is employed, the parameters for it may also require adjustment to correct errors in hydrograph shape. When calibration is complete, a simulation program should yield runoff volumes within a few percent for annual data, a reasonable monthly distribution, hydrograph shape consistent with observed hydrographs, and peak flows randomly distributed above and below the historic data. Cases of estimated input data or estimated streamflow should be excluded from comparisons.

Automatic parameter-determination routines may be incorporated into a single-segment model [8, 9]. Observed flows are input as data, and an objective function (commonly the sum of squared differences between observed and simulated mean daily flow) is calculated. A parameter is adjusted by some predetermined amount, and the simulation is repeated. By successive iterations it is possible to approach an optimum set of parameters (minimum objective function). The process is facilitated by programming a systematic search procedure. For a multisegment basin, automatic adjustment of parameters is practicable only if all segments are assumed to have the same values for L, U, INFILTRATION, and INTERFLOW. If all segments have different parameters, the number of iterations increases exponentially as the number of segments and the computation time becomes too great.

Manual adjustment is greatly enhanced by direct display of output hydrographs on a cathode-ray tube [10]. With a time-sharing computer system, direct interaction between the hydrologist and the simulation program permits rapid adjustment of parameters.

While calibration is normally limited to determining the four parameters U, L, INFILTRATION, and INTERFLOW, it sometimes becomes evident during calibration that other parameters require adjustment. This is especially true of K1 (precipitation adjustment), A (impervious area), K24L (percolation to deep groundwater), and K24EL (evapotranspiration from groundwater). The necessity for adjustment of these parameters is usually evident from an inability to obtain a water balance. If groundwater flows are consistently high but reducing INFILTRATION would yield too much surface runoff, it may be appropriate to invoke K24L. If total water balance is in error, an adjustment in K1 may be indicated. A thorough understanding of the simulation algorithm and the effects of the individual parameters is essential for good calibration.

Errors or lack of representivity in the input data and errors in the streamflow data used for calibration are a significant problem. If parameters are adjusted to bring erroneous cases to a fit, the parameters may be seriously biased. This is a special problem with automatic optimization if a squared-difference criterion is used, since some of the largest differences between observed and simulated flow may result from data errors or inadequacies. Adjustment of parameters controlling volume of runoff is fairly straightforward through comparison of simulated and observed values of monthly flow volume. Adjustment of parameters influencing hydrograph shape is

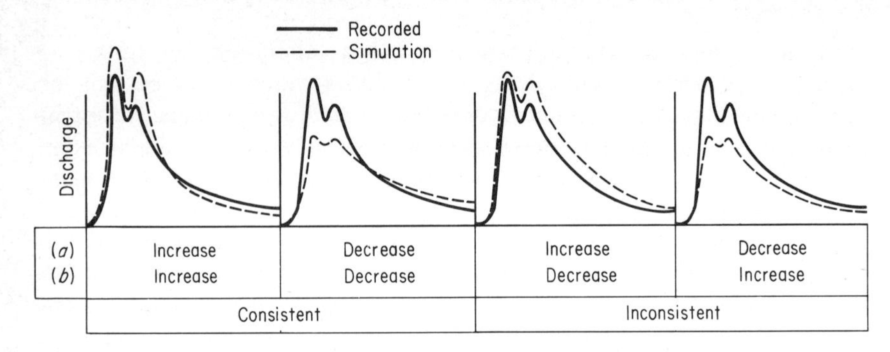

FIGURE 10-6
Some consistent and inconsistent indications for the adjustment of INFILTRATION based on (*a*) surface-runoff volume and (*b*) groundwater flow.

more difficult because differences between observed and simulated hydrographs may arise from errors in timing, estimated volume of storm runoff, or the initial assumption of shape parameters. In some cases the comparison will indicate ambiguous situations (Fig. 10-6), which cannot be corrected by parameter adjustment.

The problem of fitting is not unique to simulation but is present with all hydrologic techniques. Because simulation is capable of more precise results, and because it is tested against several years of observed data, there is a tendency to be more concerned with accuracy of calibration and the selection of optimal parameters than with conventional techniques. Inadequate input data are probably the most serious constraint on simulation. No matter how good the model is, it cannot overcome the limitations of poor data. Scattered random errors in the data are usually not serious, but data containing a significant bias will yield biased answers. The closeness of fit in calibration is controlled by the input data, and there is a tendency to revise parameters and rerun the calibration many times in an attempt to get a better fit than is possible with the data. Johanson [11] investigated the limits of accuracy in calibration in relation to the adequacy of the input data (Sec. 3-6).

10-7 Other Applications of Simulation in Hydrology

The ability to simulate streamflow hydrographs suggests that simulation should have other applications in hydrology. The flow simulation discussed in this chapter can output other information such as groundwater accretion; actual evapotranspiration; soil-moisture levels; snowpack depth, density, and water equivalent; and possibly other information of potential hydrologic value.

Since erosion is the result of raindrop impact and transport of eroded material to the stream is accomplished by overland flow, the basic structure for sediment simulation exists in a flow-simulation model (Sec. 13-8). Quality of water in streams and lakes can also be simulated in a manner paralleling flow simulation. A quality simulation model should be especially valuable because of the relative scarcity of long records of quality suitable for defining the probability of critical quality conditions. Water quality is measured by a large number of factors including temperature, dissolved oxygen, biochemical oxygen demand, dissolved salts, phytoplankton, zooplankton, sediment or turbidity, nutrients, and bacteria. Algorithms must be provided for each factor, and input data must be available for artificial inputs from waste discharges. Dissolved salts vary with the mix of surface and subsurface water in the stream, which is calculated by a flow-simulation model, and also with the addition of salts from waste outflows. Temperature is controlled by radiation, evaporation, and other natural factors and is also influenced by discharges of waste heat. Dissolved oxygen is affected by the demands of waste discharges, natural aeration of the water, and the activity of algae. Thus the programming of a quality model is necessarily complex, and the data-handling requirements are large. Diurnal variations of some factors such as temperature and dissolved oxygen are often large, so that for realistic understanding of the processes the model must operate on a short time interval [12].

REFERENCES

1. J. P. McCallister, Role of Digital Computers in Hydrologic Forecasting and Analysis, *Int. Assoc. Sci. Hydrol. Publ.* 63, pp. 68–77, 1963.
2. D. M. Rockwood, Columbia Basin Streamflow Routing by Computer, *J. Waterw. Div. ASCE*, vol. 84, p. 1874, December 1958.
3. R. K. Linsley and N. H. Crawford, Computation of a Synthetic Streamflow Record on a Digital Computer, *Int. Assoc. Sci. Hydrol. Publ.* 51, pp. 526–538, 1960.
4. N. H. Crawford and R. K. Linsley, Digital Simulation in Hydrology: Stanford Watershed Model IV, *Stanford Univ., Dept. Civ. Eng. Tech. Rep.* 39, 1966.
5. E. A. Anderson and N. H. Crawford, The Synthesis of Continuous Snowmelt Runoff Hydrographs on a Digital Computer, *Stanford Univ., Dept. Civ. Eng. Tech. Rep.* 36, 1964.
6. A. M. Lumb and R. K. Linsley, Hydrologic Consequences of Rainfall Augmentation, *J. Hydraul. Div. ASCE*, vol. 97, pp. 1065–1080, July 1971.

7. National Weather Service River Forecast System Forecast Procedures, *NOAA Tech. Mem.* NWS HYDRO-14, December 1972.
8. L. D. James, An Evaluation of Relationships Between Streamflow Patterns and Watershed Characteristics through the Use of OPSET: A Self-calibrating Version of the Stanford Watershed Model, *Univ. Ky. Water Resour. Inst. Res. Rep.* 36, 1970.
9. J. C. Monro, Direct Search Optimization in Mathematical Modeling and a Watershed Model Application, *U.S. Natl. Weather Serv., NOAA Tech. Mem.* NWS HYDRO-12, April 1971.
10. D. D. Franz and R. K. Linsley, An Interactive Time Series Plotting System for Classroom Instruction in Hydrology, *Stanford Univ., Dept. Civ. Eng. Tech. Rep.* 143, February 1971.
11. R. C. Johanson, Precipitation Network Requirements for Streamflow Estimation, *Stanford Univ., Dept. Civ. Eng. Tech. Rep.* 147, August 1971.
12. P. Lombardo, "A Critical Review of Available Water Quality Simulation Models," Hydrocomp, Inc., Palo Alto, Calif., July 1973.

BIBLIOGRAPHY

AITKEN, A. D.: Assessing Systematic Errors in Rainfall-Runoff Models, *J.Hydrol.*, vol. 20, pp. 121–137, October 1973.

ANDERSON, E. A.: Development and Testing of Snow Pack Energy Balance Equations, *Water Resour. Res.*, vol. 4, pp. 19–37, February 1968.

———: National Weather Service River Forecast System—Snow Accumulation and Ablation Model, *NOAA Tech. Mem.* NWS HYDRO-17, November 1973.

BOUGHTON, W. C.: A Mathematical Catchment Model for Estimating Runoff, *J. Hydrol. (New Zealand)*, vol. 7, pp. 75–100, 1968.

CHEN, C. W., and SHUBINSKI, R. P.: Computer Simulation of Urban Storm Water Runoff, *J. Hydraul. Div. ASCE*, vol. 97, pp. 289–301, February 1971.

CLABORN, B. J., and MOORE, W.: Numerical Simulation in Watershed Hydrology, *Univ. Tex., Hydraul. Eng. Lab. Tech. Rep.* HYD. 14-7001, 1970.

DAWDY, D. R., and BERGMANN, J. M.: Effect of Rainfall Variability on Streamflow Simulation, *Water Resour. Res.*, vol. 5, pp. 958–966, 1969.

FLEMING, G.: "Deterministic Simulation in Hydrology," American Elsevier, New York, 1974.

HOLTAN, H. N., and LOPEZ, N. C.: An Engineering-oriented Model of Watershed Hydrology, *Proc. Int. Seminar Hydrol. Prof.*, vol. 3. pp. 572–605, 1969.

HUFF, D. D.: "Simulation of the Hydrologic Transport of Radioactive Aerosols," Ph.D. thesis, Comm. Hydrology, Stanford Univ., December 1967.

IBBITT, R. P., and O'DONNELL, T.: Fitting Methods for Conceptual Catchment Models, *J. Hydraul. Div. ASCE*, vol. 97, pp. 1331–1342, September 1971.

JOHNSTON, P. R., and PILGRIM, D. H.: A Study of Parameter Optimization for a Rainfall-Runoff Model, *Water Res. Lab. Rep. no.* 131, Univ. New South Wales, February 1973.

KUCHMENT, L. S., and KOREN, V. I.: Modelling of Hydrologic Processes with the Aid of Electronic Computers, *Int. Assoc. Sci. Hydrol. Publ.* 81, pp. 616–624, 1968.

LIOU, E. Y.: OPSET: Program for Computerized Selection of Watershed Parameter Values for the Stanford Watershed Model, *Univ. Ky. Water Resour. Inst. Res. Rep.* 34, 1970.

MANDEVILLE, A. N., O'CONNELL, P. E., SUTCLIFFE, J. V., and NASH, J. E.: River Flow Forecasting through Conceptual Models, *J. Hydrol.*, vol. 10, pp. 282–290 and 317–329, 1970, and vol. 11, pp. 109–128, 1970.

NARAYANA, V. V. D., RILEY, J. P., and ISRAELSON, E. K.: "Analog Computer Simulation of the Runoff Characteristics of an Urban Watershed," Utah Water Research Laboratory, Utah State University, Logan, 1969.

SCHERMERHORN, V. P., and KUEHL, D. W.: Operational Streamflow Forecasting with the SSAR Model, *Int. Assoc. Sci. Hydrol. Publ.* 80, pp. 317–328, 1968.

SITTNER, W. T., SCHAUSS, C. E., and MONRO, J. C.: Continuous Hydrograph Synthesis with an API-Type Hydrologic Model, *Water Resour. Res.*, vol. 5, pp. 1007–1022, October 1969.

SUGAWARA, M.: On the Analysis of the Runoff Structure about Several Japanese Rivers, *Jap. J. Geophys.*, vol. 2, 1961.

PROBLEMS

10-1 In the infiltration-interflow function (Fig. 10-2) if b is 0.9 in./h, c is 1.3, and $X = 0.1$ in./h, what surface runoff and interflow will occur? Suppose $X = 0.5$ in./h. What is the result?

10-2 If $U = 1$ in., $L = 10$ in., and $S_U/U = 0.7$, what part of the surface runoff of Prob. 10-1 will be retained in surface storage? What is the increment to overland flow? Repeat with $S_U = 2.0$ in.

10-3 If $L = 10$ in. and $S_L = 7$ in., what part of the net infiltration of Prob. 10-1 will be added to groundwater storage? Repeat with $S_L = 18$ in.

10-4 If a simulation model is available, calibrate it for a small stream in your area. Discuss the results in terms of such topics as adequacy of input data, interaction between model parameters, etc.

10-5 Using the simulation model, explore its sensitivity to input data and parameters. For example, compare output when the potential evapotranspiration is increased 20 percent. What is the effect of a 10 percent change in precipitation? How sensitive is the model to changes in parameters? To what extent do you think your results are specific for the conditions you have assumed? For example, would the changes resulting from an assumed change in evapotranspiration be different if the precipitation were different?

11

PROBABILITY IN HYDROLOGY: A BASIS FOR DESIGN

Since design and planning are concerned with future events whose time or magnitude cannot be forecast, we must resort to statements of probability or frequency† with which a specified rate or volume of flow will be equaled or exceeded. Selection of the appropriate level of probability for design, i.e., the risk that will be acceptable, rests on economic or policy grounds [1]. Designing for the 10 percent or 10-y flood involves a calculated risk. If the hydrologic analysis is correct, the system will occasionally be overtaxed. The alternative of designing against the worst possible event that could happen is usually so costly that it is justified only where consequences of failure are especially grave.

This chapter deals with techniques for defining probability from a given set of data and with special methods employed for determining the spillway-design flood for major dams. Special methods for probability analysis using synthetically generated data are discussed in Chap. 12.

† Probability: a mathematical basis for prediction, which, for an exhaustive set of outcomes, is the ratio of the outcomes that will produce a given event to the total number of possible outcomes (paraphrased from "Webster's 7th New Collegiate Dictionary," 1971). Frequency: the number of cases in a class when events are classified according to differences in one or more attributes.

FLOOD PROBABILITY

The following section discusses basic concepts of probability analysis with specific reference to flood peaks. In general these methods are also applicable to other hydrologic parameters with minor differences discussed in subsequent sections.

11-1 Selection of Data

If probability analysis is to provide useful answers, it must start with a data series that is relevant, adequate, and accurate. Relevance implies that the data must deal with the problem. Most flood studies are concerned with peak flows, and the data series will consist of selected observed peaks. However, if the problem is duration of flooding, e.g., for what periods of time a highway adjacent to a stream is likely to be flooded, the data series should represent duration of flows in excess of some critical value. If the problem is one of interior drainage of a leveed area, the data required may consist of those flood volumes occurring when the main river is too high to permit gravity drainage.

Adequacy refers primarily to length of record, but sparsity of data-collecting stations is often a problem. The observed record is merely a sample of the total population of floods that have occurred and may be expected to occur again. If the sample is too small, the probabilities derived cannot be expected to be reliable. Available streamflow records are too short to provide an answer to the question: How long must a record be to define flood probabilities within acceptable tolerances? Table 11-1 gives some estimates derived from synthetic data series.

Table 11-1 suggests that extrapolation of frequency estimates beyond

Table 11-1 LENGTH OF RECORD IN YEARS REQUIRED TO ESTIMATE FLOODS OF VARIOUS PROBABILITIES WITH 95 PERCENT CONFIDENCE [2]

Design probability	Acceptable error	
	10%	25%
0.1	90	18
0.02	110	39
0.01	115	48

probability of 0.01 is extremely risky with data series generally available [3, 4]. Ott's data [5] show that with 20 y of record the probability is 80 percent that the design flow will be overestimated and that 45 percent of the overestimates will exceed 30 percent. It thus appears that records shorter than 20 y should not be used for frequency analysis.

Accuracy of the data series refers primarily to the problem of homogeneity. Most flow records are satisfactory in terms of their intrinsic accuracy. Obviously, if a station is so poor that the reported flows are unreliable, it cannot provide satisfactory data for probability analysis. Even though reported flows are accurate, they may be unsuited for probability analysis if changes in the watershed have caused a change in the hydrologic characteristics; i.e., the record is not internally homogeneous. Dams, levees, diversions, urbanization, and other land-use changes may introduce inconsistencies (Sec. 4-17). Such records should not be used without adjusting to a common set of watershed conditions, usually either natural conditions or current conditions.

If the analysis is concerned with probabilities less than 0.5, a data series composed of *annual floods*, the largest event in each year, is the best choice. For frequent events, the *partial duration series* (or simply *partial series*) is better (Sec. 11-10).

11-2 Plotting Positions

Probability analysis seeks to define the flood flow with a probability p of being equaled or exceeded in any year. *Return period*† T_r is commonly used instead of probability p to define the design flood. Return period and probability are reciprocals, i.e.,

$$p = \frac{1}{T_r} \tag{11-1}$$

There are various formulas [6, 7] for plotting positions, but the one most commonly used is Weibull's [8]:

$$p = \frac{m}{n+1} \qquad \text{or} \qquad T_r = \frac{n+1}{m} \tag{11-2}$$

where n is the number of years of record and m is the rank of the event in order of magnitude, the largest event having $m = 1$. Equation (11-2)

† *Return period* and *recurrence interval* are used interchangeably to signify the average number of years within which a given event will be equaled or exceeded.

assigns an average return period to the largest event in a data series equal to $n + 1$ years. Table 11-2 presents the theoretical distribution of the return periods for floods having specified average return periods. Note that the probability that the return period will be less than the average value is greater than 0.5. It will be seen that over a long period of time 25 percent of the intervals between floods $\geq$ 100-y flood will be less than 29 y, while an equal number will be in excess of 139 y. In other words, for a 75 percent assurance that the capacity of a structure will not be exceeded in the next 29 y, it must be designed for the 100-y (average-return-period) flood.

The probability j that the actual probability of the mth of n floods is less than p_0 can be obtained from [9]

$$j = \binom{n}{m} m \int_{1}^{1-p_0} p^{n-m}(1-p)^{m-1}\, dp \qquad (11\text{-}3)$$

which can be used to calculate Table 11-3. In this equation $\binom{n}{m}$ is the binomial coefficient [see Eq. (11-17)]. There are approximately 2 chances out of 3 that the true return period of the largest event is greater than the period of record. The table also shows the considerable uncertainty in the plotting positions assigned to the highest events in the series, but for $m = 4$ the range of uncertainty is reasonably narrow. This leads to a general rule that if the frequency analysis is intended to yield information on floods with return periods less than $n/5$, one may plot flood magnitude against plotting position and fit a curve by eye. For longer return periods, it is better to fit a theoretical distribution to the data.

Table 11-2 THEORETICAL DISTRIBUTION OF THE RETURN PERIOD

Average return period $\bar{T}_r$	Actual return period T_r exceeded various percentages of the time						
	1%	5%	25%	50%	75%	95%	99%
2	8	5	3	1	0	0	0
5	22	14	7	3	1	0	0
10	45	28	14	7	3	0	0
30	137	89	42	21	8	2	0
100	459	300	139	69	29	5	1
1,000	4,620	3,000	1,400	693	288	51	10
10,000	46,200	30,000	14,000	6,932	2,880	513	100

11-3 Theoretical Distribution of Floods

Since records are generally short, it is not possible to determine the proper distribution to use for flood-probability analysis. Various distributions [10–12] have been suggested, and the evidence adduced in their support is the ability to fit the plotted data of one or more streams. While considerable effort has been expended in attempts to define the best distribution for floods, tests [13] suggest that there is no clearly superior distribution. Intuitively there is no reason to expect that a single distribution will apply to all streams. The log-Pearson Type III distribution has been recommended [14] for use by United States federal agencies for flood analysis. The first asymptotic distribution of extreme values, commonly called the *Gumbel distribution*, has also been used widely. These two distributions are described in the following sections.

Ott [5] used stochastically generated hourly rainfall data (Chap. 12) to simulate (Chap. 10) synthetic 500-y records at two stations. Of several distributions tested, the Gumbel distribution best fitted the synthetic data for Dry Creek (California), while the log-Pearson was best for the Fisher

Table 11-3 AVERAGE RETURN PERIODS FOR VARIOUS LEVELS OF PROBABILITY

Rank from top m	Number of years of record n	Probability				
		0.01	0.25	0.50	0.75	0.99
1	2	1.11	2.00	3.41	7.46	200
	5	1.66	4.13	7.73	17.9	498
	10	2.71	7.73	14.9	35.3	996
	20	4.86	14.9	29.4	70.0	1990
	60	13.5	43.8	87.0	209.0	5970
2	3	1.06	1.48	2.00	3.06	17.0
	6	1.42	2.57	3.78	6.20	37.4
	11	2.13	4.41	6.76	11.4	71.1
	21	3.61	8.12	12.7	21.8	138
	61	9.62	23.0	36.6	63.4	408
3	4	1.05	1.32	1.63	2.19	7.10
	7	1.31	2.06	2.75	3.95	14.1
	12	1.86	3.32	4.62	6.86	25.6
	22	3.03	5.86	8.35	12.6	48.6
	62	7.76	16.1	23.3	35.8	140
4	5	1.03	1.24	1.46	1.83	4.50
	8	1.25	1.80	2.27	3.04	8.26
	13	1.70	2.77	3.63	5.02	14.4
	23	2.67	4.72	6.36	8.98	26.6
	63	6.63	12.5	17.2	24.8	75.2

River (North Carolina). In both cases, Ott's synthetic-streamflow data departed substantially from probability curves defined by the short historic records, as might be expected.

Chow [15] has shown that most frequency functions can be generalized to

$$X = \bar{X} + K\sigma_X \qquad (11\text{-}4)$$

where X is a flood of specified probability, $\bar{X}$ is the mean of the flood series, σ_X is the standard deviation of the series, and K, a frequency factor defined by a specific distribution, is a function of the probability level of X.

11-4 Log-Pearson Type III Distribution

The recommended procedure for use of the log-Pearson distribution is first to convert the data series to logarithms and then compute:†

Mean:

$$\overline{\log X} = \frac{\sum \log X}{n} \qquad (11\text{-}5)$$

Standard deviation:

$$\sigma_{\log X} = \sqrt{\frac{\sum (\log X - \overline{\log X})^2}{n-1}} \qquad (11\text{-}6)$$

Skew coefficient:

$$g = \frac{n \sum (\log X - \overline{\log X})^3}{(n-1)(n-2)(\sigma_{\log X})^3} \qquad (11\text{-}7)$$

The value of X for any probability level is computed from Eq. (11-4) modified

$$\log X = \overline{\log X} + K\sigma_{\log X} \qquad (11\text{-}8)$$

where K is taken from Table 11-4. The distribution will plot as a straight line on log-normal paper when the skew coefficient $g = 0$. Note that the plots are for a cumulative frequency distribution rather than the probability density function.

The Type III distribution is one of a widely used series devised by Pearson [16]. Although the distribution has little theoretical basis, its

† Alternatively,

$$\sigma_{\log x} = \sqrt{\frac{\Sigma (\log x)^2 - (\Sigma \log x)^2/n}{n-1}}$$

$$g = \frac{n^2 \Sigma (\log x)^3 - 3n\Sigma \log x \, \Sigma (\log x)^2 + 2(\Sigma \log x)^3}{n(n-1)(n-2)(\sigma_{\log x})^3}$$

where $\log x = \log X - \overline{\log X}$.

popularity derives from the fact that with skew = 0 it conforms to the log-normal distribution. The probability density function for Type III is (origin at the mode)

$$p(X) = p_0\left(1 - \frac{X}{a}\right)^c e^{-cX/2} \tag{11-9}$$

Table 11-4 *K* VALUES FOR THE LOG-PEARSON TYPE III DISTRIBUTION

Skew coefficient g	Recurrence interval, years							
	1.0101	1.2500	2	5	10	25	50	100
	Percent chance							
	99	80	50	20	10	4	2	1
3.0	−0.667	−0.636	−0.396	0.420	1.180	2.278	3.152	4.051
2.8	−0.714	−0.666	−0.384	0.460	1.210	2.275	3.114	3.973
2.6	−0.769	−0.696	−0.368	0.499	1.238	2.267	3.071	3.889
2.4	−0.832	−0.725	−0.351	0.537	1.262	2.256	3.023	3.800
2.2	−0.905	−0.752	−0.330	0.574	1.284	2.240	2.970	3.705
2.0	−0.990	−0.777	−0.307	0.609	1.302	2.219	2.912	3.605
1.8	−1.087	−0.799	−0.282	0.643	1.318	2.193	2.848	3.499
1.6	−1.197	−0.817	−0.254	0.675	1.329	2.163	2.780	3.388
1.4	−1.318	−0.832	−0.225	0.705	1.337	2.128	2.706	3.271
1.2	−1.449	−0.844	−0.195	0.732	1.340	2.087	2.626	3.149
1.0	−1.588	−0.852	−0.164	0.758	1.340	2.043	2.542	3.022
0.8	−1.733	−0.856	−0.132	0.780	1.336	1.993	2.453	2.891
0.6	−1.880	−0.857	−0.099	0.800	1.328	1.939	2.359	2.755
0.4	−2.029	−0.855	−0.066	0.816	1.317	1.880	2.261	2.615
0.2	−2.178	−0.850	−0.033	0.830	1.301	1.818	2.159	2.472
0	−2.326	−0.842	0	0.842	1.282	1.751	2.054	2.326
−0.2	−2.472	−0.830	0.033	0.850	1.258	1.680	1.945	2.178
−0.4	−2.615	−0.816	0.066	0.855	1.231	1.606	1.834	2.029
−0.6	−2.755	−0.800	0.099	0.857	1.200	1.528	1.720	1.880
−0.8	−2.891	−0.780	0.132	0.856	1.166	1.448	1.606	1.733
−1.0	−3.022	−0.758	0.164	0.852	1.128	1.366	1.492	1.588
−1.2	−3.149	−0.732	0.195	0.844	1.086	1.282	1.379	1.449
−1.4	−3.271	−0.705	0.225	0.832	1.041	1.198	1.270	1.318
−1.6	−3.388	−0.675	0.254	0.817	0.994	1.116	1.166	1.197
−1.8	−3.499	−0.643	0.282	0.799	0.945	1.035	1.069	1.087
−2.0	−3.605	−0.609	0.307	0.777	0.895	0.959	0.980	0.990
−2.2	−3.705	−0.574	0.330	0.752	0.844	0.888	0.900	0.905
−2.4	−3.800	−0.537	0.351	0.725	0.795	0.823	0.830	0.832
−2.6	−3.889	−0.499	0.368	0.696	0.747	0.764	0.768	0.769
−2.8	−3.973	−0.460	0.384	0.666	0.702	0.712	0.714	0.714
−3.0	−4.051	−0.420	0.396	0.636	0.660	0.666	0.666	0.667

SOURCE: Adapted from [12].

where
$$c = \frac{4}{\beta} - 1 \qquad (11\text{-}10)$$

$$a = \frac{c}{2}\frac{\mu_3}{\mu_2} \qquad (11\text{-}11)$$

$$p_0 = \frac{n}{a}\frac{c^{c+1}}{e^c\Gamma(c+1)} \qquad (11\text{-}12)$$

$$\beta = \frac{\mu_3^{\ 2}}{\mu_2^{\ 3}} \qquad (11\text{-}13)$$

where μ_2 is the variance, μ_3 is the third moment about the mean $= \sigma^6 g$, Γ is the gamma function, and e is the base of napierian logarithms.

11-5 Extreme-Value Type I Distribution

Fisher and Tippett [17] found that the distribution of the maximum (or minimum) values selected from n samples approached a limiting form as the size of the samples increased. When the initial distributions within the samples are exponential, the Type I distribution results. This distribution is given by [18]

$$p = 1 - e^{-e^{-y}} \qquad (11\text{-}14)$$

where p is the probability of a given flow being equaled or exceeded, e is the base of napierian logarithms, and y, the reduced variate, is a function of probability (Table 11-5). Then

$$X = \bar{X} + \left(\frac{y - y_n}{\sigma_n}\right)\sigma_X \qquad (11\text{-}15)$$

where $\bar{X}$ is the mean of the data series, σ_X is its standard deviation, and σ_n and y_n are functions of record length. This equation is equivalent to Eq. (11-4) with K equal to the term in parentheses. Table 11-5 gives values of K for various return periods and record lengths. Two or more computed values of X define a straight line on extreme-value probability paper. (See Figs. 11-1 and 11-2.)†

† Extrapolation of a straight line is easier than extrapolation of a curve. Hence, frequency analysis is simplified by using special plotting papers scaled so that a specific distribution will plot as a straight line. On such probability paper, the ordinate is normally flow rate, and the abscissa is probability or return period. For some distributions the logarithm of flow is the ordinate. The abscissa scale is warped to achieve the straight-line plot. Figure 11-1 illustrates a log-Pearson Type III plotting on log-normal probability paper, and Fig. 11-2 shows an extreme-value plot.

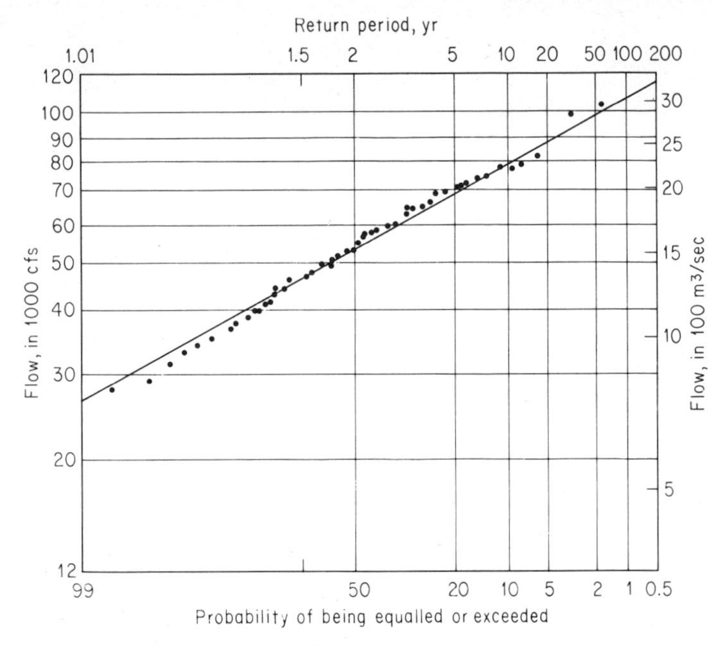

FIGURE 11-1
Flood-frequency curve for the Clearwater River at Kamiah, Idaho. The curve is plotted on log-normal probability paper and fitted by the log-Pearson Type III function.

Table 11-5 VALUES OF K [EQ. (11-4)] FOR THE EXTREME-VALUE (TYPE I) DISTRIBUTION

Return period, years	Probability	Reduced variate y	Record length, years						
			20	30	40	50	100	200	∞
1.58	0.63	0.000	−0.492	−0.482	−0.476	−0.473	−0.464	−0.459	−0.450
2.00	0.50	0.367	−0.147	−0.152	−0.155	−0.156	−0.160	−0.162	−0.164
2.33	0.43	0.579	0.052	0.038	0.031	0.026	0.016	0.010	0.001
5	0.20	1.500	0.919	0.866	0.838	0.820	0.779	0.755	0.719
10	0.10	2.250	1.62	1.54	1.50	1.47	1.40	1.36	1.30
20	0.05	2.970	2.30	2.19	2.13	2.09	2.00	1.94	1.87
50	0.02	3.902	3.18	3.03	2.94	2.89	2.77	2.70	2.59
100	0.01	4.600	3.84	3.65	3.55	3.49	3.35	3.27	3.14
200	0.005	5.296	4.49	4.28	4.16	4.08	3.93	3.83	3.68
400	0.0025	6.000	5.15	4.91	4.78	4.56	4.51	4.40	4.23

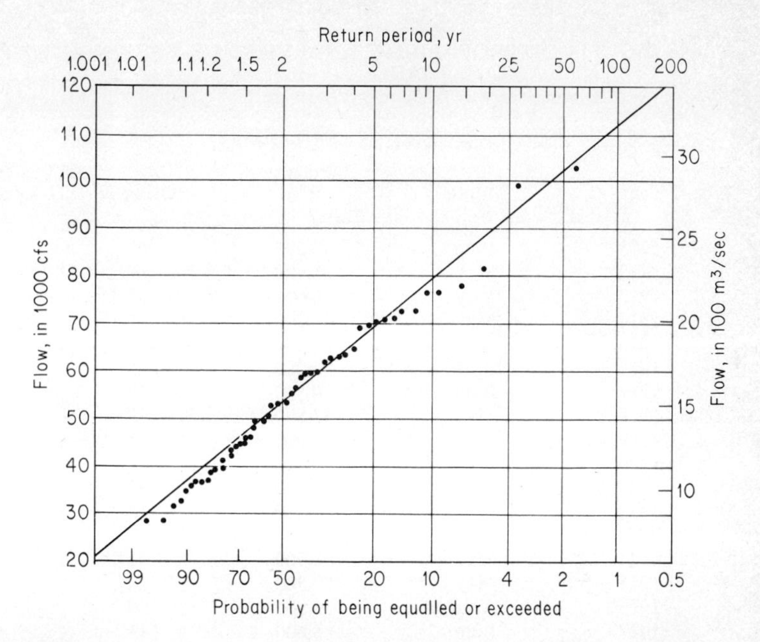

FIGURE 11-2
Flood-frequency curve for the Clearwater River at Kamiah, Idaho, plotted on extreme-value paper and fitted with the Type I (Gumbel) distribution.

Exercise in flood-probability analysis Using the data of Table 11-6, find the magnitude of the 10- and 100-y floods with the log-Pearson Type III and Gumbel distributions. For Gumbel,

$$\bar{q} = 54{,}971 \text{ ft}^3/\text{s} \qquad \sigma_q = 16{,}483 \text{ ft}^3/\text{s}$$

From Table 11-5, $K_{10} = 1.46$ and $K_{100} = 3.48$ when $n = 55$.

$$q_{10} = 54{,}971 + 1.46 \times 16{,}483 = 79{,}040 \text{ ft}^3/\text{s}$$

$$q_{100} = 54{,}971 + 3.48 \times 16{,}483 = 112{,}300 \text{ ft}^3/\text{s}$$

For log-Pearson Type III from Table 11-6

$$\overline{\log q} = 4.7212 \qquad \sigma_{\log q} = 0.1296 \qquad g = 0.0434$$

From Table 11-4, $K_{10} = 1.286$ and $K_{100} = 2.358$.

$$\log q_{10} = 4.7212 + 0.1296 \times 1.286 = 4.8879$$

$$q_{10} = \text{antilog } 4.8879 = 77{,}250 \text{ ft}^3/\text{s}$$

Table 11-6 FLOOD-PROBABILITY ANALYSIS FOR THE CLEARWATER RIVER AT KAMIAH, IDAHO
Drainage area = 4,850 mi^2 (12,560 km^2)

Year	Month	Discharge, ft^3/s	Order m	Plotting position, y
1911	June	39,500	45	1.24
1912	May	61,900	19	2.95
1913	May	76,600	5	11.20
1914	May	42,200	42	1.33
1915	May	28,200	55	1.02
1916	June	56,000	25	2.24
1917	June	70,500	10	5.60
1918	May	52,800	28	2.00
1919	May	52,000	31	1.81
1920	May	43,600	41	1.37
1921	May	69,700	12	4.67
1922	June	62,400	18	3.11
1923	May	49,600	32	1.75
1924	May	58,900	22	2.55
1925	May	59,800	20	2.80
1926	April	35,900	50	1.12
1927	June	68,600	13	4.31
1928	May	72,100	7	8.00
1929	May	52,700	29	1.93
1930	April	31,000	53	1.06
1931	May	40,800	43	1.30
1932	May	72,100	8	7.00
1933	June	81,400	3	18.67
1934	April	45,900	37	1.51
1935	May	44,000	40	1.40
1936	May	63,200	16	3.50
1937	May	34,300	51	1.10
1938	April	63,400	15	3.73
1939	May	46,000	36	1.56
1940	May	37,100	47	1.19
1941	May	28,900	54	1.04
1942	May	37,100	48	1.17
1943	May	52,200	30	1.87
1944	May	34,200	52	1.08
1945	May	44,400	38	1.47
1946	May	36,600	49	1.14
1947	May	69,900	11	5.09
1948	May	99,000	2	28.00
1949	May	76,200	6	9.33
1950	June	62,600	17	3.29
1951	May	44,200	39	1.44
1952	April	49,200	34	1.65
1953	June	53,100	27	2.07
1954	May	58,800	23	2.43
1955	June	64,100	14	4.00
1956	May	77,800	4	14.00
1957	May	71,200	9	6.22

Year	Month	Discharge, ft³/s	Order m	Plotting position, y
1958	May	59,600	21	2.67
1959	June	55,100	26	2.15
1960	May	49,600	33	1.70
1961	May	58,600	24	2.33
1962	April	39,700	44	1.27
1963	May	38,200	46	1.22
1964	June	103,000	1	56.00
1965	May	47,900	35	1.60
	Σ	3,023,400		
	Mean	54,971		
	σ_q	16,483		
	$\overline{\log q}$	4.7212		
	$\sigma_{\log q}$	0.1296		
	g	0.0434		

SOURCE: Data from *U.S. Geological Survey Water-Supply Papers.*

$$\log q_{100} = 4.7212 + 0.1296 \times 2.358 = 5.0268$$

$$q_{100} = \text{antilog } 5.0268 = 106{,}400 \text{ ft}^3/\text{s}$$

All answers should be rounded to three significant figures.

11-6 Selection of Design Frequency

Analyses described in preceding sections define average probability level or average return period. This information is useful for numerous problems. For example, in calculating average annual flood damages, the damage caused by a particular flow should be multiplied by its annual probability of occurrence to find its contribution to the average annual damage. The design of any structure with essentially unlimited life would be based on annual probabilities. Design of highway culverts to pass flows of annual probability p implies that, on the average, pN culverts will be overtopped each year, where N is the total number of culverts.

There are situations when one is concerned with the probability of a flood occurring during a specified interval of future time. For example, what flood probabilities exist during the construction period of a dam? The probability J_k that a flood with average probability of occurrence p will be exceeded exactly k times during an N-y period is given by [9]

$$J_k = \binom{N}{k} (1 - p)^{N-k} p^k \qquad (11\text{-}16)$$

where the binomial coefficient

$$\binom{N}{k} = \frac{N!}{k!\,(N-k)!} \tag{11-17}$$

The probability of one *or more* exceedances in N y is found by taking $k = 0$ and noting that the probability of exceedance is one minus the probability of non-exceedance.

$$J_{1 \text{ or more}} = 1 - (1 - p)^N \tag{11-18}$$

For example, there is a 10 percent chance that the 460-y event will be equalled or exceeded during the next 50 y (Table 11-7). *Note that it is assumed that the annual probability of occurrence is known accurately.* Since this is unlikely, some further uncertainty remains in the estimates. Thus it is important to obtain as accurate an estimate of p as possible. The procedure outlined cannot be looked upon as a certain safety factor. If p is underestimated appreciably, this procedure may lead to gross over-design, a strong possibility if the record is short (Sec. 11-1).

11-7 Regional Flood Frequency

Analysis of historic records is of little value in flood-frequency studies for ungaged watersheds or watersheds with only a few years of record. One procedure designed to overcome this problem is regional-flood-frequency analysis [19]. Records from a number of streams are assembled, and a standard period is selected. Some annual peaks may be estimated so that all records are complete for the standard period. Flow data for each station are made nondimensional by dividing by the mean annual flood $\bar{q}$ for the station. Median ratios for each order number are then plotted (omitting estimated values), and a regional curve fitted (Fig. 11-3). Statistical tests may

Table 11-7 RETURN PERIOD REQUIRED FOR SPECIFIED RISK OF OCCURRENCE WITHIN PROJECT LIFE

Permissible risk of failure	Expected life of project, years				
	1	10	25	50	100
0.01	100	910	2440	5260	9100
0.10	10	95	238	460	940
0.25	4	35	87	175	345
0.50	2	15	37	72	145
0.75	1.3	8	18	37	72
0.99	1.01	2.7	6	11	22

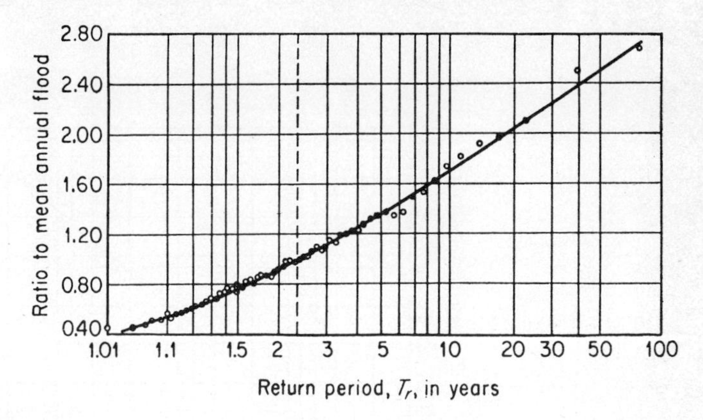

FIGURE 11-3
Regional-flood-frequency curve for selected stations in the Youghiogheny and Kiskiminetas River basins, Pennsylvania and Maryland. (*U.S. Geological Survey.*)

be used to confirm the homogeneity of the data. A relation defining the mean annual flood is then required. Often this is a simple relation between $\bar{q}$ and drainage area (Fig. 11-4), although other parameters such as basin slope, mean elevation, percentage of area in lakes, and mean annual precipitation over the watershed may be employed.

A probability curve for an ungaged watershed is then constructed by first determining the mean annual flood and then converting the ratio scale of the regional-frequency curve to flow by multiplying the ratios by the estimated $\bar{q}$. The weakness of the regional-frequency approach lies in the assumption that all streams will show the same variance of $q/\bar{q}$ (slope of the cumulative-frequency function). Statistical tests may indicate that several streams *could* have the same function, but they can never prove the functions are identical. Differences between frequency curves of several streams may be real differences arising from different regimes. They may, on the other hand, be apparent differences arising from the variability of short samples.

If the regional approach is used, considerable care should be taken to select streams as nearly similar in hydrologic characteristics as possible. They should have similar vegetal cover, land use, topographic conditions, and geologic characteristics, and large basins should not be grouped with very small basins. They should also have similar rainfall and evapotranspiration regimes. Combining data from a number of watersheds may help to reduce the uncertainty arising from short records if the records are sufficiently independent. However, if maximum events at most stations derive from the same storms and tend to plot as outliers from the station-frequency curve, the composite curve will have the same outliers.

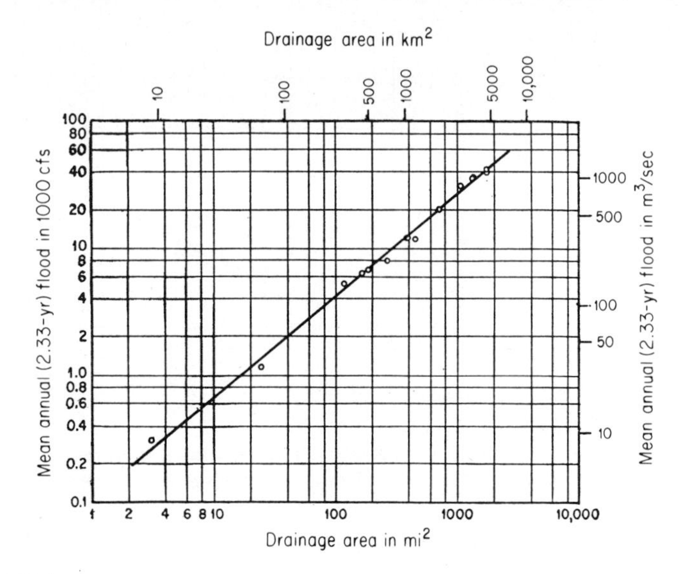

FIGURE 11-4
Variation of mean annual flood with drainage area, Youghiogheny and Kiskiminetas River basins, Pennsylvania and Maryland. (*U.S. Geological Survey.*)

11-8 Frequency Analysis from Synthetic Data

Since the major problems of frequency analysis result from short records, it is clearly desirable to use the longest possible records. Often rainfall data will be available for a longer period than streamflow data, and techniques discussed in Chaps. 7 to 10 may be used to estimate flood peaks prior to the beginning of a streamflow record. These same techniques can be used to develop data for ungaged catchments. Because there will be some error in the estimated peaks, the effective record length is not equal to the total of observed and estimated records, but an improvement in the frequency analysis can result if the estimates are made with sufficient care. It is important that there be no bias in the estimates of peaks as this will bias the frequency analysis. Random errors will tend to cancel.

Procedures for estimating peak flows may be derived, if necessary, from 2 or 3 y of observed data although a longer sample is to be preferred. Hence, records which are quite useless for frequency analysis may be extended. Simulation techniques (Chap. 10) are especially useful for record extension.

It is not necessary that rainfall stations used for estimating peak flows be in the watershed under study. The only requirement is that the stations

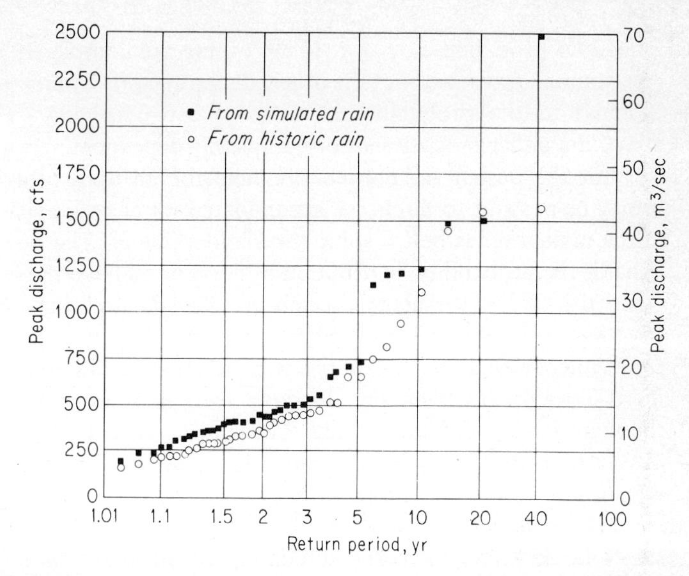

FIGURE 11-5
Comparison of flood-frequency curves for Silver Creek near San Jose, California, based on historic data and on flows simulated using stochastically generated hourly rainfall. (*From* [23].)

used experience the same precipitation-frequency regime as the study basin [20]. This may permit use of precipitation data from as far as 100 mi (160 km). If remote data are used, calculated peaks may not necessarily reproduce historic peaks on the watershed, but if the data used represent the same climatological regime, the resulting frequency curve will approximate the true curve for the study watershed. It even appears possible to use stochastically generated rainfall data in lieu of observed rainfall data to develop a flood-frequency curve for a watershed [5, 21, 22]. Figure 11-5 illustrates a frequency curve so derived [23]. If stochastic data are used, exact correspondence with historic events cannot be expected. Hence, a short stochastic extension of observed precipitation records is not advisable. Instead, a long stochastic record should be developed and utilized for simulation.

11-9 Conditional Probability

Hydrologic problems often involve occurrence of two or more events simultaneously. These situations require special treatment. If the events are independent, the probability of their joint occurrence is the product of their individual probabilities, $p_1 p_2$. If the events are completely dependent, i.e., one cannot happen without the other, the probability of their joint occurrence is equal to the probability of either event occurring. Few hydrologic events are either independent or completely dependent.

In the design of internal drainage for an area protected by levees, it may be possible to discharge accumulated water by gravity when the stage in the main river is below some threshold value h. The question might be: What is the probability distribution of flood flows from the internal drainage when the main-river stage exceeds h? The answer may be obtained by selecting only tributary flood peaks when the main-river stage is h or higher for frequency analysis. The series will probably be different from the annual flood series for the tributary. In some years there may be no occurrences if the main stream did not exceed the critical stage. If the occurrence is relatively rare, there may be few events for analysis. Extension of the record is indicated to provide a larger data base.

Some knowledge may be desired of the interrelation between high stages in the main river and floods on the tributary rather than a simple threshold analysis. In this case, the problem can be treated as one involving multiple thresholds, and the results may be expressed as a matrix (Table 11-8) showing the number of tributary floods in various magnitude classes for selected ranges of main-stream stage or flow. Alternatively, a frequency curve can be drawn for each class interval of main-stream stage or above each threshold level. In each case, n equals the record length in calculating plotting position. The data constitute several partial series and are not subject to analysis by rigorous statistical methods. The relatively limited sample in each row is the major difficulty.

Table 11-8 CONDITIONAL PROBABILITY OF FLOOD FLOWS ON A TRIBUTARY BEING EQUALED OR EXCEEDED WHEN THE MAIN STREAM IS AT VARIOUS LEVELS

Probabilities are shown in parentheses

Main-stream stage, ft	Tributary peak flow, ft^3/s						Total cases in row
	0–99	100–199	200–299	300–399	400–499	500–599	
0–5	2 (0.25)	1 (0.15)	...	1 (0.10)	1 (0.05)		5 (0.25)
5–8	1 (0.20)	1 (0.15)	2 (0.10)				4 (0.20)
8–11	...	1 (0.20)	1 (0.15)	1 (0.10)	1 (0.05)		4 (0.20)
11–14	1 (0.25)	...	1 (0.20)	2 (0.15)	1 (0.05)		5 (0.25)
14–17	...	...	1 (0.10)	...	...	1 (0.05)	2 (0.10)
Total cases in column	4 (0.20)	3 (0.15)	5 (0.25)	4 (0.20)	3 (0.15)	1 (0.05)	20 (1.00)

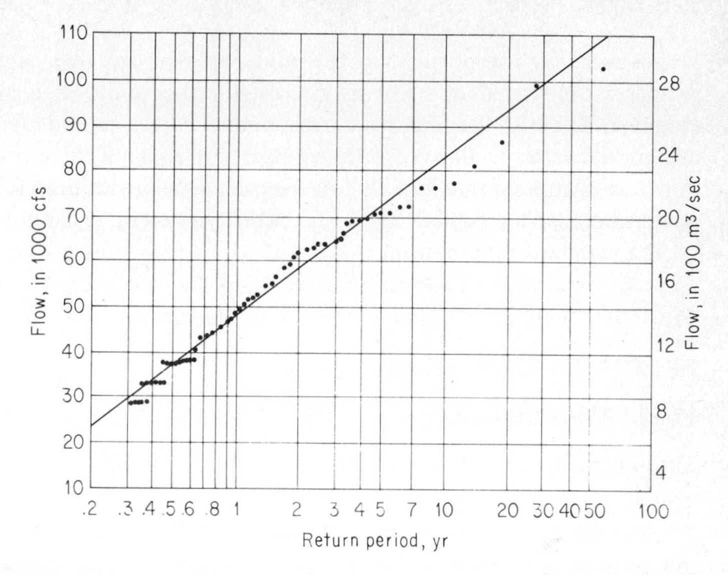

FIGURE 11-6
Flood-frequency curve for the Clearwater River at Kamiah, Idaho, based on partial-duration flow data.

Some problems can be handled by combining the several variables into a single variable which can be analyzed by conventional methods. For example, if the probability of average rainfall over a watershed is required, the observed rainfalls should be averaged and the averages treated as a data series for analysis. Similarly, one might add simultaneous flows of two or more streams to determine the probability of combined discharges into a reservoir. Unless the events are perfectly correlated, it is incorrect to add the 10-y flows to obtain the 10-y combined flow. This procedure will almost always result in an overestimate of flow for any return period.

11-10 Frequent Events

When the design problem requires consideration of events with return periods less than 5 y, a partial flood series is preferable to the annual series. Use of such short return periods might be appropriate in urban drainage, where damage is negligible. The partial series is made up of all floods above some selected base value. The base is usually chosen so that not more than two or three events are included for each year. The partial series can then indicate the probability of events being equaled or exceeded 2 or 3 times per year (Fig. 11-6). Minor peaks associated with a larger flood are usually not included in the series. However, the decision must be arbitrary and should be governed by the purpose of the study. If it is felt that each peak constitutes an independent event of interest for the study, then all should be counted. Table 11-9 compares corresponding return periods for the annual and partial series as derived on theoretical grounds. Return period for any flow rate is approximately 0.5 y less for the partial series than for the annual series. Because the partial series is arbitrarily selected, it cannot be expected to fit a standard distribution.

PRECIPITATION PROBABILITY

11-11 Distributions

The preceding discussion on flood probability applies generally to precipitation. Annual maximum hourly or daily amounts ordinarily conform to a Fisher-Tippett (extreme-value Type I), log-Pearson, log-normal, or gamma distribution. In humid areas, where the mean is high, monthly, seasonal, or annual totals will approximate a normal distribution. In drier areas a skew distribution such as the log-Pearson, log-normal, gamma, square-root, or cube-root normal may give a better fit. Various comparisons [25–28] have been made of different distributions, but the results are generally inconclusive.

11-12 Generalization of Rainfall-Frequency Data

Precipitation records are often too short to permit reliable frequency analysis except for relatively short return periods, say 10 y or less. In such cases reliable rainfall-frequency values for longer return periods can be obtained by applying ratios of rainfall magnitudes for different return periods, say 100- to 10-y rainfall, computed from long-record stations with precipitation regimes similar to those of the short-record stations for which the data are

Table 11-9 CORRESPONDING RETURN PERIODS IN YEARS FOR ANNUAL AND PARTIAL SERIES [24]

Partial series	Annual series	Partial series	Annual series
0.5	1.16	5.0	5.52
1.0	1.58	10	10.5
1.45	2.00	50	50.5
2.0	2.54	100	100.5

required. The need for such generalization is greatest for short-duration rainfall since recording-gage records are generally much shorter than those of nonrecording gages. Also, recording-gage networks are relatively sparse, and there is a greater need for using records from as many gages as possible.

Several empirical relationships have been developed [29–32] for estimating rainfall-frequency values when precipitation data are generally inadequate for frequency analysis. Rainfall-frequency values for stations with adequate data are correlated with such readily available climatological data as mean annual precipitation, mean annual number of rainy days, and mean annual number of thunderstorm days. These relationships, which are usually developed and presented in graphic form, are then used to estimate rainfall-frequency values for localities with inadequate data. Such relations have been used to estimate rainfall-frequency values for durations from 20 min to 24 h for return periods from 2 to 100 y.

11-13 Adjustment of Fixed-Interval Precipitation Amounts

Analysis of short-duration rainfall is ordinarily performed on data published as clock-hour or observational-day amounts. Since it is unlikely that an intense rainfall will occur entirely within an interval between fixed observation times, data of this type give an underestimate of true maximum amounts for the specified durations. Rainfall-frequency data based on hourly or daily amounts should be increased by 13 percent to approximate true values for 60 min or 24 h (Table 11-10). The adjustment decreases as the number of observations on which the maximum value is based increases. For example, 24-h maximum amounts based on maximum four consecutive 6-h rainfall increments require, on the average, an increase of about 3 percent. The same adjustment would apply to 96-h rainfall-frequency values computed from observational-day data.

Table 11-10 RATIO OF TRUE TO FIXED-INTERVAL MAXIMUM PRECIPITATION

Number of observational increments	Ratio
1	1.13
2	1.04
3–4	1.03
5–8	1.02
9–24	1.01

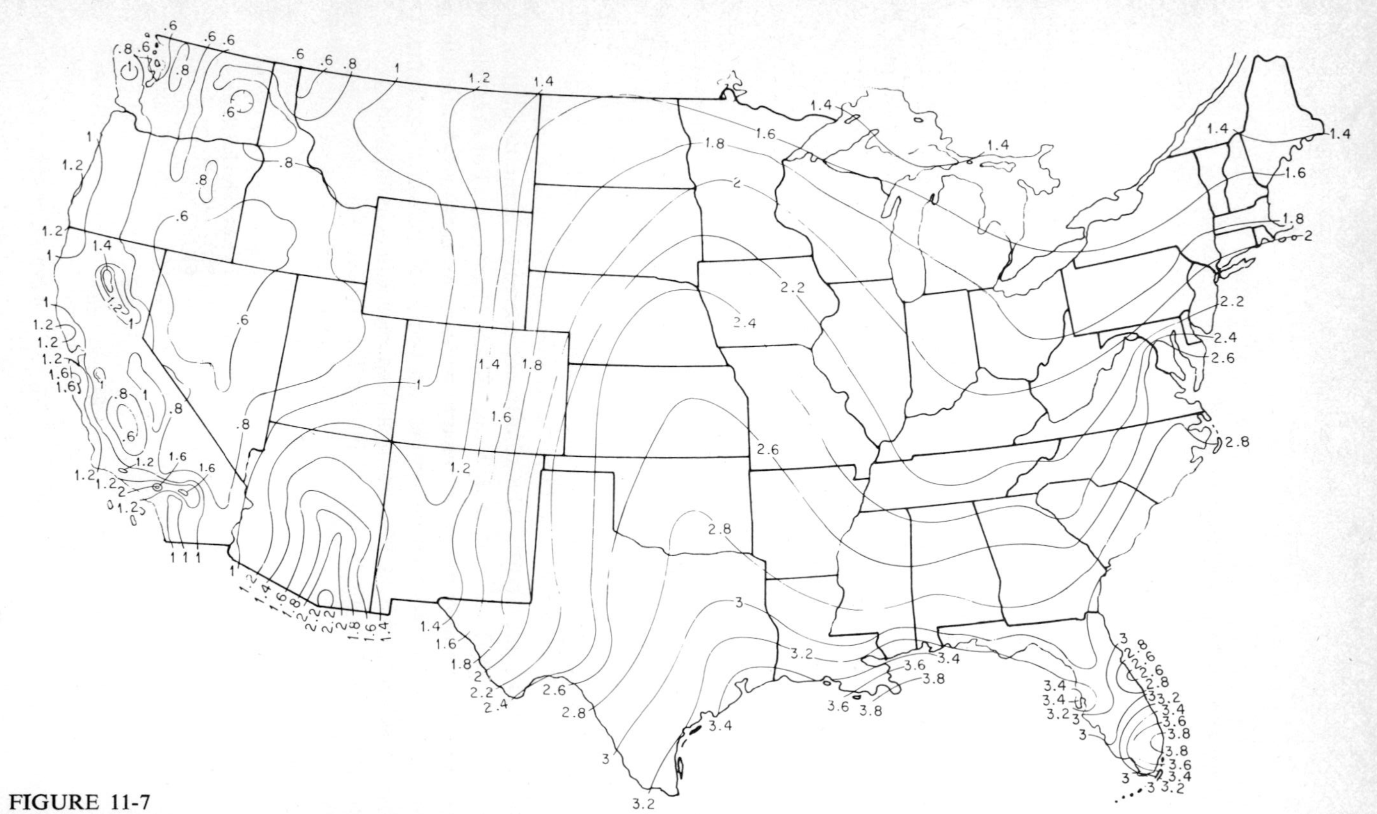

FIGURE 11-7
One-hour rainfall (inches) to be expected on the average once in 10 y. (Map scale precludes reliable depiction for mountainous region west of the 105th meridian. Detailed maps for that region are published in Precipitation-Frequency Atlas of Western United States, *NOAA Atlas* 2.) (*U.S. National Weather Service*.)

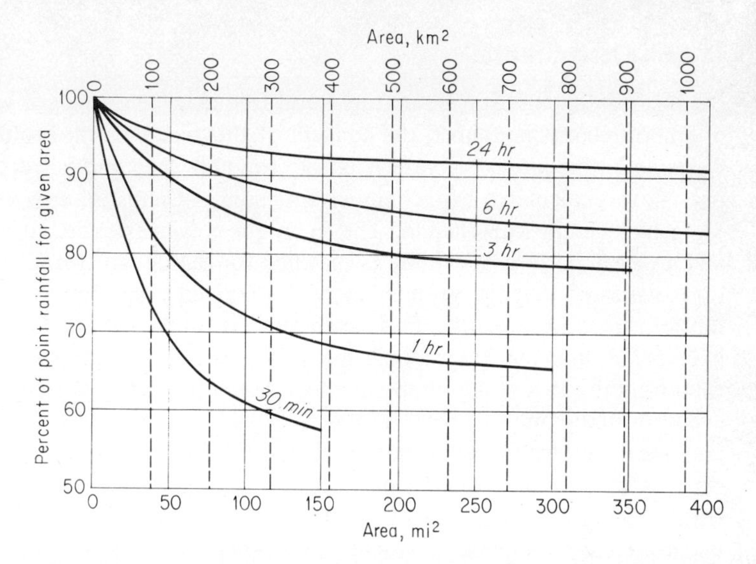

FIGURE 11-8
Reduction of point rainfall-frequency values for application to drainage areas up to 400 mi^2 (1040 km^2). (*U.S. National Weather Service.*)

11-14 Rainfall-Frequency Maps

Except in mountainous terrain, rainfall intensity-frequency variations over short distances may be expected to be small. Thus, it is practical to prepare maps of rainfall amounts for various frequencies and durations. Figure 11-7 is one of a series [33] of such maps which cover durations from 30 min to 24 h and return periods from 1 to 100 y. Similar maps are available [34] also for Alaska, Hawaii, and the West Indies and for durations up to 10 d. These maps provide a good basis for estimating intensity-frequency values except in mountainous areas, where the scale is too small for reliable interpolation. Larger-scale frequency maps of 6- and 24-h precipitation have been prepared [35] on a state-by-state basis for the mountainous western United States. These maps are based on both observed precipitation and on precipitation estimated from multiple correlations with slope, elevation, distance to moisture source, and distance and elevation of intervening barriers.

All the above precipitation-frequency maps depict point-precipitation values only. These values are often assumed to apply up to 10 mi^2 (25 km^2). Areal average precipitation must be less than the maximum point value in the area. Figure 11-8 indicates the average reduction of point precipitation of various durations for size of area [36]. It should be noted, however, that depth-area-duration relations of the type depicted vary with storm type and intensity and may exhibit regional differences.

11-15 Design Storm

Because of difficulties in estimating flood frequency for ungaged watersheds or where records are short, the concept of the design storm has developed. A rainfall time-intensity pattern is selected and the runoff hydrograph or peak flow is calculated by techniques discussed in Chaps. 7 and 8 or, in many cases, by an empirical formula. The technique is even extended to large areas, and an areal pattern of rainfall is specified for the design storm. It is sometimes assumed that the probability of the derived flood flow is the same as that of the design rainfall. This assumption is rarely right and often grossly in error. A storm includes a time-intensity pattern, an areal pattern, and the total rainfall. It is really impossible to assign a frequency to such a complex event. Usually only the total rainfall is considered. Since the time-intensity and areal patterns affect runoff volume and flood peak, storms with the same rainfall volume seldom produce the same flood peak. In addition, a storm occurs in a sequence of events which fix antecedent conditions and, in turn, runoff volume and hydrograph shape.

A single design storm, even if the frequency is known accurately, is inadequate for the economic analysis which should be made for flood mitigation, storm drainage, culvert design, etc. The preferred procedure is to synthesize the longest possible flood series and derive a frequency relation from the synthetic data. The design-storm approach is seldom warranted except when there has been a policy decision that the structure should be designed for the probable maximum event (Sec. 11-18).

PROBABILITY OF RUNOFF VOLUME

11-16 Distributions

Monthly and annual volumes of runoff seem to conform to the normal or log-normal distributions, although the gamma distribution is sometimes used. Frequency analysis of runoff volume seems to require fewer data than for flood peaks. No systematic analysis has been undertaken to define the required record length for a stable volume frequency analysis, but 30 to 50 y would seem reasonably adequate for most purposes. Arid areas, where many years or months have zero or near zero runoff, require longer records.

11-17 Drought

Hydrologic drought may be defined as a period during which streamflows are inadequate to supply established uses under a given water-management system. No more specific definition seems possible since each situation must be analyzed separately. Commonly, the hydrologist is involved in two kinds of drought studies. He may be concerned with low flows which would be limiting for a water supply served by diversion without storage and may be critical in water-pollution problems. He may also be concerned with extended periods of low flow as they may affect the yield of a storage reservoir. This latter problem is best treated by stochastic analysis (Chap. 12) since occurrences of drought conditions in recorded streamflow are generally too few for frequency analysis.

For low-flow analysis, it is necessary to select the time period of concern (1, 5, 10 d, etc.) and to abstract the annual minimum flows for this interval [37]. The data are usually plotted using Eq. (11-2), and a curve is fitted by eye. Figure 11-9 illustrates such a plot for several time intervals on log-normal probability paper. Gumbel suggested that a Type III extreme-value distribution (log–extreme-value distribution) would be appropriate. This can be treated as a Type I distribution using logarithms of the variate or plotted on extreme-value paper with a logarithmic scale of flow.

PROBABLE MAXIMUM EVENTS

In some situations where substantial risk of loss of life exists, it may be appropriate to design a facility against what appears to be the worst possible conditions. No probability can be realistically attached to such an event, and no meaningful economic evaluation is feasible. It must simply rest on a policy decision that maximum protection be provided. The probable maximum flood is accepted as the standard for design of spillways on dams where failure could lead to catastrophic loss of life. It might be appropriate also to locate water-treatment plants, wastewater-treatment facilities, and other essential public utilities above the level of the probable maximum flood (PMF). The PMF is very large and almost always beyond the possibility of control with conventional flood-mitigation works. If it were to occur, flooding would be extensive and damage would be heavy. Consideration of the PMF in design serves only to eliminate the possibility of the addition of a sudden dam failure or the loss of a potable-water supply to the already serious flood conditions.

11-18 Hydrometeorological Studies

Determination of the PMF begins with the determination of the probable maximum precipitation (PMP). There is a reasonable basis on which to analyze the basic factors of major floods, i.e., storm rainfall and snowmelt; maximize them to their upper physical limits consistent with accepted

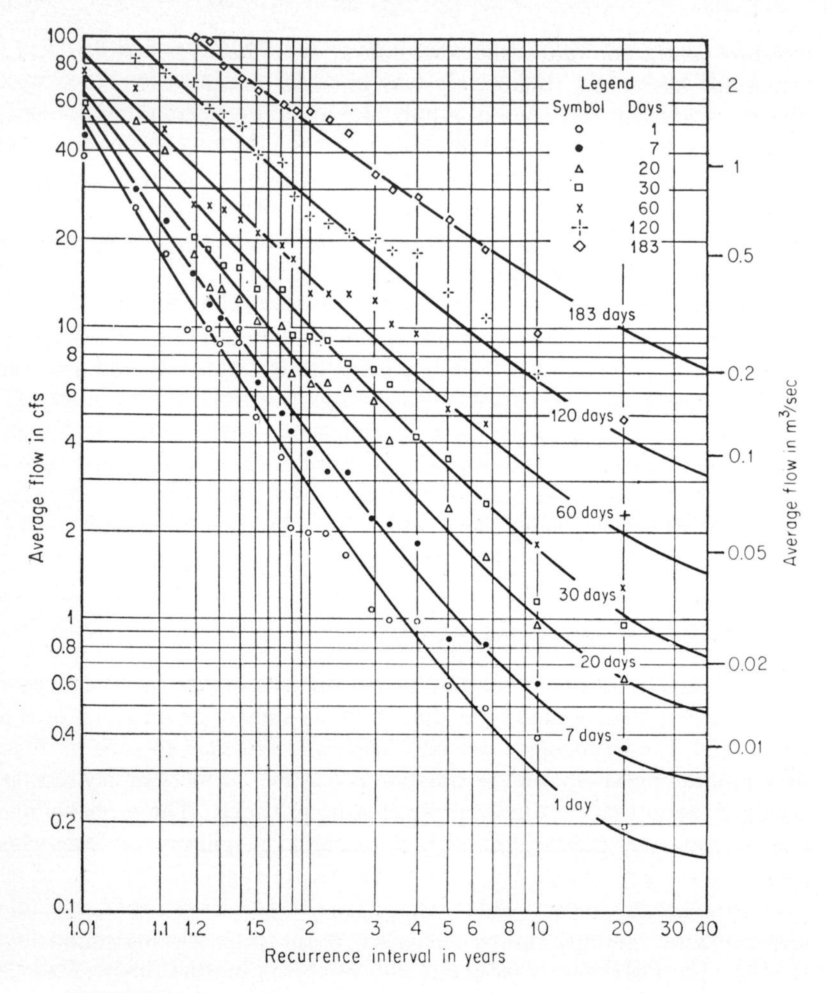

FIGURE 11-9
Frequency of minimum flows for Yellow Creek near Hammondsville, Ohio, 1915–1935. (*From W. P. Cross and E. E. Webber, Ohio Stream-Flow Characteristics, Ohio Dept. Nat. Resour. Bull.* 13, *pt.* 2, *table* 1, *December* 1950.)

meteorological knowledge; and then reassemble them into more critical but meteorologically and hydrologically acceptable combinations or chronological sequences.

PMP is usually derived [38–41] by (1) taking the results of depth-area-duration analyses (Sec. 3-12) of precipitation in major storms that have or could have occurred in the area of interest, (2) adjusting them for maximum moisture charge and rate of moisture inflow, and (3) enveloping the adjusted values for all storms to obtain the depth-area-duration curves of PMP. The use of storms outside the area of interest, called *storm transposition*, involves modification for differences in factors affecting rainfall, e.g., elevation, latitude, and distance from moisture source. Changes in shape and orientation of isohyetal patterns are also considered. Storm patterns are not transposed in mountainous regions because it is not possible to adjust rainfall accurately for orographic influences.

Storm models have also been used [41–43] to estimate PMP. A model using wind and moisture as parameters is hypothesized, tested against major observed storms, and then used to estimate PMP by introducing into the model maximum values of the parameters.

PMP can also be estimated by statistical procedures. In one approach [44–46], PMP for a point and given duration is expressed as the mean of the annual series plus K standard deviations, as in Eq. (11-4). The empirically derived coefficient K varies with rainfall duration and inversely with the mean of the series and ranges between 5 and 30. Various adjustments are made to the mean and standard deviation for outliers and length of record, and depth-area curves similar to those of Fig. 11-8 are used to adjust the estimated point PMP values for size of area.

In many areas snowmelt is an important, and sometimes predominant, factor in major floods. In such cases the PMF requires determination [39, 47, 48] of optimum snow cover, maximum melting rates, and the PMP consistent with the optimum snowmelt conditions. This requires that the seasonal variation of the PMP be determined so that the magnitudes of possible floods, with and without snowmelt, can be estimated for various times of the year and compared.

Detailed hydrometeorological studies involve a tremendous amount of work, and the cost may not be justified when only preliminary estimates of the design flood are required. For this reason generalized charts of PMP and its seasonal variation have been prepared [49] for the United States and a few other areas. Figure 11-10 shows enveloping values of PMP for 200 mi^2 (518 km^2) and 24 h for central and eastern United States [50]. Values for areas from 10 to 1000 mi^2 (26 to 2600 km^2) and for durations from 6 to 48 h can be obtained by applying the depth-area-duration percentage values of Table 11-11.

11-19 Probable Maximum Flood

The PMF must be estimated from the PMP by hydrologic techniques. The conversion may be accomplished with simulation techniques or by use of rainfall-runoff relations and unit hydrographs.

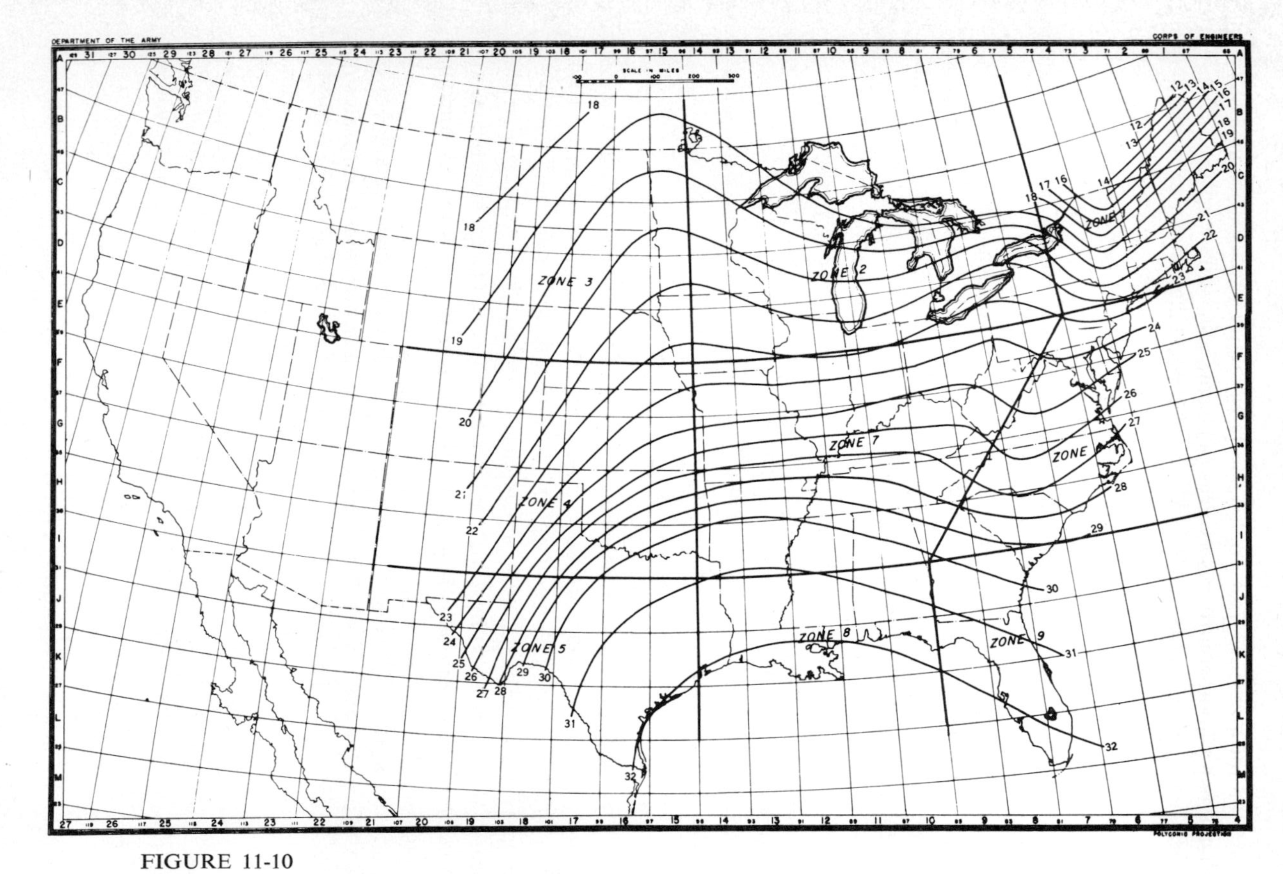

FIGURE 11-10
Probable maximum 24-h precipitation over areas of 200 mi² (518 km²). (*U.S. National Weather Service.*)

To illustrate the hydrologic aspects of spillway design, let us assume that a major flood-control reservoir is to be constructed on the Neosho River just upstream of Council Grove, Kansas. For simplicity it is assumed that the discharge record at Council Grove (250 mi², or 650 km²) is applicable to the dam site. Possible loss of life in case of a dam failure dictates a spillway design flood well beyond that obtainable through frequency analysis.

The discharge record at Council Grove began in 1939, and the maximum observed discharge before July 1951 was 69,500 ft³/s in October, 1941 (1970 m³/s). Flood marks indicate that the 1903 flood reached a stage of 37.3 ft (11.4 m), or about 0.2 ft (6 cm) higher than the 1941 flood. A peak of 121,000 ft³/s (3,420 m³/s) occurred at 8:30 A.M., July 11, 1951, and just

Table 11-11 DEPTH-AREA-DURATION RELATIONSHIPS OF PROBABLE MAXIMUM PRECIPITATION [50]
Percentage values to be applied to 200-mi² (518-km²) 24-h values of Fig. 11-10

Area		Duration, h	Zone							
mi²	km²		1	2	3	4	5	6	7	8–9
10	26	6	111	117	102	112	101	113	102	96
		12	123	127	121	124	121	123	120	108
		24	133	141	134	132	130	132	130	123
		48	142	151	155	141	144	143	140	137
20	52	6	103	108	98	105	95	106	95	90
		12	115	118	115	117	114	116	113	103
		24	126	131	126	124	124	125	122	118
		48	134	141	146	134	138	137	133	132
50	129	6	92	97	92	96	85	97	86	83
		12	105	106	107	108	104	106	104	96
		24	116	118	117	114	115	116	114	111
		48	123	129	133	125	129	128	124	125
100	259	6	83	88	87	89	78	90	79	78
		12	97	98	100	100	96	98	96	91
		24	108	109	108	107	108	108	106	105
		48	114	119	124	118	123	122	118	119
200	518	6	74	80	81	82	70	82	73	72
		12	89	90	93	93	88	90	89	86
		24	100	100	100	100	100	100	100	100
		48	105	110	115	111	116	115	111	114
500	1295	6	62	70	72	74	60	72	64	64
		12	77	79	83	84	79	79	80	78
		24	88	88	90	91	91	89	91	92
		48	94	98	104	102	107	104	102	107
1000	2590	6	52	63	64	68	52	64	58	58
		12	67	72	74	76	68	70	73	74
		24	78	80	82	85	84	79	84	86
		48	85	88	95	96	100	94	96	102

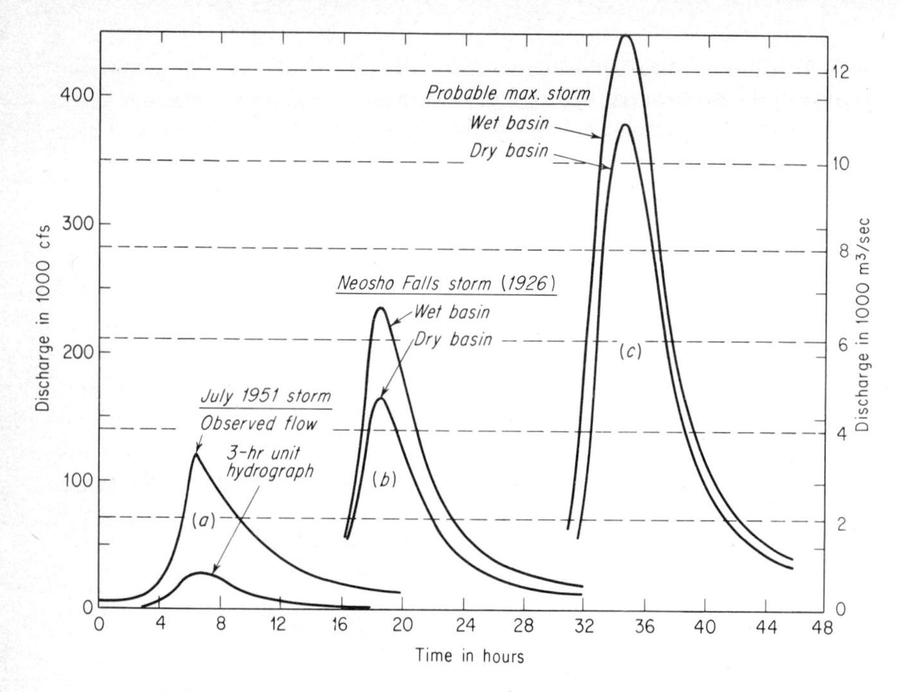

FIGURE 11-11
Observed and synthesized hydrographs for the Neosho River at Council Grove, Kansas.

24 h later a second distinct peak of 71,100 ft^3/s (2010 m^3/s) occurred [51]. The first of these two floods and the 3-h unit hydrograph derived therefrom are shown in Fig. 11-11*a*. The derived unit hydrograph fits the July 11 flood reasonably well, but it is more peaked than one derived from lesser storms of record (Sec. 9-2).

The July 1951 storm which produced the maximum observed discharge at Council Grove was not nearly as severe as the maximum observed storm in that region. On Sept. 12, 1926, a storm centered at Neosho Falls, 60 mi (100 km) to the southeast, deposited an average of 11.4 in. (290 mm) of rain over 200 mi^2 (518 km^2) in 6 h. The depth-area-duration data for this storm and the one at Council Grove in July 1951 are given [52] in Table 11-12. The 9-h, 250-mi^2 rain for the 1926 storm is about 11.2 in. (284 mm), and the time distribution of the storm was such that 3-h increments of 1.5, 9.5, and 0.2 in. (38, 241, and 5 mm) appear reasonable. From a runoff relation (similar to Fig. 8-6), this rainfall would yield runoff increments of 0.4, 8.5, and 0.2 in. (10, 216, and 5 mm) if the basin were initially very wet and 0.2, 6.0, and 0 in. (5, 152, and 0 mm) if the basin were initially very dry. Application of the unit hydrograph derived from the 1951 storm to these runoff increments results in the synthesized hydrograph shown in Fig. 11-11*b*.

From a meteorological viewpoint, there is every reason to expect that the 1926 storm is far short of the maximum which can occur over the basin. From Fig. 11-10 the probable maximum 24-h 200-mi^2 (518 km^2) rainfall is about 24.5 in. (622 mm). Interpolation from Table 11-11 gives 250-mi^2 values of 19.5, 22.0, and 24.0 in. (495, 559, and 610 mm) for durations of 6, 12, and 24 h, respectively. From the depth-duration curve defined by these three values it is estimated that the probable maximum rainfall for the problem basin is 17.0, 19.5, and 21.0 in. (432, 495, and 533 mm) for durations of 3, 6, and 9 h, respectively. Assuming that successive differences of these three values constitute reasonable 3-h increments and that the largest is most apt to occur during the second period gives a chronological sequence of 2.5, 17.0, and 1.5 in. (63, 432, and 38 mm). Storms approaching this magnitude can occur from June through September [50]. From an applicable runoff relation it is found that such a storm occurring on an initially wet basin in June or July would generate runoff increments of 1.5, 16.0, and 1.2 in. (38, 406, and 30 mm). Should it occur in August or September with the basin initially dry, the corresponding runoff increments would be 0.2, 14.0, and 1.1 in. (15, 356, and 28 mm). Applying the unit hydrograph provides the two flood hydrographs shown in Fig. 11-11*c*.

The observed and synthesized hydrographs are indicative of flow potentialities under natural conditions. If the reservoir capacity and plan of operation are such that some water could be stored at time of peak discharge, the hydrograph should be modified by routing (Sec. 9-6). Any anticipated flow through sluiceways should also be subtracted from the spillway flow.

Table 11-12 DEPTH-AREA-DURATION DATA FOR MAJOR STORMS CENTERED AT NEOSHO FALLS (SEPTEMBER 1926) AND COUNCIL GROVE, KANSAS (JULY 1951) [52]

Area, mi^2	Rainfall depth, in.					
	Neosho Falls			Council Grove		
	6 h	12 h	18 h	6 h	12 h	18 h
Point	13.6	13.8	14.0	5.8	7.5	8.2
10	13.4	13.7	13.9	5.3	7.0	7.9
100	12.2	12.5	12.7	4.7	6.4	7.4
200	11.4	11.7	11.9	4.6	6.2	7.2
500	9.5	10.0	10.2	4.3	5.8	6.7

Having determined the hydrologic potentialities, the designer must evaluate the economic and practical aspects of several assumed design values. If the consequences of failure are sufficiently severe, he may be justified in designing for the probable maximum flood (450,000 ft^3/s, or 12,700 m^3/s, in this case). If a lesser spillway-design flood is adopted, it must be done with the full realization that some risk is being accepted. Just what degree of risk is involved cannot be determined for such extreme events as the Neosho Falls storm. Even the July 1951 flood is well beyond the limits of reliable frequency analysis. Although quite meaningless, straight-line extrapolation of a regional frequency relation [53] to 1000 y yields a flow of only 40,000 ft^3/s (1130 m^3/s), a value which has been exceeded on four occasions since the turn of the century. The 1000-y flood may be well above the indicated 40,000 ft^3/s (1130 m^3/s) and could be as great as the 1951 flood, but the 235,000 ft^3/s (6650 m^3/s) flow computed from the Neosho Falls storm is an extremely remote event and is perhaps less likely to occur than other types of catastrophe facing the populace of the area. If the cost of providing spillway capacity for the largest flood is about the same as for lesser floods, the decision is simplified. In some cases a portion of the flood flow may be assumed to pass over the top of the dam with negligible damage. While called a spillway-design flood, the basic criterion is that this flow safely pass the dam by any feasible means, emergency or otherwise.

In the transfer from PMP to PMF the hydrologist has a number of choices. If he assumes very wet antecedent conditions, uses a unit hydrograph with an unusually high peak, and assumes the reservoir full at the beginning of the storm, the spillway flow will be much larger than if he were to use less extreme conditions. These decisions must be made with judgment. Because the PMP is an extraordinary event, it does not necessarily follow that it will occur when all other conditions are most favorable for maximizing the peak. Judgment based on major floods of record will usually suggest that other conditions tend to fall in a reasonably normal range.

REFERENCES

1. L. D. James and R. R. Lee, "Economics of Water Resources Planning," McGraw-Hill, New York, 1971.
2. M. A. Benson, Characteristics of Frequency Curves Based on a Theoretical 1000-Year Record, *U.S. Geol. Surv. Water-Supply Pap.* 1543-A, pp. 51–74, 1952.
3. E. Glos and R. Krause, Estimating the Accuracy of Statistical Flood Values by Means of Long-Term Discharge Records and Historical Data, *Int. Assoc. Sci. Hydrol. Publ.* 84, pp. 144–151, 1967.

4. M. Melentijevich, Estimation of Flood Flows using Mathematical Statistics, *Int. Assoc. Sci. Hydrol. Publ.* 84, pp. 164–172, 1967.
5. R. F. Ott, Streamflow Frequency Using Stochastically Generated Hourly Rainfall, *Stanford Univ. Dept. Civ. Eng. Tech Rep.* 151, December 1971.
6. M. A. Benson, Plotting Positions and Economics of Engineering Planning, *J. Hydraul. Div. ASCE*, vol. 88, pp. 57–71, November 1962.
7. M. A. Benson, Evolution of Methods for Evaluating the Occurrence of Floods, *U.S. Geol. Surv. Water-Supply Pap.* 1580-A, 1962.
8. W. Weibull, A Statistical Theory of the Strength of Materials, *Ing. Vetenskaps-akad. Handl.* (*Stockh.*), vol. 151, p. 15, 1939.
9. H. A. Thomas, Jr., Frequency of Minor Floods, *J. Boston Soc. Civ. Eng.*, vol. 35, pp. 425–442, 1948.
10. Methods of Flow Frequency Analysis, *Subcomm. Hydrol., Inter-Agency Comm. Water Resour., Washington, Notes Hydrol. Act. Bull.* 13, April 1966.
11. H. C. Riggs, Frequency Curves, *U.S. Geol. Surv. Tech. Water Resour.*, bk. 4, chap. A2, 1968.
12. G. Réméniéras, Statistical Methods of Flood Frequency Analysis, in "Assessment of the Magnitude and Frequency of Flood Flows," *Water Resour. Ser.* 30, pp. 50–108, United Nations/World Meteorological Organization, New York, 1967.
13. M. A. Benson, Uniform Flood-Frequency Estimating Methods for Federal Agencies, *Water Resour. Res.*, vol. 4, pp. 891–898, October 1968.
14. A Uniform Technique for Determining Flood Flow Frequencies, *Hydrol. Comm., Water Resour. Counc., Washington, Bull.* 15, 1967.
15. V. T. Chow, A General Formula for Hydrologic Frequency Analysis, *Trans. Am. Geophys. Union*, vol. 32, pp. 231–237, April, 1951.
16. K. Pearson, "Tables for Statisticians and Biometricians," 3d ed., Cambridge University Press, London, 1930.
17. R. A. Fisher and L. H. C. Tippett, Limiting Forms of the Frequency Distributions of the Smallest and Largest Member of a Sample, *Proc. Camb. Phil. Soc.*, vol. 24, pp. 180–190, 1928.
18. E. J. Gumbel, "Statistics of Extremes," Columbia University Press, New York, 1958.
19. T. Dalrymple, Regional Flood Frequency, *High. Res. Board Res. Rep.* 11-B, pp. 4–20, 1950. The U.S. Geological Survey has summarized flood data and presented regional frequency methods for the United States in *Water-Supply Pap.* 1671–1689, 1964.
20. J. L. H. Paulhus and J. F. Miller, Flood Frequencies Derived from Rainfall Data, *J. Hydraul. Div. ASCE*, vol. 83, pap. 1451, December 1957.
21. D. Valencia and John C. Schaake, A Disaggregation Model for Time Series Analysis and Synthesis, *Mass. Inst. Tech., Ralph M. Parsons Lab. Rep.* 149, June 1972.

22. D. D. Franz, Hourly Rainfall Synthesis for a Network of Stations, *Stanford Univ. Dept. Civ. Eng., Tech. Rep.* 126, 1970.
23. A. Pattison, Synthesis of Hourly Rainfall Data, *Water Resour. Res.*, vol. 1, pp. 489–498, 1965.
24. W. B. Langbein, Annual Floods and Partial Duration Series, *Trans. Am. Geophys. Union*, vol. 30, pp. 879–881, December 1949.
25. D. M. Hershfield, An Empirical Comparison of the Predictive Value of Three Extreme-Value Procedures, *J. Geophys. Res.*, vol. 67, pp. 1535–1542, April 1962.
26. F. A. Huff and J. C. Neill, Comparison of Several Methods for Rainfall Frequency Analysis, *J. Geophys. Res.*, vol. 64, pp. 541–548, May 1959.
27. R. D. Markovic, Probability Functions of Best Fit to Distributions of Annual Precipitation and Runoff, *Colo. State Univ. Hydrol. Pap.* 8, August 1965.
28. H. C. S. Thom, Some Methods of Climatological Analysis, *World Meteorol. Org., Tech. Note* 81, 1966.
29. D. M. Hershfield, L. L. Weiss, and W. T. Wilson, Synthesis of Rainfall Intensity-Frequency Regimes, *Proc. ASCE*, vol. 81, sep. no. 744, July 1955.
30. D. M. Hershfield and W. T. Wilson, Generalizing Rainfall-Intensity-Frequency Data, *Int. Assoc. Sci. Hydrol. Gen. Assem. Toronto*, vol. 1, pp. 499–506, 1957.
31. B. M. Reich, Short-Duration Rainfall-Intensity Estimates and Other Design Aids for Regions of Sparse Data, *J. Hydrol.*, vol. 1, pp. 3–28, March 1963.
32. F. C. Bell, Generalized Rainfall-Duration-Frequency Relationships, *J. Hydraul. Div. ASCE*, vol. 95, pp. 311–327, January 1969.
33. D. M. Hershfield, Rainfall Frequency Atlas for the United States, *U.S. Weather Bur. Tech. Pap.* 40, May 1961.
34. See United States Data Sources at end of Chap. 3.
35. J. F. Miller, R. H. Frederick, and R. J. Tracey, Precipitation-Frequency Atlas for the Conterminous Western United States (by States), *NOAA Atlas* 2, 11 vols., 1973.
36. Rainfall Intensity-Frequency Regime, pt. 2, Southeastern United States, *U.S. Weather Bur. Tech. Pap.* 29, March 1958.
37. J. B. Stall and J. C. Neill, Partial Duration Series for Low-Flow Analysis, *J. Geophys. Res.*, vol. 66, pp. 4219–4225, December 1961.
38. J. L. H. Paulhus and C. S. Gilman, Evaluation of Probable Maximum Precipitation, *Trans. Am. Geophys. Union*, vol. 34, pp. 701–708, October 1953.
39. World Meteorological Organization, Estimation of Maximum Floods, WMO no. 233, *Tech Pap.* 126, *Tech. Note* 98, pp. 1–116, Geneva, 1969.
40. World Meteorological Organization, Manual for Estimation of Probable Maximum Precipitation, *Opera. Hydrol. Rep.* 1, WMO no. 332, Geneva, 1973.
41. C. J. Wiesner, "Hydrometeorology," pp. 147–225, Chapman & Hall, London, 1970.
42. Generalized Estimates of Probable Maximum Precipitation and Rainfall: Frequency Data for Puerto Rico and Virgin Islands, *U.S. Weather Bur. Tech. Pap.* 42, 1961.

43. Interim Report: Probable Maximum Precipitation in California, *U.S. Weather Bur. Hydrometeorol. Rep.* 36, October 1961.
44. D. M. Hershfield, Estimating the Probable Maximum Precipitation, *J. Hydraul. Div. ASCE*, vol. 87, pp. 99–106, September 1961.
45. D. M. Hershfield, Method for Estimating Probable Maximum Rainfall, *J. Am. Waterw. Assoc.*, vol. 57, pp. 965–972, August 1965.
46. G. A. McKay, Statistical Estimates of Precipitation Extremes for the Prairie Provinces, PFRA Engineering Branch, Canada Dept. of Agriculture, April 1965.
47. J. P. Bruce and R. H. Clark, "Introduction to Hydrometeorology," pp. 235–239, Pergamon, New York, 1966.
48. Runoff from Snowmelt, Engineering and Design Manual EM 1110-2-1406, U.S. Corps of Engineers, Washington, January 1960.
49. See United States Data Sources at end of this chapter.
50. Seasonal Variation of the Probable Maximum Precipitation East of the 105th Meridian for Areas from 10 to 1000 Square Miles and Durations of 6, 12, 24, and 48 Hours, *U.S. Weather Bur. Hydrometeorol. Rep.* 33, 1956.
51. Kansas-Missouri Floods of July 1951, *U.S. Geol. Surv. Water-Supply Pap.* 1139, p. 180, 1952.
52. Storm Rainfall in the United States, U.S. Corps of Engineers (processed).
53. T. Dalrymple, Flood-Frequency, Kansas-Missouri Floods of July 1951, *U.S. Geol. Surv. Water-Supply Pap.* 1139, pp. 225–229, 1952.

BIBLIOGRAPHY

ALEKSEEV, G. A.: Objective Statistical Methods of Computation and Generalization of the Parameters of Maximum Rainfall Runoff, *Proc. Symp. Floods and Their Computation*, IASH-UNESCO-WMO, 1969.

CHOW, V. T.: Frequency Analysis, sec. 8-I in V. T. Chow (ed.), "Handbook of Hydrology," McGraw-Hill, New York, 1964.

GILMAN, C. S.: Rainfall, sec. 9 in V. T. Chow (ed.), "Handbook of Hydrology," McGraw-Hill, New York, 1964.

GRINGORTEN, I. I.: Fitting Meteorological Extremes by Various Distributions, *Q. J. R. Meteorol. Soc.*, vol. 88, pp. 170–176, 1962.

GUMBEL, E. J.: On the Plotting of Flood Discharge, *Trans. Am. Geophys. Union*, pt. II, pp. 699–716, 1943.

HAZEN, A.: "Flood Flows," Wiley, New York, 1930.

JENKINSON, A. F.: The Frequency Distribution of the Annual Maximum (or Minimum) Values of Meteorological Events, *Q. J. R. Meteorol. Soc.*, vol. 81, pp. 158–171, 1955.

KIMBALL, B. F.: On the Choice of Plotting Positions on Probability Paper, *J. Am. Stat. Assoc.*, vol. 55, pp. 546–560, 1960.

LINSLEY, R. K., and FRANZINI, J. B.: "Water Resources Engineering," 2d ed., McGraw-Hill, New York, 1972.

NATIONAL BUREAU OF STANDARDS: Probability Tables for the Analysis of Extreme-Value Data, *Appl. Math. Ser.*, no. 22, 1953.

WORLD METEOROLOGICAL ORGANIZATION: Guide to Hydrological Practices, 3d ed., WMO no. 168, Geneva, 1974.

UNITED STATES DATA SOURCES

Generalized maps of point-precipitation–frequency data for durations up to 10 d and return periods up to 100 y are available for the entire United States and a few foreign areas in various issues of the *U.S. Weather Bureau Technical Paper* and *NOAA Atlas* series. Area-reduction curves are provided for reducing point values to areas up to 400 mi^2 (1000 km^2). Estimates of probable maximum precipitation for various watersheds in the United States and in a few foreign countries are available in the U.S. National Weather Service's *Hydrometeorological Report* series. Generalized estimates for the entire United States can be obtained from the *Hydrometeorological Report* series and various issues of the *U.S. Weather Bureau Technical Paper* series. See also Chaps. 3 and 4 for precipitation and streamflow data.

PROBLEMS

11-1 Using the data given in Table 11-9 and the frequency curve of Fig. 11-1 construct a partial-duration frequency curve. Plot the frequency curve of Fig. 11-6 on the same chart for comparative purposes. Are the two curves of the same functional form? Are the differences between the two curves significant in view of possible sampling errors?

11-2 Assuming the curve of Fig. 11-2 to represent the true average return period, for what discharge would you design to provide 50 percent assurance that failure would not occur within the next 20 y?

11-3 Using the data in Table 11-6, compute the frequency distributions for the five 10-y periods beginning with 1911, 1921, 1931, 1941, and 1951. Plot the five curves so derived on a single sheet of extreme-probability paper for comparative purposes. What is the extreme error at the 50-y return period, assuming the curve of Fig. 11-2 to yield the true average return period? Take $K_{10} = 1.85$ and $K_{100} = 4.27$ in Eq. (11-4) when $n = 10$.

11-4 Obtain annual and partial-duration flood data for a selected gaging station and plot the two series on extreme-value paper and semilogarithmic paper, respectively. Fit a curve to each by eye, and also compute the frequency curve for annual data. How does the curve fitted by eye compare with the Gumbel curve? What is the percentage error at return periods of 10 and 100 y? How do the annual and partial-duration return periods compare relative to those given in Table 11-9?

11-5 Obtain excessive-precipitation data for a first-order weather station and derive intensity-frequency curves for durations of 5, 15, 30, 60, and 120 min. Compare your results for the 60-min duration with Fig. 11-7.

11-6 Compile an annual series of minimum flow for a selected gaging station and plot on extreme-value paper. Fit curves to the plotted data by eye and then by calculation. Do the data appear to follow the theory of extreme values? Does plotting log q help?

11-7 Using Fig. 11-10 and Table 11-11, construct depth-area-duration curves of probable maximum precipitation for a basin centered at 31°N, 91°W and for one centered at 45°N, 69°W. (Try plotting on semilogarithmic paper, using the ordinate logarithmic scale for area and the abscissa linear scale for depth, and labeling the curves with duration.)

12

STOCHASTIC HYDROLOGY

"The science of conjecture, or the stochastic science, is defined as the art of estimating as best one can, the probability of things so that in our judging and acting we may choose to follow the best, the safest, the surest, or the most soul-searching way [1]."

In statistics the word stochastic is synonymous with random, but in hydrology it has been used in a special way to refer to a time series which is partially random. Stochastic hydrology fills the gap between deterministic models (Chaps. 8 to 10) and probabilistic hydrology (Chap. 11). In deterministic hydrology the time variability is assumed to be totally explained by other variables as processed through an appropriate model. In probabilistic hydrology we are not concerned with time sequence but only with the probability, or chance, that an event will be equaled or exceeded. In stochastic hydrology, the time sequence is all-important.

A simple example of a stochastic process is that of drawing colored balls from an urn. The important feature is the actual order in which the balls are withdrawn from the urn. Valuable information is contained in the sequence of withdrawals: red, red, black, green, black, white, green, green, etc. Average probability statements, by contrast, are concerned only with the relative numbers of different colored balls withdrawn. Stochastic representation preserves sequencing of events.

Typical hydrologic time series are the quantitative description of streamflow or precipitation history at a point. There is a finite amount of information contained in any hydrologic time series. This information has its most complete description in the continuous observed time record. However, the same record can be described in terms of mechanisms (mathematical relationships) with various degrees of precision. It is possible to generate (by mathematical functions) time series that differ from the observed time series but retain many properties of the original series. Each generated sequence is so constructed that the events have the same probability of occurrence as the observed sequence. Such time series are manufactured by stochastic generation techniques.

Stochastic hydrology is meaningful only in a design or operational decision-making sense. In hydrologic design the designer usually wishes to see how the particular facility will perform for representative future hydrologic inputs. The designer is not in a position to know what future flows or future precipitation events will be, but he can assume that future events will have the same stochastic properties as the observed historical record. It is this assumption that forms the principal basis of stochastic hydrology, namely, generation of equiprobable input traces, each trace having similar statistical properties. Each input sequence yields a sequence of outputs from the system under investigation. A stochastic analysis using many input traces thus yields a probability distribution of system response. The probability distributions of system response are used for design and operation decision making.

Stochastic methods were first introduced into hydrology to cope with the problem of reservoir design. The required capacity of a reservoir depends on the sequence of flows, especially a sequence of low flows. If a reservoir operates on an annual cycle, i.e., it fills and is partially or wholly emptied each year, it may be possible to assess its *reliability*, the probability that it will deliver its intended yield each year, on the basis of an analysis of the historic flow record if this record is sufficiently long. If, however, the reservoir operates with carryover storage, i.e., it provides storage to meet required withdrawals over a period of several dry years, the historic record is unlikely to provide adequate information on reliability, since it is generally too short to define the probability of a series of subnormal years. Stochastic methods provide a means of estimating the probability of sequences of dry years during any specified period of future time. Even if the historic record suggests that a reservoir will operate on an annual cycle, the possibility of two or more dry years in sequence exists and thus stochastic analysis should be part of the hydrologic study for all reservoirs dependent on natural stream inflows.

Stochastic methods in combination with deterministic methods also offer the prospect of improving estimates of flood frequency (Chap. 11) since this task also depends on long records for reliable estimates.

Attempts to solve the problem of short record length by various statistical means were probably first initiated by Hazen [2], who suggested combining records from several stations into a single long record. Sudler [3] wrote historic flows on cards and dealt the cards randomly from the deck to construct a synthetic record of 1000 y. This procedure produces a variety of sequences of historic flows for storage analysis. The advent of the computer has made more sophisticated techniques [4] possible, collectively known as *stochastic hydrology*, the generation of synthetic hydrologic time series.

12-1 The Stochastic Generator

Basic to stochastic analysis is the assumption that the process is stationary, i.e., that the statistical properties of the process do not vary with time. Thus the properties of the historic record can be used to derive long synthetic sequences which can be more effectively used in planning than the short historic sequence. The synthetic sequences must resemble the historical sequences, i.e., must display similar statistical properties.

Some properties of hydrologic time series can be investigated in a time domain through correlogram analysis. In some situations it is convenient to work in a frequency domain and use the tools of spectral analysis to identify the principal harmonics contained in the series. However, the short length of hydrologic time series limits the usefulness of spectral analysis. Correlogram and spectral analysis enable some deterministic trends to be located [5]. When "trends" have been detected and subtracted from the original series, the residual series is examined. Of usual interest are the probability distributions of the elements of the residual series. For example, if a monthly time scale is the basic unit used in analysis, the probability distributions of the flow volumes (or residual flow volumes) for each month are of interest.

Basically a time series can be modeled mathematically as the combination of a deterministic and residual random component. One purpose for analyzing a time series is to determine the particular forms of the deterministic and random residual terms. The form of a stochastic generation equation can be very simple (preserving mean, variance, and lag-1 serial correlation) or more sophisticated. More sophisticated generators [6, 7] attempt to preserve low-frequency (as well as high-frequency) fluctuations in the time series; the simpler generators [8] are restricted to preserving high-frequency fluctuations.

In stochastic design applications, the designer is interested in the total system response. Although many properties are needed to describe a historic sequence fully, the stochastic analysis need consider only those which are important for the system under study. In fact, this is paramount in any type of mathematical system simulation, and it reflects the important coupling of inputs, system demands, and system operation. It is therefore important to identify the most suitable generation scheme necessary for the problem at hand.

In most streamflow-volume generation schemes it suffices to assume a first-order Markov structure, i.e., that any event is dependent only on the preceding event. A simple Markov generating function for annual flow volume Q is

$$Q_i = \underbrace{\bar{Q} + \rho(Q_{i-1} - \bar{Q})}_{\text{Deterministic}} + \underbrace{t_i \sigma \sqrt{1 - \rho^2}}_{\text{Random}} \qquad \text{(12-1)}$$

Components

where t is a random variate from an appropriate distribution (Sec. 12-2) with a mean of zero and unit variance, σ is the standard deviation [Eq. (12-4)] of Q, ρ is the lag-1 serial correlation coefficient, and $\bar{Q}$ is the mean of Q. The subscript i designates flows in series from year 1 to year n. If the parameters $\bar{Q}$, σ, and ρ can be determined from the historic record and a starting value of Q_{i-1} is assumed, a simple computer algorithm can be used to generate a long series of values of Q using values of the random variate t derived sequentially by the computer. The $[Q_i]$ are obtained by Monte Carlo sampling from the probability distribution t. The computation, of course, could be made by hand using a table of random numbers as a source of t but would be too time-consuming to be useful.

If seasonal or monthly values of Q are desired, the procedure must allow for the characteristic seasonal variation as follows:

$$Q_{i,j} = \bar{Q}_j + \rho_j \frac{\sigma_j}{\sigma_{j-1}} (Q_{i-1,j-1} - \bar{Q}_{j-1}) + t_i \sigma_j \sqrt{1 - \rho_j^2} \qquad \text{(12-2)}$$

where the subscript j defines the seasons or months. For monthly synthesis j varies from 1 to 12 through the year. The subscript i is a serial designation from month 1 to month n, as in Eq. (12-1). ρ_j is the serial correlation coefficient between Q_j and Q_{j-1}. Other symbols are as for Eq. (12-1). Equation (12-2) is used by determining $\bar{Q}$, σ, and ρ for each month or season. A starting value of $Q_{i-1,j-1}$ is assumed. It is advisable to start at the beginning of the local water year, when the flows are low, although this is not necessary. To avoid the influence of start-up bias, sufficient flow values should be generated to remove the influence of the assumed initial flow. Usually 12 to 20 time increments suffice: the sequence beyond the warm-up period is suitable for use as input to the system under investigation. The coding of

Eq. (12-2) for a digital computer is relatively simple with values of t_i produced by a random-number generator having the appropriate probability distribution.

The simple single-season generator [Eq. (12-1)] is appropriate for large reservoirs where seasonal variation of flow cannot materially affect the storage requirement or in cases where flow is highly seasonal with the high-flow season preceding the high-use season. In all other cases the multiseason generator [Eq. (12-2)] should be used.

12-2 Distribution of t

Unless the random variate t is drawn from an appropriate distribution, Eq. (12-2) cannot reproduce the historical distribution even though it may satisfactorily duplicate the mean and variance. If the historic distribution is normal, the problem is simple; one draws from a normal random-number generator. If streamflow volumes are distributed log-normally, generation is effected by using transformed variables that are normally distributed. Exponentiation yields log-normal sequences.

There is no a priori basis for choosing a distribution, and no single distribution can be applicable to all streams. In addition, the relatively short historic flow records usually available do not clearly define the properties (parameters) of the distribution. One therefore selects a distribution that fits the observed data within the bounds of acceptance criteria. This selection is influenced by the fact that certain distributions are amenable to generation techniques while others are extremely difficult to work with. For example a three-parameter, log-normal distribution might fit the observed data as satisfactorily as a more complicated general gamma distribution. One would elect the former because it can be easily and economically programmed. Probability distributions can be fitted using bayesian or classical methods. For example, if there is good physical reason to expect that flow volumes are of an additive nature, a normal distribution could be anticipated.

Figure 12-1 shows the distributions of annual flows for three streams representing widely varied climatic regimes. The data suggest distributions ranging from nearly normal (Delaware and Flathead Rivers) to approximately exponential (Rio Grande). Figure 12-2 shows a similar comparison of distributions for selected months at a single station. Distribution selection is most difficult for streams where zero flows are a frequent occurrence. There is little alternative to selecting the distribution to which the historic data seem to conform. If there is serious doubt as to the best distribution, it is wise to generate two (or more) synthetic sequences from different distributions to determine the effect on the project plan resulting from differences in the sequences.

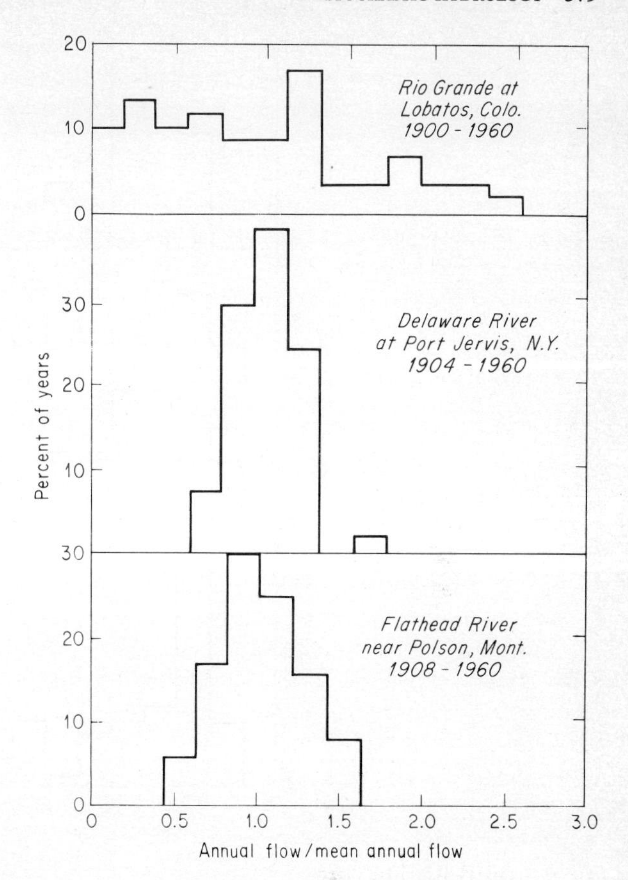

FIGURE 12-1
Distribution of annual streamflows for three different watersheds.

12-3 Definition of Parameters

A probability distribution can be defined by a few parameters. For example, a normal distribution can be completely defined by two parameters, the mean and variance. Most probability distributions of importance in hydrology require two or more parameters to define them. These parameters can be directly related to such quantities as the arithmetic mean, variance, and skew of the observed data. At best, however, it is only possible to estimate the true value of the parameters. The observed time series is only a short sample of the total time series. If the true arithmetic mean of the population is μ, we can only approximate μ by the sample mean $\bar{X}$. Thus for a normal distribution $N(\mu, \sigma_1^2)$ having population mean μ and population variance σ_1^2 we approximate the distribution by $N(\bar{X}, \sigma^2)$.

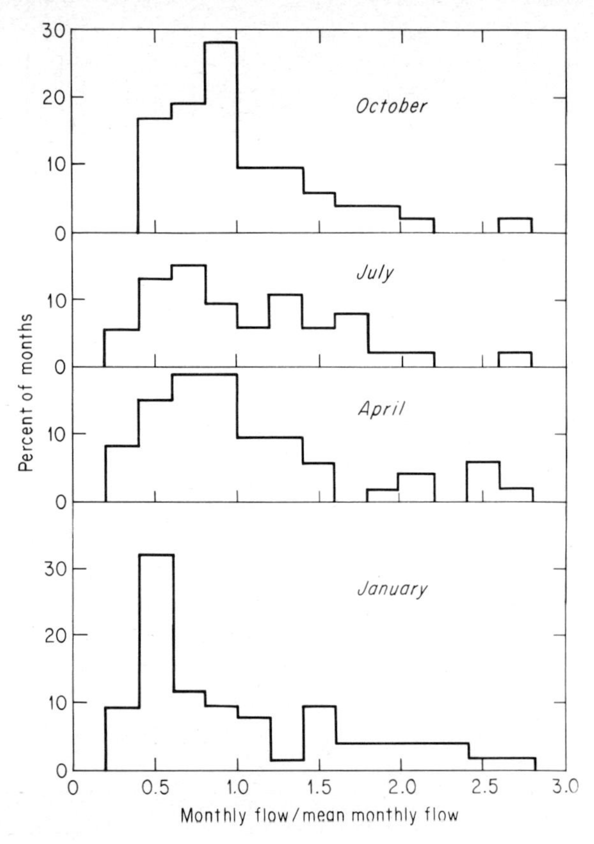

FIGURE 12-2
Comparison of selected monthly-flow distributions for the Flathead River at Polson, Montana [7096 mi^2 (18,379 km^2); data from 1908–1960].

There are two principal methods in use for computation of parameters. The most commonly used is the method of moments [9]. A second method, the method of maximum likelihood, has received increasing attention with the wide availability of digital computers. Because most generation schemes in use employ parameters calculated by the method of moments, only moment-derived parameters are discussed in this chapter.

For any probability distribution the arithmetic mean is approximated by

$$\overline{X} = \sum_{i=1}^{n} \frac{X_i}{n} \tag{12-3}$$

where n is the number of items in the sample. The variance must be corrected for serial correlation if present as

$$\sigma^2 = \psi \frac{\sum_{i=1}^{n} X_i^2 - n\overline{X}^2}{n-1} \tag{12-4}$$

where ψ is a function of serial correlation ρ and length of record (Table 12-1).

$$\psi = \frac{\dfrac{n-1}{n}}{1 - \dfrac{1-\rho^2}{n(1-\rho)} + \dfrac{2\rho(1-\rho^n)}{n^2(1-\rho)^2}} \tag{12-5}$$

If the sequences $\{Y_i\} = \{X_i\}$, $i = 1, 2, \ldots, n-1$, and $\{Z_{i-1}\} = \{X_i\}$, $i = 2, 3, \ldots, n$, are formed and have means $\overline{Y}$ and $\overline{Z}$ and variances σ_y^2 and σ_z^2, respectively, then the serial correlation coefficient of the series is

$$\rho = \frac{1}{n-2} \frac{\sum_{j=1}^{n-1} (Y_j Z_j) - (n-1)\overline{Y}\overline{Z}}{\sigma_y \sigma_z} \tag{12-6}$$

The calculated value of ρ should not be accepted without some thought. Anderson [11] suggested a test for the significance of the correlation coefficient which is displayed in Fig. 12-3. With short records, values of

Table 12-1 VALUES OF ψ IN EQ. (12-4) FOR COMPUTING VARIANCE

Corr. coeff.	Size of sample n						
	10	20	30	40	60	80	100
0.00	1.000	1.000	1.000	1.000	1.000	1.000	1.000
0.10	1.011	1.006	1.004	1.003	1.002	1.001	1.001
0.20	1.025	1.013	1.008	1.006	1.004	1.003	1.003
0.30	1.043	1.022	1.014	1.011	1.007	1.005	1.004
0.40	1.068	1.034	1.022	1.017	1.011	1.008	1.007
0.50	1.103	1.051	1.034	1.025	1.017	1.013	1.010
0.60	1.155	1.077	1.051	1.038	1.025	1.019	1.015
0.70	1.241	1.121	1.080	1.060	1.040	1.030	1.024
0.80	1.403	1.210	1.140	1.104	1.069	1.051	1.041
0.90	1.818	1.461	1.316	1.238	1.158	1.118	1.093
0.99	5.275	3.865	3.220	2.833	2.377	2.110	1.932

SOURCE: Adapted from [10].

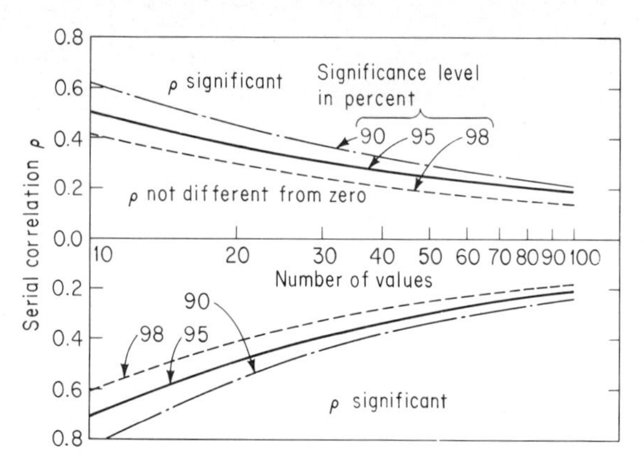

FIGURE 12-3
Significance of serial correlation coefficient as a function of sample size. (*Adapted from* [11].)

$\rho \leq 0.3$ are not statistically different from zero. However, even a small correlation reduces the effect of the random term and increases the persistence term of the generating function. At the same time the variance is increased [Eq. (12-5)]. If the value of ρ is too large, these effects lead to an overestimate of the storage required.

There is no obvious physical explanation of negative values of ρ. If such negative values are statistically not distinguishable from zero, they should be set to zero. However, if negative correlations are thought to be significant, the analyst should immediately check the flow record for deterministic causes. If such a cause is found, it must be incorporated into the generation model. A scattergram is helpful for examining correlation coefficients. A plot of Q_i versus Q_{i-1} may indicate (Fig. 12-4) that only a few chance flow sequences materially influence the correlation. For Yegua Creek, 26 y of data yield $\rho = 0.29$, which is not significantly different from zero by Anderson's test. If point A is omitted, $\rho = -0.10$, while with 35 y (omitting point A) $\rho = -0.007$. Similarly, a chance occurrence of two high years in sequence may lead to a large positive value of ρ. Test calculations can be made omitting outliers from the main body of data to define the probable values of ρ more accurately. Fairly high values of intermonthly correlation may often occur, but interannual correlations in excess of 0.4 are most unlikely, and if the last months of the year are poorly correlated, the interannual correlation must be near zero (under the assumption of a Markov relationship).

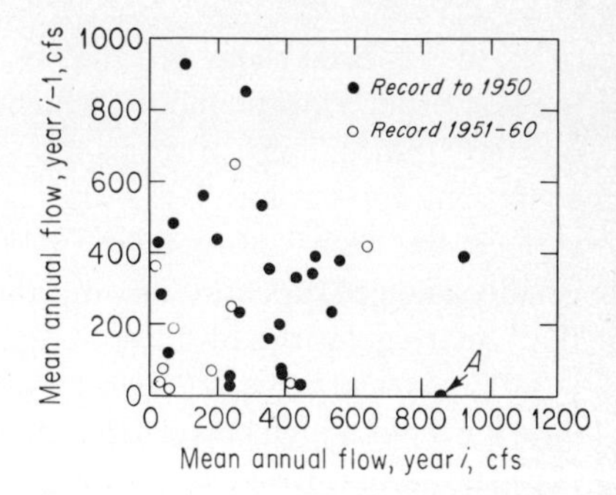

FIGURE 12-4
Serial plot of annual flows for Yegua Creek, near Somerville, Texas. (*Data from U.S. Geological Survey.*)

If the log-normal distribution is selected, the parameters should be calculated from.

$$\mu_X{}^2 = \exp\left(\sigma_y{}^2 + 2\mu_y\right) \tag{12-7}$$

$$\sigma_X{}^2 = \exp\left(2\sigma_y{}^2 + 2\mu_y\right) - \exp\left(\sigma_y{}^2 + 2\mu_y\right) \tag{12-8}$$

$$\rho_X = \frac{\exp\left(\sigma_y{}^2\rho_y - 1\right)}{\exp\left(\sigma_y{}^2\right) - 1} \tag{12-9}$$

where μ_X, σ_X, and ρ_X are calculated with Eqs. (12-3), (12-4), and (12-6) using the observed data and μ_y, σ_y, and ρ_y are the parameters to be used. The common procedure of transforming the data to logarithms and calculating the parameters of the transformed data can lead to substantial bias in the generated flows because the statistical parameters of the streamflow are not preserved. The transformed parameters $\sigma_y{}^2$ and μ_y are calculated by solving Eqs. (12-7) and (12-8) simultaneously. Then ρ_y can be found from Eq. (12-9).

It is unwise to go to detailed analysis when parameters are estimated from a very short record because gross errors could be made in selecting the underlying distribution. Furthermore there is considerable uncertainty in parameter estimates. If an analysis is necessary, the record should be extended as far as possible by a deterministic approach such as simulation (Chap. 10). Table 12-2 compares parameters for Arroyo Seco near Paso Robles, California, as derived from a 65-y record and from two 32-y portions of the record. The data suggest that record length should exceed 65 y for adequate definition of parameters. It should be noted, however, that Arroyo Seco exhibits high variability in runoff volumes. Variability should be examined when deciding what data-base length is needed to yield sufficient accuracy for the stochastic analysis to be performed. Selection of a data sample size must be viewed in terms of the total system under consideration.

It has been suggested that multi-lag models should be used in lieu of the first-order Markov model, which assumes that any flow is dependent only on the immediately previous flow. A correlogram (plot of correlation coefficient for various lags, Fig. 12-5) for a Markov process is a smooth exponential declining function. A correlogram calculated from a record is usually a jagged function showing abrupt changes. While it might be assumed that the irregularities of the correlogram are real features of the regime, there is little physical basis for such an assumption. If, for example, with annual data lag 5 has a high correlation, then lag 4 should be at least as high. Burges [10] shows that irregular correlograms can be obtained from short samples drawn from a randomly generated flow record. The irregular correlogram results from either chance variations in a short sample or from some underlying trends or from both.

There are instances of significant multiperiod carry-over effects, primarily as the result of groundwater storage. In the short term, soil-moisture carry-over may also have a significant effect. Thus, use of multiple lags may sometimes be appropriate in monthly models and occasionally in annual models. However, considerable thought should be given to use of multiple lags, and a reasonable physical understanding of their significance should be available before their adoption.

Table 12-2 PARAMETERS OF THE HISTORIC STREAMFLOW RECORD FOR ARROYO SECO, NEAR SOLEDAD, CALIFORNIA [10]

	Mean†			Variance			Serial correlation		
Month	1903–1967	1903–1934	1936–1967	1903–1967	1903–1934	1936–1967	1903–1967	1903–1934	1936–1967
Oct.	0.040	0.050	0.032	0.004	0.005	0.003	0.539	0.736	0.172
Nov.	0.224	0.233	0.216	0.259	0.300	0.233	−0.003	−0.012	0.003
Dec.	0.824	0.717	0.952	1.269	0.799	1.776	0.028	−0.034	0.078
Jan.	1.693	1.925	1.460	5.014	7.577	2.661	0.260	0.187	0.437
Feb.	2.360‡	2.412	2.370	5.802	5.389	6.456	0.209	0.280	0.129
Mar.	1.979	2.123	1.864	5.007	5.959	4.316	0.442	0.140	0.770
Apr.	1.191	1.035	1.296	2.246	0.798	3.715	0.470	0.382	0.626
May	0.427	0.423	0.429	0.127‡	0.133	0.130	0.625	0.541	0.732
June	0.172	0.173	0.171	0.025	0.030	0.021	0.947	0.926	0.981
July	0.058	0.058	0.058	0.005	0.005	0.004	0.966	0.960	0.978
Aug.	0.020	0.022	0.019	0.001	0.001	0.001	0.963‡	0.964	0.965
Sept.	0.018	0.022	0.014	0.001	0.001	0.001	0.684	0.660	0.732

† Data are expressed as inches of runoff over the catchment (244 mi² or 632 km²).

‡ These values fall outside the range of the shorter periods because the water year 1934–1935 is included in the full period and not in the short periods.

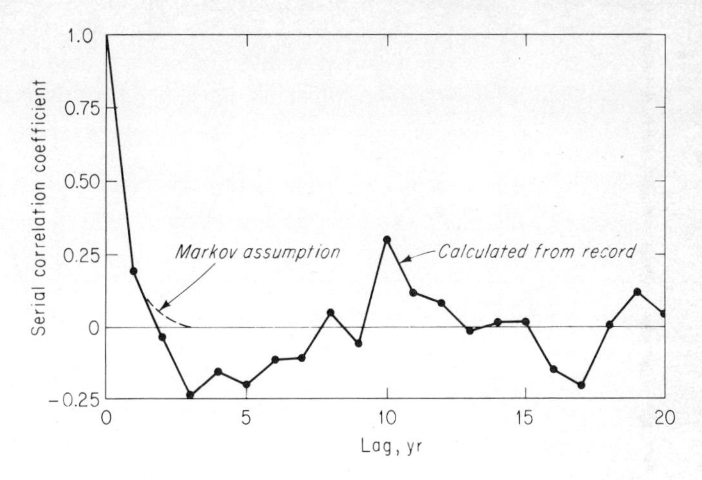

FIGURE 12-5
Correlogram of annual flows for the Susquehanna River at Danville, Pennsylvania ($n = 51$, $\rho_1 = 0.19$) and the correlogram for a first-order Markov process with $\rho = 0.19$.

12-4 The Hurst Phenomenon

While studying long-term storage requirements on the Nile, Hurst [12] found that the range R_n can be expressed as

$$R_n = \sigma_n \left(\frac{n}{2}\right)^h \tag{12-10}$$

where σ_n is the standard deviation and n the length of the series. The range R_n is defined as the difference between the greatest cumulative excess inflow above the mean flow and the greatest cumulative deficiency below the mean inflow. The exponent h is called the *Hurst coefficient*. For a Markov process $h = 0.5$. Hurst tested some 800 time series (streamflow, varves, tree rings, precipitation, and temperature) and found $0.5 < h < 1$ with a mean value of $h = 0.73$ and a standard deviation of 0.09. One implication of $h > 0.5$ is that there is long-term persistence in natural time series and that the Markov process is not a valid model.

Mandelbrot and Wallis [6] suggest the use of fractional gaussian noise to generate sequences with $h > 0.5$. Autoregressive integrated moving averages (ARIMA) [5] and a broken-line process [7] have also been suggested. It is not practical to give an adequate description of these processes here, and interested readers should investigate the references.

Long-term trends in climate are clearly indicated by geologic evidence, but no physical explanation for the kind of persistence suggested by Hurst's results exists. The cause must be other than the carry-over processes of

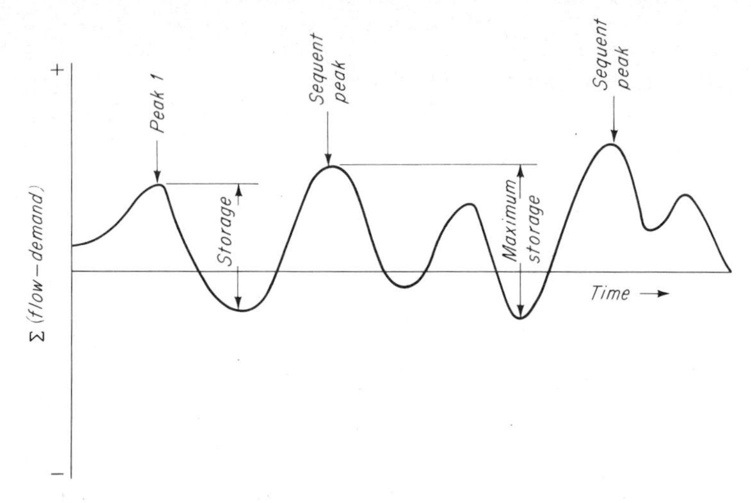

FIGURE 12-6
Schematic description of the sequent-peak algorithm.

hydrology and must be sought in terms of long-term climatic change. Currently available evidence does not suggest significant climatic change in the time scale of 100 y or less, within which project-planning horizons normally fall. Since the purpose of stochastic analysis is usually to generate many short sequences representing possible alternative short-term futures, it can be argued that the matter of long-term persistence is of no importance. For most purposes it would be incorrect to generate a series of several thousand years with persistence effects on the same scale. If there were identifiable trends expected to continue into the planning period of a project, the only correct treatment would be to add a trend term to the generating function. It surely would not be proper to attempt to treat a known trend by a random function. Since a Markov process will generate sequences that are either more critical or less critical than the historic sample, and since the difficulty of defining the correct value of h from a short historic record is recognized, there is considerable argument for continued use of the Markov assumption until clear proof of a better method is available.

12-5 The Storage Model

Numerous methods have been suggested to determine reservoir storage capacity required to meet a specified demand. All are based on the storage equation. For a simple storage reservoir a sequent-peak analysis [13] is simple and convenient. Values of the cumulative sum of flow minus demand are calculated (Fig. 12-6). The first inflow peak (local maximum of the cumulative net inflow) and the sequent peak (next following peak which is greater than the first peak) are identified. The required storage for the interval is the difference between the initial peak and the lowest trough in the interval. The process is repeated through the complete record, and the largest computed value of required storage governs.

For multiple reservoirs, a multipurpose reservoir, or a single-purpose reservoir with special release requirements, a more complex operation study or simulation may be necessary. Table 12-3 illustrates the computations for a short period for a hypothetical reservoir. Precipitation on the reservoir (column 6) and evaporation from it (column 7) are calculated by multiplying the rates (columns 2 and 3) by the average reservoir area for each month (taken as the average of the areas at the beginning and end of each month from column 11). A variable monthly demand (column 5) is given, and a special release to meet downstream water rights defined as 400 ft^3/s or the natural inflow, whichever is less, is required. The maximum difference between a storage maximum and a succeeding minimum is the storage required to deliver the specified demand.

Alternatively, a simulation can be devised to determine the output of a system works of specified size. In this case the reservoir, pipeline, and generating capacities, as well as a set of operating rules, must be specified. The flows are run through the system using a simulation procedure similar to Table 12-3 but usually more complicated by virtue of the more complex system. The outputs in acre-feet of irrigation water, kilowatt-hours of power, etc., are calculated, and the monetary value of these outputs is computed. By successive trials, the system configuration supplying the maximum benefits can be determined [14]. If money values can be assigned to all outputs, their combined value can be computed.

Any of the storage analyses described above can be readily performed on a computer. The more complex system simulation requires a computer solution because of the many calculations involved.

12-6 Storage Requirement from Stochastic Data

If a large number of traces of streamflow are routed through a reservoir algorithm and the maximum storage requirement for each trace is determined, data for a probability analysis of storage requirements are available. Burges [10] showed that when approximately 1000 traces of annual flows are used, the probability distribution conforms to the extreme-value Type I distribution (Gumbel) and that a replication of a second 1000 traces reproduces a virtually identical probability curve. On the other hand, when fewer traces are used, the curves do not necessarily coincide (Fig. 12-7). When annual flows are

Table 12-3 OPERATION STUDY FOR A STORAGE RESERVOIR

Month (1)	Precip-itation, ft (2)	Evap-oration, ft (3)	Inflow, acre-ft (4)	With-drawal, acre-ft (5)	Precip-itation, acre-ft (6)	Evap-oration, acre-ft (7)	Release, acre-ft (8)	Storage change, acre-ft (9)	Storage,† acre-ft (10)	Area,† acres (11)	Storage required
1931											
Dec.	1.062	0.060	3440	100	82	5	400	3017	3017	153	
1932											
Jan.	0.379	0.079	2030	102	65	14	400	1579	4596	190	
Feb.	0.390	0.111	4470	133	97	27	400	4007	8603	304	
Mar.	0.042	0.199	96	143	13	60	96	−190	8413	298	
Apr.	0.062	0.294	11	255	18	86	11	−323	8090	286	
May	0.018	0.411	11	448	5	115	11	−558	7532	276	
June	0	0.482	0	448	0	130	0	−578	6954	266	
July	0	0.500	0	448	0	127	0	−575	6379	245	
Aug.	0	0.450	0	408	0	107	0	−515	5864	229	
Sept.	0	0.385	0	377	0	66	0	−443	5421	217	
Oct.	0.034	0.252	0	255	7	54	0	−302	5119	213	
Nov.	0.066	0.137	0	132	14	29	0	−147	4972	210	
Dec.	0.381	0.074	0	102	80	16	0	−38	4934	210	
1933											
Jan.	0.745	0.083	1090	104	160	18	400	728	5662	220	
Feb.	0.100	0.102	33	136	22	22	33	−136	5526	218	
Mar.	0.280	0.210	347	146	61	46	347	−130	5395	216	
Apr.	0.021	0.274	12	260	45	59	12	−274	5121	212	
May	0.155	0.398	11	458	31	79	11	−506	4615	198	8462 acre-ft

June	0	0.499	3	458	0	95	3	−553	4062	185
July	0	0.547	0	458	0	96	0	−554	3508	168
Aug.	0	0.465	0	416	0	75	0	−491	3017	155
Sept.	0	0.384	0	385	0	57	0	−442	2575	144
Oct.	0.158	0.240	0	260	22	34	0	−272	2303	136
Nov.	0	0.140	0	135	0	19	0	−154	2149	132
Dec.	0.577	0.079	3	104	75	10	3	−39	2110	131
1934										
Jan.	0.110	0.080	566	106	14	10	400	64	2174	132
Feb.	0.455	0.098	1332	138	65	14	400	845	3019	154
Mar.	0	0.214	14	148	0	32	14	−180	2839	148
Apr.	0.056	0.266	8	265	8	39	8	−296	2543	144
May	0.054	0.392	5	466	7	53	5	−512	2031	128
June	0.046	0.480	0	466	5	56	0	−517	1514	105
July	0	0.507	0	466	0	47	0	−513	1001	81
Aug.	0	0.450	0	424	0	32	0	−456	545	63
Sept.	0.079	0.370	0	392	3	15	0	−404	141	18
Oct.	0.084	0.250	700	265	1	5	400	30	171	25
Nov.	0.434	0.126	666	138	14	6	400	126	307	41
Dec.	0.304	0.079	515	106	14	4	400	19	326	50
1935										
Jan.	0.755	0.068	1820	108	91	6	400	1387	1713	120
Feb.	0.108	0.088	18	140	13	10	18	−136	1576	110
Mar.	0.491	0.213	1620	151	62	27	400	1112	2688	146
Apr.	0.486	0.258	3680	270	89	47	400	3044	5732	222

† At end of month.

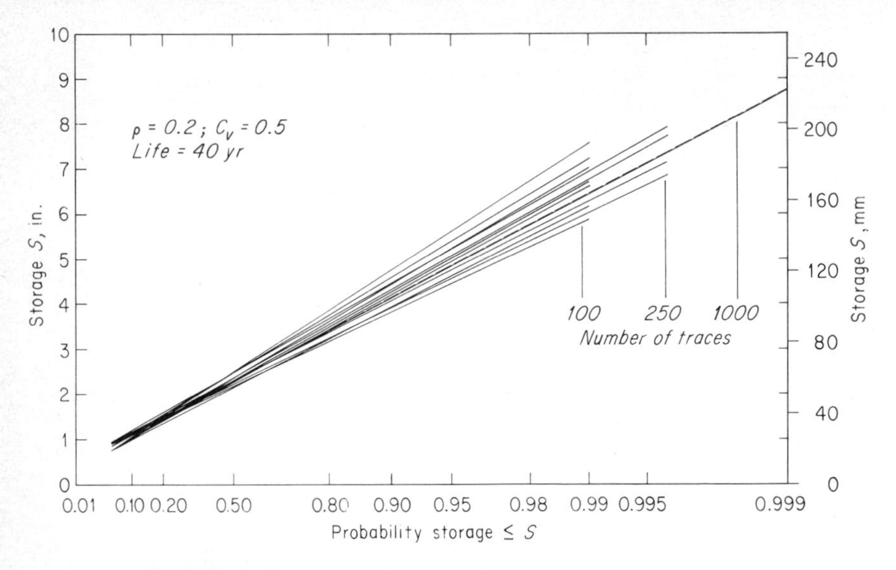

FIGURE 12-7
Influence of the number of traces used to define the distribution of storage assuming a first-order Markov process with normally distributed annual data and a demand equal to 90 percent of mean flow. (*From* [10].)

used, reservoir analysis should normally involve determination of at least 1000 traces, each as long as the expected useful economic life of the reservoir under study. If monthly flows arc uscd, it is likely that fewer traces will be adequate. Because many trials are likely to be made in a particular design situation, it is wise to determine by trial the number of simulated traces that are needed to replicate the storage distribution. This number may vary from about 300 traces for streams that exhibit low variability to the order of 1000 traces for streams with high variability.

Burges also showed that the assumption about the initial condition of the reservoir plays an important part in determining reliability. The example illustrated in Table 12-3 assumes zero initial storage, but any other initial value could have been chosen. The sequent-peak algorithm does not specify an initial condition. When the sequent-peak algorithm is used to analyze a large number of traces, a random distribution of initial storage values is likely to result. This may be the best assumption, since the actual initial condition is often unknown. However, if there is some other basis for fixing the initial condition, it may be used by starting the summation with the predetermined initial storage value.

When a reservoir is expected to supply a progressively increasing

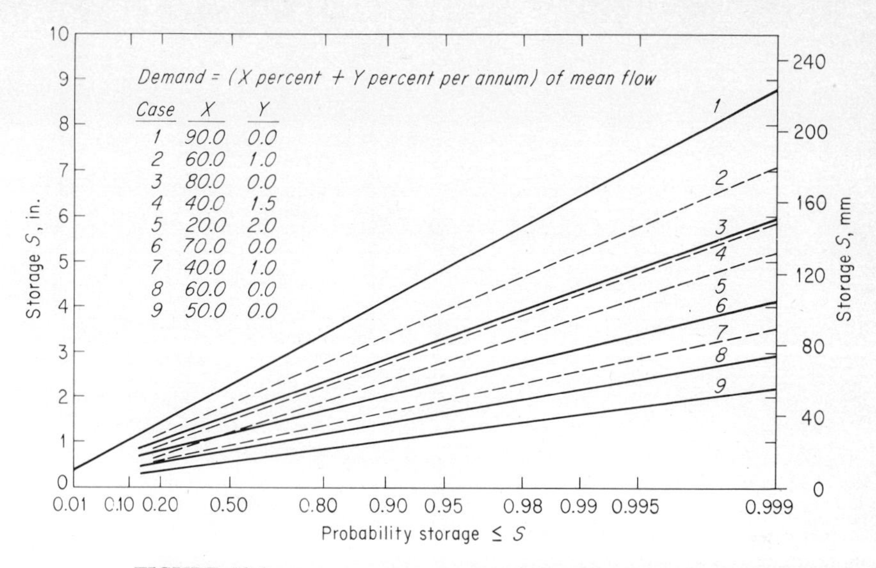

FIGURE 12-8
Comparison of storage requirements for a constant and linearly increasing demand assuming a first-order Markov process with normally distributed annual flow, $\sigma/\bar{Q} = 0.5$, $\rho = 0.2$. (*From* [10].)

demand, the storage algorithm should provide for demand to increase appropriately through each trace. Use of a constant demand at the ultimate level will indicate a requirement for a reservoir considerably larger than may actually be needed (Fig. 12-8).

12-7 Reservoir Reliability

It has been customary to refer to the safe yield of a reservoir as if this were a guaranteed minimum yield. Determinations based on analysis of the historic record provide no evidence regarding the reliability of a reservoir. When a stochastic analysis is accepted as a realistic example of what may happen in the future, a storage-probability curve like Fig. 12-7 provides useful information. Such a curve indicates the probability that flows during any future period equal to the trace length will prove adequate to sustain the desired demand without deficiency. A reliability of 0.99 indicates that only 1 out of every 100 traces showed a deficiency, i.e., a reservoir with the indicated capacity offers a 99 percent assurance of successful operation during the project life.

The stochastic data will also define the magnitude of the deficiency in

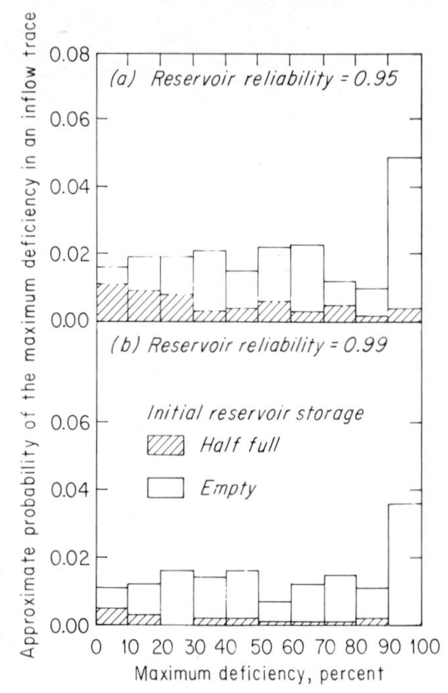

FIGURE 12-9
Approximate probability distribution of deficiencies for Markov process with normally distributed annual flows and $\sigma/\bar{Q} = 0.5$. Traces with zero deficiency not included. (*From* [10].)

each trace so that probability statements can be formulated (Fig. 12-9). A 1-percent risk of deficiency lacks meaning unless something can be said about the magnitude of the shortfall. Many deficiencies computed in the stochastic series will be negligibly small; some may be catastrophically large. Note that the stochastic approach does not seek primarily to generate extraordinary events—either high or low. Such events will be generated with the probability appropriate to their magnitude in the assumed distribution. Basically, however, stochastic analysis seeks to find sequences of more ordinary events which in combination may create shortages.

12-8 Time Trends

Stochastic analysis usually assumes a stationary series; i.e., there are no time trends. In the light of our present knowledge of secular trends in weather, this assumption is probably the wisest one to use. We are concerned with a fairly short period in the future—rarely more than 100 y, commonly less. No established secular trends will cause a significant change in such a short span of time. Changes anticipated as a result of man's activities must be incorporated in the parameters used for stochastic generation. That is, if runoff is expected to increase or decrease as a result of other activities in the watershed, some adjustment must be made. Increased evaporation loss from the reservoir should be included in the storage simulation.

Time trends in the historic data on which the stochastic generation is based are still another problem. Many streamflow records display obvious trends as a result of changes in land use in the watershed, diversions for irrigation or other purposes, construction of reservoirs, etc. Adjustment for these trends should be made before parameters are calculated. It is probably best to adjust to full natural flow to define parameters for the stochastic generation and then to incorporate reservoirs or diversions into the storage simulation. However, if the change is permanent and uncontrolled, such as the effects of urbanization, it may be wiser to adjust to current or future conditions. The adjustment to natural flow is usually accomplished by adding known diversions and reservoir-storage changes to the observed flows (Sec. 4-17). Adjustment to current or future conditions is probably best accomplished by simulation from precipitation data. An alternative is to fit a trend line to the time series of historic data. This approach achieves only an approximate correction and should be resorted to only if other alternatives are impractical because of data deficiencies.

12-9 Multistation Models

In large watersheds with several storage projects, it becomes necessary to develop stochastic data at each project site. The generating process must preserve the statistical properties of the several locations and at the same time generate compatible flow data. Flows for concurrent time periods at the several sites must represent flows which could reasonably have occurred at the same time; i.e., there cannot be a major flood at one point on the stream concurrent with extreme low flows at some other point on the same stream.

One approach is to generate the synthetic data at a *base station* and then construct the records at the other stations by a simple cross-correlation equation much like Eq. (12-2), i.e.,

$$Q_{2,i} = \bar{Q}_2 + \rho_{12}(Q_{1,i} - \bar{Q}_1) + t_i\sigma_2\sqrt{1 - \rho_{12}^2} \qquad (12\text{-}11)$$

where the subscript 1 refers to the base station and 2 to the dependent station.

Equation (12-11) preserves the cross correlation between stations 1 and 2 but does not preserve the serial correlation in station 2 or cross correlations between station 2 and any other station in the system. Consequently Eq. (12-11) should be used only when two or three stations are involved. When more stations are involved, a multivariate solution has been suggested by Matalas [15].

12-10 Stochastic Analysis of Precipitation

Precipitation data may be treated by stochastic methods although the procedures must be somewhat different from those for flow. The advantages of stochastic analysis of precipitation are primarily (1) the possibility of stochastic analysis in situations where there is no flow record or the available record is inadequate and (2) the possibility of stochastic treatment of short-interval data. Both cases can be handled by stochastically generated rainfall data converted to flow with a deterministic simulation procedure.

Monthly flow volumes are usually satisfactory for reservoir studies, but for very small reservoirs or for flood problems greater detail is required. The techniques described earlier in this chapter for deriving stochastic streamflow data are not applicable to daily data because the sequential relations between daily flows are not wholly random and cannot be described by a simple function. For example, from point *A* to point *B* (Fig. 12-10) flow on successive days is related to flow on the previous day by the recession equation [Eq. (7-1)]. If a random component is present, it is very small and represented by the variability of the evapotranspiration from day to day. Between points *B* and *C* the relation between successive days is quite poor, since the controlling factor is a separate stochastic element—precipitation—filtered by the catchment. The flow on day $B + 1$ is only partially determined by the flow on day *B*. A stochastic daily-flow-generating procedure would have to recognize these differing conditions and include an algorithm that would determine the occurrence of streamflow rises. Once a rise is under way, a function which constrained the flows to a pattern consistent with the watershed characteristics would be used.

Such a daily-flow-generating function could probably be devised, but the use of rainfall data offers several advantages. It seems more logical to work with the causal stochastic variable since precipitation displays relatively low serial correlation and the generating functions can be more simple and direct. Usually the precipitation records are longer than streamflow records, and hence the parameters are more stable. Finally, extremes of precipitation for short time intervals are relatively less important in determining streamflow than the sequences of precipitation intensities in the moderate to heavy range, and the generation of sequences of data is the primary function of stochastic analysis.

If stochastically generated precipitation is to be used to establish long flow records, the generated precipitation data must meet the requirements of the simulation system to be used in the transformation. In addition, it could be argued that stochastic potential evapotranspiration (PET) data should also be utilized, although experience shows that the variability of PET

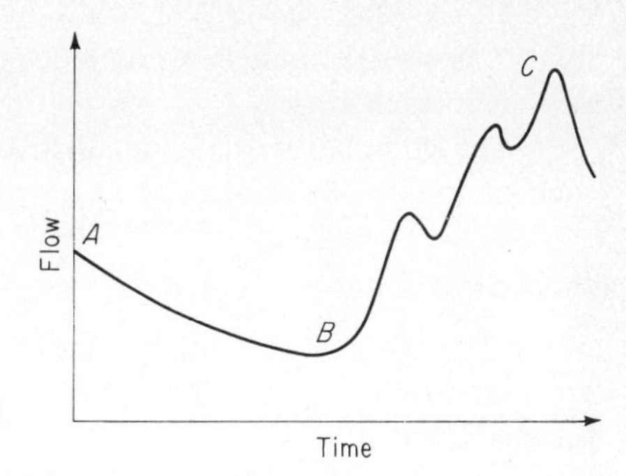

FIGURE 12-10
Mean daily flow hydrograph showing the differing regimes.

is much less than that of precipitation and generally has little effect on the results. If snow is involved in the hydrology of a watershed, temperature data will also be required along with the other parameters utilized in the snowmelt routines of the simulation program to be employed. Finally, for large watersheds it will be necessary to generate several rainfall records to represent the variability in space over the watershed properly.

Pattison [16] demonstrated the feasibility of developing stochastic flow data for a small watershed by generation of a single-station record of precipitation and transformation with the Stanford Watershed Model. He used a Markov model and a transition-probability matrix between successive hours of rain and a sixth-order model to determine the onset of rain after a period of no rainfall. The procedure was cumbersome, and difficulty was experienced in generating the proper number of hours with 0.01 in. (0.25 mm). The transition matrix had numerous empty boxes, and transitions were limited to those which had actually occurred in the historic record. Despite these limitations, he successfully reproduced the flood-frequency curve of a stream (Fig. 11-5). His experience suggests that something less than perfect reproduction of rainfall is a satisfactory input to a simulation model for flow generation.

Franz [17] used a multivariate normal approach to multistation hourly rainfall generation. To account for the variation in rainfall characteristics throughout the year, he dealt independently with four seasons. More seasons might better define the annual cycle, but as the seasons are shortened, the amount of data for defining the parameters diminishes. Selection of the number and extent of the seasons must be based on judgment.

Franz defined a *wet hour* as an hour in which at least one station in the network had rain and a *dry hour* as an hour in which no rain occurred at any station. A *storm* was an interval of wet hours beginning after a dry hour and terminating with a dry hour. An *interstorm* was a sequence of dry hours. Separate models were used to synthesize storm and interstorm periods in each season.

The cumulative frequency distribution of observed hourly rainfall at each station was transformed to a normal distribution by fitting the equation

$$Y = a + b(X + 0.5)^c \qquad (12\text{-}12)$$

where Y is a standard normal variable, X is the rainfall value, and a, b, and c are parameters derived in the fitting process. Rainfall amounts were expressed as integers with 0.01 as 1. The normal distribution was assumed to be truncated at zero. The generating equations for the storm model are then

$$\mathbf{S}_0 = E(x_i x_i^T) \qquad (12\text{-}13)$$

$$\mathbf{S}_1 = E(x_{i+1} x_i^T) \qquad (12\text{-}14)$$

$$\mathbf{LL}^T = \mathbf{S}_0 - \mathbf{S}_1 \mathbf{S}_0^{-1} \mathbf{S}_1^T \qquad (12\text{-}15)$$

$$X_{i+1} = \bar{X} + \mathbf{S}_1 \mathbf{S}_0^{-1}(X_i - \bar{X}) + \mathbf{L}Z_{i+1} \qquad (12\text{-}16)$$

where $\mathbf{S}_0$ is the lag-0 covariance matrix for transformed rainfall, $\mathbf{S}_1$ is the corresponding lag-1 covariance matrix, X_i is the vector of transformed rainfall values in hour i, $\bar{X}$ is the vector of means for X_i, Z_i is the vector of standard normal independent variables, x_i is the vector of deviations of X_i from $\bar{X}$, E denotes the expected value, superscript T denotes transposition, and $\mathbf{L}$ is the lower triangular matrix.

Empirically defined distributions of the interstorm interval were used to generate interstorm periods. Thus, starting with a dry hour, Eq. (12-16) is used to generate a series of wet hours. When this series is finally terminated by a dry hour, the interstorm model is called to define the length of the following interstorm interval, after which the storm model is called again. Since lag-1 dependence is assumed, the generating process is markovian.

When tested with data for three stations in the Russian River basin, California, the procedure seemed to generate satisfactory values for such factors as the probability of zero rain, mean storm length, mean hourly intensity, and annual rainfall. Generated data utilized in a simulation model closely approximated the flood frequency curve of Dry Creek. Ott [18] subsequently utilized the Franz model on another stream with apparently good results. As for stochastic streamflow, there is no positive measure of success. Reproduction of the original parameters only verifies that the generating algorithm is satisfactory. In the case of stochastic rainfall a further factor is that slight errors in the generated rainfall probably have little effect on the frequency of simulated streamflow, which is the end product desired.

REFERENCES

1. J. Bernoulli, "Ars Conjectandi," p. 213, Basel, 1713.
2. A. Hazen, Storage to Be Provided in Impounding Reservoirs for Municipal Water Supply, *Trans. ASCE*, vol. 77, pp. 1539–1669, 1914.
3. C. E. Sudler, Storage Required for the Regulation of Streamflow, *Trans. ASCE*, vol. 91, pp. 622–704, 1927.
4. H. A. Thomas, Jr., and M. B. Fiering, Mathematical Synthesis of Streamflow Sequences for the Analysis of River Basins by Simulation, chap. 12 in A. Maass et al., "Design of Water Resources Systems," Harvard University Press, Cambridge, Mass., 1962.
5. G. E. P. Box and G. M. Jenkins, "Time Series Analysis: Forecasting and Control," Holden-Day, San Francisco, 1970.
6. B. B. Mandelbrot and J. P. Wallis, Computer Experiments with Fractional Gaussian Noises, *Water Resour. Res.*, vol. 5, pp. 228–267, February 1969.
7. J. M. Mejia, I. Rodriguez-Iturbe, and D. R. Dawdy, Streamflow Simulation: 2, The Broken Line Process as a Potential Model for Hydrologic Simulation, *Water Resour. Res.*, vol. 8, pp. 931–941, August 1972.
8. M. Fiering, "Streamflow Synthesis," Harvard University Press, Cambridge, Mass., 1967.
9. J. R. Benjamin and C. A. Cornell, "Probability, Statistics, and Decision for Civil Engineers," McGraw-Hill, New York, 1970.
10. S. J. Burges, Use of Stochastic Hydrology to Determine Storage Requirements for Reservoirs: A Critical Analysis, *Stanford Univ. Progr. Eng. Econ. Plann. Rep.* EEP-34, September 1970.
11. R. L. Anderson, Distribution of the Serial Correlation Coefficient, *Ann. Math. Stat.*, vol. 13, pp. 1–13, 1962.
12. H. E. Hurst, Long-Term Storage Capacity of Reservoirs, *Trans. ASCE*, vol. 116, pp. 770–808, 1951.
13. H. A. Thomas, Jr., and M. B. Fiering, The Nature of the Storage Yield Function in "Operations Research in Water Quality Management," Harvard University Water Program, 1963.
14. M. Hufschmidt and M. Fiering, "Simulation Techniques for the Design of Water-Resource Systems," Harvard University Press, Cambridge, Mass., 1962.
15. N. C. Matalas, Autocorrelation of Rainfall and Streamflow Minimums, *U.S. Geol. Surv. Prof. Pap.* 434-B, 1963.
16. A. Pattison, Synthesis of Rainfall Data, *Stanford Univ. Dept. Civ. Eng., Tech. Rep.* 40, 1964.
17. D. D. Franz, Hourly Rainfall Synthesis for a Network of Stations, *Stanford Univ. Dept. Civ. Eng. Tech. Rep.* 126, 1969.
18. R. F. Ott, Streamflow Frequency Using Stochastically Generated Hourly Rainfall, *Stanford Univ. Dept. Civ. Eng. Tech. Rep.* 151, December 1971.

BIBLIOGRAPHY

FIERING, M. B., and JACKSON, B. B.: Synthetic Streamflows, *Water Resour. Monogr.* 1, American Geophysical Union, 1971.

KARTVELISHVILI, N. A.: "Teorija Veroyatnostnych Processov v Gidrologii i Regulirovane Stoka," in Russian (Theory of Stochastic Processes in Hydrology and Regulation of Flow), Gidrometeoizdat, Leningrad, 1967.

KLEMES, V.: Applications of Hydrology to Water Resources Management, *Opera. Hydrol. Rep.*, no. 4, World Meteorological Organization, Geneva, 1973.

MORAN, P. A. P.: "The Theory of Storage," Methuen, London, 1959.

PROBLEMS

12-1 Using a historic-data series of 25 y or more and the sequent-peak method, determine the storage required to satisfy a demand of 30, 50, 70, and 90 percent of the mean flow. If your classmates analyze other streams, express the storage as a percent of mean flow and compare. Are the variations between streams as you might expect?

12-2 For the stream of Prob. 12-1, calculate mean, variance, and serial correlation. Plot the flow data and inspect the plot for occurrences which might bias the computed serial correlation.

12-3 Use the parameters of Prob. 12-2 to generate a record of annual flows. Develop a reliability curve like Fig. 12-7. What probability is applicable to the results of Prob. 12-1? How do the reliability curves constructed from the two halves of your synthetic record agree?

12-4 Analyze the parameters for the monthly data for a stream. Do all the months seem to exhibit the same distribution? Can you explain the variation in the serial correlation throughout the year? In the variance?

13

SEDIMENTATION

Soil is continually being removed from the land surfaces of the earth and transported downstream in rivers until it is deposited in lakes, estuaries, and the oceans. Since water is the primary agent of erosion and the principal vehicle for transport of the eroded material, the process is of interest to the hydrologist. Of concern to the hydrologist are the rates of deposition of sediment in reservoirs, harbors, and estuaries and the methods of controlling erosion at its source, both to conserve the soil in place and to minimize the accumulation in reservoirs and harbors.

13-1 The Erosion Process

Erosion may be viewed as starting with the *detachment* of soil particles by the impact of raindrops [1]. The kinetic energy of the drops can splash soil particles into the air. On level ground the particles are distributed more or less uniformly in all directions, but on a slope there is a net transport downslope (Fig. 13-1). If overland flow is occurring, the falling particles will be entrained in the flowing water and moved even farther downslope before settling to the soil surface. Overland flow is predominantly laminar and

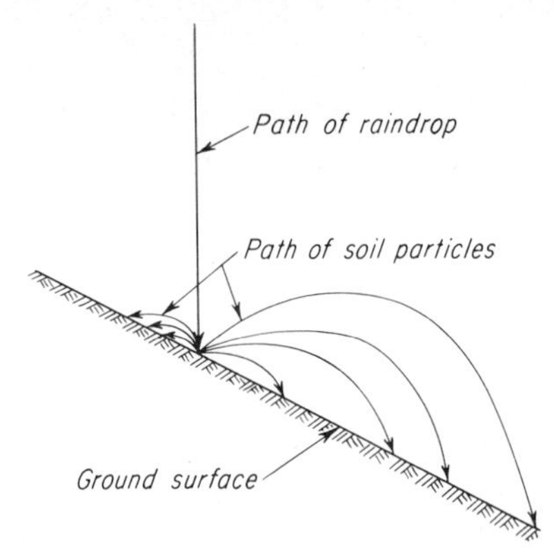

FIGURE 13-1
Downhill transport of soil particles by splash.

cannot detach soil particles from the soil mass, but it can move loose particles already on the soil surface. The splash and overland-flow processes are responsible for *sheet erosion*, the relatively uniform degradation of the soil surface. Sheet erosion is difficult to detect except as the soil surface is lowered below old soil marks on fence posts, tree roots are exposed, or small pillars of soil capped by stones remain.

Raindrops vary in diameter d from 0.02 to 0.25 in. (0.5 to 6 mm), and terminal velocity v varies with diameter from about 7 to 30 ft/s (2 to 9 m/s). Since kinetic energy is proportional to d^3v^2, the erosive power of the largest drops may be 10,000 times that of the smaller. This conforms with the observation that a few intense storms account for most of the erosion. The effect is augmented by the fact that overland flow is more likely to occur during intense rains.

At some point on the slope sufficient overland flow may accumulate to cause a small rivulet. If turbulence in the flow is strong enough to dislodge particles from the bed and banks of the channel, *gully erosion* may occur. As the gully deepens, its profile is steepest near its head (Fig. 13-2). Erosion is most rapid in this region, and the gully tends to grow headward.

The third factor in erosion is *mass movement*, which may take the form of a slow downslope *creep* of the soil mass or a rapid collapse of a large soil mass as a *landslide*. Some landslides enter the stream channels, delivering a considerable volume of soil directly to the stream.

13-2 Factors Controlling Erosion

Many factors affect the rate of erosion. The most important among these are the rainfall regime, vegetal cover, soil type, and land slope. The significance of the rainfall regime is discussed in the previous section. Because

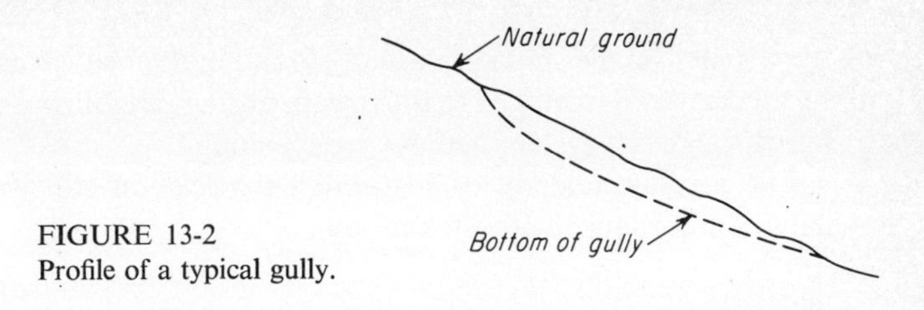

FIGURE 13-2
Profile of a typical gully.

of the important role of raindrop impact, vegetation provides significant protection against erosion by absorbing the energy of the falling drops and generally reducing the drop sizes which reach the ground. Vegetation may also provide mechanical protection to the soil against gully erosion. In addition, a good vegetal cover generally improves infiltration capacity through the addition of organic matter to the soil. Higher infiltration capacity means less overland flow and consequently less erosion.

A well-cemented soil will resist splash erosion more readily than loose soils. Generally, splash erosion increases with an increasing fraction of sand in the soil because of the loss of cohesion. Splash erosion decreases with an increasing percentage of water-stable aggregates. A soil whose individual grains do not tend to form aggregates will erode more readily than one in which aggregates are plentiful.

Rates of erosion are greater on steep slopes than on flat slopes. The steeper the slope the more effective splash erosion is in moving soil downslope. Overland-flow velocities are also greater on steep slopes, and mass movements are more likely to occur in steep terrain. Length of slope is also important. The shorter the slope length, the sooner the eroded material reaches the stream, but this is offset by the fact that overland-flow discharge and velocity increase with length of slope.

A number of equations relating the factors discussed above have been proposed as a basis for estimating erosion rates, but it is difficult to define properly such factors as the precipitation regime by a simple index number, and the equations are at best approximate.

13-3 Suspended-Sediment Transport

Sediment moves in the stream as *suspended sediment* in the flowing water and as *bed load*, which slides or rolls along the channel bottom. A third term, *saltation*, is used to describe the movement of particles which seem to bounce along the bed. The processes are not independent, for material which appears as bed load at one section may be in suspension at another. Another useful distinction is between *bed-material load*, represented by those particles of grain size normally found in the streambed, and *wash load*, made up of particles smaller than are usually found in the bed. Wash load consists of the fine material washed into the stream during rainfall and which normally travels through the system without redepositing.

The settling velocity of suspended particles in still water is approximately in accordance with Stokes' law:

$$v_s = \frac{2(\rho_g - \rho)gr^2}{9\mu} \tag{13-1}$$

where ρ_g and ρ are densities of the particle and the liquid, respectively, r is the radius of the particle, and μ is the absolute viscosity of the water. Generally considered applicable to particles from 0.0002 to 0.2 mm, the equation assumes that viscosity offers the only resistance to settling, that the particles are rigid and spherical, and that their fall is not impeded by other particles.

In turbulent flow the gravitational settling of particles is counteracted by upward transport in turbulent eddies. Since the concentration of suspended material is greatest near the bottom of the stream, upward-moving eddies carry more sediment than downward-moving eddies. The system is in equilibrium if gravity movement and turbulent transport are in balance and the amount of suspended material remains constant.

The general two-dimensional nonequilibrium equation for suspended-sediment transport is

$$v\frac{\partial c_s}{\partial_x} = v_s\frac{\partial c_s}{\partial_y} + \frac{\partial \varepsilon_x}{\partial_x}\frac{\partial c_x}{\partial_x} + \frac{\partial \varepsilon_y}{\partial_y}\frac{\partial c_s}{\partial_y} + \varepsilon_x\frac{\partial^2 c_s}{\partial x^2} + \varepsilon_y\frac{\partial^2 c_s}{\partial y^2} \tag{13-2}$$

where c_s is the sediment concentration for a particular size of particle, v_s is settling velocity, ε is a mixing coefficient, and x and y are longitudinal and vertical dimensions respectively. No solution of this equation has been achieved without so many simplifying assumptions that the solution is of little value in dealing with natural streams. The literature is replete with analytic and experimental studies of suspended-load transport. Most effort has been directed toward deriving a function which would describe the vertical variation of sediment concentration in the stream. Such a function combined with a vertical distribution of flow velocity would permit calculation of suspended-load transport. Alternatively, given sediment samples at one or two depths, the total sediment load could be calculated in a manner analogous to current-meter gaging of flow. Several functions were derived which conformed reasonably well with observed sediment-concentration variations in the vertical [2]. However, such functions can apply only to a limited particle-size range and must be summed over the total range of particle size.

When these functions are applied to the range of particle sizes in the bed material, a fair approximation to the transport of suspended bed material seems possible, but a large and variable wash-load component precludes a reliable computation of total suspended load. Hence, methods of suspended-load measurement independent of a knowledge of the sediment-concentration gradient were developed (Sec. 13-5).

13-4 Bed-Material Transport

For many years analysis of bed-load transport has been based on the classical equation of du Boys [3]

$$G_i = \Upsilon \frac{\tau_0}{w} (\tau_0 - \tau_c) \qquad (13\text{-}3)$$

where G_i is the rate of bed-load transport per unit width of stream, Υ is an empirical coefficient depending on the size and shape of the sediment particles, w is the specific weight of water, τ_0 is the shear at the streambed, and τ_c is the magnitude of shear at which transport begins. Numerous variations on the original du Boys formula have been proposed [2], all using the concept of a critical tractive force to initiate motion. This approach ignores the modern concepts of turbulence and the boundary layer as they affect entrainment of bed particles. The successful application of Eq. (13-3) lies in the proper selection of the coefficient Υ. Most available values are determinations from studies with small flumes. Table 13-1 summarizes values given by Straub [4].

Accuracy of instruments for bed-load measurement is so uncertain (Sec. 13-5) that field comparison of bed-load formulas is very difficult.

Table 13-1 FACTORS IN EQ. (13-3) FOR BED-LOAD MOVEMENT [4]

Particle diameter, mm	Υ		τ_c	
	ft^6/lb^2-s	m^6/kg^2-s	lb/ft^2	kg/m^2
$\frac{1}{8}$	0.81	0.0032	0.016	0.078
$\frac{1}{4}$	0.48	0.0019	0.017	0.083
$\frac{1}{2}$	0.29	0.0011	0.022	0.107
1	0.17	0.0007	0.032	0.156
2	0.10	0.0004	0.051	0.249
4	0.06	0.0002	0.090	0.439

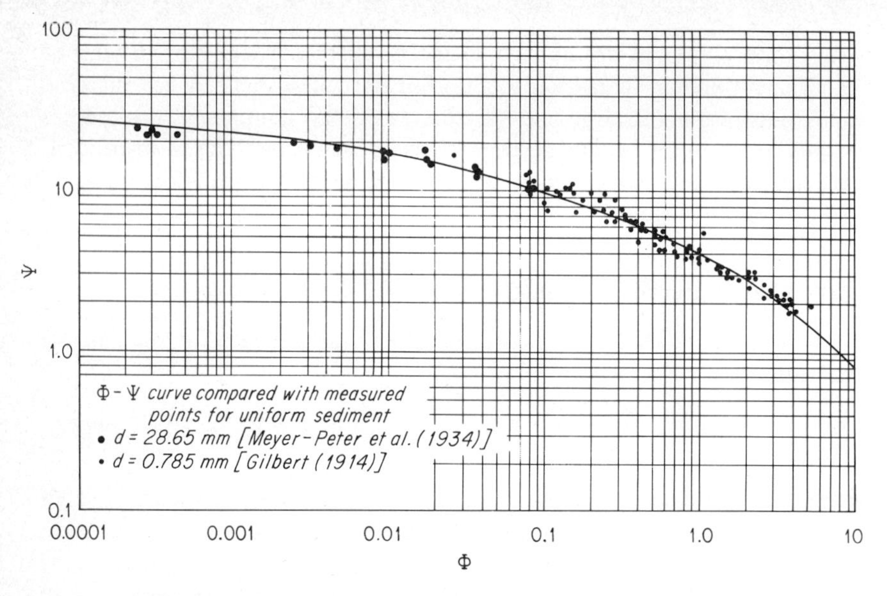

FIGURE 13-3
Plot of Einstein's function Φ versus Ψ. (*From* [5].)

The validity of bed-load formulas is therefore quite indefinite. Recent work on the bed-load problem has utilized the concepts of turbulent flow and statistical variation of fluid forces at a point. A widely used approach is that of Einstein [5], who defines the *intensity* of bed-load transport as

$$\Phi = \frac{G_i}{w}\sqrt{\frac{\rho}{\rho_s - \rho}\frac{1}{gd^3}} \qquad (13\text{-}4)$$

and the *flow intensity* as

$$\psi = \frac{\rho_s - \rho}{\rho}\frac{d}{sR} \qquad (13\text{-}5)$$

where w is the specific weight of water, ρ is the density of water, ρ_s is the density of the bed material, d is the grain diameter, s is the channel slope, and R is the hydraulic radius. An empirical relation between Φ and ψ (Fig. 13-3) permits the solution of Eqs. (13-4) and (13-5) for G_i. The method applies to a specific narrow range of particle sizes and must be repeated to cover the entire range of the bed-material sizes. However, if the range of particle size is small, Einstein suggests a single solution using d_{35}, the particle size for which 35 percent of the bed material is finer.

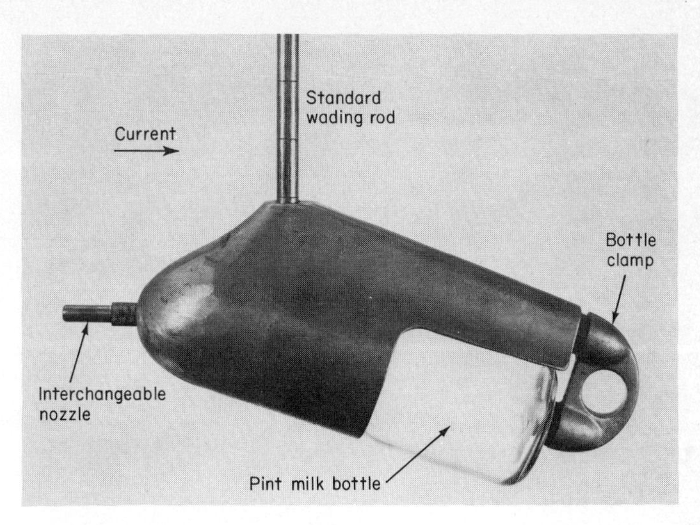

FIGURE 13-4
The DH-48 depth-integrating hand sampler for use on small streams.

13-5 Sediment Measurement

Early suspended-sediment observations were made with open bottles or complex grab samplers, which failed to provide adequate data for a number of reasons. A good sampler must cause minimum disturbance of streamflow, avoid errors from short-period fluctuations in sediment concentration, and give results which can be related to velocity measurements. These requirements seem to be met in a series of samplers [6] designed at the Iowa Hydraulic Laboratory under the sponsorship of several federal agencies. The samplers (Fig. 13-4) consist of a streamlined shield enclosing a glass milk bottle as a sample container. A vent permits escape of air as water enters the bottle and controls the inlet velocity so that it is approximately equal to the local stream velocity. Nozzle tips of various sizes are available to control the rate at which the bottle fills. The large models have the bottle fully enclosed and are fitted with tail vanes to keep the sampler headed into the current when cable-supported.

The sampler is lowered through the stream at constant vertical speed until the bottom is reached and is then raised to the surface at constant speed. The result is an integrated sample with the relative quantity collected at any depth in proportion to the velocity (or discharge) at that depth. The duration of the traverse is determined by the time required to nearly fill the sample bottle and can be computed from the filling-rate curves for the particular nozzle when the stream velocity is known. A number of traverses are made at intervals across the section to determine the total suspended-sediment

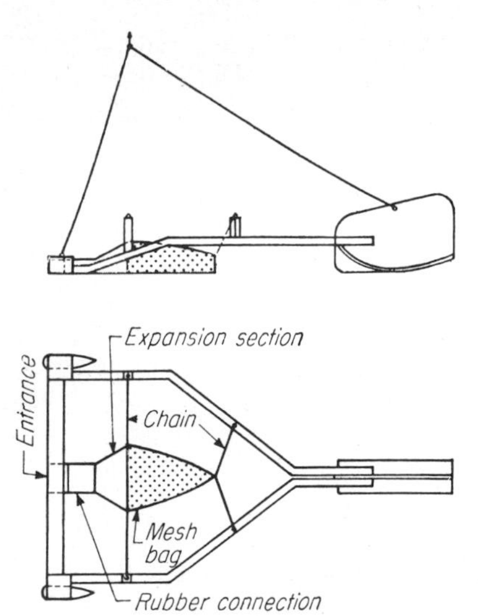

FIGURE 13-5
A bed-load sampler of the pressure-difference type used in the Netherlands.

load for the section. Thus there is no problem of whether a point sample is representative of sediment load in the section. Point samplers are used only where it is impossible to use the depth-integrating type because of great depths or high velocity or for studies of sediment distribution in streams. Because of the shape of the sampler, the nozzle cannot be lowered to the streambed, and consequently a few inches of the depth near the bed is not sampled. This may represent a large error [7] in shallow streams.

The collected samples are filtered and the sediment dried. The ratio of dry weight of sediment to total weight of the sample is the sediment concentration, usually expressed in parts per million. Other analyses which may be performed include determination of grain-size distribution, fall velocity, and, occasionally, heavy-mineral or chemical analysis. The latter tests may be useful in tracing the original source of the sediment.

There is as yet no wholly satisfactory bed-load sampler. Portable samplers consist of a container lowered to the streambed. In the pressure-difference sampler (Fig. 13-5), the expansion section causes a pressure drop which encourages inflow at the same rate as the prevailing flow while at the same time the reduced exit velocity encourages deposition of sediment in the mesh bag [8]. Fine sediment particles can escape through the mesh, and a calibration in a flume is desirable. The difficulty of designing a sampler which does not disturb the flow and at the same time effectively traps all sizes of bed load should be apparent.

One type of permanent bed-load trap for research purposes on small streams consists of a grated opening in the streambed into which the bed material falls. The trapped material is later excavated and measured. Turbulence-producing weirs have also been designed which throw the bed load into suspension locally so that it can be sampled with a suspended-sediment sampler. Comparison of the samples thus obtained with those from a section upstream of the weir indicates the quantity of bed load. This method is suitable only when the bed material is relatively fine.

Because of the considerable variability of sediment load in streams, a continuous monitoring device would be very useful, and a number of such devices have been tested. Pumping samplers which draw water from a fixed point in the stream at preset time intervals and store the samples in bottles for later analysis have been tested, as have samplers which pass the water through a cell where the attenuation of a light beam by the suspended sediment is measured with a photocell [9]. Photocell probes which are immersed in the stream [10, 11] and an x-ray probe [12] have also been tested. A significant limitation of all such devices is that they sample only one point in the cross section. An intensive investigation with conventional samplers must precede their installation to determine a representative location within the stream. At best this will be an approximation since the most representative location will vary with stage and possibly other factors. Thus it is difficult to test a monitoring device in the field, and the absolute accuracy of such devices is unknown. Nevertheless, they do provide a means of obtaining a record of the variation of sediment load with time, and if used in conjunction with a regular program of sampling, should provide better information than is obtained from occasional samples alone.

13-6 Sediment-Rating Curves

Sediment measurements, like current-meter measurements, give only occasional samples of the sediment discharge. A *sediment-rating curve* relating suspended-sediment discharge and water discharge (Fig. 13-6) is commonly used to estimate sediment load on days when no measurements are available. The figure clearly shows that such relations are approximate. A given flow rate may result from melting snow or rains of differing intensity, and a different sediment load would result in each case. Areal distribution of runoff may be a factor if different portions of the basin are more prolific sediment sources than others. Sediment-rating curves should be used with caution and where possible applied to small and relatively homogeneous basins. However, when they are used to estimate mean annual sediment yield, the errors in the sediment rating will tend to compensate and the resulting answer should be reasonably satisfactory if a sufficiently long record is used.

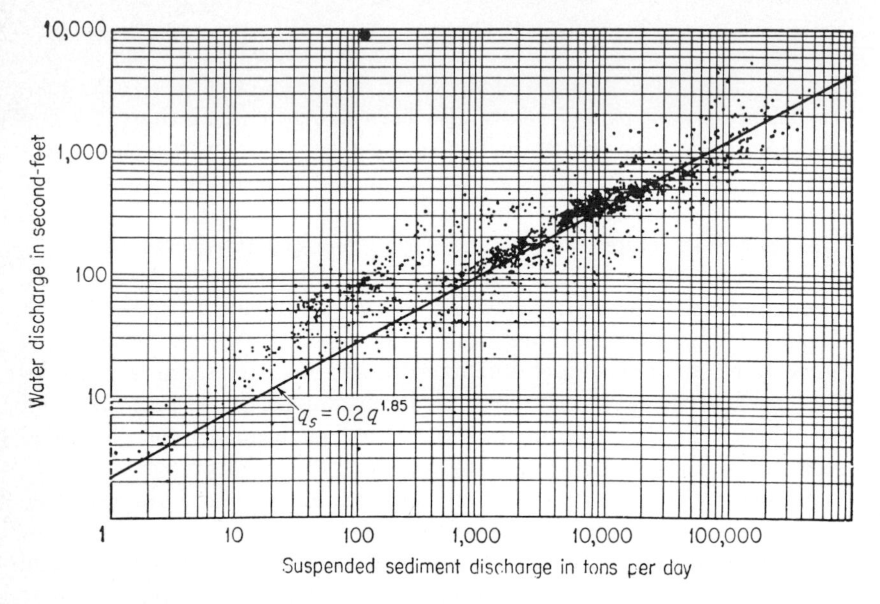

FIGURE 13-6
Sediment-rating curve for the Powder River at Arvada, Wyoming. (*From L. B. Leopold and T. Maddock, Jr., The Hydraulic Geometry of Stream Channels and Some Physiographic Implications, U.S. Geol. Surv. Prof. Pap.* 252, 1953.)

13-7 Sediment Yield of a Watershed

The average annual sediment production from a watershed is dependent on many factors such as climate, soil type, land use, topography, and the presence of reservoirs. Adequate data for a complete analysis of all factors are difficult to obtain. Langbein and Schumm [13] used data from a number of watersheds to construct the curve of Fig. 13-7 which relates average annual sediment production per unit area to mean annual precipitation. Maximum production rates occur at about 12 in. (305 mm) of mean annual precipitation because such areas usually have little protective vegetal cover. With higher rainfalls, vegetal cover reduces the erosion, and with lesser rainfalls the erosion also decreases.

Fleming [14] utilized data from over 250 catchments about the world to derive relations [Eq. (13-6) and Table 13-2] for mean annual suspended load Q_s in tons as a function of mean annual discharge in cubic feet per second for various vegetal covers:

$$Q_s = aQ^n \qquad (13\text{-}6)$$

Errors of ± 50 percent may be expected from these relations.

For watersheds without sediment records the relations presented above

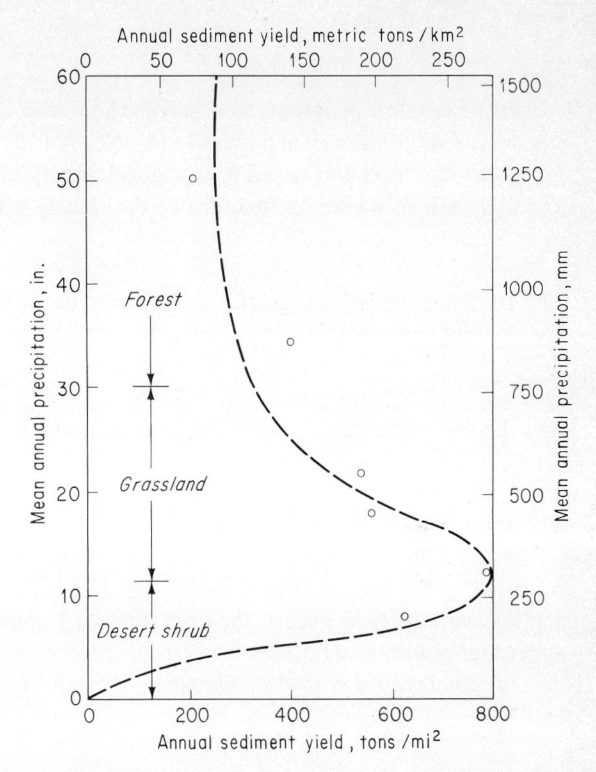

FIGURE 13-7
Sediment yield as a function of mean annual precipitation. (*From* [13].)

may be viewed as offering an order-of-magnitude estimate of sediment yield. If possible, such estimates should be compared with sediment data on similar watersheds in the same region.

Table 13-2 VALUES OF a AND n IN EQ. (13-6) FOR VARIOUS COVER TYPES

Vegetal cover	n	a: For Q_s in tons	a: For Q_s in metric tons
Mixed broadleaf and coniferous	1.02	117	106
Coniferous forest and tall grassland	0.82	3,523	3,196
Short grassland and scrub	0.65	19,260	17,472
Desert and scrub	0.72	37,730	34,228

SOURCE: Adapted from [14].

13-8 Sediment Simulation

A flow-simulation program, as described in Chap. 10, includes the essential factors for simulation of sediment load. Negev [15] developed such a model for suspended load and tested it with good results. In this model the amount of soil splash R is made a function of the hourly precipitation amount i

$$R = K_1 i^j \tag{13-7}$$

The transport of splash residue S is a function of the residue in storage on the ground surface R_s and the overland flow rate q_o

$$S = K_2 R_s q_o^{\,k} \tag{13-8}$$

and the sediment washed from impervious areas E is

$$E = K_3 R \tag{13-9}$$

The total wash load then becomes

$$W = R + S + E \tag{13-10}$$

Unless overland flow occurs, the only source of wash load is from impervious areas and is very small.

Gully erosion G is also related to overland flow

$$G = K_4 q_o^{\,m} \tag{13-11}$$

Negev divided the sediment from gully erosion into two portions. The first B has particle size substantially thc same as the bed-material load in the stream

$$B = (1 - K_5)G \tag{13-12}$$

and the second portion, which he called *interload* I, represents material finer than about 95 percent of the bed-material load:

$$I = G - B = K_5 G \tag{13-13}$$

The total suspended-load transport q_s is then

$$q_s = K_6 I_s q^n + K_7 q^r \tag{13-14}$$

where q is the mean daily streamflow and I_s is the quantity of interload in storage in the streambed calculated by maintaining a running balance of the input from erosion and the outflow of suspended interload material.

The procedure involves a number of coefficients which must be found by calibration (Chap. 10). The exponent j can be assumed as 3.0, and k and m as 2.5. Negev assumed $n = r =$ the slope of a sediment-rating curve for the station. The coefficient K_5 can be estimated from particle-size data.

The overland-flow rate is calculated with a flow-simulation model.

By generating a fairly long sediment record, a better estimate of the mean annual sediment yield can be obtained than from a short observed record because of the great variability in sediment transport from year to year. A limitation on the simulation approach is that it requires calibration against at least a short but continuous sediment record. Occasional, infrequent sediment samples do not provide a very good basis for calibration, just as they fail to give a firm base for estimating sediment yield. Thus a program of daily sediment samples or continuous monitoring is important in improving our ability to deal with sediment problems. The Negev model does not attempt to estimate bed-load transport because of the virtually total lack of bed-load data available for development of algorithms.

13-9 Reservoir Sedimentation

The rate at which the capacity of a reservoir is reduced by sedimentation depends on (1) the quantity of sediment inflow, (2) the percentage of this inflow trapped in the reservoir, and (3) the density of the deposited sediment. The quantity of sediment inflow may be estimated by any of the methods discussed in Secs. 13-6 to 13-8, or if data are available, by reference to mean-annual-yield data per unit area from similar basins in the region. Table 13-3 presents some selected values of sediment yield derived from reservoir surveys. These data are generally obtained by surveying the reservoir with sounding lines or echo-sounding equipment and are published periodically [16].

Table 13-3 SELECTED DATA ON RATES OF SEDIMENT PRODUCTION

	Drainage area		Annual sediment production	
Location	mi^2	km^2	Tons/mi^2	Metric tons/km^2
Bayview Reservoir, Ala.	72	186	1,769	620
San Carlos Reservoir, Ariz.	12,900	33,411	389	136
Morena Reservoir, Calif.	112	290	3,340	1,170
Black Canyon Reservoir, Idaho	2,540	6,579	172	60
Pittsfield Reservoir, Ill.	1.8	4.7	3,090	1,082
Mission Lake, Kans.	11.4	29.5	2,705	947
High Point Reservoir, N.C.	63	163	544	191
Tygart Reservoir, W.Va.	1,182	3,061	51	18

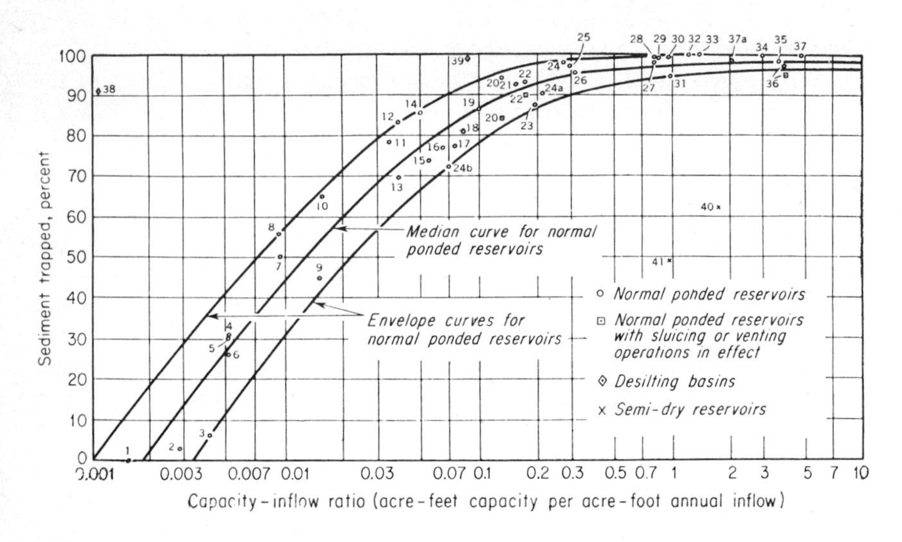

1 Williams Reservoir
2 Lake Halbert (Rock Reservoir No. 1)
3 Lake Halbert (Rock Reservoir No. 3)
4 Hales Bar Reservoir (1935 and 1936)
5 Hales Bar Reservoir (1938)
6 Hales Bar Reservoir (1937)
7 Keokuk Reservoir
8 Lake Taneycomo
9 Wilson Lake
10 Lake Marinuka
11 Lake Decatur
12 Bullard's Bar Reservoir
13 Lake Halbert (Earth Reservoir No. 1)
14 Lake Rockwell
15 Corpus Christi Reservoir (1942–1943)
16 Corpus Christi Reservoir (1934–1942)
17 Lexington Reservoir
18 Lloyd Shoals Reservoir
19 Lake Michie
20 Lake Issaqueena
21 Guernsey Reservoir
22 Arrowrock Reservoir
23 T. and P. Reservoir
24 Hiwassee Reservoir
24a Imperial Dam Reservoir (1938–1942)
24b Imperial Dam Reservoir (1943–1947)
25 Lake of the Ozarks
26 Pardee Reservoir
27 Possum Kingdom Lake
28 White Rock Reservoir
29 Buchanan Lake
30 Norris Reservoir
31 Senecaville Reservoir (1939–1943)
32 H; Lage Pond
33 Denison Reservoir
34 Lake Mead
35 San Carlos Reservoir
36 Conchas Reservoir
37 Fort Peck Reservoir
37a Elephant Butte Reservoir
38 All American Canal Desilting Basin
39 Hadley Creek New Desilting Basin
40 John Martin Reservoir
41 Senecaville Reservoir (1936–1939)

FIGURE 13-8
Reservoir trap efficiency as a function of capacity-inflow ratio. (*From* [17].)

By comparing sediment accumulation in reservoirs with estimates of sediment inflow based on measured sediment transport, Brune [17] derived a relationship between reservoir *trap efficiency*, the percent of incoming sediment retained in the reservoir, and the ratio of reservoir capacity to mean annual water inflow. A reservoir with a small capacity-inflow ratio will experience substantial spillway discharge, and much of the sediment load will be discharged. A reservoir with a very high capacity-inflow ratio may discharge very little water over the spillway and hence retain most of the inflowing sediment. Figure 13-8 can be used to estimate the fraction of the sediment inflow which is trapped. As the reservoir is filled with sediment, the trap efficiency will decrease so that it may be necessary to make the computation for short time intervals with appropriate adjustment of the trap efficiency. The curves apply to total load, and an increment of bed load should be added to suspended-load data.

The volume† occupied by the sediment in the reservoir will depend on the specific weight of the deposited material. The specific weight varies with the kind of sediment and the age of the deposits. Older sediments have more time to consolidate and are under a superimposed load from the more recent deposits. Lane and Koelzer [18] found that dry specific weight w_t at time t can be defined by

$$w_t = w_1 + K \log t \qquad (13\text{-}15)$$

where w_1 is the initial specific weight and K is a consolidation coefficient (Table 13-4). If a mixture of materials is present in the sediment, a weighted average specific weight should be calculated. Equation (13-15) applies to each annual accumulation of sediment, and the average weight of the total sediment accumulation at time t must be found by integrating from year 1 to year t. Table 13-5 presents average specific weights after 50 y used by the U.S. Soil Conservation Service for general design purposes.

† Gross volume of sediment is generally used in computing rate of filling of reservoirs, but part of the water in the pore spaces of the sediment may be recoverable.

Table 13-4 CONSTANTS IN EQ. (13-15) FOR THE SPECIFIC WEIGHT OF SEDIMENT IN LB/FT³ [18]. (FOR KG/M³ MULTIPLY BY 16.1.)

Reservoir condition	Sand		Silt		Clay	
	w_1	K	w_1	K	w_1	K
Sediment always submerged	93	0	65	5.7	30	16.0
Moderate reservoir drawdown	93	0	74	2.7	46	10.7
Considerable reservoir drawdown	93	0	79	1.0	60	6.0
Reservoir normally empty	93	0	82	0.0	78	0.0

Table 13-5 RANGE OF SPECIFIC WEIGHTS OF RESERVOIR SEDIMENTS AFTER 50-y ACCUMULATION [19]

Material	Permanently submerged		Aerated	
	lb/ft³	kg/m³	lb/ft³	kg/m³
Clay	40–60	640–960	60–80	960–1280
Silt	55–75	880–1200	75–85	1200–1360
Sand	85–100	1360–1600	85–100	1360–1600
Poorly sorted sand and gravel	95–130	1520–2080	95–130	1520–2080

REFERENCES

1. W. D. Ellison, Studies of Raindrop Erosion, *Agric. Eng.*, vol. 25, pp. 131–136, 181, 182, 1944.
2. W. H. Graf, "The Hydraulics of Sediment Transport," McGraw-Hill, New York, 1971.
3. P. du Boys, Le Rhône et les rivières à lit affouillable, *Ann. Ponts Chaussées*, ser. 5, vol. 18, pp. 141–145, 1879.
4. L. G. Straub, H.R. Doc. no. 238, 73d Cong., 2d Sess., p. 1135, 1935.
5. H. A. Einstein, The Bed-Load Function for Sediment Transportation in Open Channel Flows, *U.S. Dept. Agric. Soil Conserv. Serv. Tech. Bull.* 1026, 1950.
6. U.S. Interagency Committee on Water Resources, A Study of Methods Used in Measurement and Analysis of Sediment Loads in Streams, *Rep.* 14, Minneapolis, Minn., 1963.
7. B. R. Colby, Relationship of Unmeasured Discharge to Mean Velocity, *Trans. Am. Geophys. Union*, vol. 38, pp. 708–717, October 1957.
8. E. M. Schaak, Eine Vorrichtung zur Messung des Geschiebes an der Flußsohle, *Int. Assoc. Hydr. Struct. Res., 1st Congr., Berlin, 1937.*
9. G. B. Thorpe, A New Suspended Solids Recorder, *Ind. Electron.*, vol. 2, pp. 415–418, September 1964.
10. G. Fleming, Suspended Solids Monitoring: A Comparison between Three Instruments, *Water Water Eng.*, vol. 73, pp. 377–382, September 1969.
11. W. H. Jackson, An Investigation into Silt in Suspension in the River Humber, *Dock Harbour Auth.*, vol. 45, August 1964.
12. C. E. Murphree, G. C. Bolton, J. R. McHenry, and D. A. Parsons, Field Test of an X-Ray Sediment Concentration Gage, *J. Hydraul. Div. ASCE*, vol. 94, pp. 515–528, March 1968.
13. W. B. Langbein and S. A. Schumm, Yield of Sediment in Relation to Mean Annual Precipitation, *Trans. Am. Geophys. Union*, vol. 39, pp. 1076–1084, December 1958.

14. G. Fleming, Design Curves for Suspended Load Estimation, *Proc. Inst. Civ. Eng.*, vol. 43, pp. 1–9, 1969.
15. M. Negev, A Sediment Model on a Digital Computer, *Stanford Univ. Dept. Civ. Eng. Tech. Rep.* 76, March 1967.
16. U.S. Dept. of Agriculture, Summary of Reservoir Sediment Deposition Surveys Made in the United States through 1970, *Misc. Publ.* 1266, July 1973.
17. G. M. Brune, Trap Efficiency of Reservoirs, *Trans. Am. Geophys. Union*, vol. 34, pp. 407–418, June 1953.
18. E. W. Lane and V. A. Koelzer, Density of Sediments Deposited in Reservoirs, *U.S. Corps Eng. St. Paul Dist. Rep.* 9, 1943.
19. L. C. Gottschalk, Sedimentation, sec. 17 in V. T. Chow (ed.), "Handbook of Applied Hydrology," McGraw-Hill, New York, 1964.

BIBLIOGRAPHY

BOGARDI, J.: "Vizgolkasok Horddalekszallitasa," in Hungarian (Sediment Transport in Alluvial Streams), Akademiai Kiado, Budapest, 1970. (Translated into English by International Courses in Hydrology at Budapest.)

BROWN, C. B.: Sediment Transportation, chap. 12 in H. Rouse (ed.), "Engineering Hydraulics," Wiley, New York, 1949.

LEOPOLD, L. B., WOLMAN, M. G., and MILLER, J. P.: "Fluvial Processes in Geomorphology," Freeman, San Francisco, 1964.

TRASK, P. D.: "Applied Sedimentation," Wiley, New York, 1950.

PROBLEMS

13-1 Find the equation of a sediment rating passing through the points $q = 10$, $q_s = 4$ and $q = 1000$, $q_s = 8000$ as in Fig. 13-6.

13-2 (*a*) Find the equation applicable to the graph of Fig. 13-6 with sediment load expressed in acre-feet per day. Assume an in-place density of 100 lb/ft^3.

(*b*) The drainage area of the Powder River at Arvada is 6050 mi^2. Write the equation in terms of average depth of erosion in inches.

(*c*) What is the approximate extreme departure (error) of the plotted points from the line in percent?

13-3 A proposed reservoir has a capacity of 3000 acre-ft and a tributary area of 50 mi^2. If the annual streamflow averages 5 in. of runoff and the sediment production is 0.69 acre-ft/mi^2, what is the probable life of the reservoir before its capacity is reduced to 500 acre-ft? Use the median curve from Fig. 13-8. Repeat the computations using the two envelope curves.

13-4 A reservoir has a capacity of 50,000 acre-ft, and the annual inflow averages 78,000 acre-ft. The estimated sediment production of the area is 950 tons/mi^2, and the drainage area is 1120 mi^2. Sediment samples indicate that the grain-size distribution is 24 percent sand, 33 percent silt, 43 percent clay. When will 80 percent of the reservoir capacity be filled with sediment? Use line 3 of Table 13-4.

13-5 A reservoir has a capacity of 6 hm^3 and a drainage area of 200 km^2. The annual streamflow averages 350 mm of runoff, and the sediment production is 1100 metric tons per square kilometer. The sediment has an average in-place specific weight of 1500 kg/m^3. Using the median curve of Fig. 13-8, find the time required to reduce the reservoir capacity to 1 hm^3.

13-6 A reservoir has a capacity of 50 hm^3 and an annual inflow of 75 hm^3. Sediment production of the watershed is 275 metric tons per square kilometer. Area of the watershed is 1000 km^2. The grain-size distribution of the sediment is 8 percent sand; 41 percent silt; and 51 percent clay. The sediment is expected to be always submerged. How long will it take for the reservoir to become 80 percent filled with sediment?

14

MORPHOLOGY OF RIVER BASINS

Numerous references to the influence of the physical features of a watershed on its hydrologic response have been made in preceding chapters. Conversely, the hydrologic character of a watershed to a considerable degree shapes its physical character. One might assume that this interrelationship could be the basis for useful quantitative tools for predicting hydrologic response from measurable physical features. Some useful relations have evolved, but to date the results are more qualitative than quantitative. This chapter briefly summarizes some of the morphologic concepts that seem significant in hydrology and some of the more firmly established relationships that have been developed.

The inability to relate closely the physical and hydrologic features of watersheds stems from a number of causes. Accurate determination of physical characteristics is limited by available maps, which are of different scales and are compiled to differing cartographic standards, so that some physical descriptors may not be equivalent when determined from different maps. For other descriptors definitions must be arbitrary, leaving the possibility that a satisfactory definition has not yet been found. Finally, it is clear that the relationship between the essentially static physical features of a basin and the highly stochastic hydrologic characteristics are likely to be relatively complex. Hence, it is quite possible that the proper relationships have not yet been explored.

14-1 Physical Descriptors of Watershed Form

Many numerical measures describing various characteristics of a watershed and its stream system have been proposed. The following paragraphs summarize but a few which seem to have special relevance to hydrology.

Stream order Horton [1] suggested a classification of *stream order* as a measure of the amount of branching within a basin. A first-order stream is a small, unbranched tributary (Fig. 14-1). A second-order stream has only first-order tributaries. A third-order stream has only first- and second-order tributaries. The order of a particular drainage basin is determined by the order of the principal stream.

Order is extremely sensitive to the map scale used. A careful study of aerial photographs will often show three or four orders of streams (mostly ephemeral rills and channels) not indicated on a standard 1:24,000 scale topographic map. The 1:24,000 scale map will show one or two orders more than a 1:62,500 scale map. Even standard maps are not consistent in delineation of streams. Thus if order is to be used as a comparative parameter, it must be carefully defined. For some uses it may be desirable to make an adjustment to estimates of order on the basis of a detailed field survey of a few small tributary basins.

Horton also introduced the *bifurcation ratio* to describe the ratio of the number of streams of any order to the number in the next lowest order. The bifurcation ratios within a basin tend to be about the same magnitude. Generally, bifurcation ratios are found to be between 2 and 4 with a mean value near 3.5. This observation led to the *law of stream numbers*

$$N_u = r_b^{k-u} \tag{14-1}$$

where N_u is the number of streams of order u, r_b is the bifurcation ratio, and k is the order of the main stream. Similarly, Horton suggested the law of stream lengths.

$$\bar{L}_u = \bar{L}_1 r_e^{u-1} \tag{14-2}$$

where $\bar{L}$ is the average length of streams of order u and r_e is the length ratio. An equivalent equation also applies to the average area $\bar{A}$ of basins of order u

$$\bar{A}_u = \bar{A}_1 r_a^{u-1} \tag{14-3}$$

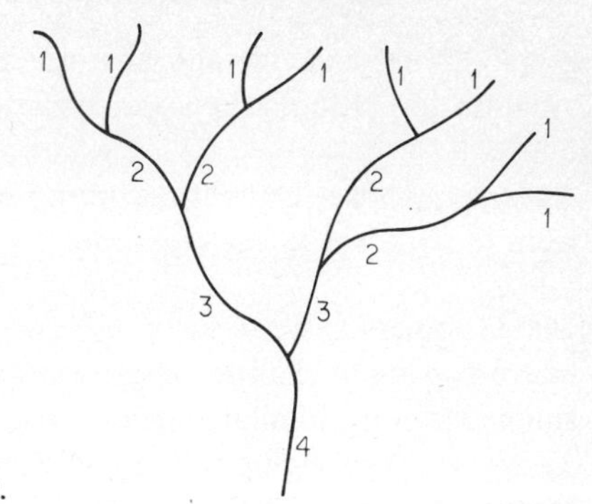

FIGURE 14-1
Definition sketch for stream order.

Equations (14-1) to (14-3) indicate a geometric progression of number, length, and area. Graphically, they suggest a linear plot of order number versus the logarithms of number, length, or area. The relationships have been confirmed under a wide range of conditions. The equations can be used by measuring N, L, and A for the two highest orders in the basin and then estimating these values for all lower orders.

Drainage density The total length of streams within a watershed divided by the drainage area defines the *drainage density*, the length of channels per unit area. A high drainage density reflects a highly dissected watershed, which should respond relatively rapidly to a rainfall input, while low drainage density reflects a poorly drained watershed with slow hydrologic responses. Observed values of drainage density range from as low as 3 in parts of the Appalachian area of the United States to 400 or more in Badlands National Monument, South Dakota. Low drainage densities are observed where soil materials are resistant to erosion or very permeable and where the relief is small. High values may be expected where soils are easily eroded or relatively impermeable, slopes are steep, and vegetal cover is scanty.

Length of overland flow The average length of overland flow $\bar{L}_o$ may be approximated by

$$\bar{L}_o = \frac{1}{2D} \qquad (14\text{-}4)$$

where D is the drainage density. This approximation ignores the effects of ground and channel slope, which make the actual overland path longer than the estimate from Eq. (14-4). The error is probably of little significance. Horton suggested that the denominator be multiplied by $\sqrt{1 - s_c/s_g}$, where

s_c and s_g are the average channel and ground slopes, respectively. This modification reduces the approximation inherent in Eq. (14-4).

Area relations Data for a number of the larger rivers of the world [2] seem to conform to the equation

$$L = 1.4A^{0.6} \qquad (14\text{-}5)$$

where L is main-channel length (mi) and A is drainage area (mi^2). Data by Hack [3] show a similar relation. The exponent generally seems to be between 0.6 and 0.7, suggesting that a basin tends to elongate as it grows larger. The coefficient in Eq. (14-5) becomes 1.27 with dimensions in kilometers.

Basin shape The shape of a watershed affects the streamflow hydrograph and peak-flow rates. Numerous efforts to develop a factor which describes basin shape by a single numerical index have been reported. Watersheds tend to the form of a pear-shaped ovoid, but geologic controls result in many substantial deviations from this shape. Horton [4] suggested the dimensionless form factor R_f as an index of shape, or

$$R_f = \frac{A}{L_b^2} \qquad (14\text{-}6)$$

where A is basin area and L_b is the length of the basin measured from outlet to divide near the head of the longest stream along a straight line. This index or its reciprocal has been used as an indicator of unit hydrograph shape.

Equation (14-6) implies no special assumption of basin shape. For a circle $R_f = \pi/4 = 0.79$; for a square with the outlet at the midpoint of one side $R_f = 1$; and for a square with the outlet at one corner $R_f = 0.5$. Values for other geometrical shapes are easily calculated. Numerous writers have suggested the use of a circle [5] or lemniscate [6] as a reference shape. The resulting shape factors reduce to Horton's R_f multiplied by a constant.

14-2 Descriptors of Watershed Relief

The topography or relief of a watershed may have more influence on its hydrologic response than catchment shape, and numerous descriptors of relief have been advanced by various writers. Some of the more useful descriptors are discussed in this section.

Channel slope The slope of a channel affects velocity of flow and must play a role in hydrograph shape. Typical channel profiles (Fig. 14-2) are concave upward. In addition all but the very smallest watersheds contain several channels, each with its own profile. Thus the definition of average

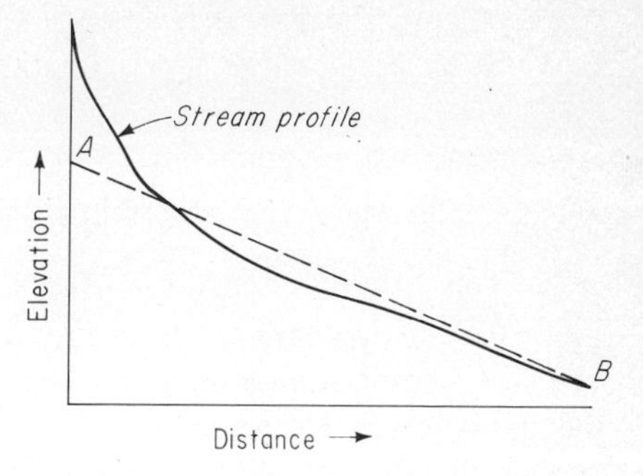

FIGURE 14-2
One method of defining mean channel slope.

channel slope of a watershed is difficult. Commonly, only the main stream is considered in describing the channel slope of a watershed. A simple and widely used measure of channel slope is the slope of a line (*AB*, Fig. 14-2) drawn such that the area under it equals the area under the main-channel profile. Taylor and Schwarz [7] calculated the slope of a uniform channel having the same length and time of flow as the main channel. Since the velocity is proportional to square root of slope, the procedure used by Taylor and Schwarz is equivalent to weighting channel segments by the square root of their slope, which gives relatively less weight to the steep upstream reaches of the stream. Thus, if the channel were divided into n equal segments, each of slope s_i, a simple index of slope would be

$$R_s = \left(\frac{\sum_{i=1}^{i=n} \sqrt{s_i}}{n} \right)^2 \qquad (14\text{-}7)$$

Land slope The slope of the ground surface is a factor in the overland-flow process and hence a parameter of hydrologic interest, especially on small watersheds, where the overland-flow process may be a dominant factor in determining hydrograph shape. Because of the considerable variation in land-surface slope in the usual watershed, a method of defining an average or index value is required.

The distribution of land-surface slope can be determined by establishing a grid or a set of randomly located points over a map of the watershed. The slope of a short segment of line normal to the contours is determined at each grid intersection or random point. The mean, median, and variance of the resulting distribution can be calculated. The accuracy of the results depends on the adequacy of the map used.

Area-elevation data When one or more of the factors of interest in a hydrologic study vary with elevation, it is useful to know how the watershed

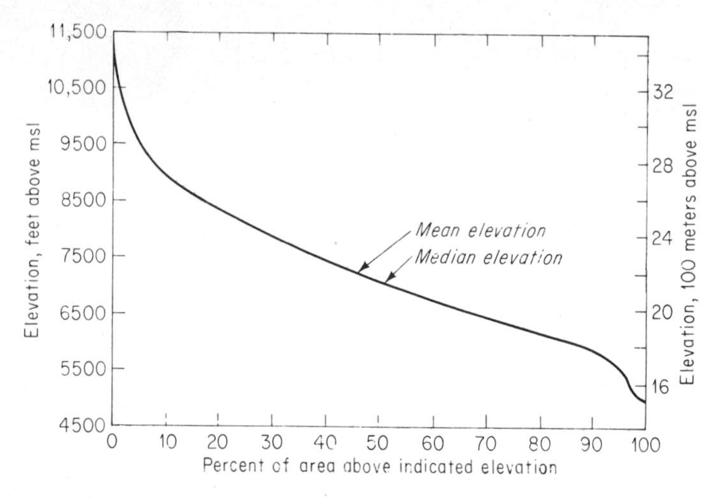

FIGURE 14-3
Area-elevation curve for the Teton River watershed above St. Anthony, Idaho.

area is distributed with elevation. An area-elevation (or *hypsometric*) curve can be constructed by planimetering the area between contours on a topographic map and plotting the cumulative area above (or below) a given elevation versus that elevation (Fig. 14-3). In some cases it is convenient to use percentage of area instead of actual area, particularly if a comparison between watersheds is desired. If a grid with about 100 or more intersections is superimposed on the watershed and the number of grid intersections in each elevation range is noted, an area-elevation curve can be constructed which will be about as accurate as one derived by planimetry and the effort may be substantially less.

The curve of Fig. 14-3 is typical of area-elevation curves of geologically mature watersheds. Very small watersheds, however, may show quite different characteristics. Snowmelt computations in mountainous areas must usually be made for elevation zones because both snow depth and temperature are functions of elevation. Precipitation in mountainous areas may sometimes be weighted by elevation in calculations of basin average precipitation.

Aspect The aspect of a slope is the direction toward which the slope faces; i.e., it is the bearing of a line normal to the slope read in the direction away from the slope. Because of the influence of insolation on snowmelt, aspect may be of interest in dealing with snow. Precipitation amounts are often influenced by the aspect of a slope relative to the direction of the wind. Customarily, aspect is used as a characteristic of a particular point or at most of a specific hillside. The distribution of aspect may be determined in manner similar to that described for land slope. A grid or set of random points is superimposed over a map of the watershed, and the bearing of a line normal to the contours at each intersection is recorded. From these data a polar diagram can be plotted with aspect shown as angle from north and percentage area as distance from the origin. Typically, such plots appear nearly circular, indicating a uniform distribution of aspect. However, very small basins may show a substantial departure from uniformity.

14-3 Hydraulic Geometry

Hydraulic geometry describes the character of the channels of a basin: the variation of mean depth, width, and velocity at a particular cross section and between cross sections. These relations apply to alluvial channels, where the cross section is readily adapted to the flows which occur, but are less reliable where rock outcrops control the channel characteristics.

The basic equations of hydraulic geometry are [8]

$$w = aq^b \tag{14-8}$$

$$d = cq^f \tag{14-9}$$

$$v = kq^m \tag{14-10}$$

where q is discharge, w is channel width, d is mean depth, v is mean velocity, and a, b, c, f, k, and m are numerical coefficients. Since $q = wdv$, it follows that $ack = 1$ and $b + f + m = 1$. The equations plot as straight lines on log-log plots, the exponents representing the slope of the lines and the coefficients the intercept when $q = 1$. Plotting data for a fairly large number of streams indicates marked regional conformity in the values of the exponents, and agreement between regions is close enough to suggest the possibility of some degree of universality to the values.

Table 14-1 presents values of the exponents b, f, and m determined by various investigators. At-station values describe the variation of w, d, and v with q at a given cross section, while the between-station values reflect the change in the values of w, d, and v as q increases in the downstream direction along a stream. Figure 14-4 illustrates the interrelation between the two sets of equations. In Fig. 14-4 the solid lines represent the between-station relations and the dashed lines the at-station relations with slopes plotted at approximately the mean values as given in Table 14-1. Note that Eq. (14-9) is the equation of a rating curve and that the exponent f remains

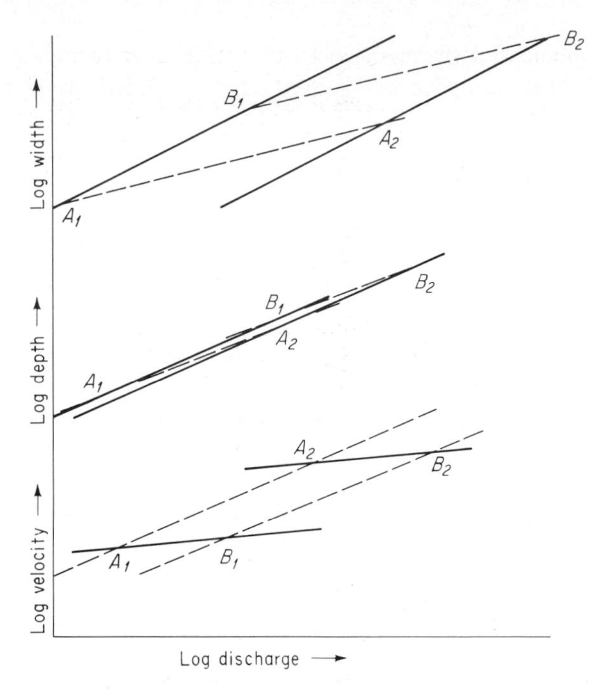

FIGURE 14-4
Interrelationships in hydraulic geometry between stations A and B and between high and low flows at each station.

approximately the same at a station and between stations. The width, however, tends to increase more rapidly in a downstream direction than it does at a station, while velocity increases rapidly with q at a station but remains relatively constant in the downstream direction.

To be meaningful, the between-station relations must be defined for flows of the same frequency. Most of the data have been derived for a flow rate equal to the mean annual discharge, which represents a flow at about one-third bankfull depth and is equaled or exceeded about 25 percent of the time (about 90 d/y). However, the bankfull flow with a return period of about 1.5 y and the 10-y flow have also been used.

Much additional work needs to be done to establish the quantitative relations of hydraulic geometry and to explain fully the reasons underlying the relations. In part, they are related to the sediment-transport characteristics of the stream [12]. However, it is clear that for most streams a reasonable relationship can be derived and used, for example, to supplement cross-section data as a basis for routing flow from the watershed.

14-4 Stream Patterns

When viewed in plan, stream channels may be described as meandering, braided, or straight. A *meandering stream* flows in large, more or less symmetrical loops, or bends (Fig. 14-5). The median length of meandering streams appears to be about 1.5 times the valley length [13]; i.e., the *sinuosity* averages about 1.5. The wavelength of meanders ranges from 7 to 11 times the channel width, and the radius of curvature of the bend usually ranges between 2 and 3 times the channel width. The amplitude of the meanders or width of the meander belt varies considerably and seems to be controlled more by the characteristics of the bank material than other factors. Amplitude usually ranges from 10 to 20 times channel width.

A *braided* stream (Fig. 14-6) consists of many intertwined channels (*anabranches*) separated by islands. Braided streams tend to be very wide and relatively shallow with coarse bed material. No formal statements about the geometry of braided streams are possible. Few long straight channels exist in nature, but many lack sufficient curvature to be called a meandering stream. A *straight* stream is commonly defined as one with a sinuosity of less than 1.25.

The important question for the engineering hydrologist is to explain why a channel adopts one of the patterns described above. Braided channels are usually found in reaches where the banks are easily erosible—sandy material with little vegetal protection. Bed material is relatively coarse and of heterogenous particle sizes. The slope of the braided reach is greater than that of adjacent unbraided reaches. Hydraulically, the braided reach is less efficient than the unbraided reach. The total width of branches in a braided reach may be 1.5 to 2 times that of an undivided channel, and the depth of

Table 14-1 VALUES OF THE EXPONENTS IN EQS. (14-8) TO (14-10)

	At station			Between stations		
	b	f	m	b	f	m
Average, midwestern states [8]	0.26	0.40	0.34	0.5	0.4	0.1
Ephemeral streams in semi-arid United States [9]	0.29	0.36	0.34	0.5	0.3	0.2
Average, 158 United States stations [2]	0.12	0.45	0.43			
10 stations on the Rhine River	0.13	0.41	0.43			
Appalachian streams [10]	...	...	...	0.55	0.36	0.09
Kaskaskia River, Ill. [11]	...	...	...	0.51	0.39	0.14

FIGURE 14-5
A meandering stream. (*U.S. Geological Survey.*)

flow is correspondingly less. Braiding is thus a way of dissipating energy when stream slope steepens. Velocity increases that would otherwise lead to erosion are thus avoided.

An initially straight alluvial channel, either in a laboratory flume or in the field, will usually develop meanders as water flows through it. Friedkin [14] feels that the only requirement for meandering is erosible banks. Others are less sure of the mechanism, but whatever it is, meanders also serve to dissipate energy. A meandering channel may be 1.5 to 2 times as long as a nonmeandering channel, Its slope is correspondingly reduced, but head losses are increased both because of the longer channel length and because of the bend losses. Without these losses, velocities would be higher, with corresponding tendency to downcut the channel. Many meandering streams cannot downcut because they discharge into a water body with fixed elevation. If downcutting cannot occur, some other device is required to dissipate the available energy.

Thus both braiding and meandering can be explained as means of energy dissipation. Braiding will occur when bed material is coarse and heterogenous and banks very easily erosible. Meandering is likely to occur

FIGURE 14-6
A braided stream. (*U.S. Geological Survey.*)

on flatter slopes where the material is finer and the banks somewhat more cohesive.

In either case the stream is in a kind of equilibrium—equilibrium in the sense that it will maintain its grade but obviously not in the sense that there will be no channel changes. In the braided channel there is continual shifting and changing between individual anabranches, and the meanders undergo a more or less continuous process of erosion in the concave bend and deposition at the subsequent point bar, so that the meanders seem to be constantly moving downstream. Any attempt by man to change the natural pattern of a stream requires careful planning and usually costly revetment work to prevent erosion of banks and return to the original pattern.

Yang [15] defines *unit stream power* as the time rate of expenditure of potential energy per unit weight of water. Thus

$$\frac{dy}{dt} = \frac{dx}{dt}\frac{dy}{dx} = vs \qquad (14\text{-}11)$$

where y is elevation above datum (equals potential energy), x is distance, t is time, v is mean velocity, and s is energy slope. The law of least time rate of energy expenditure requires that unit stream power decrease in a downstream direction [16]. Meanders, riffles and pools, and braided streams all seem to conform to this law.

14-5 Floodplains

The floodplain of a river is the valley floor adjacent to the incised channel, which may be inundated during high water. The river tends to swing back and forth across the valley bottom, reworking the floodplain deposits and eroding first one valley side and then the other. Floodplains are built up primarily from deposition of sediment in the river channel and deposition of fine sediments on the floodplain when flooded. Additionally, organic materials may accumulate in cutoff meander loops (*oxbow lakes*). Often a natural levee will form along the banks of the incised channel caused by the deposit of coarse sediment as the water from the stream invades the floodplain. Sediment deposition in the channel plus natural levees on the bank can lead to a situation in which the stream flows at a higher elevation than its floodplain. This condition develops quite frequently in streams flowing across an alluvial cone.

As suggested earlier, floodplains tend to be flooded at fairly low recurrence intervals. Leopold et al. [17] report numerous evaluations of the flow magnitude required to overflow the floodplain. Return periods generally range between 1 and 2 y, and a general statement that floodplains of the eastern and central United States are inundated by floodwaters in 2 out of 3 y is quite reasonable. It is not clear whether this is a general rule or not. Nixon [18] made a similar analysis of British streams and found that flooding occurred on the average about twice each year. There is some difficulty in defining precisely what the floodplain is and a problem in defining precisely the bankfull stage. Thus it is not clear whether Nixon's data indicate conditions substantially different from those in the United States. In any case it is clear that the floodplain is subject to frequent flooding, and hence its use for buildings and other purposes should be carefully regulated. Transverse slope of a floodplain is usually quite small, and it is often difficult to detect natural levees by visual inspection. Consequently in studies for which the floodplain characteristics are important, one must have either adequately detailed maps or special field surveys which satisfactorily define the information needed.

REFERENCES

1. R. E. Horton, Erosional Development of Streams, *Geol. Soc. Am. Bull.*, vol. 56, pp. 281–283, 1945.
2. L. B. Leopold, M. G. Wolman, and J. P. Miller, "Fluvial Processes in Hydrology," p. 145, Freeman, San Francisco, 1964.
3. J. T. Hack, Studies of Longitudinal Stream Profiles in Virginia and Maryland, *U.S. Geol. Surv. Prof. Pap.* 294-B, pp. 45–97, 1957.

4. R. E. Horton, Drainage Basin Characteristics, *Trans. Am. Geophys. Union*, vol. 13, pp. 350–361, 1932.
5. S. A. Schumm, Evolution of Drainage Systems and Slopes in Badlands at Perth Amboy, New Jersey, *Bull. Geol. Soc. Am.*, vol. 67, pp. 597–646, 1956.
6. R. J. Chorley, D. E. G. Malm, and H. A. Pogorzelski, A New Standard for Estimating Drainage Basin Shape, *Am. J. Sci.*, vol. 255, pp. 138–141, 1957.
7. A. B. Taylor and H. E. Schwarz, Unit-Hydrograph Lag and Peak Flow Related to Drainage Basin Characteristics, *Trans. Am. Geophys. Union*, vol. 33, pp. 235–246, April 1952.
8. L. B. Leopold and T. G. Maddock, The Hydraulic Geometry of Stream Channels and Some Physiographic Implications, *U.S. Geol. Surv. Prof. Pap.* 252, 1953.
9. L. B. Leopold, Downstream Change of Velocity of Rivers, *Am. J. Sci.*, vol. 251, pp. 606–624, 1953.
10. L. M. Bush, Jr., Drainage Basins, Channels and Flow Characteristics of Selected Streams in Central Pennsylvania, *U.S. Geol. Surv. Prof. Pap.* 282-F, pp. 145–181, 1961.
11. J. B. Stall and Y. S. Fok, Hydraulic Geometry of Illinois Streams, *Univ. Ill. Water Resour. Cent. Res. Rep.* 15, July 1968.
12. See [2], pp. 248–291.
13. See [2], p. 296.
14. J. F. Friedkin, A Laboratory Study of the Meandering of Alluvial Rivers, *U.S. Waterw. Eng. Exp. Stn.*, 1945.
15. C. T. Yang, Potential Energy and Stream Morphology, *Water Resour. Res.*, vol. 7, pp. 311–322, April 1971.
16. C. T. Yang, On River Meanders, *J. Hydrol.*, vol. 13, pp. 231–253, July 1971.
17. See [2], pp. 319–322.
18. M. Nixon, A Study of the Bankfull Discharges of Rivers in England and Wales, *Proc. Inst. Civ. Eng.*, pp. 157–174, 1959.

BIBLIOGRAPHY

BLENCH, T.: "Regime Behaviour of Rivers and Canals," Butterworth's, London, 1957.

LELIAVSKY, S.: "Introduction to Fluvial Hydraulics," Constable, London, 1955.

LOBECK, A. K.: "Geomorphology: An Introduction to the Study of Landscapes," McGraw-Hill, New York, 1939.

NĚMEC, J.: "Engineering Hydrology," McGraw-Hill, London, 1972.

STRAHLER, A. N.: Dynamic Basis for Geomorphology, *Bull. Geol. Soc. Am.*, pp. 923–928, 1952.

STRAHLER, A. N.: Quantitative Geomorphology of Drainage Basins and Channel Networks, sec. 4-II in V. T. Chow (ed.), "Handbook of Applied Hydrology," McGraw-Hill, New York, 1964.

PROBLEMS

14-1 Secure a topographic map (preferably 1:24,000 scale) and define the watershed of a small stream on it. From the map determine the stream order, bifurcation ratio, length ratio, and area ratio. Does the stream appear to conform to Eqs. (14-1) to (14-3)? If your classmates used other streams in the same area, is there any regional consistency in the ratios?

14-2 If it is possible to obtain a 1:62,500 or other small-scale map of the same area you used in Prob. 14-1, repeat the analysis. What differences in values occur when the map scale changes?

14-3 For the watershed of Prob. 14-1 (or similar) determine the drainage density and length of overland flow.

14-4 Collect data on area and channel length for at least 10 streams. Do these data appear to conform to Eq. (14-5)? If not, can you suggest why they do not conform?

14-5 For the watershed of Prob. 14-1 (or other similar watershed) determine average channel slope as given by line *AB*, Fig. 14-2, and by Eq. (14-7). How well do you think these estimates actually represent the true slope of the stream?

14-6 Construct an area-elevation curve for the watershed you studied in Prob. 14-1 or other similar watershed.

APPENDIX A

GRAPHICAL CORRELATION

Hydrology is largely an empirical science, and a preponderance of the problems confronting the hydrologist involves correlation analysis or the application of a relation derived through such analysis. Most computers have statistical programs which perform least-squares and other types of analytical correlations very rapidly. Usually, however, a mathematical correlation requires that the form of the relationship (linear, exponential, additive, multiplicative, etc.) be assumed in advance, and in many hydrologic studies it is difficult to specify the form in advance. In some cases, it is most helpful to the analyst to study the data point by point as he plots a graphical relationship. Hence, a brief discussion of methods of graphical correlation is included in this text.

A-1 Two-Variable Correlation

If a linear relation is to be used, the line of best fit must pass through the point defined by the means ($\bar{X}$ and $\bar{Y}$) of the observed values of the two variables X and Y. This is true not only for graphical correlation but also for the least-squares line. With one point $(\bar{X},\bar{Y})$ on the line determined, the proper slope can be estimated by first plotting the data (Fig. A-1) and then determining average co-ordinates for groups of points. The selected groups should comprise all points

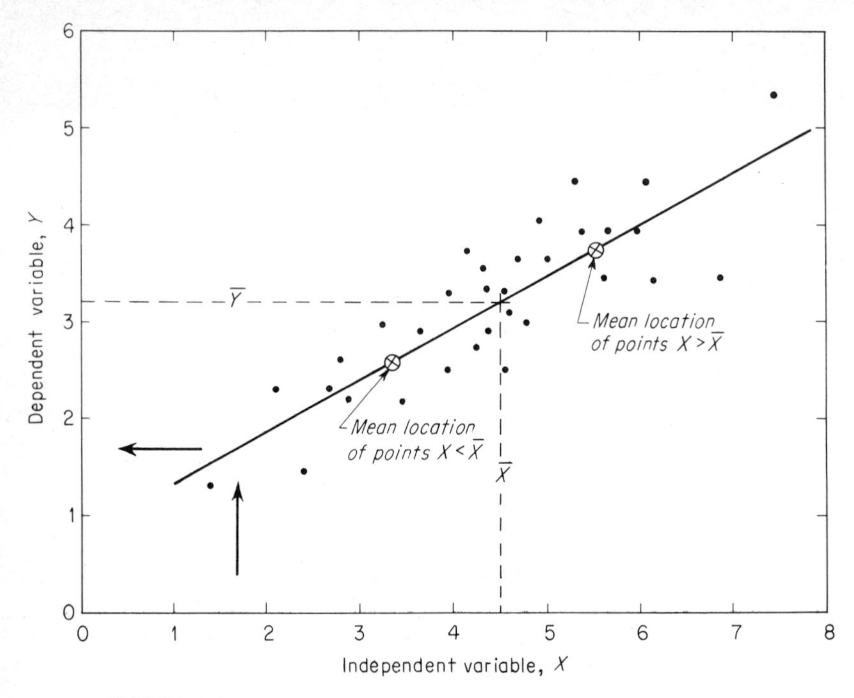

FIGURE A-1
Two-variable linear correlation by graphical method.

falling within specified values of the independent variable (X) irrespective of the value of Y, the factor to be estimated. If the total number of points is divided into two groups of approximately equal size, then the line connecting their means will pass through the mean of all points and this is the best line that can be readily determined graphically. The relation determined by *group averages* usually has a slightly steeper slope (dy/dx) than that determined by least squares. As the degree of correlation increases, the difference between the two lines diminishes, and for perfect correlation they are coincident. The relation of averages tends to minimize the absolute sum of the deviations, while the least-squares relation minimizes the sum of the squares of the deviations.

The group averages can be estimated graphically by first estimating successive two-point averages (halfway between the plotted points). The four-point averages are then halfway between the two-point averages, etc. (Fig. A-2). The points should be grouped with respect to values of the independent variable. Unless the correlation is perfect, a different line will result if points are grouped according to the dependent variable, the difference increasing as the correlation decreases.

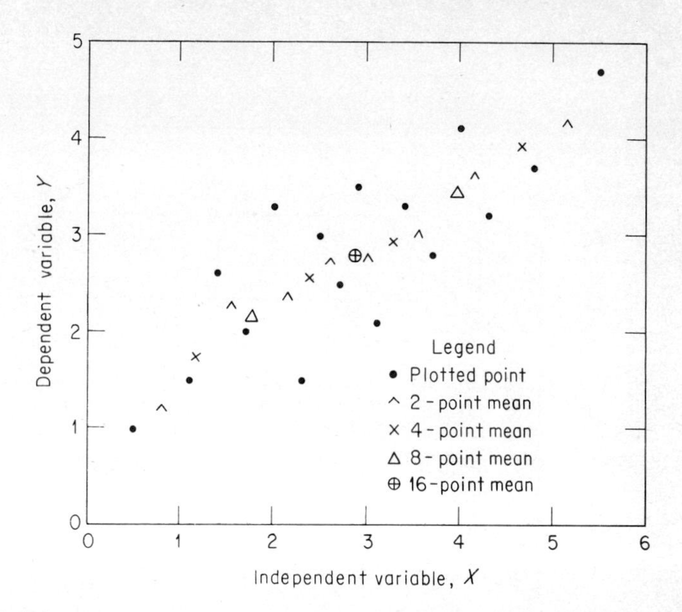

FIGURE A-2
Graphical determination of group means of points classified with respect to the independent variable X.

If upon examination of the group averages (two-point, four-point, etc.) it is decided that the relation is curvilinear in form, then a curve can be fitted to the group means. A curve does not necessarily pass through the mean of all the data.

A-2 Three-Variable Correlation

The most logical method of presenting a three-variable relation is by means of a three-dimensional sketch (Fig. A-3). The *contours* of X_3, that is, curves connecting equal values of X_3, are parallel straight lines and are equally spaced for equal increments of X_3. This is true of any plane surface.

Given the values of any two of the variables, the third can be estimated from the chart of Fig. A-3. A close examination of the figure will disclose that once the three families of lines have been constructed on the surface *ABCD* represented by the equation, all remaining portions of the sketch become superfluous. In practice, the surface *ABCD* is projected onto one of the coordinate planes and shown by a family of curves on cross-section paper.

A chart similar to *ABCD* of Fig. A-3 can be developed from a series of simultaneous observations of three related variables. By plotting values of Y versus X_1 and labeling the points with corresponding values of X_2, contours of X_2 can be fitted in much the same manner as elevation contours are drawn in the preparation

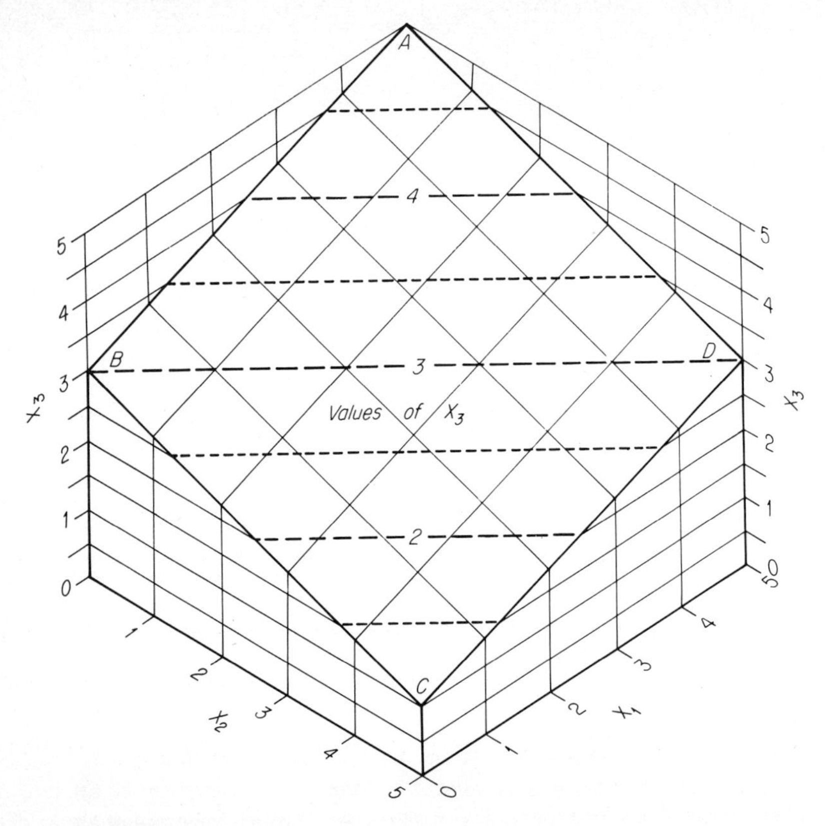

FIGURE A-3
Three-dimensional figure for the solution of $X_1 = X_2 + 2.5X_3 - 7.5$

of a topographic map. In correlation analysis, however, we are attempting to derive a logical relationship between three variables from observations subject to random error. Moreover, the true relationship frequently involves additional variables. For these reasons, no attempt is made to fit each observation exactly. If the degree of correlation displayed by the plotted data is not particularly high, averages of points grouped with respect to the independent variables (X_1 and X_2) should be determined before any attempt is made to construct X_2 contours (Fig. A-4). The location of the mean points is determined graphically, and the label is the average of the X_2 values for the group.

Sharp discontinuities, in either spacing or curvature of the lines, should not be embodied in a family of curves based on limited data unless they can be logically explained. The complexity of the derived curve family should be held to a minimum

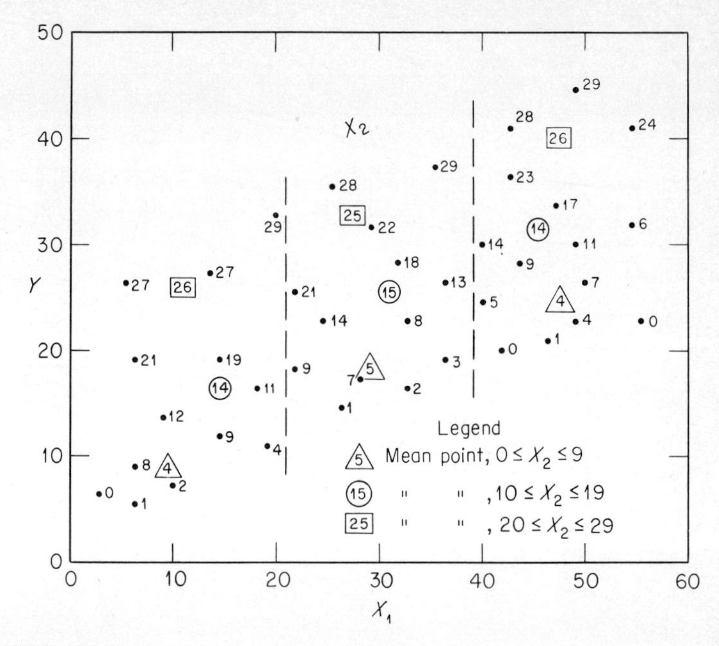

FIGURE A-4
Graphical determination of group means of points classified with respect to two independent variables X_1 and X_2.

consistent with available data and a priori knowledge of the relationship. With the general character of the curve family decided upon, the method [1] of construction (Fig. A-5) is as follows:

1 Estimate the optimum position of the drafting curve for various X_2 curves, noting in each case the location of the hole in the drafting curve (or any other identifiable point) and the value of Y for some fixed value of X_1.
2 Fit a smooth curve (A) through the points defined by the locations of the hole in the drafting curve.
3 Replot the points used to define curve A by shifting them horizontally to the proper value of X_2 on the auxiliary scale (for example, $X_2 = 48$ and $Y = 68$) and fit construction curve B. Curves A and B together fix the position of the hole in the drafting curve for any desired X_2 curve.
4 Plot the values of Y (for $X_1 = 0$) derived in step 1 at the corresponding values of X_2 on the auxiliary scale and fit construction curve C. This curve fixes the orientation of the drafting curve for any selected X_2 curve.
5 Construct selected X_2 curves as illustrated for $X_2 = 10$. The hole in

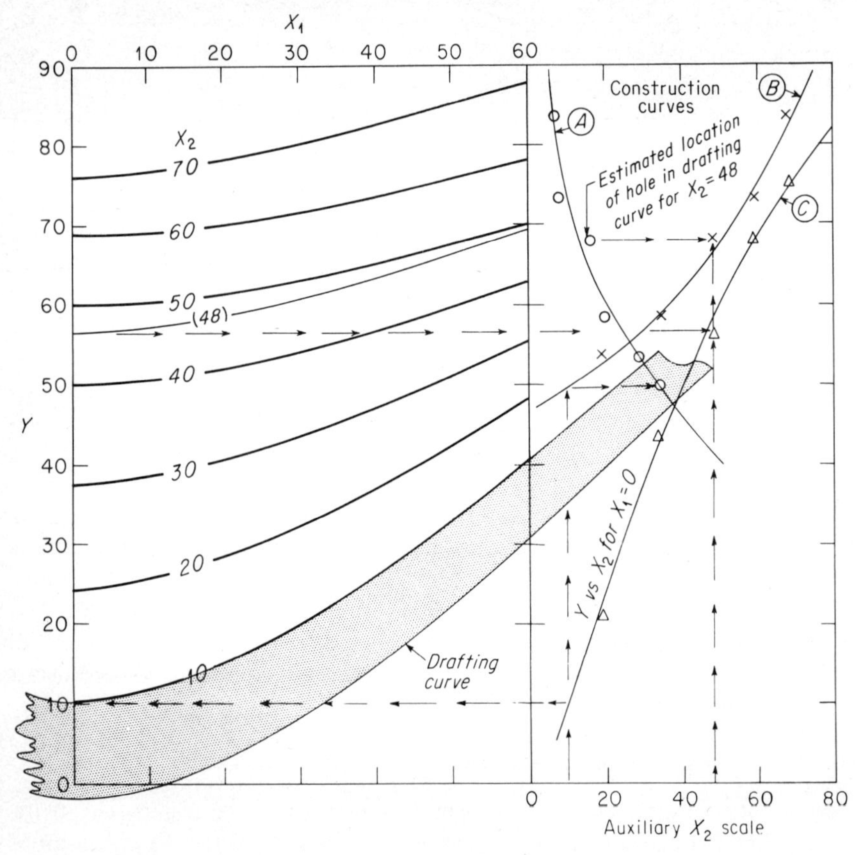

FIGURE A-5
Construction of a smooth-curve family.

the drafting curve is placed on curve A at a value of Y corresponding to $X_2 = 10$ on curve B. The drafting curve is then rotated to fit the point $X_1 = 0$ and Y corresponding to $X_2 = 10$ on curve C.

If the curve family were to consist of equally spaced straight parallel lines, construction curve C (also a straight line) would determine the spacing, and the line for $\bar{X}_2$ drawn through the point $(\bar{X}_1, \bar{Y})$ would determine the slope. If variable spacing and/or convergence were to be incorporated into a family of straight lines, two construction curves analogous to curve C (for two selected values of X_1) would be required.

Any three-variable relation can be expressed by a chart of the type just discussed. For linear functions, the contours are equally spaced, straight, and parallel; for curvilinear functions, they may be curved, unequally spaced, or both. Contours which converge indicate a *joint function*, i.e., a function in which the effect of one independent variable is dependent upon the value of a second (for example, $Y = X_1X_2$).

A-3 Coaxial Correlation of Four or More Variables

The development of a runoff relation by coaxial correlation is described in Sec. 8-6. The treatment of the method in this section is general in character, but detailed plottings and tabular data are provided. Although minor duplication between the two sections is intentional, they are largely supporting and should be studied jointly. The *coaxial method of graphical correlation* [1] is based on the premise that if any important factor is omitted from a relation, the scatter of points in a plotting of observed values of the dependent variable versus values computed from the relation will be at least partially explained by the omitted factor. In other words, if the points on such a plotting are labeled with the corresponding values of the new factor, a family of curves can be drawn to modify the values computed from the original relation. It will be seen that a coaxial relation is in reality a series of three-variable relations arranged with common axes to facilitate plotting and computing (Figs. 8-5 and A-6).

For purposes of illustration, assume that we wish to develop a relation for predicting Y from X_1, X_2, and X_3 and that the available data have been tabulated as shown in Table A-1. The analysis proceeds as follows:

1 For the entire series of observations, plot Y against X_1 and label each point with the value of X_2, as shown in chart *A* of Fig. A-6. Fit a family of curves to the plotted data (Sec. A-2). In this example it appeared that straight, parallel lines would adequately fit the data; more complex functions should be used only when clearly indicated or when they can be logically explained.

2 Next plot observed values of Y against those computed from chart *A* (entering with X_1 and interpolating within the X_2 curves); label each point with the value of X_3, as shown in chart *B* of Fig. A-6; and again fit a family of curves.

3 Entering with X_1, X_2, and X_3 in sequence (Fig. A-6), compute the value of Y for each observation in the series and tabulate the corresponding error (Table A-1).

If additional variables were to be introduced, step 2 of the process would be repeated in essence. In each case, Y as computed from the derived chart sequence would be plotted against observed values of Y, and the points would be labeled with values of the variable to be introduced.

Since chart *A* of Fig. A-6 was derived without consideration of X_3, the subsequent introduction of this factor may necessitate a revision of chart *A*, particularly

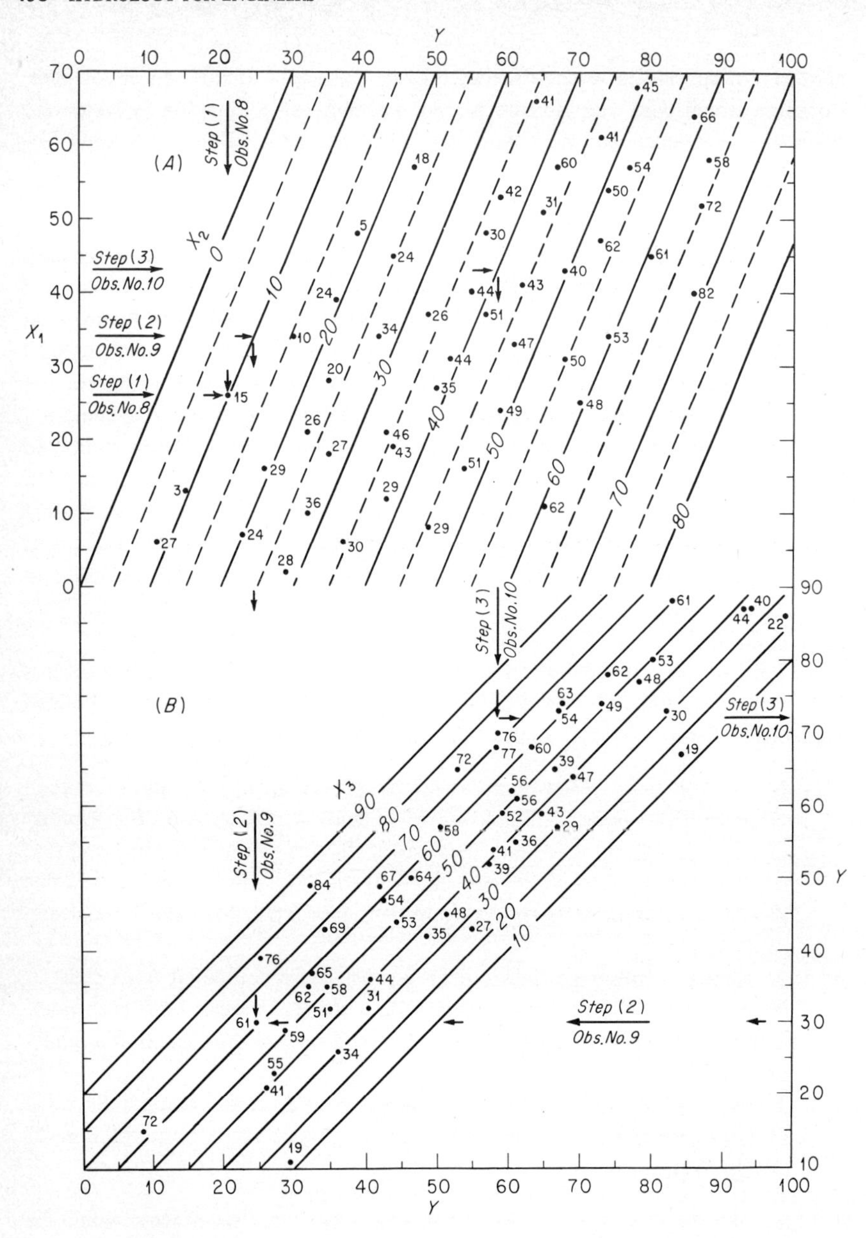

FIGURE A-6
Development of first approximation curves by coaxial method (data from Table A-1).

if appreciable correlation exists between X_3 and either X_1 or X_2. In other words, the coaxial technique usually requires two or more successive approximations to achieve best results. Although several techniques are available [2, 3], only one method of developing second and succeeding approximations is discussed here. With this method the analysis continues as follows:

4 Reconstruct the first-approximation X_3 curves (Fig. A-6) on a second sheet of cross-section paper, as in chart *B* of Fig. A-7.

5 Next, in chart *A* of Fig. A-7, plot values of X_1 against coordinate values derived by entering chart *B* with observed *Y* (on the vertical scale) and X_3, labeling each point with X_2. Fit the second-approximation X_2 curves.

6 In chart *B* (Fig. A-7), plot observed values of *Y* against those computed from chart *A* in exactly the same manner as under step 2 above. Revise the

Table A-1 DATA FOR COAXIAL-CORRELATION ILLUSTRATION (FIGS. A-6 AND 7)

Observation No.	Y	X_1	X_2	X_3	Absolute error: First approximation	Absolute error: Second approximation	Observation No.	Y	X_1	X_2	X_3	Absolute error: First approximation	Absolute error: Second approximation
1	88	58	58	61	1	0	27	44	45	24	53	2	1
2	32	21	26	51	3	1	28	78	68	45	62	2	2
3	74	34	53	63	0	0	29	50	27	35	64	4	2
4	73	47	62	30	1	1	30	36	39	24	44	2	0
5	86	40	82	22	1	0	31	54	16	51	41	0	0
6	29	2	28	59	4	0	32	64	66	41	47	4	4
7	35	18	27	58	4	1	33	61	33	47	56	3	3
8	21	26	15	41	1	3	34	62	41	43	56	1	1
9	30	34	10	61	0	2	35	65	11	62	39	4	1
10	68	43	40	77	4	4	36	68	31	50	60	0	0
11	57	48	30	58	2	2	37	55	40	44	36	1	1
12	67	57	60	19	2	2	38	35	28	20	62	3	1
13	87	64	66	44	3	2	39	59	53	42	43	2	3
14	49	8	29	84	0	3	40	73	61	41	54	4	3
15	65	51	31	72	1	1	41	32	10	36	31	1	3
16	70	25	48	76	2	1	42	39	48	5	76	1	0
17	15	13	3	72	5	2	43	74	54	50	49	1	1
18	52	31	44	39	0	1	44	45	19	43	48	5	4
19	80	45	61	53	2	2	45	43	21	46	27	0	0
20	26	16	29	34	2	1	46	49	37	26	67	2	0
21	37	6	30	65	3	1	47	87	52	72	40	2	2
22	11	6	27	19	3	1	48	42	34	34	35	1	2
23	23	7	24	55	7	2	49	47	57	18	54	2	1
24	59	24	49	52	1	1	50	43	12	29	69	1	2
25	57	37	51	29	1	0	Sum	...	...	...	...	101	71
26	77	57	54	48	0	0	Mean	...	...	...	...	2.0	1.4

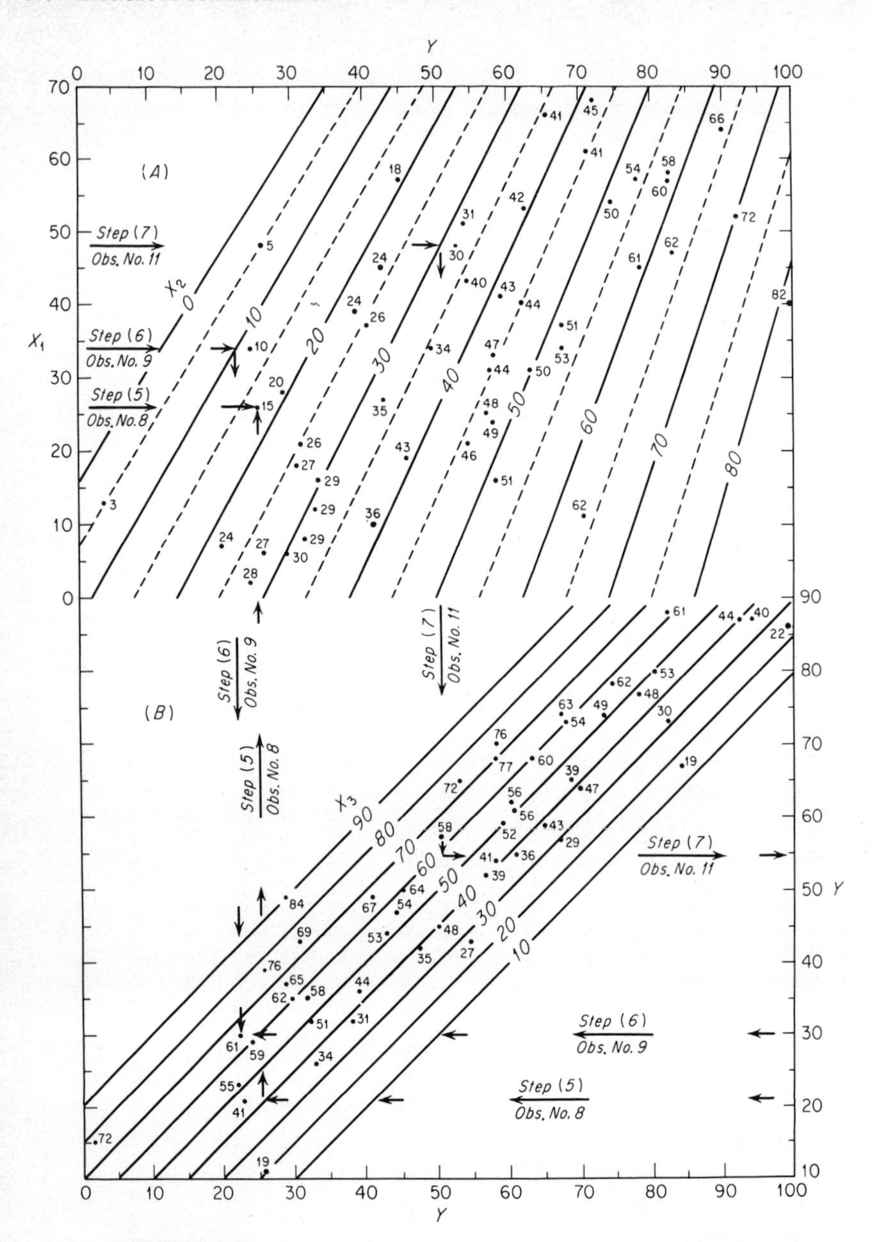

FIGURE A-7
Development of second approximation curves by coaxial method (data from Table A-1).

X_3 curves as indicated by the plotted points. In this example it was found that the first-approximation X_3 curves were entirely satisfactory.

7 Repeat step 3 above, using the second-approximation curves of Fig. A-7, and tabulate the corresponding error (Table A-1). If any material change is made in any of the curve families, the average error of the second approximation should be appreciably smaller than that of the first approximation. To avoid possible bias in the final relation that might result from inconsistent curve fitting, the algebraic sum of the errors should be computed. If this sum is not essentially zero, one of the curve families should be shifted as required.

Third and subsequent approximations are derived by simple repetition of steps 4 to 7. If the analysis involved five or more variables, there would be three or more charts in the sequence. In any event, the second-approximation plotting for any chart is accomplished by entering the chart sequence from both ends with corresponding values of all factors (including the dependent variable). The point so determined is labeled with the value of the factor for which the chart is being developed, and a revised curve family is fitted.

This method of multiple graphical correlation usually yields good results for many problems involving joint functions, such as rainfall-runoff and crest-stage relations, where three or more independent factors are significant. If the approximate shape and spacing of the curve families are known, the first-approximation curves for *all but one* variable can be sketched without plotting the data. The curves representing the remaining variable can then be developed by plotting in the prescribed manner. In general, it is advisable to determine the curves for the most important factor by plotting. The final position of all curve families can then be determined by the process outlined. This approach constitutes the substitution of an estimated effect for each variable in preference to the assumed no effect for those variables subsequently introduced. The number of approximations required to achieve the final solution is accordingly reduced.

REFERENCES

1. M. A. Kohler, The Use of Crest Stage Relations in Forecasting the Rise and Fall of the Flood Hydrograph, U.S. Weather Bureau, 1944 (mimeo.).
2. R. K. Linsley, M. A. Kohler, and J. L. H. Paulhus, "Applied Hydrology," pp. 652–655, McGraw-Hill, New York, 1949.
3. M. A. Kohler and R. K. Linsley, Predicting the Runoff from Storm Rainfall, *U.S. Weather Bur. Res. Pap.* 34, 1951.

APPENDIX B

TABLES OF PHYSICAL CONSTANTS AND EQUIVALENTS AND PSYCHROMETRIC TABLES

Table B-1 VOLUME EQUIVALENTS

Unit	Equivalent						
	in.^{3}	gal	Imperial gal	ft^3	m^3	acre-ft	sfd
Cubic inch	1	0.00433	0.00361	5.79×10^{-4}	1.64×10^{-5}	1.33×10^{-8}	6.70×10^{-9}
U.S. gallon	231	1	0.833	0.134	0.00379	3.07×10^{-6}	1.55×10^{-6}
Imperial gallon	277	1.20	1	0.161	0.00455	3.68×10^{-6}	1.86×10^{-6}
Cubic foot	1728	7.48	6.23	1	0.0283	2.30×10^{-5}	1.16×10^{-5}
Cubic meter	61,000	264	220	35.3	1	8.11×10^{-4}	4.09×10^{-4}
Acre-foot	7.53×10^{7}	3.26×10^{5}	2.71×10^{5}	43,560	1230	1	0.504
Second-foot-day	1.49×10^{8}	6.46×10^{5}	5.38×10^{5}	86,400	2450	1.98	1

Table B-2 DISCHARGE EQUIVALENTS

Unit	Equivalent						
	gal/d	ft^3/d	gal/min	Imperial gal/min	acre-ft/d	ft^3/s	m^3/s
U.S. gallon per day	1	0.134	6.94×10^{-4}	5.78×10^{-4}	3.07×10^{-6}	1.55×10^{-6}	4.38×10^{-8}
Cubic foot per day	7.48	1	5.19×10^{-3}	4.33×10^{-3}	2.30×10^{-5}	1.16×10^{-5}	3.28×10^{-7}
U.S. gallon per minute	1440	193	1	0.833	4.42×10^{-3}	2.23×10^{-3}	6.31×10^{-5}
Imperial gallon per minute	1728	231	1.20	1	5.31×10^{-3}	2.67×10^{-3}	7.57×10^{-5}
Acre-foot per day	3.26×10^{5}	43,560	226	188	1	0.504	0.0143
Cubic foot per second	6.46×10^{5}	86,400	449	374	1.98	1	0.0283
Cubic meter per second	2.28×10^{7}	3.05×10^{6}	15,800	13,200	70.0	35.3	1

Table B-3 MISCELLANEOUS EQUIVALENTS AND PHYSICAL CONSTANTS

1 second-foot-day per square mile	=	0.03719 inch
1 inch of runoff per square mile	=	26.9 second-foot-days
	=	53.3 acre-feet
	=	2,323,200 cubic feet
1 miner's inch	=	0.025 cubic foot per second in Arizona, California, Montana, and Oregon
	=	0.020 cubic foot per second in Idaho, Kansas, Nebraska, New Mexico, North and South Dakota, and Utah
	=	0.026 cubic foot per second in Colorado
	=	0.028 cubic foot per second in British Columbia
1 cubic foot per second	=	0.000214 cubic mile per year
	=	0.9917 acre-inch per hour
1 pound of water	=	0.5507 inch over 8-inch circle
	=	0.3524 inch over 10-inch circle
	=	0.2448 inch over 12-inch circle
1 horsepower	=	0.746 kilowatt
	=	550 foot-pounds per second
e	=	2.71828
$\log e$	=	0.43429
$\ln 10$	=	2.30259

Table B-4 VALUES OF n FOR THE MANNING FORMULA [EQ. (4-7)]

Channel condition	n
Plastic, glass, drawn tubing	0.009
Neat cement, smooth metal	0.010
Planed timber, asbestos pipe	0.011
Wrought iron, welded steel, canvas	0.012
Ordinary concrete, asphalted cast iron	0.013
Unplaned timber, vitrified clay, glazed brick	0.014
Cast-iron pipe, concrete pipe	0.015
Riveted steel, brick, dressed stone	0.016
Rubble masonry	0.017
Smooth earth	0.018
Firm gravel	0.020
Corrugated metal pipe and flumes	0.023
Natural channels:	
Clean, straight, full stage, no pools	0.029
As above with weeds and stones	0.035
Winding, pools and shallows, clean	0.039
As above at low stages	0.047
Winding, pools and shallows, weeds and stones	0.042
As above, shallow stages, large stones	0.052
Sluggish, weedy, with deep pools	0.065
Very weedy and sluggish	0.112

Values quoted above are averages of many determinations; variations of as much as 20 percent must be expected, especially in natural channels.

Table B-5 PROPERTIES OF WATER IN ENGLISH UNITS

Temp., °F	Specific gravity	Specific weight, lb/ft³	Heat of vaporization, Btu/lb	Viscosity		Vapor pressure		
				Absolute, lb-s/ft²	Kinematic, ft²/s	in. Hg	Millibars	lb/in.²
32	0.99986	62.418	1075.5	3.746×10^{-5}	1.931×10^{-5}	0.180	6.11	0.089
40	0.99998	62.426†	1071.0	3.229	1.664	0.248	8.39	0.122
50	0.99971	62.409	1065.3	2.735	1.410	0.362	12.27	0.178
60	0.99902	62.366	1059.7	2.359	1.217	0.522	17.66	0.256
70	0.99798	62.301	1054.0	2.050	1.058	0.739	25.03	0.363
80	0.99662	62.216	1048.4	1.799	0.930	1.032	34.96	0.507
90	0.99497	62.113	1042.7	1.595	0.826	1.422	48.15	0.698
100	0.99306	61.994	1037.1	1.424	0.739	1.933	65.47	0.950
120	0.98856	61.713	1025.6	1.168	0.609	3.448	116.75	1.693
140	0.98321	61.379	1014.0	0.981	0.514	5.884	199.26	2.890
160	0.97714	61.000	1002.2	0.838	0.442	9.656	326.98	4.742
180	0.97041	60.580	990.2	0.726	0.386	15.295	517.95	7.512
200	0.96306	60.121	977.9	0.637	0.341	23.468	794.72	11.526
212	0.95837	59.828	970.3	0.593	0.319	29.921	1013.25	14.696

† Maximum specific weight is 62.427 lb/ft³ at 39.2°F.

Table B-6 PROPERTIES OF WATER IN METRIC UNITS

Temp., °C	Specific gravity	Density, g/cm³	Heat of vaporization, cal/g	Viscosity		Vapor pressure		
				Absolute centipoises‡	Kinematic, centistokes§	mm Hg	Millibars	g/cm²
0	0.99987	0.99984	597.3	1.79	1.79	4.58	6.11	6.23
5	0.99999	0.99996†	594.5	1.52	1.52	6.54	8.72	8.89
10	0.99973	0.99970	591.7	1.31	1.31	9.20	12.27	12.51
15	0.99913	0.99910	588.9	1.14	1.14	12.78	17.04	17.38
20	0.99824	0.99821	586.0	1.00	1.00	17.53	23.37	23.83
25	0.99708	0.99705	583.2	0.890	0.893	23.76	31.67	32.30
30	0.99568	0.99565	580.4	0.798	0.801	31.83	42.43	43.27
35	0.99407	0.99404	577.6	0.719	0.723	42.18	56.24	57.34
40	0.99225	0.99222	574.7	0.653	0.658	55.34	73.78	75.23
50	0.98807	0.98804	569.0	0.547	0.554	92.56	123.40	125.83
60	0.98323	0.98320	563.2	0.466	0.474	149.46	199.26	203.19
70	0.97780	0.97777	557.4	0.404	0.413	233.79	311.69	317.84
80	0.97182	0.97179	551.4	0.355	0.365	355.28	473.67	483.01
90	0.96534	0.96531	545.3	0.315	0.326	525.89	701.13	714.95
100	0.95839	0.95836	539.1	0.282	0.294	760.00	1013.25	1033.23

† Maximum density is 0.999973 g/cm³ at 3.98°C. ‡ centipoise: g/(cm) (sec) (10^2) § centistokes: cm²/sec × 10^2

Table B-7 VARIATION OF RELATIVE HUMIDITY IN PERCENT WITH TEMPERATURE AND WET-BULB DEPRESSION ON THE FAHRENHEIT SCALE

Pressure = 30.00 in. = 1015.9 millibars

Air temp., °F	Wet-bulb depression, deg														
	0	1	2	3	4	6	8	10	12	14	16	18	20	25	30
0	84	56	27												
5	86	63	40	16											
10	89	69	50	30	11										
15	91	74	58	42	26										
20	94	79	65	51	37	10									
25	96	84	71	59	47	24	1								
30	99	88	77	66	56	35	15								
35	100	91	81	72	63	45	27	10							
40	100	92	84	76	68	52	37	22	7						
45	100	93	85	78	71	57	44	31	19	6					
50	100	93	87	80	74	61	49	38	27	16	5				
55	100	94	88	82	76	65	54	43	33	24	14	5			
60	100	94	89	83	78	68	58	48	39	30	21	13	5		
65	100	95	90	85	80	70	61	52	44	35	28	20	13		
70	100	95	90	86	81	72	64	55	48	40	33	26	19	3	
75	100	95	91	87	82	74	66	58	51	44	37	31	24	10	
80	100	96	91	87	83	75	68	61	54	47	41	35	29	15	3
85	100	96	92	88	84	77	70	63	56	50	44	38	33	20	8
90	100	96	92	89	85	78	71	65	58	53	47	41	36	24	13
95	100	96	93	89	86	79	72	66	60	55	49	44	39	28	17
100	100	96	93	89	86	80	74	68	62	57	51	46	42	31	21

SOURCE: U.S. Weather Bureau, Relative Humidity and Dew Point Table, TA 454-O-3E, September 1965.

Table B-8 VARIATION OF RELATIVE HUMIDITY IN PERCENT WITH TEMPERATURE AND WET-BULB DEPRESSION ON THE CELSIUS SCALE

Pressure = 990 millibars = 29.24 in.

Air temp., °C	Wet-bulb depression, deg															
	0	1	2	3	4	5	6	7	8	9	10	11	12	13	14	15
−10	91	60	31	2												
−8	93	65	39	13												
−6	94	70	46	23	0											
−4	96	74	53	32	11											
−2	98	78	58	39	21	3										
0	100	81	63	46	29	13										
2	100	84	68	52	37	22	7									
4	100	85	71	57	43	29	16									
6	100	86	73	60	48	35	24	11								
8	100	87	75	63	51	40	29	19	8							
10	100	88	77	66	55	44	34	24	15	6						
12	100	89	78	68	58	48	39	29	21	12	4					
14	100	90	79	70	60	51	42	34	26	18	10	3				
16	100	90	81	71	63	54	46	38	30	23	15	8				
18	100	91	82	73	65	57	49	41	34	27	20	14	7			
20	100	91	83	74	66	59	51	44	37	31	24	18	12	6		
22	100	92	83	76	68	61	54	47	40	34	28	22	17	11	6	
24	100	92	84	77	69	62	56	49	43	37	31	26	20	15	10	5
26	100	92	85	78	71	64	58	51	46	40	34	29	24	19	14	10
28	100	93	85	78	72	65	59	53	48	42	37	32	27	22	18	13
30	100	93	86	79	73	67	61	55	50	44	39	35	30	25	21	17
32	100	93	86	80	74	68	62	57	51	46	41	37	32	28	24	20
34	100	93	87	81	75	69	63	58	53	48	43	39	35	30	26	23
36	100	94	87	81	75	70	64	59	54	50	45	41	37	33	29	25
38	100	94	88	82	76	71	66	61	56	51	47	43	39	35	31	27
40	100	94	88	82	77	72	67	62	57	53	48	44	40	36	33	29

SOURCE: From "Radiosonde Observation Computation Tables," Dept. of Commerce–Dept. of Defense, Washington, June 1972.

Table B-9 VARIATION OF DEWPOINT WITH TEMPERATURE AND WET-BULB DEPRESSION AND OF SATURATION VAPOR PRESSURE OVER WATER WITH TEMPERATURE ON THE FAHRENHEIT SCALE

Pressure = 30.00 in. = 1015.9 millibars

Air temp., °F	Saturation vapor pressure		Wet-bulb depression, deg														
	Millibars	in. Hg	0	1	2	3	4	6	8	10	12	14	16	18	20	25	30
0	1.52	0.045	−4	−12	−26												
5	1.91	0.056	2	−5	−14	−31											
10	2.40	0.071	7	2	−5	−15	−34										
15	2.99	0.088	13	8	3	−4	−14										
20	3.71	0.110	18	15	10	5	−2	−27									
25	4.58	0.135	24	21	17	13	8	−7									
30	5.63	0.166	30	27	24	20	16	6	−12								
35	6.89	0.203	35	33	30	27	24	16	5	−16							
40	8.39	0.248	40	38	35	33	30	24	16	4	−18						
45	10.17	0.300	45	43	41	39	36	31	24	16	5	−18					
50	12.27	0.362	50	48	46	44	42	37	32	25	17	5	−17				
55	14.75	0.436	55	53	51	50	48	43	39	33	27	18	7	−15			
60	17.66	0.522	60	58	57	55	53	49	45	40	35	29	20	9	−11		
65	21.07	0.622	65	63	62	60	59	55	51	47	42	37	31	23	12		
70	25.03	0.739	70	69	67	66	64	61	57	53	49	45	39	33	26	−14	
75	29.63	0.875	75	74	72	71	69	66	63	59	56	52	47	42	36	14	
80	34.96	1.032	80	79	77	76	74	72	68	65	62	58	54	50	45	29	−9
85	41.10	1.214	85	84	82	81	80	77	74	71	68	64	61	57	53	39	18
90	48.15	1.422	90	89	87	86	85	82	79	76	73	70	67	63	60	48	33
95	56.24	1.661	95	94	93	91	90	87	85	82	79	76	73	70	66	56	44
100	65.47	1.933	100	99	98	96	95	93	90	87	85	82	79	76	73	64	53

SOURCE: U.S. Weather Bureau, Relative Humidity and Dew Point Table, TA 454-0-3E, September 1965.

Table B-10 VARIATION OF DEWPOINT WITH TEMPERATURE AND WET-BULB DEPRESSION AND OF SATURATION VAPOR PRESSURE OVER WATER WITH TEMPERATURE ON THE CELSIUS SCALE

Pressure = 1013.2 millibars = 29.92 in.

Air temp., °C	Saturation vapor pressure		Wet-bulb depression, deg															
	Millibars	in. Hg	0	1	2	3	4	5	6	7	8	9	10	11	12	13	14	15
−10	2.86	0.085	−11	−16	−24													
−8	3.35	0.099	−9	−13	−20	−33												
−6	3.91	0.115	−7	−11	−16	−24												
−4	4.55	0.134	−5	−8	−12	−19	−32											
−2	5.28	0.156	−2	−5	−9	−14	−22											
0	6.11	0.180	0	−3	−6	−11	−16	−27										
2	7.05	0.208	2	−1	−3	−7	−12	−19	−33									
4	8.13	0.240	4	2	−1	−4	−8	−13	−21	−47								
6	9.35	0.276	6	4	2	−1	−5	−9	−14	−23								
8	10.72	0.317	8	6	4	1	−2	−5	−9	−15	−26							
10	12.27	0.362	10	8	6	4	1	−2	−5	−10	−17	−29						
12	14.02	0.414	12	10	8	6	4	1	−2	−6	−11	−18	−34					
14	15.98	0.472	14	12	11	9	6	4	1	−2	−6	−11	−19					
16	18.17	0.532	16	14	13	11	9	7	4	1	−2	−6	−11					
18	20.63	0.609	18	16	15	13	11	9	7	4	2	−2	−6					
20	23.37	0.690	20	19	17	15	14	12	10	7	5	2	−1					
22	26.43	0.780	22	21	19	17	16	14	12	10	8	5	2	−1	−5			
24	29.83	0.881	24	23	21	20	18	16	15	13	11	8	6	3	−1	−5	−10	
26	33.61	0.992	26	25	23	22	20	19	18	15	13	11	9	6	4	0	−4	−9
28	37.80	1.116	28	27	25	24	22	21	19	18	16	14	12	10	7	4	1	−3
30	42.43	1.253	30	29	27	26	25	23	22	20	18	17	15	13	10	8	5	2
32	47.55	1.404	32	31	29	28	27	25	24	22	21	19	17	15	13	11	9	6
34	53.20	1.571	34	33	32	30	29	28	26	25	23	21	20	17	16	14	12	10
36	59.42	1.755	36	35	34	32	31	30	28	27	25	24	22	21	19	17	15	13
38	66.26	1.957	38	37	36	34	33	32	30	29	28	26	25	23	21	20	18	16
40	73.78	2.179	40	39	38	36	35	34	33	31	30	28	27	25	24	22	20	19

SOURCE: U.S. National Weather Service, Marine Surface Observations, *Weather Bur. Handb.* 1, 1969.

Table B-11 RATIO OF SATURATION VAPOR PRESSURE OVER ICE TO THAT OVER WATER AT THE SAME TEMPERATURE

Deg	0	1	2	3	4	5	6	7	8	9
					Celsius					
−0	1.000	0.990	0.981	0.971	0.962	0.953	0.943	0.934	0.925	0.916
−10	0.907	0.899	0.890	0.881	0.873	0.864	0.856	0.847	0.839	0.831
−20	0.823	0.815	0.807	0.799	0.791	0.784	0.776	0.769	0.761	0.754
−30	0.746	0.739	0.732	0.725	0.718	0.711	0.704	0.698	0.691	0.685
−40	0.678	0.672	0.666	0.660	0.654	0.648	0.642	0.636	0.630	0.625
					Fahrenheit					
30	0.989	0.994	1.000							
20	0.937	0.942	0.947	0.953	0.958	0.963	0.968	0.973	0.979	0.984
10	0.888	0.893	0.897	0.902	0.907	0.912	0.917	0.922	0.927	0.932
0	0.841	0.845	0.850	0.855	0.859	0.864	0.868	0.873	0.878	0.883
−0	0.841	0.836	0.832	0.827	0.823	0.818	0.814	0.809	0.805	0.801
−10	0.796	0.792	0.788	0.784	0.779	0.775	0.771	0.767	0.763	0.759
−20	0.755	0.750	0.746	0.742	0.739	0.734	0.731	0.727	0.723	0.719
−30	0.715	0.711	0.708	0.704	0.700	0.696	0.693	0.690	0.686	0.682
−40	0.678	0.675	0.672	0.668	0.665	0.661	0.658	0.654	0.651	0.648
−50	0.644	0.641	0.638	0.635	0.631	0.628	0.625	0.622	0.619	0.616

SOURCE: From "Smithsonian Meteorological Tables," 6th ed., p. 370, Smithsonian Institution, Washington, 1966.

Table B-12 VARIATION OF PRESSURE, TEMPERATURE, DENSITY, AND BOILING POINT WITH ELEVATION
U.S. standard atmosphere

Elevation from mean sea level, m	Pressure			Air temp., °C	Air density, kg/m^3	Boiling point, °C
	mm Hg	Millibars	cm H_2O			
−500	806.15	1074.78	1096.0	18.2	1.285	101.7
0	760.00	1013.25	1033.2	15.0	1.225	100.0
500	716.02	954.61	973.4	11.8	1.167	98.3
1000	674.13	898.76	916.5	8.5	1.112	96.7
1500	634.25	845.60	862.3	5.3	1.058	95.0
2000	596.31	795.01	810.7	2.0	1.007	93.4
2500	560.23	746.92	761.6	−1.2	0.957	91.7
3000	525.95	701.21	715.0	−4.5	0.909	90.0
3500	493.39	657.80	670.8	−7.7	0.863	88.3
4000	462.49	616.60	628.8	−11.0	0.819	86.7
4500	433.18	577.52	588.9	−14.2	0.777	85.0
5000	405.40	540.48	551.1	−17.5	0.736	83.3

SOURCE: "U.S. Standard Atmosphere, 1962," National Aeronautics and Space Administration, U.S. Air Force, and U.S. Weather Bureau.

Table B-13 VALUES OF k^t FOR VARIOUS VALUES OF k AND t [EQ. (8-7)]

	k									
t	0.80	0.82	0.84	0.86	0.88	0.90	0.92	0.94	0.96	0.98
1	0.800	0.820	0.840	0.860	0.880	0.900	0.920	0.940	0.960	0.980
2	0.640	0.672	0.706	0.740	0.774	0.810	0.846	0.884	0.922	0.960
3	0.512	0.551	0.593	0.636	0.681	0.729	0.779	0.831	0.885	0.941
4	0.410	0.452	0.498	0.547	0.600	0.656	0.716	0.781	0.849	0.922
5	0.328	0.371	0.418	0.470	0.528	0.590	0.659	0.734	0.815	0.904
6	0.262	0.304	0.351	0.405	0.464	0.531	0.606	0.690	0.783	0.886
7	0.210	0.249	0.295	0.348	0.409	0.478	0.558	0.648	0.751	0.868
8	0.168	0.204	0.248	0.299	0.360	0.430	0.513	0.610	0.721	0.851
9	0.134	0.168	0.208	0.257	0.316	0.387	0.472	0.573	0.693	0.834
10	0.107	0.137	0.175	0.221	0.279	0.349	0.434	0.539	0.665	0.817
11	0.086	0.113	0.147	0.190	0.245	0.314	0.400	0.506	0.638	0.801
12	0.069	0.092	0.123	0.164	0.216	0.282	0.368	0.476	0.613	0.785
13	0.055	0.076	0.104	0.141	0.190	0.254	0.338	0.447	0.588	0.769
14	0.044	0.062	0.087	0.121	0.167	0.229	0.311	0.421	0.565	0.754
15	0.035	0.051	0.073	0.104	0.147	0.206	0.286	0.395	0.542	0.739
16	0.028	0.042	0.061	0.090	0.129	0.185	0.263	0.372	0.520	0.724
17	0.023	0.034	0.052	0.077	0.114	0.167	0.242	0.349	0.500	0.709
18	0.018	0.028	0.043	0.066	0.100	0.150	0.223	0.328	0.480	0.695
19	0.014	0.023	0.036	0.057	0.088	0.135	0.205	0.309	0.460	0.681
20	0.012	0.019	0.031	0.049	0.078	0.122	0.189	0.290	0.442	0.668
21	0.009	0.015	0.026	0.042	0.068	0.109	0.174	0.273	0.424	0.654
22	0.007	0.013	0.022	0.036	0.060	0.098	0.160	0.256	0.407	0.641
23	0.006	0.010	0.018	0.031	0.053	0.089	0.147	0.241	0.391	0.628
24	0.005	0.009	0.015	0.027	0.047	0.080	0.135	0.227	0.375	0.616
25	0.004	0.007	0.013	0.023	0.041	0.072	0.124	0.213	0.360	0.603
26	0.003	0.006	0.011	0.020	0.036	0.065	0.114	0.200	0.346	0.591
27	0.002	0.005	0.009	0.017	0.032	0.058	0.105	0.188	0.332	0.579
28	0.002	0.004	0.008	0.015	0.028	0.052	0.097	0.177	0.319	0.568
29	0.002	0.003	0.006	0.013	0.025	0.047	0.089	0.166	0.306	0.557
30	0.001	0.003	0.005	0.011	0.022	0.042	0.082	0.156	0.294	0.545
40	...	...	...	0.002	0.006	0.015	0.036	0.084	0.195	0.446
50	...	...	...	...	...	0.005	0.015	0.045	0.130	0.364
60	...	...	...	...	...	...	0.007	0.024	0.086	0.298

Table B-14 MAP-SCALE EQUIVALENTS

Ratio	in./mi	mi/in.	mi²/in.²	cm/km	km/cm	km²/cm²
1:1,000,000	0.0634	15.7828	249.097	0.1000	10.0000	100.000
1:500,000	0.1267	7.8914	62.274	0.2000	5.0000	25.000
1:250,000	0.2534	3.9457	15.569	0.4000	2.5000	6.250
1:126,720	0.5000	2.0000	4.000	0.7891	1.2672	1.606
1:125,000	0.5069	1.9728	3.892	0.8000	1.2500	1.562
1:90,000	0.7040	1.4205	2.018	1.1111	0.9000	0.8100
1:63,360	1.0000	1.0000	1.000	1.5783	0.6336	0.4014
1:62,500	1.0138	0.9864	0.9730	1.6000	0.6250	0.3906
1:45,000	1.4080	0.7102	0.5044	2.2222	0.4500	0.2025
1:31,680	2.0000	0.5000	0.2500	3.1566	0.3168	0.1004
1:30,000	2.1120	0.4735	0.2242	3.3333	0.3000	0.0900
1:24,000	2.6400	0.3788	0.1435	4.1667	0.2400	0.0576
1:12,000	5.2800	0.1894	0.0358	8.3333	0.1200	0.0144
1:2,300	26.4000	0.0379	0.001435	41.6667	0.0240	0.00576
1:1,200	52.8000	0.0189	0.000358	83.3333	0.0120	0.000144

APPENDIX C

EVAPORATION COMPUTATION CHARTS IN METRIC UNITS

The following figures (C-1 through C-4) are conversions of Figs. 5-1, 5-2, 5-5, and 5-6 to metric units.

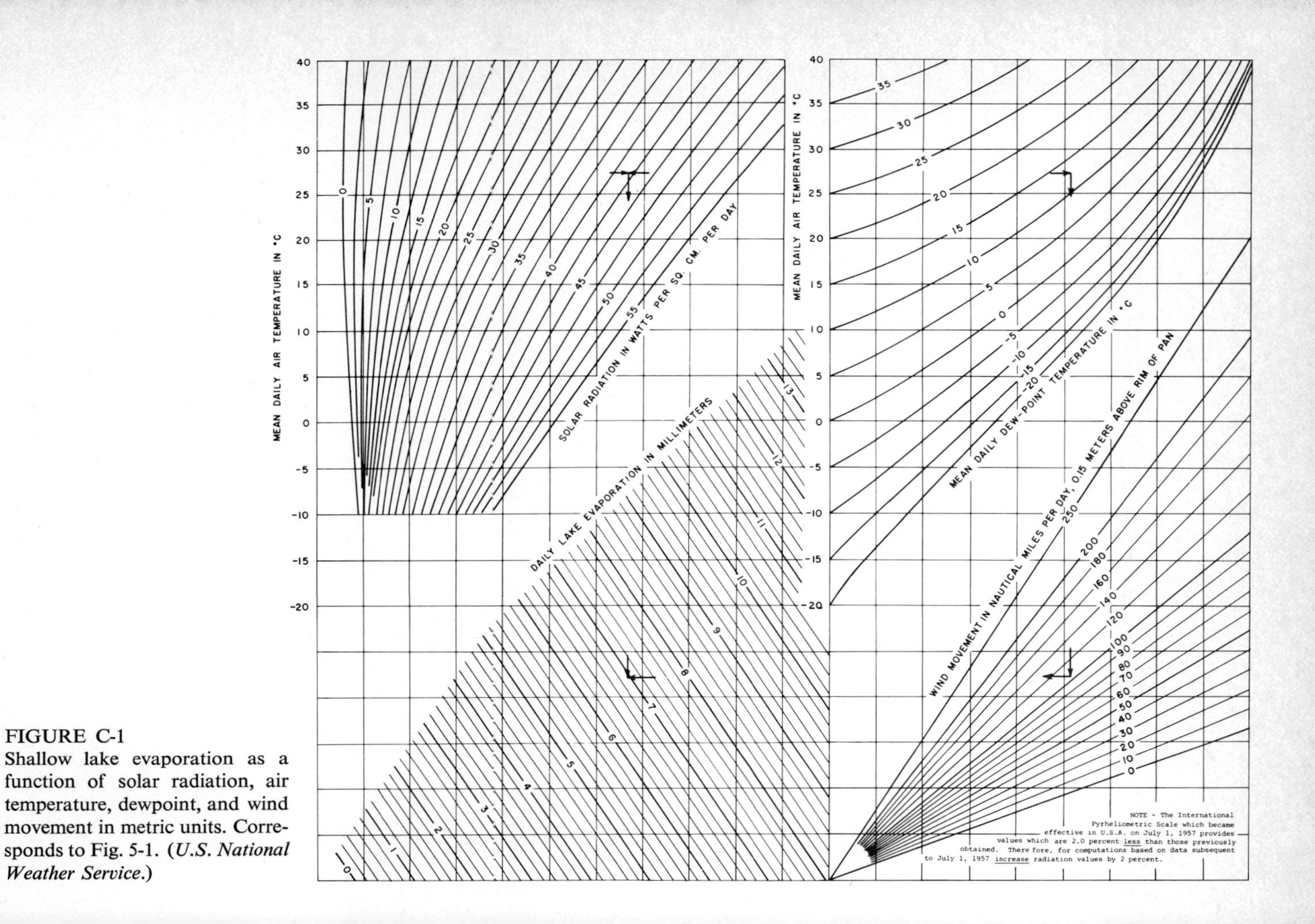

FIGURE C-1
Shallow lake evaporation as a function of solar radiation, air temperature, dewpoint, and wind movement in metric units. Corresponds to Fig. 5-1. (*U.S. National Weather Service.*)

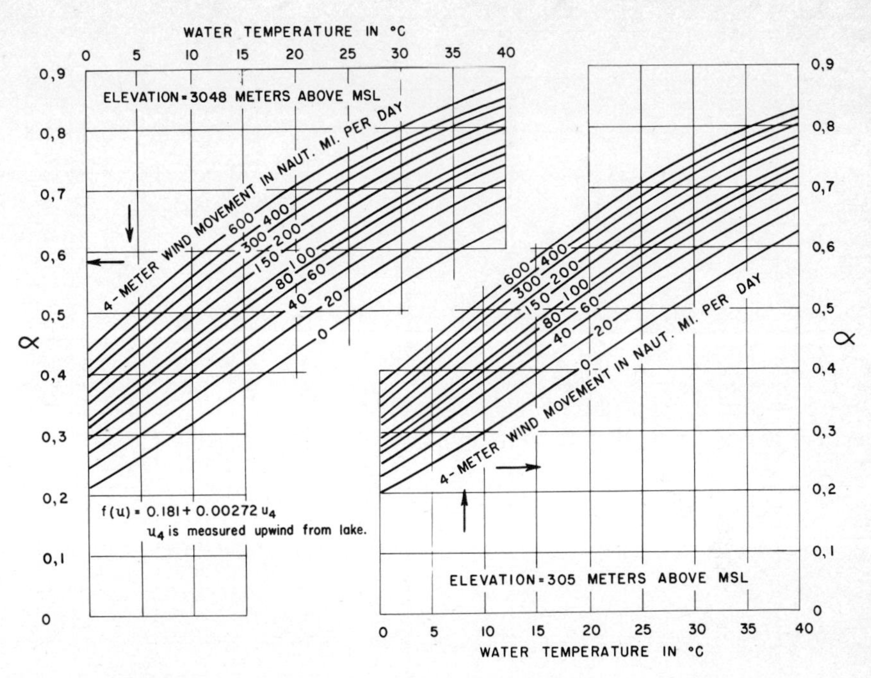

FIGURE C-2
Variation of α with water temperature, wind movement, and elevation in metric units. Corresponds to Fig. 5-2. (*U.S. National Weather Service.*)

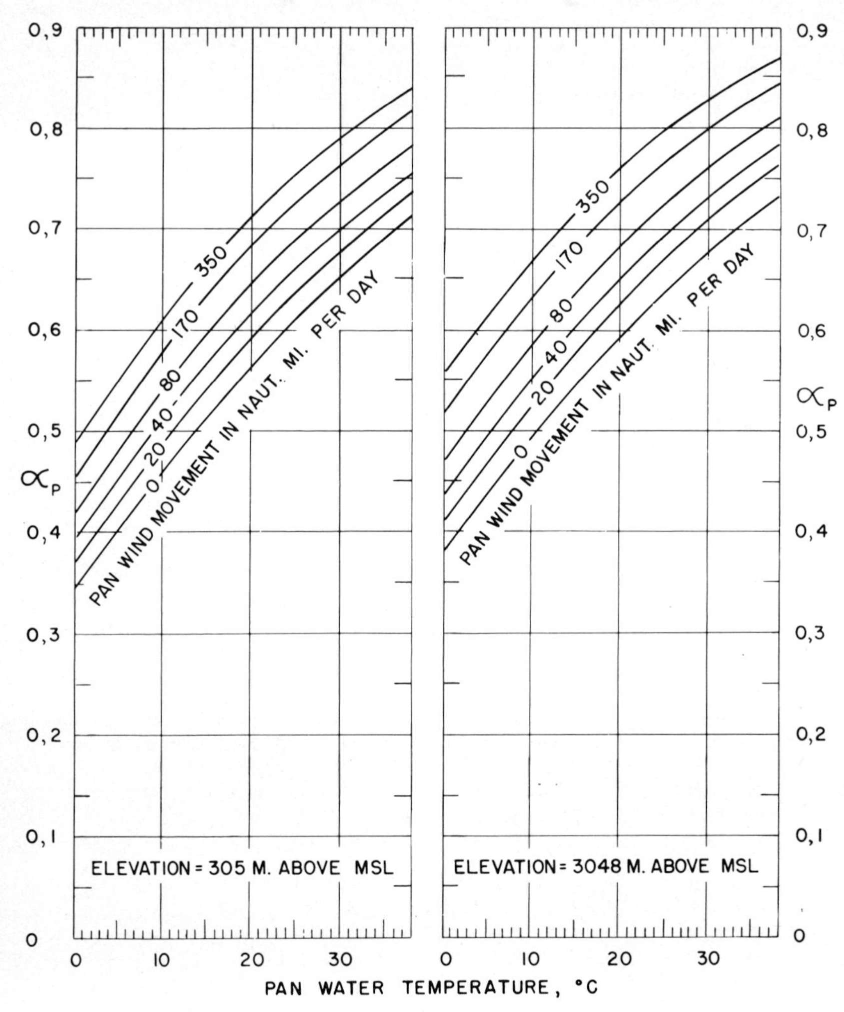

FIGURE C-3
Portion of advected energy (into a Class A pan) utilized for evaporation in metric units. Corresponds to Fig. 5-5. (*U.S. National Weather Service.*)

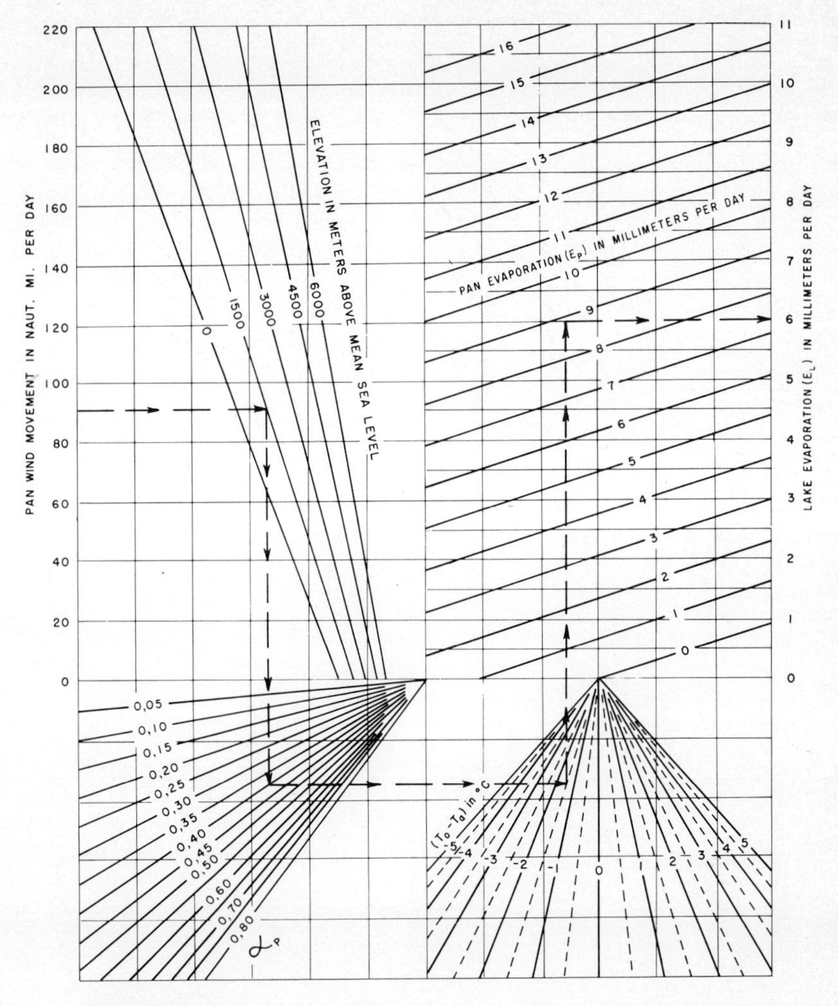

FIGURE C-4
Shallow lake evaporation as a function of Class A pan evaporation and heat transfer through the pan in metric units per day. Corresponds to Fig. 5-6. (*U.S. National Weather Service.*)

NAME INDEX

SUBJECT INDEX